Strong Interactions in the Standard Model: Massless Bosons to Compact Stars

Strong Interactions in the Standard Model: Massless Bosons to Compact Stars

Editors

Minghui Ding
Craig Roberts
Sebastian M. Schmidt

Basel • Beijing • Wuhan • Barcelona • Belgrade • Novi Sad • Cluj • Manchester

Editors

Minghui Ding
Helmholtz-Zentrum
Dresden-Rossendorf
Dresden
Germany

Craig Roberts
Nanjing University
Nanjing
China

Sebastian M. Schmidt
Helmholtz-Zentrum
Dresden-Rossendorf
Dresden
Germany

Editorial Office
MDPI AG
Grosspeteranlage 5
4052 Basel, Switzerland

This is a reprint of articles from the Special Issue published online in the open access journal *Particles* (ISSN 2571-712X) (available at: https://www.mdpi.com/journal/particles/special_issues/strong_interactions_standard_model).

For citation purposes, cite each article independently as indicated on the article page online and as indicated below:

Lastname, A.A.; Lastname, B.B. Article Title. *Journal Name* **Year**, *Volume Number*, Page Range.

ISBN 978-3-7258-1501-2 (Hbk)
ISBN 978-3-7258-1502-9 (PDF)
doi.org/10.3390/books978-3-7258-1502-9

© 2024 by the authors. Articles in this book are Open Access and distributed under the Creative Commons Attribution (CC BY) license. The book as a whole is distributed by MDPI under the terms and conditions of the Creative Commons Attribution-NonCommercial-NoDerivs (CC BY-NC-ND) license.

Contents

About the Editors vii

Preface ix

Daniel S. Carman, Ralf W. Gothe, Victor I. Mokeev and Craig D. Roberts
Nucleon Resonance Electroexcitation Amplitudes and Emergent Hadron Mass
Reprinted from: *Particles* **2023**, *6*, 416–439, doi:10.3390/particles6010023 1

Kai-Bao Chen, Tianbo Liu, Yu-Kun Song and Shu-Yi Wei
Several Topics on Transverse Momentum-Dependent Fragmentation Functions
Reprinted from: *Particles* **2023**, *6*, 515–545, doi:10.3390/particles6020029 25

Alexandre Deur, Volker Burkert, Jian-Ping Chen and Wolfgang Korsch
Experimental Determination of the QCD Effective Charge $\alpha_{g_1}(Q)$
Reprinted from: *Particles* **2022**, *5*, 171–179, doi:10.3390/particles5020015 56

Minghui Ding, Craig D. Roberts and Sebastian M. Schmidt
Emergence of Hadron Mass and Structure
Reprinted from: *Particles* **2023**, *6*, 57–120, doi:10.3390/particles6010004 65

Mauricio N. Ferreira and Joannis Papavassiliou
Gauge sector dynamics in QCD
Reprinted from: *Particles* **2023**, *6*, 312–363, doi:10.3390/particles6010017 129

Thomas Klähn, Lee C. Loveridge and Mateusz Cierniak
Chaos in QCD? Gap Equations and Their Fractal Properties
Reprinted from: *Particles* **2023**, *6*, 470–484, doi:10.3390/particles6020026 181

Cédric Mezrag
Generalised Parton Distributions in Continuum Schwinger Methods: Progresses, Opportunities and Challenges
Reprinted from: *Particles* **2023**, *6*, 262–296, doi:10.3390/particles6010015 196

Armen Sedrakian
Impact of Multiple Phase Transitions in Dense QCD on Compact Stars
Reprinted from: *Particles* **2023**, *6*, 713–730, doi:10.3390/particles6030044 231

Hans Ströher, Sebastian M. Schmidt, Paolo Lenisa and Jörg Pretz
Precision Storage Rings for Electric Dipole Moment Searches: A Tool En Route to Physics Beyond-the-Standard-Model
Reprinted from: *Particles* **2023**, *6*, 385–398, doi:10.3390/particles6010020 249

Rico Zöllner, Minghui Ding and Burkhard Kämpfer
Masses of Compact (Neutron) stars with Distinguished Cores
Reprinted from: *Particles* **2023**, *6*, 217–238, doi:10.3390/particles6010012 263

About the Editors

Minghui Ding

Dr. Minghui Ding is currently a Member of the High Potential Programme at Helmholtz-Zentrum Dresden-Rossendorf (HZDR) in Germany. She obtained her bachelor's degree in physics from Central China Normal University in 2011, and her doctoral degree in Theoretical Physics from Peking University in 2016. After completing her PhD, she was made a postdoctoral researcher at Nankai University, China; Argonne National Laboratory, USA; and the European Centre for Theoretical Studies in Nuclear Physics and Related Areas (ECT*), Italy. Her primary areas of research include the development and application of continuum Schwinger function methods and the theory and phenomenology of strong interactions in the Standard Model of particle physics (quantum chromodynamics) (QCD). An early career researcher in these areas, she has published 27 articles in peer-reviewed journals that have attracted over 1300 total citations.

Craig Roberts

Prof. Craig Roberts is the International Distinguished Professor and Head of the Institute for Nonperturbative Physics at Nanjing University. He received his PhD in theoretical physics from the Flinders University of South Australia in 1988. Following a postdoctoral fellowship at the University of Melbourne, Victoria, Australia, he joined the Theory Group in the Physics Division at Argonne National Laboratory in 1989, first as a postdoctoral fellow, then a staff member. Additionally, from 2001-2017, he was Leader of the Theory Group.

Prof. Roberts has held numerous visiting and joint appointments at research centres and universities worldwide. Amongst other recognitions, he was elected to APS Fellowship in 2001; won the Friedrich Wilhelm Bessel Research Award of the Alexander von Humboldt Foundation in 2003; was awarded the Flinders University Convocation Medal in 2009; was selected as an International Fellow of Germany's Helmholtz Association in 2012; awarded the University of Chicago/Argonne LLC Board of Governors Distinguished Performance Award in 2014; was honored by the Chinese Ministry of Education as an International Distinguished Professor in 2015; and was selected into the Jiangsu Province 100 Talents Plan for Professionals in 2018. Since accepting his positions at Nanjing University in 2019, he has won many additional honors and awards.

Prof. Roberts conducts a broad-ranging research programme discussing modern high-energy nuclear and particle theory, pursuing the development and refinement of novel theoretical approaches to hadroparticle physics and strong-coupling quantum field theory. He is a leading practitioner of nonperturbative continuum Schwinger function methods in quantum chromodynamics. Prof. Roberts has published 271 articles in peer-reviewed journals with over 23,500 citations, and he has delivered 450 colloquia, seminars, and lectures at research centres worldwide, including more than 200 invited presentations at conferences and workshops.

Sebastian M. Schmidt

Prof. Sebastian M. Schmidt has been Scientific Director at the Helmholtz-Zentrum Dresden-Rossendorf since 1 April 2020. Prof. Dr. Dr. hc. mult. Sebastian M. Schmidt, started his scientific career at the University of Rostock, Germany, before joining the Joint Institute for Nuclear Research in Dubna, Russia. He completed his doctorate in theoretical physics in Rostock in 1995. After a fellowship at Tel Aviv University, he won an Alexander von Humboldt Foundation Research Fellowship (Lynen-Award) in association with the Theory Group at Argonne National Laboratory (USA). Subsequently, Prof.

Schmidt took on leadership of an Emmy Noether junior research group, sponsored by the German Research Foundation (DFG), at the Eberhard Karls University of Tübingen in 2000. In 2001, Prof. Schmidt completed his habilitation at the Universities of Tübingen and Rostock. From 2002 to 2006, Prof. Schmidt was first a staff member, then managing director of the Helmholtz Association's office in Berlin. From 2007 to 2020, he served as a member of the Board of Directors at the Forschungszentrum Jülich, working in the areas of matter and information. In 2012, Prof. Schmidt was appointed a full university professor at RWTH Aachen University; in 2024, he began work at the Technical University of Dresden. In October 2023, Prof. Schmidt was elected to the Germany National Academy of Science and Engineering—acatech. In addition to his overviewing responsibilities for national and international scientific projects, Prof. Schmidt has conducted research focused on strong interactions in quantum field theory, publishing over 100 articles in peer-reviewed journals that have attracted more than 11,000 citations.

Preface

The Standard Model of particle physics (SM) was formulated roughly fifty years ago, and, with discovery of the Higgs boson at CERN in 2012, it became complete. However, despite the SM's enormous body of successes, it still presents an array of unsolved problems. Primary among them is the following question: can the SM explain the origin of nuclear size masses? This is the puzzle of emergent hadron mass (EHM), whose solution is supposed to lie within quantum chromodynamics (QCD), the SM's strong interaction component. EHM could provide the unifying explanation for all the SM's remarkable nonperturbative phenomena, including confinement and absolute stability of the proton, the proton's mass and radii, the lepton-like scale of the pion mass and its hadron-like radius, and so much more, running up to the character and composition of dense astrophysical objects. Of course, as a source of mass, EHM interferes constructively with a range of Higgs boson effects. For instance, such feedback sets the kaon apart from the pion and separates heavy quark systems from those containing only light quarks. Presented with such a plethora of interrelated phenomena, whose implications reach throughout nature, the world has responded with huge investments of personnel and resources in strong interaction experiment and theory.

Reflecting the scope of associated endeavors, this volume collects a diverse range of perspectives on the problem of EHM, its observable manifestations, and the approaches and tools that are today being employed to deliver an insightful understanding and, perhaps, finally, a solution. Some of the contributions are reviews, but all may be described as feature articles, containing novel perspectives on topical problems and original material that highlights promising new directions in strong interaction physics. The collection was compiled with a view to stressing the importance of openness in basic research, with authors drawn from a wide range of backgrounds and operational environments. All contributions recognize and emphasize the vitality of international collaborative endeavors, whether it be through experimentation, phenomenology, or theory, and they aim, ultimately, to explain the emergence of nuclei from the remarkably simple Lagrangian of QCD. This, indeed, is one of the greatest challenges facing science today. Moreover, if QCD is, as some of the contributions argue, the first mathematically well-defined quantum field theory in four dimensions that science has produced, then it might serve as the paradigm for extending physics beyond the SM. With such coverage, the volume is likely to be of interest to all students and researchers working in high-energy nuclear physics, hadron physics, and particle physics.

We, the guest editors, and the Members of the Editorial Board of Particles are grateful to all those who contributed to the collection, as well as to the referees who donated their time in order to ensure that the project was a success.

Minghui Ding, Craig Roberts, and Sebastian M. Schmidt
Editors

Review

Nucleon Resonance Electroexcitation Amplitudes and Emergent Hadron Mass

Daniel S. Carman [1,*,†], Ralf W. Gothe [2,*,†], Victor I. Mokeev [1,*,†] and Craig D. Roberts [3,4,*,†]

1. Jefferson Laboratory, 12000 Jefferson Ave., Newport News, VA 23606, USA
2. Department of Physics and Astronomy, University of South Carolina, 712 Main St., Columbia, SC 29208, USA
3. School of Physics, Nanjing University, Nanjing 210093, China
4. Institute for Nonperturbative Physics, Nanjing University, Nanjing 210093, China
* Correspondence: carman@jlab.org (D.S.C.); gothe@sc.edu (R.W.G.); mokeev@jlab.org (V.I.M.); cdroberts@nju.edu.cn (C.D.R.)
† These authors contributed equally to this work.

Abstract: Understanding the strong interaction dynamics that govern the emergence of hadron mass (EHM) represents a challenging open problem in the Standard Model. In this paper we describe new opportunities for gaining insight into EHM from results on nucleon resonance (N^*) electroexcitation amplitudes (i.e., $\gamma_v p N^*$ electrocouplings) in the mass range up to 1.8 GeV for virtual photon four-momentum squared (i.e., photon virtualities Q^2) up to 7.5 GeV^2 available from exclusive meson electroproduction data acquired during the 6-GeV era of experiments at Jefferson Laboratory (JLab). These results, combined with achievements in the use of continuum Schwinger function methods (CSMs), offer new opportunities for charting the momentum dependence of the dressed quark mass from results on the Q^2-evolution of the $\gamma_v p N^*$ electrocouplings. This mass function is one of the three pillars of EHM and its behavior expresses influences of the other two, *viz.* the running gluon mass and momentum-dependent effective charge. A successful description of the $\Delta(1232)3/2^+$ and $N(1440)1/2^+$ electrocouplings has been achieved using CSMs with, in both cases, common momentum-dependent mass functions for the dressed quarks, for the gluons, and the same momentum-dependent strong coupling. The properties of these functions have been inferred from nonperturbative studies of QCD and confirmed, e.g., in the description of nucleon and pion elastic electromagnetic form factors. Parameter-free CSM predictions for the electrocouplings of the $\Delta(1600)3/2^+$ became available in 2019. The experimental results obtained in the first half of 2022 have confirmed the CSM predictions. We also discuss prospects for these studies during the 12-GeV era at JLab using the CLAS12 detector, with experiments that are currently in progress, and canvass the physics motivation for continued studies in this area with a possible increase of the JLab electron beam energy up to 22 GeV. Such an upgrade would finally enable mapping of the dressed quark mass over the full range of distances (i.e., quark momenta) where the dominant part of hadron mass and N^* structure emerge in the transition from the strongly coupled to perturbative QCD regimes.

Keywords: exclusive meson photo- and electroproduction; exclusive reactions with the CLAS and CLAS12 detectors; nucleon resonance photo- and electroexcitation amplitudes; nucleon resonance spectrum and structure; emergence of hadron mass; continuum Schwinger function methods; hadron structure and interactions

Citation: Carman, D.S.; Gothe, R.W.; Mokeev, V.I.; Roberts, C.D. Nucleon Resonance Electroexcitation Amplitudes and Emergent Hadron Mass. *Particles* 2023, 6, 416–439. https://doi.org/10.3390/particles6010023

Academic Editor: Armen Sedrakian

Received: 19 January 2023
Revised: 1 March 2023
Accepted: 2 March 2023
Published: 15 March 2023

Copyright: © 2023 by the authors. Licensee MDPI, Basel, Switzerland. This article is an open access article distributed under the terms and conditions of the Creative Commons Attribution (CC BY) license (https://creativecommons.org/licenses/by/4.0/).

1. Introduction

Studies of the strong interaction dynamics that govern the generation of hadron ground and excited states in the regime where the running coupling of quantum chromodynamics (QCD) is large, i.e., $\alpha_s/\pi \approx 1$, known as the strong QCD (sQCD) regime, represent a crucial challenge in modern hadron physics [1]. The rapid growth of α_s in the transition from the perturbative to sQCD domains and particularly its saturation, driven by gluon self-interactions, are predicted by CSMs [2,3] and supported by recent experimental results on the Bjorken sum rule [4]. These trends suggest that the generation of hadron structure in the sQCD regime is defined by emergent degrees of freedom that are related to the partons of QCD's Lagrangian in a non-trivial manner. While this evolution with distance is determined by the QCD Lagrangian, it cannot be analyzed by employing perturbative QCD (pQCD) when α_s/π becomes comparable with unity. The active degrees of freedom seen in hadron structure and their interactions change substantially with distance at the scales where the transition from sQCD to pQCD takes place, and the structure of hadron ground and excited states emerges. Understanding how the active degrees of freedom emerge from the QCD Lagrangian and how their interactions evolve with distance requires the development of nonperturbative methods capable of making predictions, both in the meson and baryon sectors, that can be confronted with empirical results on hadron structure extracted using electromagnetic and hadronic probes.

A decade of rapid progress in the development and application of CSMs in hadron physics [5–15], complemented by advances in and results from lattice QCD (lQCD) [16–25], have delivered numerous predictions for properties of mesons and baryons within a common theoretical framework. Studies of hadron structure from data obtained in experiments with electromagnetic probes at JLab [1,26–29], MAMI [30–35], and Babar and Belle [36,37], have provided experimental results that can be confronted with predictions from the QCD-connected approaches to hadron structure. More results are expected from experiments in the ongoing 12-GeV era at JLab [1,38–40] and from planned research programs at the US electron ion collider (EIC) [39,41–43], the electron ion collider in China (EicC) [44,45], and experiments with hadronic probes conducted by the AMBER Collaboration at CERN [46].

Studies of exclusive meson electroproduction in the nucleon resonance excitation region using data from 6-GeV-era experiments at JLab have provided the first and still only available comprehensive information on the electroexcitation amplitudes (i.e., $\gamma_v p N^*$ electrocouplings) of most nucleon excited states in the mass range up to 1.8 GeV for photon virtualities $Q^2 < 5$ GeV^2 (or $Q^2 < 7.5$ GeV^2 for the $\Delta(1232)3/2^+$ and $N(1535)1/2^-$) [47–51]. Analyses of these results have revealed many facets of strong interactions in the sQCD regime seen in the generation of N^* states of different quantum numbers with different structural features [28,39,52–61]. These results also enable the evaluation of the resonant contributions to inclusive electron scattering observables [62–64], substantially expanding the capability to explore both polarized and unpolarized parton distribution functions (PDFs) of the nucleon for fractional parton light-front momenta close to unity. Analyses of the results on $\gamma_v p N^*$ electrocouplings within CSMs [1,28,48,59,65–75] have demonstrated a new and promising potential for elucidation of the sQCD dynamics that are responsible for the generation of >98% of the visible mass in the Universe.

Explaining the emergence of hadron mass represents one of the most challenging open problems in the Standard Model (SM). The emergent nature of hadron mass is made manifest by a comparison between the measured proton and neutron masses and the sum of the current masses of their valence quark constituents. Protons and neutrons are bound systems of three light u- and d-quarks. The sum of the current masses of these quarks, which is generated by Higgs couplings into QCD, accounts for less than 2% of the measured nucleon masses (see Table 1). This accounting clearly indicates that the overwhelmingly dominant component of the nucleon mass is created by mechanisms other than those associated with the Higgs boson [8–15].

The past decade of progress using CSMs to study the evolution of hadron structure with distance, maintaining a traceable and often direct connection to the QCD Lagrangian, has conclusively demonstrated that the dominant part of each hadron's mass is generated by strong interactions in the regime of large QCD running coupling [1,8–15,73]. Solving QCD's equations-of-motion for the gluon and quark fields has revealed the emergence of quasiparticles, with the quantum numbers of the Lagrangian partons but carrying momentum-dependent masses that are large in the sQCD domain. It is the presence of these quasiparticles within mesons and baryons that explains the greatest part of the visible mass in the Universe.

Table 1. Comparison between the measured masses of the proton and neutron, $m_{p,n}$, and the sum of the current-quark masses of their three u- and d-quark constituents [76]. (Current quark masses are listed at a scale of 2 GeV, but the comparison remains qualitatively unchanged if renormalization group invariant current masses are used.)

	Proton	Neutron
Measured masses (MeV)	$938.2720813 \pm 0.0000058$	$939.5654133 \pm 0.0000058$
Sum of the current quark masses (MeV)	$8.09^{+1.45}_{-0.65}$	$11.50^{+1.45}_{-0.60}$
Contribution of the current quark masses to the measured nucleon mass (%)	<1.1	<1.4

Herein we describe advances in the exploration of the structure of nucleon excited states, using data from the 6-GeV-era experiments at JLab, and discuss the impact of these results on the understanding of EHM. A successful description of JLab results on the $\Delta(1232)3/2^+$ and $N(1440)1/2^+$ electrocouplings has been achieved using CSMs [65,66,68,69,71]. The CSM calculations are distinguished by (a) having employed common momentum-dependent mass functions for the dressed quarks, whose behavior is intimately connected with the running of the gluon mass and QCD's effective charge [77] and (b) thereby unifying the description of these electrocouplings with kindred studies of, inter alia, nucleon and pion elastic form factors [68,78,79]. Such successes provide a sound foundation for arguments supporting the potential of experimental results on the Q^2-dependence of nucleon resonance electrocouplings to deliver new information on the running quark mass.

Parameter-free CSM predictions for the Q^2-evolution of the $\Delta(1600)3/2^+$ electrocouplings became available in 2019 [74]. At that time, there were no experimental results for the electroexcitation amplitudes of this resonance. Herein, too, we present the first preliminary experimental results on these amplitudes, obtained from analysis of $\pi^+\pi^- p$ electroproduction data off protons in the W-range up to 1.7 GeV and $2 < Q^2/\text{GeV}^2 < 5$ using the JLab-Moscow State University (JM) meson–baryon reaction model [67,80–83]. The comparison between the CSM predictions and these experimental results represents a further sensitive test of the capability for validating the EHM paradigm [8–15].

Our discussion is organized as follows. In Section 2, the basic features of the EHM paradigm are outlined, with special emphasis on the dressed-quark and gluon running masses and their evolution in the transition from the weak to strong coupling domains of the strong interaction. We also emphasize the complementarity and critical role of combined studies of both meson and baryon structure in validating any understanding of the generation of the dressed masses of gluon and quark quasiparticles. We outline the analysis framework used for extraction of the $\gamma_v p N^*$ electrocouplings from data on exclusive meson electroproduction available from experiments during the 6-GeV era at JLab in Section 3. The impact of these results on the understanding of EHM is presented in Section 3.2. In Section 4, plans for future studies and their prospects in ongoing experiments in the 12-GeV era at JLab are highlighted, along with physics motivations for a possible increase of the JLab electron beam energy up to 22 GeV. Such an upgrade would offer the only foreseeable opportunities to explore QCD dynamics in the full range of distances over which the dominant part of hadron mass and structure emerge, particularly reaching, for the first time, into the kinematic region where perturbative and nonperturbative QCD calculations overlap.

2. Basics for Insight into EHM Using CSMs

The notable progress in developing an understanding of EHM via CSMs has conclusively demonstrated that the dominant part of hadron mass is generated by strong interactions at momentum scales $k \lesssim 2$ GeV. We now sketch the EHM paradigm and discuss how studies of the meson and baryon structure of both ground and excited states offer complementary and crucial information that will enable the elucidation of the sQCD dynamics responsible for EHM and its manifold corollaries.

2.1. CSMs and the EHM Paradigm

Every scheme proposed for the solution of QCD reveals that the current-quark masses, which are generated by Higgs boson couplings into the Lagrangian, acquire momentum-dependent corrections owing to gluon emission and absorption, as illustrated in Figure 1 (top row). Gluons, too, come to be dressed by the analogous processes shown in Figure 1 (lower rows). Treated in a weak coupling expansion, these "gap equations" generate every diagram in perturbation theory. On the other hand, nonperturbative analyses can reveal emergent features of the strong interaction, such as dynamical chiral symmetry breaking (DCSB) and intimations of confinement [14] (Section 5).

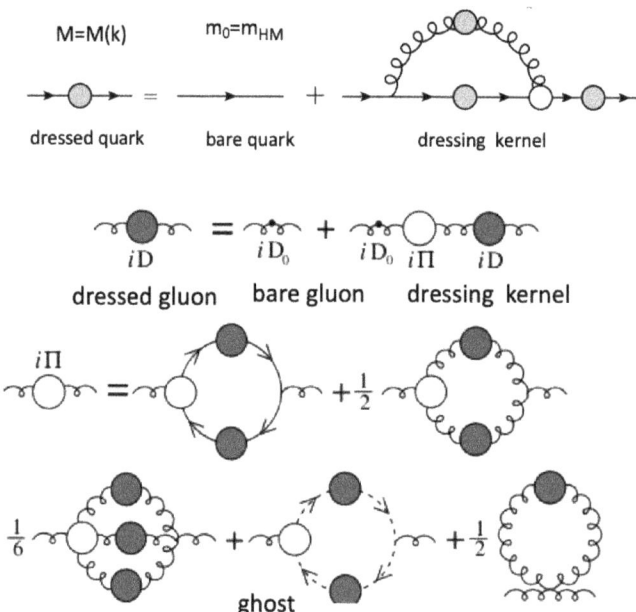

Figure 1. Integral equations for the dressed quark and gluon two-point functions [84] (Section 2.2), drawn in terms of the Feynman diagrams that govern the emergence of gluon and quark quasiparticles from the partons used to express the QCD Lagrangian. (Total momentum k flows from left to right in each diagram, being conserved in passing through the loop integrals.) These quasiparticles are the active components in hadron structure at low resolving scales. Their parton content is revealed at higher resolutions. (Unbroken lines—quarks; spring-like lines—gluons; short-dashed lines—ghosts; filled circles—dressed propagators; open circles—two-point = self-energies and three/four-point = dressed vertices. The vertices satisfy their own Dyson–Schwinger equations, involving higher n-point functions [84]).

As explained elsewhere [84], the integral equations in Figure 1 and their analogs for higher-n-point functions can be understood as QCD's Euler–Lagrange equations, *viz.* QCD's equations of motion. The solutions of those shown explicitly for the dressed quark and gluon two-point functions predict the emergence of gluon and quark quasiparticles. Each is a superposition of enumerably many gluon and quark (and ghost) partons, is characterized by its own momentum-dependent mass function—

$$S(k) = Z(k^2)/[i\gamma \cdot k + M(k^2)], D(k) = 1/[k^2 J(k^2) + m_g^2(k^2)], \qquad (1)$$

drawn in the left panel of Figure 2 (the wave function renormalizations, $Z(k^2)$ and $J(k^2)$, are not displayed and, here, only the gluon's scalar dressing function is written—see References [3,85] for details— and evolves with distance $1/k$, where k is the momentum scale flowing through the diagram, in a well-defined manner that reproduces perturbative results on $m_p/k \simeq 0$.

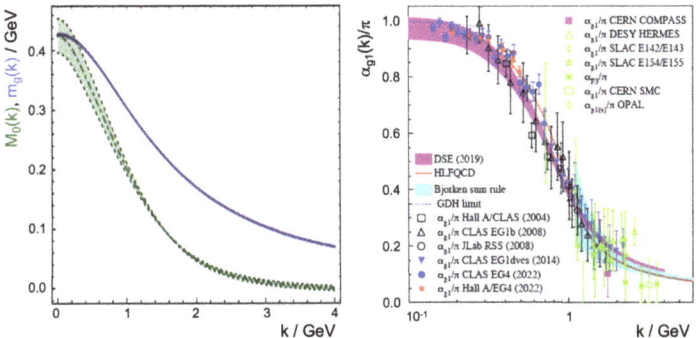

Figure 2. (**Left**): CSM predictions for the momentum dependence of the dressed-gluon (blue solid) and quark (green dot-dashed) masses [9–11]. The associated like-colored bands express the uncertainties in the CSM predictions. (N.B. Since the Poincaré-invariant kinetic energy operator for a vector boson has mass–dimension two and that for a spin-half fermion has mass–dimension unity, then for $m_p^2/k^2 \to 0$, $M_0(k) \propto 1/k^2$ and $m_g^2(k) \propto 1/k^2$, up to $\ln k^2$ corrections). (**Right**) CSM prediction [3] (magenta band) for the process-independent QCD running coupling $\hat{\alpha}(k)$ compared with the empirical results [4] for the process-dependent effective charge defined via the Bjorken sum rule, which is prominent in deep inelastic scattering.

Of primary significance is the dressing of gluons, described by the lower three rows in Figure 1, with effects driven by the three-gluon vertex being most prominent. It was realized long ago [86] that this led to the emergence of a running gluon mass, like that in Figure 2 (left panel), through the agency of a Schwinger mechanism [87,88] in QCD, the details of which have steadily been unfolded during the past fifteen years [89–93]. This essentially nonperturbative consequence of gauge sector dynamics, revealed in both continuum and lattice-regularized studies of QCD, is the first pillar of EHM.

Capitalizing on such progress in understanding gauge sector dynamics, a unique QCD analog of the Gell-Mann–Low effective charge has been defined and calculated [2,3], $\hat{\alpha}(k)$, with the result shown in Figure 2 (right panel). For $k \gtrsim 2$ GeV, this charge matches the pQCD coupling, but it also supplies an infrared completion of the running coupling, which is free of a Landau pole and saturates to the value $\hat{\alpha}(k = 0) = 0.97(4)$. Both these latter features are direct consequences of the emergence of a gluon mass function, whose infrared value is characterized by the renormalization-group-invariant mass-scale $\hat{m} = 0.43(1)$ GeV $\approx m_p/2$. This effective charge is the second pillar of EHM.

As highlighted in Figure 2 (right panel), the pointwise behavior of $\hat{\alpha}(k)$ is almost identical to that of the process-dependent charge [94,95] defined via the Bjorken sum rule [96,97] for reasons that are explained in Reference [14] (Section 4). The form of $\hat{\alpha}(k)$—in particular, its being defined and smooth on the entire domain of spacelike momentum transfers—provides strong support for the conjecture that QCD is a mathematically well-defined quantum gauge field theory. As such, it can serve as a template for extensions of the SM using the notion of compositeness for seemingly pointlike objects.

Turning to the quark gap equation, Figure 1 (top row), and constructing its kernel using the first two pillars of EHM, one obtains a dressed-quark propagator that is characterized by the mass-function shown in Figure 2 (left panel). Critically, for $k \lesssim 2$ GeV, the behavior of this mass function is practically unchanged in the absence of Higgs boson couplings into QCD, i.e., in the chiral limit. Such an outcome is impossible in pQCD. The emergence of

$M(k)$ is the principal manifestation of DCSB in QCD; and this dressed-quark mass function, again a prediction common to both continuum and lattice-regularized QCD, is the third pillar of EHM.

The appearance of dressed-gluon and -quark quasiparticles following the transition into the domain of sQCD, whose strong mutual- and self-interactions are described by a process-independent momentum-dependent effective charge, form the basis for the EHM paradigm and its explanation of hadron mass and structure. Indeed, it is worth reiterating that the dressed-quark mass function in Figure 2 (left panel) shows how the almost massless current-quark partons, which are the degrees-of-freedom best suited for the description of truly high-energy phenomena, are transmogrified, by a nonperturbative accumulation of interactions, into fully dressed quarks. It is these quark quasiparticles, to which is attached an infrared mass-scale $M(k \simeq 0) \approx 0.4\,\text{GeV}$, that provide a link between QCD and the long line of quark potential models developed in the past sixty years [98].

Insofar as the light u- and d-quarks are concerned, Higgs boson couplings into QCD are almost entirely irrelevant to the size of their infrared mass, contributing $< 2\%$ (Reference [11] (Figure 2.5)); hence, equally irrelevant to the masses of the nucleon and its excited states. The dominant component of the masses of all light-quark hadrons is that deriving from $M(k \simeq 0)$, viz. EHM.

Since the quark quasiparticles carry the same quantum numbers as the seed quark-partons, then N^* electroexcitation processes can be used to chart $M(k)$ by exploiting the dependence of the associated N^* electroexcitation amplitudes on the momentum transfer squared. Sketched simply, owing to the quasielastic nature of the transition, the momentum transferred in the process, Q, is shared equally between the three bound quark quasiparticles in the initial and, subsequently, the final states. This means that the quark mass function is predominantly sampled as $M(Q/3)$ because bound-state wave functions are peaked at zero relative momentum. Hence, increasing Q takes the reaction cross section smoothly from the sQCD (constituent-quark) domain into the pQCD domain. Any Poincaré-invariant, QCD-connected calculational framework can then relate the Q^2-dependence of the electroexcitation amplitudes to the momentum dependence of the quark mass function and, crucially, when it comes to predictions, vice versa. Examples are provided in References [28,59,65,66,68–71,73,74,99]. Regarding $M(k)$ in Figure 2, one enters the perturbative domain for $k \gtrsim 2\,\text{GeV}$; hence, a comprehensive mapping of the nonperturbative part of the dressed-quark mass requires

$$0 \leq Q^2/\text{GeV}^2 \lesssim 20 - 30. \qquad (2)$$

Experiments at JLab during the 6-GeV era provided a beginning, their progeny during the 12-GeV era will extend the map further, but only an upgrade of the JLab accelerator energy to beyond 20 GeV will deliver near exhaustive coverage of the full EHM domain.

2.2. Some Highlights from the EHM Experiment-Theory Connection

Charting the dressed-quark mass function using results from hadron structure experiments is a principal goal of modern hadron physics. As always, there are challenges to overcome, but the potential rewards are great. Empirical verification of the EHM paradigm will pave the way to understanding the origin of the vast bulk of the visible mass in the Universe. As an illustration, we note that CSMs have supplied a large body of results for meson and baryon structure observables; some examples are shown in Figure 3. Each of these predictions was obtained within a common theoretical framework and expresses different observable consequences of the dressed-gluon and -quark mass functions shown in Figure 2; hence, each draws a clear connection between observation and the QCD Lagrangian. Notably, here we have only highlighted results for ground-state hadrons because CSM predictions for the $\gamma_v p N^*$ electrocouplings and their comparison with experimental results are discussed below.

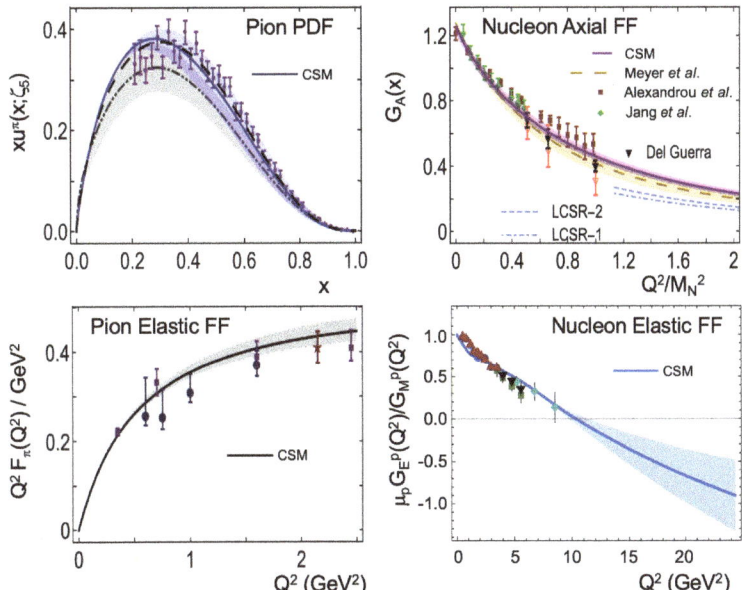

Figure 3. CSM predictions for observables of the structure for the ground state hadrons in comparison with experimental results (points with error bars) or comparable theory. (**Upper left**)—pion valence quark PDF [100]; (**Upper right**)—nucleon axial form factor G_A [101]; (**Lower left**)—pion elastic form factor [11]; and (**Lower right**)—ratio of nucleon elastic electric and magnetic form factors [79]. Sources for comparison curves and points are listed in References [11,79,100,101].

Owing to the pattern of DCSB in QCD, a quark-level Goldberger–Treiman identity [85,102–105]:

$$f_{NG} E_{NG}(k^2) = B(k^2), \quad (3)$$

relates the leading term in the bound-state amplitude of all Nambu–Goldstone (NG) bosons, $E_{NG}(k^2)$, to the scalar piece of the dressed-quark self energy, $B(k^2)$, with the NG boson leptonic decay constant, f_{NG}, providing the constant of proportionality. This exact relationship in chiral-limit QCD is Poincaré-invariant, gauge-covariant, and renormalization-scheme independent. It is also the SM's most fundamental expression of the Nambu–Goldstone theorem [106,107]. Equation (3) explains the seeming dichotomy of massless NG bosons being composites built from massive quark and antiquark quasiparticles, ensuring that all one-body dressing effects that give rise to the quasiparticle masses are canceled exactly by binding energy within the bound states so that they emerge as massless composite objects in the chiral limit [108].

Equation (3) expresses other remarkable facts. It is also a precise statement of equivalence between the pseudoscalar-meson two-body and matter-sector one-body problems in chiral-QCD. These problems are usually considered to be essentially independent. Moreover, it reveals that the cleanest expressions of EHM in the SM are located in the properties of the massless NG bosons. It is worth stressing here that π- and K-mesons are indistinguishable in the absence of Higgs couplings into QCD. Furthermore, as noted above, Equation (3) entails that they are entirely massless in this limit: the π and K mesons are the NG bosons that emerge as a consequence of DCSB. At realistic Higgs couplings, however, π and K observables are windows onto both EHM and its modulation by Higgs boson couplings into QCD.

It is now widely recognized [27,41,43–46] that the quark-level Goldberger–Treiman identity, Equation (3), and its corollaries lift studies of π and K structure to the highest level of importance. CSM calculations are available for a broad range of such observables; e.g., in a challenge for future high-luminosity, high-energy facilities, a prediction for the elastic electromagnetic pion form factor is now available out to $Q^2 = 40 \, \text{GeV}^2$ (see Reference [41] (Figure 9)).

The peculiar character of NG bosons is further highlighted by the mass budgets drawn in Figure 4, which identify that component of the given hadron's mass that is generated by (*i*) EHM; (*ii*) constructive interference between EHM and the Higgs-boson (HB) mass contribution; and (*iii*) that part generated solely by the Higgs. The proton annulus depicts information already presented in Table 1 and highlights again that the proton mass owes almost entirely to the mechanisms of EHM. New information is expressed in the second annulus, which is the ρ-meson mass budget. Plainly, the ρ-meson and proton mass budgets are qualitatively and semi-quantitatively identical, despite one being a meson and the other a baryon.

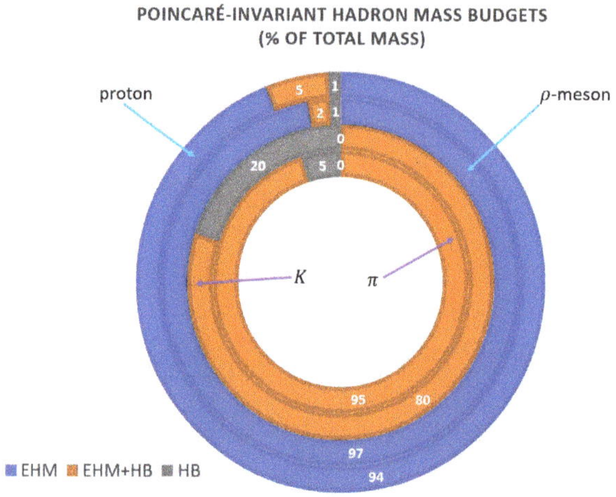

Figure 4. Mass budgets for the proton (outermost annulus), ρ-meson, kaon, and pion (innermost annulus). Each annulus is drawn using a Poincaré-invariant decomposition. The separation is made at a renormalization scale $\zeta = 2 \, \text{GeV}$, calculated using information from References [76,109–111].

The π and K mass budgets in Figure 4 are completely different. For these (near) NG bosons, there is no pure EHM component—no blue part of the ring—because they are massless in the chiral limit. On the other hand, the HB contribution to the pion mass is commensurate with the kindred component of the proton and ρ-meson masses. The biggest contribution for the π is EHM+HB interference: the small HB-only contribution is magnified by a huge, latent EHM component. The K-meson mass budget is similar. However, the larger current mass of the s-quark entails that the HB-alone contribution is four times larger in the K than in the π, but it is not $\sim$15-times larger, as a simple counting of current masses would suggest. Evidently, there is some subtlety in EHM+HB interference effects.

This discussion summarizes what others have explained in detail [8–15], namely, that studies of NG bosons on one hand and the nucleon and its excited states on the other provide complementary information about the mechanisms behind EHM: NG bosons reveal much about EHM+HB interference, whereas the other systems are directly and especially sensitive to EHM-only effects. It follows that consistent results on the dressed-quark mass function and, therefrom, indirectly, on the gluon mass and QCD effective coupling, obtained from experimental studies of these complementary systems—NG bosons and the nucleon and its excitations—will shine the brightest light on the many facets and expressions of emergent hadron mass and structure in Nature. Such a broad approach is the best (only?) way to properly verify the EHM paradigm.

3. Nucleon Resonance Electrocouplings and Their Impact on the Insight into EHM

The contemporary application of CSMs provides a QCD-connected framework that enables the development of an understanding of EHM [8–15] from the comparison of theory predictions with experimental results on the Q^2-evolution of nucleon elastic form factors and nucleon resonance electroexcitation amplitudes [48,67,72,75]. In this Section, we provide an overview of experimental $\gamma_v p N^*$ results where comparisons with CSM predictions exist.

3.1. Extraction of Electrocouplings from Exclusive Meson Electroproduction Data

Nucleon resonance electroexcitations can be fully described in terms of three electroexcitation amplitudes or $\gamma_v p N^*$ electrocouplings. $A_{1/2}(Q^2)$ and $A_{3/2}(Q^2)$ describe resonance production in the process $\gamma_v p \to N^*, \Delta^*$ by transversely polarized photons of helicity $+1$ (-1) and target proton helicities $\pm 1/2$ ($\mp 1/2$) in the center-of-mass (CM) frame, with the resonance spin projection, directed parallel (antiparallel) to the γ_v momentum, equal to $1/2$ ($-1/2$) and $3/2$ ($+3/2$), respectively. The resonance electroexcitation amplitudes of the other (flipped) helicities of the initial photon and target proton and the resonance spin projections are related by parity transformations. $S_{1/2}(Q^2)$ describes accordingly the resonance electroexcitation by a longitudinal virtual photon of zero helicity and target proton helicities $\pm 1/2$, with the absolute value of the resonance spin projection equal to $1/2$ [49]. Since parity is conserved in both electromagnetic and strong interactions, the $A_{1/2}(Q^2)$, $A_{3/2}(Q^2)$, and $S_{1/2}(Q^2)$ electrocouplings describe all possible N^* electroexcitation amplitudes. These electrocouplings are unambiguously determined through their relation with the resonance electromagnetic decay widths, Γ_γ^T and Γ_γ^L, to the final state for transversely and longitudinally polarized photons:

$$\Gamma_\gamma^T(W = M_r, Q^2) = \frac{q_{\gamma,r}^2(Q^2)}{\pi} \frac{2M_N}{(2J_r + 1)M_r} \left(|A_{1/2}(Q^2)|^2 + |A_{3/2}(Q^2)|^2 \right), \quad (4a)$$

$$\Gamma_\gamma^L(W = M_r, Q^2) = \frac{q_{\gamma,r}^2(Q^2)}{\pi} \frac{2M_N}{(2J_r + 1)M_r} |S_{1/2}(Q^2)|^2, \quad (4b)$$

with $q_{\gamma,r} = q_\gamma|_{W=M_r}$, the absolute value of the γ_v three momentum at the resonance point, M_r and J_r being the resonance mass and spin, respectively, and M_N the nucleon mass. W is the sum of the energies of the γ_v and target proton in their CM frame.

Alternatively, the resonance electroexcitation can be described by three transition form factors, $G_{1,2,3}(Q^2)$ or $G^*_{M,E,C}(Q^2)$, which represent Lorentz invariant functions in the most general expressions for the $N \to N^*$ electromagnetic transition currents. For spin $1/2$ resonances, the $F^*_{1,2}(Q^2)$ Dirac and Pauli transition form factors can also be used instead of $G_{1,2,3}(Q^2)$ or $G^*_{M,E,C}(Q^2)$. The description of resonance electroexcitation in terms of the electrocouplings and the electromagnetic transition form factors is completely equivalent, since they are unambiguously related, as described in References [49,112].

The $\gamma_v p N^*$ electrocouplings have been determined from data on exclusive meson electroproduction for most relevant channels in the resonance excitation region, including πN, ηp, and $\pi^+\pi^- p$. The extractions for KY channels are still awaiting the development of a reaction model capable of describing electroproduction observables with accuracy sufficient for the reliable separation of the resonant/non-resonant contributions [113,114]. The full amplitude for any exclusive electroproduction channel can be described as the coherent sum of the N^* electroexcitations in the s-channel for the virtual photon–proton interaction and a complex set of non-resonant mechanisms, as depicted in Figure 5. The electrocouplings determined from all exclusive meson electroproduction channels should be the same for a given N^* state since they should be independent of their hadronic decays, while the non-resonant amplitudes are different for each exclusive meson electroproduction channel. Hence consistent results on the Q^2-evolution of the electrocouplings extracted from different decay channels enable evaluation of the systematic uncertainties related to the use of the reaction models employed in the analysis.

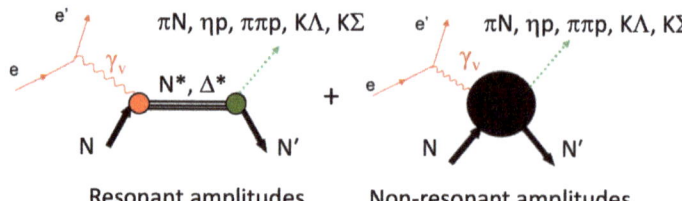

Figure 5. Resonant and non-resonant amplitudes contributing to exclusive meson electroproduction channels in the resonance region.

Systematic studies of N^* electroexcitation from the data became feasible only after experiments during the 6-GeV era with the CLAS detector in Hall B at JLab. This detector has collected the dominant part of available world data on most single- and multi-meson electroproduction channels off protons in the resonance region for Q^2 up to 5 GeV^2 [47–49,72,75]. The data are stored in the CLAS Physics Database [115,116]. For the first time, a large body of data ($\approx$ 150k points) on differential cross sections and polarization asymmetries has become available with nearly complete coverage for the final state hadron CM emission angle, which is important for the reliable extraction of electrocouplings.

Several reaction models have been developed for the extraction of electrocouplings from independent studies of the πN [117–126], ηN [126–130], and $\pi^+\pi^- p$ [67,80–83] electroproduction channels off protons. Coupled-channel approaches [131–133] are making steady progress toward determining the electrocouplings from global multichannel analyses of the combined data for exclusive meson photo-, electro-, and hadroproduction. These analyses will allow for the explicit incorporation of final state interactions between all open channels for the strong interactions between the final state hadrons. Application of such advanced coupled-channel approaches will also enable the restrictions imposed on the photo-, electro-, and hadroproduction amplitudes by the general unitarity condition to be consistently taken into account. An important extension of the database on the exclusive meson hadroproduction channels is expected from the JPARC experimental program [134,135]. These data will be of particular importance in extending the extraction of the electrocouplings within global multichannel analyses toward $W > 1.6$ GeV.

Analyses of CLAS results from the exclusive πN, ηp, and $\pi^+\pi^- p$ electroproduction channels have provided the first and still only available comprehensive information on the electrocouplings of most excited proton states in the range of $W < 1.8$ GeV and $Q^2 < 5$ GeV^2 (see Table 2). The experiments of the 6-GeV era in Halls A/C at JLab further extended this information, providing $\Delta(1232)3/2^+$ and $N(1535)1/2^-$ electrocouplings for $Q^2 < 7$ GeV^2 [50,51].

As representative examples, the transverse $A_{1/2}(Q^2)$ electrocouplings versus Q^2 for the $N(1440)1/2^+$ and $N(1520)3/2^-$, obtained from independent studies of the πN [121,123] and $\pi^+\pi^- p$ [67,83] channels, are shown in Figure 6. The electrocouplings inferred from data on the two major πN and $\pi^+\pi^- p$ electroproduction channels, with different non-resonant contributions, are consistent. This success, reproduced for all available electrocouplings and reaction channels (see Table 2), has demonstrated the capabilities of these reaction models, developed by the CLAS Collaboration, for the credible extraction of the $\gamma_v p N^*$ electrocouplings from independent studies of different electroproduction channels.

Table 2. Summary of the results for the $\gamma_v p N^*$ electrocouplings from the πN, ηp, and $\pi^+\pi^- p$ electroproduction channels measured with the CLAS detector in Hall B at JLab.

Meson Electroproduction Channels	Excited Proton States	Q^2 Ranges for Extracted $\gamma_v p N^*$ Electrocouplings, GeV^2
$\pi^0 p$, $\pi^+ n$	$\Delta(1232)3/2^+$	0.16–6
	$N(1440)1/2^+$, $N(1520)3/2^-$	0.30–4.16
	$N(1535)1/2^-$	0.30–4.16
$\pi^+ n$	$N(1675)5/2^-$, $N(1680)5/2^+$	1.6–4.5
	$N(1710)1/2^+$	
ηp	$N(1535)1/2^-$	0.2–2.9
$\pi^+\pi^- p$	$N(1440)1/2^+$, $N(1520)3/2^-$	0.25–1.50
	$\Delta(1600)3/2^+$, $\Delta(1620)1/2^-$	2.0–5.0
	$N(1650)1/2^-$, $N(1680)5/2^+$,	
	$\Delta(1700)3/2^-$	0.50–1.50
	$N(1720)3/2^+$, $N'(1720)3/2^+$	0.50–1.50

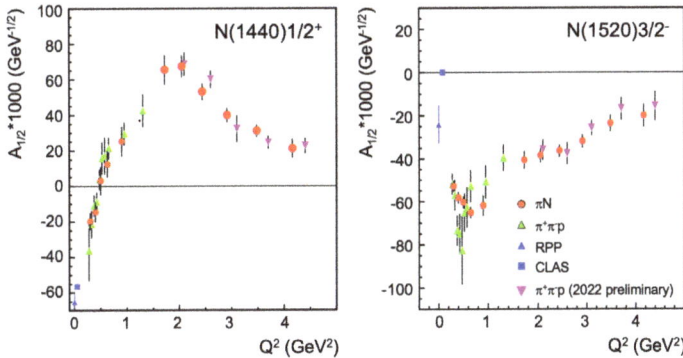

Figure 6. $N(1440)1/2^+$ and $N(1520)3/2^-$ electrocouplings extracted from the πN [121,123] and $\pi^+\pi^- p$ [67,83,136,137] electroproduction channels. The photocouplings from the Review of Particle Properties (RPP) [76] and from Reference [138] are shown by the blue squares and triangles, respectively.

3.2. Insights into the Dressed-Quark Mass Function from the $\gamma_v p N^*$ Electrocouplings

Results on the Q^2-evolution of the $\gamma_v p N^*$ electrocouplings available from experiments performed during the 6-GeV era at JLab have already had a substantial impact on understanding the sQCD dynamics responsible for the saturation of the running coupling $\hat{\alpha}$, N^* structure, and the generation of a significant portion of hadron mass [1,28]. Analyses of these results have revealed N^* structure to emerge from a complex interplay between the inner core of three dressed quarks and an outer meson–baryon cloud [1,28,67,139]. Successful descriptions of the data on the dominant $N \rightarrow \Delta(1232)3/2^+$ magnetic transition form factor [50,121] and the electrocouplings of the $N(1440)1/2^+$ [75,121,123] have been achieved using CSMs [65,66,68,69] for $Q^2 > 1.0$ GeV^2 and $Q^2 > 2.0$ GeV^2, respectively (see Figure 7). These Q^2 ranges correspond to the distance scales where contributions from the quark core to the resonance structure come to dominate. Since the CSM evaluations account for the contributions from only the quark core, they can only reasonably be confronted with experimental results in the higher-Q^2 range, where the quark core contributions to N^* structure dominate over those from the meson–baryon cloud.

It is worth noting here that the character of this separation between the inner core and outer cloud is detailed, e.g., in Ref. [140] (Section 4.2). It posits that all hadron quark cores are the same, remaining practically unaffected by exterior meson–baryon dynamics. Verification of this perspective must await an exact solution of the coupled core+cloud many-body problem in quantum field theory. Meanwhile, the fact that it delivers agreement with form factor data on numerous ground- and excited-state hadrons (mesons and baryons) provides strong empirical evidence in support of the position.

The sensitivity of the electroexcitation amplitudes to the momentum-dependence of the quark mass function is dramatically illustrated by Figure 7, which deliberately shows results obtained with $M(k) = $ constant [65,66] and $M(k)$ from Figure 2 [68,69]. The $M(k) = 0.36$ GeV results were computed first in order to provide a "straw-man" benchmark against which the subsequent realistic $M(k)$ results could be contrasted. The constant-mass results (dotted red curves in Figure 7) overestimate the data on the $N \to \Delta$ magnetic transition form factor for $Q^2 \gtrsim 1$ GeV^2. The discrepancy increases with Q^2, approaching an order of magnitude difference in the ratio at 5 GeV^2. Moreover, whilst reproducing the zero in $A_{1/2}$ for the $N(1440)1/2^+$, the frozen mass result is otherwise incompatible with the data. Plainly, therefore, the data speak against dressed quarks with a frozen mass. On the other hand, in both cases, the transition form factors are well described by an internally consistent CSM calculation built upon the mass function in Figure 2—see the solid blue curves in Figure 7. These observations confirm the statements made above, $viz.$ nucleon resonance electroexcitation amplitudes are keenly sensitive to the form of the running quark mass. Moreover, the agreement with the larger-Q^2 data clearly points to a dominance of the dressed-quark core of the nucleon resonances in the associated domains.

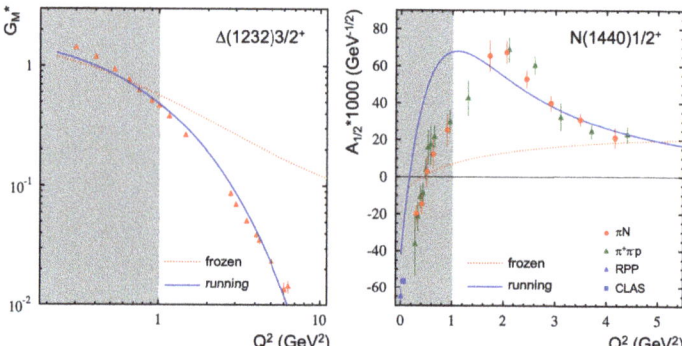

Figure 7. Description of the results for the $N \to \Delta$ magnetic transition form factor G_M^* (**left**) and the electrocoupling amplitude $A_{1/2}$ for the $N \to N(1440)1/2^+$ (**right**) achieved using CSMs [65,66,68,69]. Results obtained with a momentum-independent (frozen) dressed-quark mass [65,66] (dotted red curves) are compared with QCD-kindred results (solid blue curves) obtained with the momentum-dependent quark mass function in Figure 2. The electrocoupling data were taken from References [50,121,123]—πN electroproduction, and References [67,83,136,137]—$\pi^+\pi^- p$ electroproduction. The photocouplings for the $N(1440)1/2^+$ are from the RPP [76] and from Reference [138]—blue square and triangle, respectively. The ranges of Q^2 where the contributions from the meson–baryon cloud remain substantial are highlighted in gray.

It is worth stressing that the CSM results for the $\Delta(1232)3/2^+$ and $N(1440)1/2^+$ electroexcitation amplitudes were obtained using the same dressed-quark mass function, i.e., $M(k)$ in Figure 2: indeed, the theoretical analyses of both transitions used precisely the same framework. The common quark mass function matches that obtained by solving the quark gap equation in Figure 1 with a kernel built from the best available inputs for [77]: the gluon two-point function, running coupling, and dressed gluon–quark vertex. Moreover, the same mass function was also used in the successful description of the experimental results on nucleon elastic electromagnetic form factors [68,79], and axial and pseudoscalar form factors [101,141]. Such a mass function is also a key element in an ab $initio$ treatment of pion electromagnetic elastic and transition form factors [78,142,143].

These CSM results for meson and baryon properties, both ground and excited states, are part of a large body of mutually consistent predictions. Their success in describing and explaining data relating to such a diverse array of systems provides strong evidence in support of the position that dressed quarks, with dynamically generated running masses, are the appropriate degrees-of-freedom for use in the description of the mass and structure of all hadrons. This realization is one of the most important achievements of hadron physics during the past decade, and it was only accomplished through numerous synergistic interactions between experiment, phenomenology, and theory.

3.3. Novel Tests of CSM Predictions

In 2019, CSM predictions became available for the electrocouplings of the $\Delta(1600)3/2^+$ [74]. This baryon may be interpreted in quantum field theory as a state with aspects of the character of a first radial excitation of the $\Delta(1232)3/2^+$ [56,61], for which CSM electrocoupling results became available earlier [68] and are discussed above.

No relevant experimental results were available when the $\Delta(1600)3/2^+$ predictions were made. The first (and still preliminary) results for the $\Delta(1600)3/2^+$ electrocouplings only became available in the first half of 2022 [136,137,144]. They were extracted from the analysis of $\pi^+\pi^-p$ electroproduction off protons measured with the CLAS detector [145,146] for W from 1.4–2.1 GeV and Q^2 from 2–5 GeV^2. Nine independent one-fold differential cross sections were analyzed in each (W,Q^2) bin. The final-state hadron kinematics is fully determined by the five-fold differential cross sections. The one-fold differential cross sections were obtained by integrating the five-fold differential cross sections over different sets of four kinematic variables [82,83]. For extraction of the electrocouplings, it was necessary to fit the data for the three invariant mass distributions for the different pairs of final-state hadrons, the distributions of the final-state hadrons over the CM polar angles θ_i ($i = \pi^+, \pi^-, p_f$), and the distributions over the three CM angles $\alpha_{[i],[j]}$ between the two planes: one of which [i] is the reaction plane defined by the three-momentum of the γ_v and one of the final state hadrons, and the second [j] is determined by the three-momenta of the other two final-state hadrons for the three possible choices of the hadron pairs. Representative examples of the data measured are shown in Figure 8 at the W-bins closest to the Breit–Wigner mass of the $\Delta(1600)3/2^+$ and in different bins of Q^2.

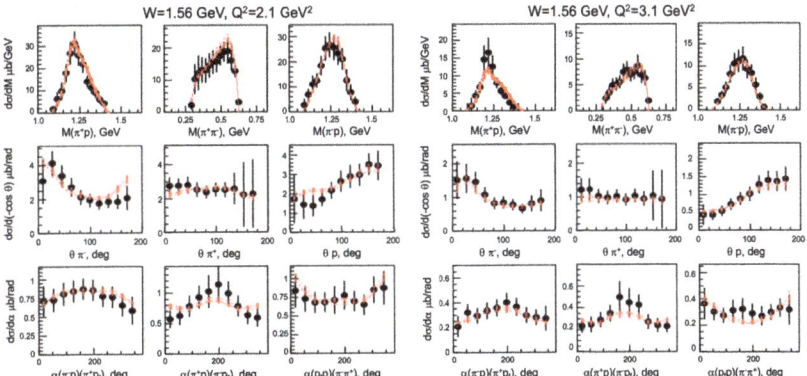

Figure 8. Regarding extraction of $\Delta(1600)3/2^+$ electrocouplings, representative examples of the nine independent one-fold differential cross sections available from the $\pi^+\pi^-p$ measurements with CLAS [145,146] at two different Q^2 values, along with the data fits within the data-driven meson–baryon JM reaction model [75,82,83].

The N^* electrocouplings on the domain $W < 1.65$ GeV were obtained from the fit of the differential $\pi^+\pi^-p$ photo- and electroproduction cross sections carried out within the framework of the data-driven JM meson–baryon reaction model [67,80–83]. This model has been developed by the CLAS Collaboration for the extraction of nucleon resonance electrocouplings and their partial hadronic decay widths to the $\pi\Delta$ and ρp final states. Within the

JM model, the full 3-body $\pi^+\pi^-p$ electroproduction amplitude includes the contributions from $\pi^-\Delta^{++}$, ρp, $\pi^+\Delta^0$, $\pi^+N(1520)3/2^-$, and $\pi^+N(1685)5/2^+$, with subsequent decays of the unstable intermediate hadrons. It also contains direct 2π photo-/electroproduction processes, where the final $\pi^+\pi^-p$ state is created without the generation of unstable intermediate hadrons. Here the nucleon resonances contribute to the $\pi^-\Delta^{++}$, $\pi^+\Delta^0$, and ρp channels.

Modeling of the non-resonant contributions is described in References [80–83]. For the resonant contributions, the JM model includes all four-star Particle Data Group (PDG) N^* states with observed decays to $\pi\pi N$, as well as the new $N'(1720)3/2^+$ resonance [147,148] observed in the combined analysis of $\pi^+\pi^-p$ photo- and electroproduction data. The resonant amplitudes are described within the unitarized Breit–Wigner Ansatz [83], thereby ensuring consistency with restrictions imposed by the general unitarity condition. The JM model offers a good description of the $\pi^+\pi^-p$ differential cross sections in the entire kinematic area covered by the data at $W < 2.1$ GeV and $Q^2 < 5$ GeV^2. All of the electrocouplings extracted from the $\pi^+\pi^-p$ data (published, in part, also in the PDG) have been determined using the JM reaction model.

In the analyses of the $\pi^+\pi^-p$ data [145,146], the following quantities were varied: the $\gamma_v p N^*$ electrocouplings for the resonances in the mass range < 1.75 GeV, their partial hadronic decay widths into the $\pi\Delta$ and ρp final states, their total decay widths, and the non-resonant parameters of the JM model. For each trial attempt at a data description, $\chi^2/d.p.$ (d.p. = data point) was computed using the comparison between the measured and computed nine one-fold differential cross sections. In the fits, the computed cross sections closest to the data were selected by requiring $\chi^2/d.p.$ to be below a predetermined threshold, ensuring that the spread of the selected phenomenological fit cross sections lies within the data uncertainties for most experimental data points. Representative examples are shown by the family of curves in Figure 8.

The electrocouplings for the computed cross sections selected from the data fits were averaged together, and their means were treated as the experimental value. The RMS width of the determined electrocouplings was assigned as the corresponding uncertainty. The preliminary results of this extraction for the $\Delta(1600)3/2^+$ [136,137,144] are shown in Figure 9, wherein they are compared with the CSM predictions obtained three years earlier [74]. These results were determined for overlapping W-intervals: 1.46–1.56 GeV, 1.51–1.61 GeV, and 1.56–1.66 GeV for Q^2 from 2–5 GeV^2. The non-resonant contributions in these W-intervals are different. The electrocouplings determined from the independent fits of the data within the three W-intervals are consistent, establishing their reliability and confirming the CSM predictions. This success has markedly strengthened the body of evidence that indicates that detailed information can be obtained on the momentum dependence of the dressed-quark mass function from sound data on N^* electroproduction and, therefrom, deep insights into the character of EHM in the SM.

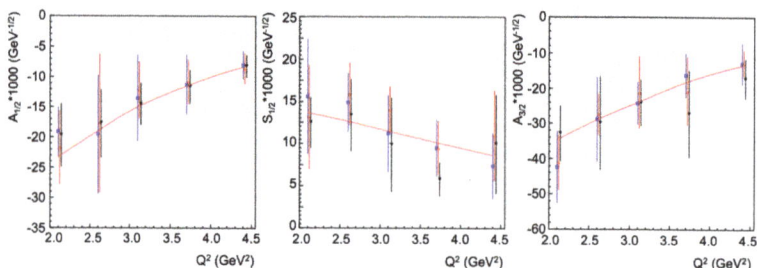

Figure 9. Preliminary $\Delta(1600)3/2^+$ electrocouplings with their assigned uncertainties, determined from independent analysis of the $\pi^+\pi^-p$ differential cross sections in three overlapping W intervals: 1.46–1.56 GeV (filled blue squares), 1.51–1.61 GeV (filled red triangles), and 1.56–1.66 GeV (filled black triangles) [136,137,144]. CSM predictions [74] are drawn as solid red curves.

4. Studies of N^* Structure in Experiments with CLAS12 and Beyond

Most results on the N^* electrocouplings have been obtained for $Q^2 < 5\,\text{GeV}^2$. Detailed comparison of these results with the CSM predictions allows for exploration of the quark mass function within the range of quark momenta $< 0.75\,\text{GeV}$, assuming equal sharing of the virtual-photon momentum transfer between the three dressed quarks in the transition between the ground and excited nucleon states. The results on the resonance electrocouplings in this range of quark momentum, shown in Figure 10 (top), cover distances over which less than 30% of hadron mass is generated [9,10].

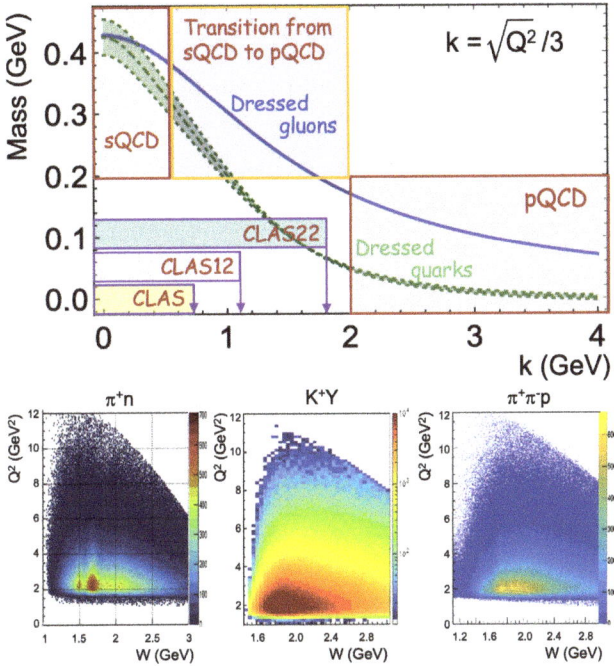

Figure 10. (Top) Momentum ranges accessible in the exploration of the momentum dependence of the dressed-quark mass function using results on the Q^2-evolution of $\gamma_v p N^*$ electrocouplings. The range of k covered by available data is mostly from experiments with CLAS, shown in yellow. The expected reach of CLAS12 experiments is shown in purple, and that achievable after a proposed increase of the JLab beam energy to 22 GeV in cyan. (Bottom) Yields of representative exclusive meson electroproduction channels available from the experiments with the CLAS12 detector.

In the northern spring of 2018, after completion of the 12-GeV-upgrade project, measurements with the CLAS12 detector in Hall B at JLab commenced [47,48,149]. Currently, CLAS12 is the only facility in the world capable of exploring exclusive meson electroproduction in the resonance region, exploiting the highest Q^2 ever achieved for these processes. Ongoing experiments with electron beam energies up to 11 GeV with CLAS12 offer a unique opportunity to obtain information on the electrocouplings of the most prominent N^* states in the mass range up to 2.5 GeV at Q^2 up to 10 GeV^2 from the exclusive πN, KY ($Y = \Lambda$ or Σ), K^*Y, KY^*, and $\pi^+\pi^- p$ channels [113,114,150–152]. Q^2 versus W event distributions for these exclusive reaction channels measured with CLAS12 at a beam energy of $\sim$11 GeV are shown in Figure 10 (bottom). The first results from the CLAS12 N^* program (at lower beam energies of 6.5/7.5 GeV) have recently been published on the beam-recoil hyperon transferred polarization in K^+Y electroproduction [153]. The increase in the Q^2-coverage for results on the electrocouplings from CLAS12 will enable exploration of the dressed-quark mass within the range of quark momenta where roughly 50% of hadron mass is expected to be generated (see Figure 10 (top)).

In order to solve the challenging SM problems relating to EHM, the dressed quark mass function should be charted over the entire quark momentum range up to ≈2 GeV. This is the domain of transition from strong to perturbative QCD (see Figure 10 (top)) and where dressed quarks and gluons become the relevant degrees-of-freedom as $\hat{\alpha}/\pi \to 1$, approaching the sQCD saturation regime (see Figure 2). This objective requires a further extension of the electrocoupling measurements up to $Q^2 \approx 30\,\text{GeV}^2$. Discussions and planning are currently underway, focusing on an energy increase of the JLab accelerator to a beam energy of 22 GeV, after the completion of the experiments planned for the 12-GeV program. Initial simulations of πN, KY, and $\pi^+\pi^- p$ electroproduction at 22 GeV using the existing CLAS12 detector at a luminosity up to $(2-5) \times 10^{35}$ cm^{-2}s^{-1} have shown that a measurement program of 1–2 years duration would enable measurements of sufficient statistical accuracy to determine the $\gamma_v pN^*$ electrocouplings of the most prominent N^* states over this full kinematic range.

Figure 11 shows the luminosity versus CM energy in lepton–proton collisions for existing and foreseeable facilities capable of exploring hadron structure in measurements with large-acceptance detectors. The luminosity requirements for the extraction of electrocouplings within the Q^2 range 10-30 GeV^2 exceed by more than an order-of-magnitude the maximum luminosity planned for experiments with the EIC [42] and EicC [45] ep colliders and even more for other facilities. The combination of a high duty-factor JLab electron beam at 22 GeV with the capacity to measure exclusive electroproduction reactions at luminosities of $(2-5) \times 10^{35}$ cm^{-2}s^{-1} using a large-acceptance detector, would make a 22 GeV JLab unique. It would be the only facility in the world able to explore the evolution of hadron structure over essentially the full range of distances where the transition from strong-coupling QCD to the weak-field domain is expected to occur.

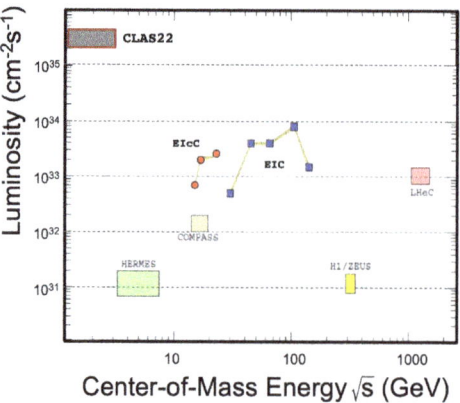

Figure 11. Luminosity versus CM energy in lepton–proton collisions for existing and foreseeable facilities capable of exploring hadron structure in measurements with large-acceptance detectors.

The increase of the JLab energy to 22 GeV, pushing the current CLAS12 detector capabilities to measure exclusive electroproduction to the highest possible luminosity and extending the available reaction models used for the extraction of the electrocouplings, will offer the only foreseeable opportunity to explore how the dominant part of hadron mass (up to 85%) and N^* structure emerge from QCD. This would make an energy-upgraded JLab at 22 GeV the ultimate QCD facility at the luminosity frontier.

5. Conclusions and Outlook

Baryons are the most fundamental three-body systems in Nature. If we do not understand how QCD generates these bound states of three dressed quarks, then our understanding of

Nature is incomplete. Remarkable progress has been achieved in recent decades through the studies of the structure of the ground and excited nucleon states in experiments at JLab during the 6-GeV era [1,28,47–49,72,154,155]. These experiments have provided a large array of new opportunities for QCD-connected hadron structure theory by opening a door to the exploration of many hitherto unseen facets of the strong interaction in the regime of large running coupling, i.e., $\alpha_s/\pi \gtrsim 0.2$, by providing the results on the $\gamma_v p N^*$ electrocouplings for numerous N^* states, with different quantum numbers and structural features.

High-quality meson electroproduction data from the 6-GeV era at JLab have enabled the determination of the electrocouplings of most nucleon resonances in the mass range up to 1.8 GeV for $Q^2 < 5$ GeV^2 (up to 7.5 GeV^2 for the $\Delta(1232)3/2^+$ and $N(1535)1/2^-$). Consistent results on the Q^2-evolution of these electrocouplings from analyses of $\pi^+ n$, $\pi^0 p$, ηp, and $\pi^+\pi^- p$ electroproduction have demonstrated the capability of the reaction models employed to extract the electrocouplings in independent studies of all of these different exclusive channels. Above, we have sketched how comparisons between the experimental results on the Q^2-evolution of the $\gamma_v p N^*$ electroexcitation amplitudes and QCD-connected theory have vastly improved our understanding of the momentum dependence of the dressed-quark mass function, which is one of the three pillars of EHM. The remaining two pillars are the running gluon mass and the QCD effective charge, and these entities, too, are constrained by the electroexcitation data.

A good description of the $\Delta(1232)3/2^+$ and $N(1440)1/2^+$ electrocouplings has been achieved using CSMs in the full range of photon virtualities where the structure of these excited states is principally determined by contributions from a core of three dressed quarks. The successful description of the electrocouplings for nucleon resonances of different structures, spin+isospin flip for the $\Delta(1232)3/2^+$, and the first radial excitation of three dressed quarks for the $N(1440)1/2^+$, was achieved with the same dressed quark mass function. This mass function is determined with QCD dynamics, and such a running mass has also been used in the successful description of data on elastic electromagnetic nucleon and pion form factors, as well as for the description of the nucleon axial form factor G_A. In thereby arriving at a unification of diverse observables, one obtains compelling evidence in support of the momentum dependence of the dressed quark mass used to describe the results on the Q^2-evolution of the electrocouplings.

This impressive hadron physics achievement in the past decade was accomplished through synergistic efforts between experiment, phenomenology, and QCD-connected hadron structure theory. In 2019, CSMs provided parameter-free predictions for the electrocouplings of the $\Delta(1600)3/2^+$. There were no experimental results available at that time. The first, preliminary results on the $\Delta(1600)3/2^+$ electrocouplings extracted from the data on $\pi^+\pi^- p$ electroproduction are reported herein. They have strikingly confirmed the CSM predictions.

Most results for the $\gamma_v p N^*$ electrocouplings are currently available for $Q^2 < 5$ GeV^2, allowing for the exploration of the dressed quark mass within the limited range of quark momenta where less than 30% of hadron mass is expected to be generated. Experiments on exclusive meson electroproduction in the resonance region are now in progress with the CLAS12 detector in Hall B at JLab, following the completion of the 12-GeV-upgrade project. CLAS12 is the only facility in the world capable of obtaining the electrocouplings of all prominent N^* states in the still unexplored Q^2 range from 5–10 GeV^2 from measurements of πN, ηp, $\pi^+\pi^- p$, and KY electroproduction. These data will probe the dressed-quark mass function at quark momenta up to $\approx$1.1 GeV, a domain where up to 50% of hadron mass is generated.

In order to solve the problem of EHM, a key challenge within the SM, the dressed-quark mass function should be mapped over the entire range of quark momenta up to $\approx$2 GeV, where the transition from strong to perturbative QCD takes place and where gluon and quark quasiparticles with dynamically generated running masses emerge as $\hat{\alpha}/\pi \to 1$. This requires an extension of existing and anticipated data so that it covers the Q^2-domain from 10–30 GeV^2. Explorations of the possibility to increase the JLab beam energy to 22 GeV are now in progress. Such a machine would enable coverage over the desired Q^2 range within the region of $W < 2.5$ GeV. Simulations with the existing CLAS12 detector configuration for the exclusive πN, KY, and $\pi^+\pi^- p$ electroproduction channels at 22 GeV beam energy and a luminosity of $(2-5) \times 10^{35}$ cm^{-2}s^{-1} show that, with beam-times of 1–2 years, differential cross section and polarization asymmetry measurements of sufficient statistical precision can be achieved to extract electrocouplings of all prominent resonances up to 30 GeV^2. Both the EIC and EicC *ep* colliders would need much higher and foreseeably unreachable luminosities than currently envisaged in order to carry out such a program. The combination of a high duty-factor 22 GeV JLab electron beam and the capability to measure exclusive electroproduction events at high luminosities with a large-acceptance detector would make JLab the ultimate QCD facility at the luminosity frontier. It would be unique in possessing the capacity to explore the evolution of hadron mass and structure over the full range of distances where the transition from sQCD to pQCD is expected.

Drawing a detailed map of proton structure is important because the proton is Nature's only absolutely stable bound state. However, understanding how QCD's simplicity explains the emergence of hadron mass and structure requires investment in a facility that can deliver precision data on much more than one of Nature's hadrons. An energy-upgraded JLab complex is the only envisaged facility that could enable scientists to produce a sufficient quantity of precise structure data on a wide range of hadrons with distinctly different quantum numbers and thereby move into a new realm of understanding. There is elegance in simplicity and beauty in diversity. If QCD possesses both, then it presents a very plausible archetype for taking science beyond the Standard Model. In that case, nuclear physics at JLab 20+ has the potential to deliver an answer that takes science far beyond its current boundaries.

Funding: Work supported by: U.S. Department of Energy (contract no. DE-AC05-06OR23177) (DSC, VIM), National Science Foundation (NSF grant PHY 10011349) (RWG), and National Natural Science Foundation of China (grant no. 12135007) (CDR).

Data Availability Statement: Not applicable.

Acknowledgments: This contribution is based on results obtained and insights developed through collaborations with many people, to all of whom we are greatly indebted.

Conflicts of Interest: The authors declare no conflict of interest.

Abbreviations

The following abbreviations are used in this manuscript:

CM	center-of-mass
CSM	continuum Schwinger function method
DCSB	dynamical chiral symmetry breaking
d.p.	data point
EHM	emergence of hadron mass
EIC	Electron-Ion Collider (at Brookhaven National Laboratory)
EicC	Electron-ion collider China
HB	Higgs boson
JLab	Thomas Jefferson National Accelerator Facility (Jefferson Laboratory)
JM	JLab-Moscow State University
lQCD	lattice-regularized quantum chromodynamics
NG (mode/boson)	Nambu-Goldstone (mode/boson)
PDFs	Particle Distribution Functions
PDG	Particle Data Group (and associated publications)
pQCD	perturbative QCD
QCD	quantum chromodynamics
RMS	root mean square
RPP	Review of Particle Properties (and associated publications)
sQCD	strong QCD
SM	Standard Model of particle physics

References

1. Brodsky, S.J.; Burkert, V.D.; Carman, D.S.; Chen, J.P.; Cui, Z.F.; Döring, M.; Dosch, H.G.; Draayer, J.; Elouadrhiri, L.; Glazier, D.I.; et al. Strong QCD from Hadron Structure Experiments. *Int. J. Mod. Phys. E* **2020**, *29*, 2030006. [CrossRef]
2. Binosi, D.; Mezrag, C.; Papavassiliou, J.; Roberts, C.D.; Rodriguez-Quintero, J. Process-Independent Strong Running Coupling. *Phys. Rev. D* **2017**, *96*, 054026. [CrossRef]
3. Cui, Z.F.; Zhang, J.L.; Binosi, D.; de Soto, F.; Mezrag, C.; Papavassiliou, J.; Roberts, C.D.; Rodríguez-Quintero, J.; Segovia, J.; Zafeiropoulos, S. Effective Charge from Lattice QCD. *Chin. Phys. C* **2020**, *44*, 083102. [CrossRef]
4. Deur, A.; Burkert, V.; Chen, J.P.; Korsch, W. Experimental Determination of the QCD Effective Charge $\alpha_{g_1}(Q)$. *Particles* **2022**, *5*, 171–179. [CrossRef]
5. Eichmann, G.; Sanchis-Alepuz, H.; Williams, R.; Alkofer, R.; Fischer, C.S. Baryons as Relativistic Three-Quark Bound States. *Prog. Part. Nucl. Phys.* **2016**, *91*, 1–100. [CrossRef]
6. Fischer, C.S. QCD at Finite Temperature and Chemical Potential from Dyson–Schwinger Equations. *Prog. Part. Nucl. Phys.* **2019**, *105*, 1–60. [CrossRef]
7. Qin, S.X.; Roberts, C.D. Impressions of the Continuum Bound State Problem in QCD. *Chin. Phys. Lett.* **2020**, *37*, 121201. [CrossRef]
8. Roberts, C.D.; Schmidt, S.M. Reflections upon the Emergence of Hadronic Mass. *Eur. Phys. J. ST* **2020**, *229*, 3319–3340. [CrossRef]
9. Roberts, C.D. Empirical Consequences of Emergent Mass. *Symmetry* **2020**, *12*, 1468. [CrossRef]
10. Roberts, C.D. On Mass and Matter. *AAPPS Bull.* **2021**, *31*, 6. [CrossRef]
11. Roberts, C.D.; Richards, D.G.; Horn, T.; Chang, L. Insights into the Emergence of Mass from Studies of Pion and Kaon Structure. *Prog. Part. Nucl. Phys.* **2021**, *120*, 103883. [CrossRef]
12. Binosi, D. Emergent Hadron Mass in Strong Dynamics. *Few Body Syst.* **2022**, *63*, 42. [CrossRef]
13. Papavassiliou, J. Emergence of Mass in the Gauge Sector of QCD. *Chin. Phys. C* **2022**, *46*, 112001. [CrossRef]
14. Ding, M.; Roberts, C.D.; Schmidt, S.M. Emergence of Hadron Mass and Structure. *Particles* **2023**, *6*, 57–120. [CrossRef]
15. Ferreira, M.N.; Papavassiliou, J. Gauge Sector Dynamics in QCD. *Particles* **2023**, *6*, 312–363. [CrossRef]
16. Blum, T.; Boyle, P.A.; Christ, N.H.; Frison, J.; Garron, N.; Hudspith, R.J.; Izubuchi, T.; Janowski, T.; Jung, C.; Jüttner, A.; et al. Domain Wall QCD with Physical Quark Masses. *Phys. Rev. D* **2016**, *93*, 074505. [CrossRef]
17. Boyle, P.A.; Christ, N.H.; Garron, N.; Jung, C.; Jüttner, A.; Kelly, C.; Mawhinney, R.D.; McGlynn, G.; Murphy, D.J.; Ohta, S.; et al. Low Energy Constants of SU(2) Partially Quenched Chiral Perturbation Theory from $N_f = 2 + 1$ Domain Wall QCD. *Phys. Rev. D* **2016**, *93*, 054502. [CrossRef]
18. Boyle, P.A.; Del Debbio, L.; Jüttner, A.; Khamseh, A.; Sanfilippo, F.; Tsang, J.T. The Decay Constants $\mathbf{f_D}$ and $\mathbf{f_{D_s}}$ in the Continuum Limit of $\mathbf{N_f = 2 + 1}$ Domain Wall Lattice QCD. *JHEP* **2017**, *12*, 008. [CrossRef]
19. Gao, F.; Qin, S.X.; Roberts, C.D.; Rodríguez-Quintero, J. Locating the Gribov Horizon. *Phys. Rev. D* **2018**, *97*, 034010. [CrossRef]
20. Oliveira, O.; Silva, P.J.; Skullerud, J.I.; Sternbeck, A. Quark Propagator with Two Flavors of O(a)-Improved Wilson Fermions. *Phys. Rev. D* **2019**, *99*, 094506. [CrossRef]
21. Boucaud, P.; De Soto, F.; Raya, K.; Rodríguez-Quintero, J.; Zafeiropoulos, S. Discretization Effects on Renormalized Gauge-Field Green's Functions, Scale Setting, and the Gluon Mass. *Phys. Rev. D* **2018**, *98*, 114515. [CrossRef]
22. Zafeiropoulos, S.; Boucaud, P.; De Soto, F.; Rodríguez-Quintero, J.; Segovia, J. Strong Running Coupling from the Gauge Sector of Domain Wall Lattice QCD with Physical Quark Masses. *Phys. Rev. Lett.* **2019**, *122*, 162002. [CrossRef] [PubMed]
23. Aguilar, A.C.; De Soto, F.; Ferreira, M.N.; Papavassiliou, J.; Rodríguez-Quintero, J. Infrared Facets of the Three-Gluon Vertex. *Phys. Lett. B* **2021**, *818*, 136352. [CrossRef]
24. Aguilar, A.C.; Ambrósio, C.O.; De Soto, F.; Ferreira, M.N.; Oliveira, B.M.; Papavassiliou, J.; Rodríguez-Quintero, J. Ghost Dynamics in the Soft Gluon Limit. *Phys. Rev. D* **2021**, *104*, 054028. [CrossRef]
25. Pinto-Gómez, F.; De Soto, F.; Ferreira, M.N.; Papavassiliou, J.; Rodríguez-Quintero, J. Lattice Three-Gluon Vertex in Extended Kinematics: Planar Degeneracy. *arXiv* **2022**, arXiv:2208.01020.
26. Aznauryan, I.G.; Bashir, A.; Braun, V.M.; Brodsky, S.J.; Burkert, V.D.; Chang, L.; Chen, C.; El-Bennich, B.; Cloët, I.C.; Cole, P.L.; et al. Studies of Nucleon Resonance Structure in Exclusive Meson Electroproduction. *Int. J. Mod. Phys. E* **2013**, *22*, 1330015. [CrossRef]
27. Horn, T.; Roberts, C.D. The Pion: An Enigma within the Standard Model. *J. Phys. G* **2016**, *43*, 073001. [CrossRef]
28. Burkert, V.D.; Roberts, C.D. Colloquium: Roper Resonance: Toward a Solution to the Fifty-Year Puzzle. *Rev. Mod. Phys.* **2019**, *91*, 011003. [CrossRef]
29. Burkert, V.D. Nucleon Resonances and Transition Form Factors. *arXiv* **2022**, arXiv:2212.08980. Available online: https://arxiv.org/abs/2212.08980 (accessed on 1 March 2023).
30. Beričič, J.; Correa, L.; Benali, M.; Achenbach, P.; Gayoso, C.A.; Bernauer, J.C.; Blomberg, A.; Böhm, R.; Bosnar, D.; Debenjak, L.; et al. New Insight in the Q^2-Dependence of Proton Generalized Polarizabilities. *Phys. Rev. Lett.* **2019**, *123*, 192302. [CrossRef]
31. Blomberg, A.; Atac, H.; Sparveris, N.; Paolone, M.; Achenbach, P.; Benali, M.; Beričič, J.; Böhm, R.; Correa, L.; Distler, M.O.; et al. Virtual Compton Scattering Measurements in the Nucleon Resonance Region. *Eur. Phys. J. A* **2019**, *55*, 182. [CrossRef]
32. Mihovilovič, M.; Weber, A.B.; Achenbach, P.; Beranek, T.; Beričič, J.; Bernauer, J.C.; Böhm, R.; Bosnar, D.; Cardinali, M.; Correa, L.; et al. First Measurement of Proton's Charge Form Factor at Very Low Q^2 with Initial State Radiation. *Phys. Lett. B* **2017**, *771*, 194–198. [CrossRef]

33. Friščić, I.; Achenbach, P.; Gayoso, C.A.; Baumann, D.; Böhm, R.; Bosnar, D.; Debenjak, L.; Denig, A.; Ding, M.; Distler, M.O.; et al. Measurement of the $p(e, e'\pi^+)n$ Reaction Close to Threshold and at Low Q^2. *Phys. Lett. B* **2017**, *766*, 301–305. [CrossRef]
34. Bernauer, J.C.; Distler, M.O.; Friedrich, J.; Walcher, T.; Achenbach, P.; Gayoso, C.A.; Böhm, R.; Bosnar, D.; Debenjak, L.; Doria, L.; et al. Electric and Magnetic Form Factors of the Proton. *Phys. Rev. C* **2014**, *90*, 015206. [CrossRef]
35. Sparveris, N.; Stave, S.; Achenbach, P.; Gayoso, C.A.; Baumann, D.; Bernauer, J.; Bernstein, A.M.; Bohm, R.; Bosnar, D.; Botto, T.; et al. Measurements of the $\gamma^* p \to \Delta$ Reaction at Low Q^2. *Eur. Phys. J. A* **2013**, *49*, 136. [CrossRef]
36. Bevan, A.J.; Golob, B.; Mannel, Th.; Prell, S.; Yabsley, B.D.; Aihara, H.; Anulli, F.; Arnaud, N.; Aushev, T.; Beneke, M.; et al. The Physics of the B Factories. *Eur. Phys. J. C* **2014**, *74*, 3026. [CrossRef]
37. Kou, E.; Urquijo, P.; Altmannshofer, W.; Beaujean, F.; Bell, G.; Beneke, M.; Bigi, I.I.; Bishara, F.; Blanke, M.; Bobeth, C.; et al. The Belle II Physics Book. *PTEP* **2019**, *2019*, 123C01; Erratum in *PTEP* **2020**, *2020*, 029201. [CrossRef]
38. Burkert, V.D. Jefferson Lab at 12 GeV: The Science Program. *Ann. Rev. Nucl. Part. Sci.* **2018**, *68*, 405–428. [CrossRef]
39. Barabanov, M.Y.; Bedolla, M.A.; Brooks, W.K.; Cates, G.D.; Chen, C.; Chen, Y.; Cisbani, E.; Ding, M.; Eichmann, G.; Ent, R.; et al. Diquark Correlations in Hadron Physics: Origin, Impact and Evidence. *Prog. Part. Nucl. Phys.* **2021**, *116*, 103835. [CrossRef]
40. Accardi, A.; Afanasev, A.; Albayrak, I.; Ali, S.F.; Amaryan, M.; Annand, J.R.M.; Arrington, J.; Asaturyan, A.; Atac, H.; Avakian, H.; et al. An Experimental Program with High Duty-Cycle Polarized and Unpolarized Positron Beams at Jefferson Lab. *Eur. Phys. J. A* **2021**, *57*, 261. [CrossRef]
41. Aguilar, A.C.; Ahmed, Z.; Aidala, C.; Ali, S.; Andrieux, V.; Arrington, J.; Bashir, A.; Berdnikov, V.; Binosi, D.; Chang, L.; et al. Pion and Kaon Structure at the Electron-Ion Collider. *Eur. Phys. J. A* **2019**, *55*, 190. [CrossRef]
42. Khalek, R.A.; Accardi, A.; Adam, J.; Adamiak, D.; Akers, W.; Albaladejo, M.; Al-bataineh, A.; Alexeev, M.G.; Ameli, F.; Antonioli, P.; et al. Science Requirements and Detector Concepts for the Electron-Ion Collider: EIC Yellow Report. *Nucl. Phys. A* **2022**, *1026*, 122447. [CrossRef]
43. Arrington, J.; Gayoso, C.A.; Barry, P.C.; Berdnikov, V.; Binosi, D.; Chang, L.; Diefenthaler, M.; Ding, M.; Ent, R.; Frederico, T. Revealing the Structure of Light Pseudoscalar Mesons at the Electron–Ion Collider. *J. Phys. G* **2021**, *48*, 075106. [CrossRef]
44. Chen, X.; Guo, F.K.; Roberts, C.D.; Wang, R. Selected Science Opportunities for the EicC. *Few Body Syst.* **2020**, *61*, 43. [CrossRef]
45. Anderle, D.P.; Bertone, V.; Cao, X.; Chang, L.; Chang, N.; Chen, G.; Chen, X.; Chen, Z.; Cui, Z.; Dai, L.; et al. Electron-Ion Collider in China. *Front. Phys.* **2021**, *16*, 64701. [CrossRef]
46. Quintans, C. The New AMBER Experiment at the CERN SPS. *Few Body Syst.* **2022**, *63*, 72. [CrossRef]
47. Mokeev, V.I.; Carman, D.S. Photo- and Electrocouplings of Nucleon Resonances. *Few Body Syst.* **2022**, *63*, 59. [CrossRef]
48. Carman, D.S.; Joo, K.; Mokeev, V.I. Strong QCD Insights from Excited Nucleon Structure Studies with CLAS and CLAS12. *Few Body Syst.* **2020**, *61*, 29. [CrossRef]
49. Aznauryan, I.; Burkert, V. Electroexcitation of Nucleon Resonances. *Prog. Part. Nucl. Phys.* **2012**, *67*, 1–54. [CrossRef]
50. Villano, A.N.; Stoler, P.; Bosted, P.E.; Connell, S.H.; Dalton, M.M.; Jones, M.K.; Kubarovsky, V.; Adams, G.S.; Ahmidouch, A.; Arrington, J.; et al. Neutral Pion Electroproduction in the Resonance Region at High Q^2. *Phys. Rev. D* **2009**, *80*, 035203.
51. Dalton, M.M.; Adams, G.S.; Ahmidouch, A.; Angelescu, T.; Arrington, J.; Asaturyan, R.; Baker, O.K.; Benmouna, N.; Bertoncini, C.; Boeglin, W.U.; et al. Electroproduction of η Mesons in the $S_{11}(1535)$ Resonance Region at High Momentum Transfer. *Phys. Rev. C* **2009**, *80*, 015205. [CrossRef]
52. Anikin, I.V.; Braun, V.M.; Offen, N. Electroproduction of the $N^*(1535)$ Nucleon Resonance in QCD. *Phys. Rev. D* **2015**, *92*, 014018. [CrossRef]
53. Giannini, M.M.; Santopinto, E. The Hypercentral Constituent Quark Model and its Application to Baryon Properties. *Chin. J. Phys.* **2015**, *53*, 020301.
54. Braun, V.M. Hadron Wave Functions from Lattice QCD. *Few Body Syst.* **2016**, *57*, 1019. [CrossRef]
55. Aznauryan, I.G.; Burkert, V.D. Electroexcitation of Nucleon Resonances in a Light-Front Relativistic Quark Model. *Few Body Syst.* **2018**, *59*, 98. [CrossRef]
56. Qin, S.X.; Roberts, C.D.; Schmidt, S.M. Poincaré-Covariant Analysis of Heavy-Quark Baryons. *Phys. Rev. D* **2018**, *97*, 114017. [CrossRef]
57. Ramalho, G. Quark Model Calculations of Transition Form Factors at High Photon Virtualities. *EPJ Web Conf.* **2020**, *241*, 02007. [CrossRef]
58. Lyubovitskij, V.E.; Schmidt, I. Nucleon Resonances with Higher Spins in Soft-Wall AdS/QCD. *Phys. Rev. D* **2020**, *102*, 094008. [CrossRef]
59. Raya, K.; Gutiérrez-Guerrero, L.X.; Bashir, A.; Chang, L.; Cui, Z.F.; Lu, Y.; Roberts, C.D.; Segovia, J. Dynamical Diquarks in the $\gamma^{(*)} p \to N(1535)\frac{1}{2}^-$ Transition. *Eur. Phys. J. A* **2021**, *57*, 266. [CrossRef]
60. Liu, L.; Chen, C.; Roberts, C.D. Wave Functions of $(I, J^P) = (\frac{1}{2}, \frac{3}{2}^{\mp})$ Baryons. *arXiv* 2022, arXiv:2208.12353.
61. Liu, L.; Chen, C.; Lu, Y.; Roberts, C.D.; Segovia, J. Composition of Low-Lying $J = 3/2^{\pm}$ Δ-Baryons. *Phys. Rev. D* **2022**, *105*, 114047. [CrossRef]
62. Blin, A.N.H.; Melnitchouk, W.; Mokeev, V.I.; Burkert, V.D.; Chesnokov, V.V.; Pilloni, A.; Szczepaniak, A.P. Resonant Contributions to Inclusive Nucleon Structure Functions from Exclusive Meson Electroproduction data. *Phys. Rev. C* **2021**, *104*, 025201. [CrossRef]
63. Hiller Blin, A.N.; Mokeev, V.; Albaladejo, M.; Fernández-Ramírez, C.; Mathieu, V.; Pilloni, A.; Szczepaniak, A.; Burkert, V.D.; Chesnokov, V.V.; Golubenko, A.A.; et al. Nucleon Resonance Contributions to Unpolarised Inclusive Electron Scattering. *Phys. Rev. C* **2019**, *100*, 035201. [CrossRef]

64. Klimenko, V. Inclusive Electron Scattering in the Resonance Region from a Hydrogen Target with CLAS12. 2023, *in progress*.
65. Wilson, D.J.; Cloet, I.C.; Chang, L.; Roberts, C.D. Nucleon and Roper Electromagnetic Elastic and Transition Form Factors. *Phys. Rev. C* **2012**, *85*, 025205. [CrossRef]
66. Segovia, J.; Chen, C.; Cloet, I.C.; Roberts, C.D.; Schmidt, S.M.; Wan, S.L. Elastic and Transition Form Factors of the $\Delta(1232)$. *Few Body Syst.* **2014**, *55*, 1–33. [CrossRef]
67. Mokeev, V.I.; Burkert, V.D.; Carman, D.S.; Elouadrhiri, L.; Fedotov, G.V.; Golovatch, E.N.; Gothe, R.W.; Hicks, K.; Ishkhanov, B.S.; Isupov, E.L.; et al. New Results from the Studies of the $N(1440)1/2^+$, $N(1520)3/2^-$, and $\Delta(1620)1/2^-$ Resonances in Exclusive $ep \to e'p'\pi^+\pi^-$ Electroproduction with the CLAS Detector. *Phys. Rev. C* **2016**, *93*, 025206. [CrossRef]
68. Segovia, J.; Cloet, I.C.; Roberts, C.D.; Schmidt, S.M. Nucleon and Δ Elastic and Transition Form Factors. *Few Body Syst.* **2014**, *55*, 1185–1222. [CrossRef]
69. Segovia, J.; El-Bennich, B.; Rojas, E.; Cloet, I.C.; Roberts, C.D.; Xu, S.S.; Zong, H.S. Completing the Picture of the Roper Resonance. *Phys. Rev. Lett.* **2015**, *115*, 171801. [CrossRef]
70. Segovia, J.; Roberts, C.D. Dissecting Nucleon Transition Electromagnetic Form Factors. *Phys. Rev. C* **2016**, *94*, 042201(R). [CrossRef]
71. Chen, C.; Lu, Y.; Binosi, D.; Roberts, C.D.; Rodríguez-Quintero, J.; Segovia, J. Nucleon-to-Roper Electromagnetic Transition Form Factors at Large Q^2. *Phys. Rev. D* **2019**, *99*, 034013. [CrossRef]
72. Burkert, V.D.; Mokeev, V.I.; Ishkhanov, B.S. The Nucleon Resonance Structure from the $\pi^+\pi^-p$ Electroproduction Reaction off Protons. *Moscow Univ. Phys. Bull.* **2019**, *74*, 243; *Vestn. Mosk. Univ. Ser. III Fiz. Astron.* **2019**, *74*, 28–38. [CrossRef]
73. Roberts, C.D. Resonance Electroproduction and the Origin of Mass. *EPJ Web Conf.* **2020**, *241*, 02008. [CrossRef]
74. Lu, Y.; Chen, C.; Cui, Z.F.; Roberts, C.D.; Schmidt, S.M.; Segovia, J.; Zong, H.S. Transition Form Factors: $\gamma^* + p \to \Delta(1232)$, $\Delta(1600)$. *Phys. Rev. D* **2019**, *100*, 034001. [CrossRef]
75. Mokeev, V.I. Two Pion Photo- and Electroproduction with CLAS. *EPJ Web Conf.* **2020**, *241*, 03003. [CrossRef]
76. Workman, R.L.; Burkert, V.D.; Crede, V.; Klempt, E.; Thoma, U.; Tiator, L.; Agashe, K.; Aielli, G.; Allanach, B.C.; Amsler, C. Review of Particle Physics. *PTEP* **2022**, *2022*, 083C01.
77. Binosi, D.; Chang, L.; Papavassiliou, J.; Qin, S.X.; Roberts, C.D. Natural Constraints on the Gluon-Quark Vertex. *Phys. Rev. D* **2017**, *95*, 031501(R). [CrossRef]
78. Chang, L.; Cloet, I.C.; Roberts, C.D.; Schmidt, S.M.; Tandy, P.C. Pion Electromagnetic Form Factor at Spacelike Momenta. *Phys. Rev. Lett.* **2013**, *111*, 141802. [CrossRef]
79. Cui, Z.F.; Chen, C.; Binosi, D.; de Soto, F.; Roberts, C.D.; Rodríguez-Quintero, J.; Schmidt, S.M.; Segovia, J. Nucleon Elastic Form Factors at Accessible Large Spacelike Momenta. *Phys. Rev. D* **2020**, *102*, 014043. [CrossRef]
80. Ripani, M.; Mokeev, V.; Anghinolfi, M.; Battaglieri, M.; Fedotov, G.; Golovach, E.; Ishkhanov, B.; Osipenko, M.; Ricco, G.; Sapunenko, V.; et al. A Phenomenological Description of $\pi^-\Delta^{++}$ Photoproduction and Electroproduction in Nucleon Resonance Region. *Nucl. Phys. A* **2000**, *672*, 220–248. [CrossRef]
81. Burkert, V.D.; Mokeev, V.I.; Shvedunov, N.V.; Boluchevskii, A.A.; Battaglieri, M.; Golovach, E.N.; Elouardhiri, L.; Joo, K.; Isupov, E.L.; Ishkhanov, B.S.; et al. Isobar Channels in the Production of $\pi^+\pi^-$ Pairs on a Proton by Virtual Photons. *Phys. Atom. Nucl.* **2007**, *70*, 427. [CrossRef]
82. Mokeev, V.I.; Burkert, V.D.; Lee, T.S.H.; Elouadrhiri, L.; Fedotov, G.V.; Ishkhanov, B.S. Model Analysis of the $p\pi^+\pi^-$ Electroproduction Reaction on the Proton. *Phys. Rev. C* **2009**, *80*, 045212. [CrossRef]
83. Mokeev, V.I.; Biselli, A. Experimental Study of the $P_{11}(1440)$ and $D_{13}(1520)$ Resonances from CLAS Data on $ep \to e'\pi^+\pi^-p'$. *Phys. Rev. C* **2012**, *86*, 035203. [CrossRef]
84. Roberts, C.D.; Williams, A.G. Dyson-Schwinger Equations and their Application to Hadronic Physics. *Prog. Part. Nucl. Phys.* **1994**, *33*, 477–575. [CrossRef]
85. Maris, P.; Roberts, C.D. π- and K Meson Bethe-Salpeter Amplitudes. *Phys. Rev. C* **1997**, *56*, 3369. [CrossRef]
86. Cornwall, J.M. Dynamical Mass Generation in Continuum QCD. *Phys. Rev. D* **1982**, *26*, 1453. [CrossRef]
87. Schwinger, J.S. Gauge Invariance and Mass. *Phys. Rev.* **1962**, *125*, 397. [CrossRef]
88. Schwinger, J.S. Gauge Invariance and Mass. 2. *Phys. Rev.* **1962**, *128*, 2425. [CrossRef]
89. Aguilar, A.C.; Binosi, D.; Papavassiliou, J. Gluon and Ghost Propagators in the Landau Gauge: Deriving Lattice Results from Schwinger-Dyson Qquations. *Phys. Rev. D* **2008**, *78*, 025010. [CrossRef]
90. Boucaud, P.; Leroy, J.P.; Le-Yaouanc, A.; Micheli, J.; Pene, O.; Rodríguez-Quintero, J. The Infrared Behaviour of the Pure Yang-Mills Green Functions. *Few Body Syst.* **2012**, *53*, 387–436. [CrossRef]
91. Aguilar, A.C.; Binosi, D.; Papavassiliou, J. The Gluon Mass Generation Mechanism: A Concise Primer. *Front. Phys. China* **2016**, *11*, 111203. [CrossRef]
92. Aguilar, A.C.; Ferreira, M.N.; Papavassiliou, J. Exploring Smoking-Gun Signals of the Schwinger Mechanism in QCD. *Phys. Rev. D* **2022**, *105*, 014030. [CrossRef]
93. Aguilar, A.C.; De Soto, F.; Ferreira, M.N.; Papavassiliou, J.; Pinto-Gómez, F.; Roberts, C.D.; Rodríguez-Quintero, J. Schwinger Mechanism for Gluons from Lattice QCD. *arXiv* **2022**, arXiv:2211.12594. Available online: https://arxiv.org/abs/2211.12594 (accessed on 1 March 2023).
94. Grunberg, G. Renormalization Scheme Independent QCD and QED: The Method of Effective Charges. *Phys. Rev. D* **1984**, *29*, 2315. [CrossRef]

95. Grunberg, G. On Some Ambiguities in the Method of Effective Charges. *Phys. Rev. D* **1989**, *40*, 680. [CrossRef]
96. Bjorken, J.D. Applications of the Chiral U(6) × (6) Algebra of Current Densities. *Phys. Rev.* **1966**, *148*, 1467. [CrossRef]
97. Bjorken, J.D. Inelastic Scattering of Polarized Leptons from Polarized Nucleons. *Phys. Rev. D* **1970**, *1*, 1376. [CrossRef]
98. Gell-Mann, M. A Schematic Model of Baryons and Mesons. *Phys. Lett.* **1964**, *8*, 214. [CrossRef]
99. Eichmann, G.; Ramalho, G. Nucleon Resonances in Compton Scattering. *Phys. Rev. D* **2018**, *98*, 093007. [CrossRef]
100. Ding, M.; Raya, K.; Binosi, D.; Chang, L.; Roberts, C.D.; Schmidt, S.M. Symmetry, Symmetry Breaking, and Pion Parton Distributions. *Phys. Rev. D* **2020**, *101*, 054014. [CrossRef]
101. Chen, C.; Roberts, C.D. Nucleon Axial Form Factor at Large Momentum Transfers. *Eur. Phys. J. A* **2022**, *58*, 206. [CrossRef]
102. Maris, P.; Roberts, C.D.; Tandy, P.C. Pion Mass and Decay Constant. *Phys. Lett. B* **1998**, *420*, 267–273. [CrossRef]
103. Höll, A.; Krassnigg, A.; Roberts, C.D. Pseudoscalar Meson Radial Excitations. *Phys. Rev. C* **2004**, *70*, 042203(R). [CrossRef]
104. Brodsky, S.J.; Roberts, C.D.; Shrock, R.; Tandy, P.C. Confinement Contains Condensates. *Phys. Rev. C* **2012**, *85*, 065202. [CrossRef]
105. Qin, S.X.; Roberts, C.D.; Schmidt, S.M. Ward–Green–Takahashi Identities and the Axial-Vector Vertex. *Phys. Lett. B* **2014**, *733*, 202–208. [CrossRef]
106. Nambu, Y. Quasiparticles and Gauge Invariance in the Theory of Superconductivity. *Phys. Rev.* **1960**, *117*, 648. [CrossRef]
107. Goldstone, J. Field Theories with Superconductor Solutions. *Nuovo Cim.* **1961**, *19*, 154–164. [CrossRef]
108. Roberts, C.D. Perspective on the Origin of Hadron Masses. *Few Body Syst.* **2017**, *58*, 5. [CrossRef]
109. Flambaum, V.V.; Holl, A.; Jaikumar, P.; Roberts, C.D.; Wright, S.V. Sigma Terms of Light-Quark Hadrons. *Few Body Syst.* **2006**, *38*, 31–51. [CrossRef]
110. Ruiz de Elvira, J.; Hoferichter, M.; Kubis, B.; Meißner, U.G. Extracting the σ-Term from Low-Energy Pion-Nucleon Scattering. *J. Phys. G* **2018**, *45*, 024001. [CrossRef]
111. Aoki, S.; Aoki, Y.; Bečirević, D.; Blum, T.; Colangelo, G.; Collins, S.; Della Morte, M.; Dimopoulos, P.; Dürr, S.; Fukaya, H.; et al. FLAG Review 2019. *Eur. Phys. J. C* **2020**, *80*, 113. [CrossRef]
112. Obukhovsky, I.T.; Faessler, A.; Fedorov, D.K.; Gutsche, T.; Lyubovitskij, V.E. Transition Form Factors and Helicity Amplitudes for Electroexcitation of Negative- and Positive Parity Nucleon Resonances in a Light-Front Quark Model. *Phys. Rev. D* **2019**, *100*, 094013. [CrossRef]
113. Carman, D.S. Nucleon Resonance Structure Studies Via Exclusive KY Electroproduction. *Few Body Syst.* **2016**, *57*, 941–948. [CrossRef]
114. Carman, D.S. CLAS N^* Excitation Results from Pion and Kaon Electroproduction. *Few Body Syst.* **2018**, *59*, 82. [CrossRef]
115. CLAS Physics Database. Available online: https://clasweb.jlab.org/physicsdb/ (accessed on 1 March 2023).
116. Chesnokov, V.V.; Golubenko, A.A.; Ishkhanov, B.S.; Mokeev, V.I. CLAS Database for Studies of the Structure of Hadrons in Electromagnetic Processes. *Phys. Part. Nucl.* **2022**, *53*, 184–190. [CrossRef]
117. Aznauryan, I.G. Multipole Amplitudes of Pion Photoproduction on Nucleons up to 2-GeV Within Dispersion Relations and Unitary Isobar Model. *Phys. Rev. C* **2003**, *67*, 015209. [CrossRef]
118. Aznauryan, I.G.; Burkert, V.D.; Fedotov, G.V.; Ishkhanov, B.S.; Mokeev, V.I. Electroexcitation of Nucleon Resonances at $Q^2 = 0.65$ (GeV/c)^2 from a Combined Analysis of Single- and Double-Pion Electroproduction Data. *Phys. Rev. C* **2005**, *72*, 045201. [CrossRef]
119. Arndt, R.; Briscoe, W.; Strakovsky, I.; Workman, R. Partial-Wave Analysis And Spectroscopy: From πN Scattering To Pion-Electroproduction. *eConf* **2007**, *C070910*, 166.
120. Drechsel, D.; Kamalov, S.S.; Tiator, L. Unitary Isobar Model - MAID2007. *Eur. Phys. J. A* **2007**, *34*, 69–97. [CrossRef]
121. Aznauryan, I.G.; Burkert, V.D.; Biselli, A.S.; Egiyan, H.; Joo, K.; Kim, W.; Park, K.; Smith, L.C.; Ungaro, M.; Adhikari, K.P.; et al. Electroexcitation of Nucleon Resonances from CLAS Data on Single Pion Electroproduction. *Phys. Rev. C* **2009**, *80*, 055203. [CrossRef]
122. Tiator, L.; Drechsel, D.; Kamalov, S.S.; Vanderhaeghen, M. Electromagnetic Excitation of Nucleon Resonances. *Eur. Phys. J. ST* **2011**, *198*, 141–170. [CrossRef]
123. Park, K.; Aznauryan, I.G.; Burkert, V.D.; Adhikari, K.P.; Amaryan, M.J.; Pereira, S.A.; Avakian, H.; Battaglieri, M.; Badui, R.; Bedlinskiy, I.; et al. Measurements of $ep \to e'\pi^+ n$ at W = 1.6 − 2.0 GeV and Extraction of Nucleon Resonance Electrocouplings at CLAS. *Phys. Rev. C* **2015**, *91*, 045203. [CrossRef]
124. Tiator, L.; Döring, M.; Workman, R.L.; Hadžimehmedović, M.; Osmanović, H.; Omerović, R.; Stahov, J.; Švarc, A. Baryon Transition Form Factors at the Pole. *Phys. Rev. C* **2016**, *94*, 065204. [CrossRef]
125. Tiator, L.; Workman, R.L.; Wunderlich, Y.; Haberzettl, H. Amplitude Reconstruction from Complete Electroproduction Experiments and Truncated Partial-Wave Expansions. *Phys. Rev. C* **2017**, *96*, 025210. [CrossRef]
126. Tiator, L. The MAID Legacy and Future. *Few Body Syst.* **2018**, *59*, 21. [CrossRef]
127. Knochlein, G.; Drechsel, D.; Tiator, L. Photoproduction and Electroproduction of η Mesons. *Z. Phys. A* **1995**, *352*, 327–343. [CrossRef]
128. Chiang, W.T.; Yang, S.N.; Tiator, L.; Drechsel, D. An Isobar Model for η Photoproduction and Electroproduction on the Nucleon. *Nucl. Phys. A* **2002**, *700*, 429–453. [CrossRef]

129. Chiang, W.T.; Yang, S.N.; Drechsel, D.; Tiator, L. An Isobar Model Study of η Photoproduction and Electroproduction. *PiN Newslett.* **2002**, *16*, 299.
130. Aznauryan, I. Resonance Contributions to η Photoproduction on Protons Found Using Dispersion Relations and an Isobar Model. *Phys. Rev. C* **2003**, *68*, 065204. [CrossRef]
131. Kamano, H. Electromagnetic N^* Transition Form Factors in the ANL-Osaka Dynamical Coupled-Channels Approach. *Few Body Syst.* **2018**, *59*, 24. [CrossRef]
132. Mai, M.; Döring, M.; Granados, C.; Haberzettl, H.; Hergenrather, J.; Meißner, U.G.; Rönchen, D.; Strakovsky, I.; Workman, R. Coupled-Channels Analysis of Pion and η Electroproduction within the Jülich-Bonn-Washington Model. *Phys. Rev. C* **2022**, *106*, 015201. [CrossRef]
133. Mai, M.; Döring, M.; Granados, C.; Haberzettl, H.; Meißner, U.G.; Rönchen, D.; Strakovsky, I.; Workman, R. Jülich-Bonn-Washington Model for Pion Electroproduction Multipoles. *Phys. Rev. C* **2021**, *103*, 065204. [CrossRef]
134. Hwang, S.; Ahn, J.K.; Bassalleck, B.; Fujioka, H.; Guo, L.; Han, Y.; Hasegawa, S.; Hicks, K.; Honda, R.; Hosomi, K.; et al. Measurement of 3-Body Hadronic Reactions with HypTPC at J-PARC. *JPS Conf. Proc.* **2015**, *8*, 022008.
135. Hicks, K.H.; Sako, H.; Imai, K.; Hasegawa, S.; Sato, S.; Shirotori, K.; Chandavar, S.; Goetz, J.; Tang, W.; Ahn, J.K.; et al. 3-Body Hadronic Reactions for New Aspects of Baryon Spectroscopy. Available online: https://j-parc.jp/researcher/Hadron/en/pac_1207/pdf/P45_2012-3.pdf (accessed on 1 March 2023).
136. Mokeev, V.I. New Opportunities for Insight into the Emergence of Hadron Mass from Studies of Nucleon Resonance Electroexcitation. In Proceedings of the Fall 2022 Meeting of the APS Division of Nuclear Physics, New Orleans, LA, USA, 27–30 October 2022.
137. Mokeev, V.I. Nucleon Resonance Electrocouplings and the Emergence of Hadron Mass. In Proceedings of the Baryons 2022—International Conference on the Structure of Baryons, Andalusia, Spain, 7–11 November 2022.
138. Dugger, M.; Ritchie, B.G.; Ball, J.P.; Collins, P.; Pasyuk, E.; Arndt, R.A.; Briscoe, W.J.; Strakovsky, I.I.; Workman, R.L.; Amaryan, M.J.; et al. π^+ Photoproduction on the Proton for Photon Energies from 0.725 to 2.875 GeV. *Phys. Rev. C* **2009**, *79*, 065206. [CrossRef]
139. Aznauryan, I.G.; Burkert, V.D. Extracting Meson-Baryon Contributions to the Electroexcitation of the $N(1675)\frac{5}{2}^-$ Nucleon Resonance. *Phys. Rev. C* **2015**, *92*, 015203. [CrossRef]
140. Roberts, H.L.L.; Chang, L.; Cloet, I.C.; Roberts, C.D. Masses of Ground and Excited-state Hadrons. *Few Body Syst.* **2011**, *51*, 1. [CrossRef]
141. Chen, C.; Fischer, C.S.; Roberts, C.D.; Segovia, J. Nucleon Axial-Vector and Pseudoscalar Form Factors and PCAC Relations. *Phys. Rev. D* **2022**, *105*, 094022. [CrossRef]
142. Chang, L.; Cloet, I.C.; Cobos-Martinez, J.J.; Roberts, C.D.; Schmidt, S.M.; Tandy, P.C. Imaging Dynamical Chiral Symmetry Breaking: Pion Wave Function on the Light Front. *Phys. Rev. Lett.* **2013**, *110*, 132001. [CrossRef]
143. Raya, K.; Chang, L.; Bashir, A.; Cobos-Martinez, J.J.; Gutiérrez-Guerrero, L.X.; Roberts, C.D.; Tandy, P.C. Structure of the Neutral Pion and its Electromagnetic Transition Form Factor. *Phys. Rev. D* **2016**, *93*, 074017. [CrossRef]
144. Mokeev, V.I. Insight into EHM from Results on Electroexcitation of $\Delta(1600)3/2^+$ Resonance. In Proceedings of the Workshop on perceiving the Emergence of Hadron Mass through AMBER @ CERN—VII, Geneve, Switzerland, 10–13 May 2022.
145. Isupov, E.L.; Burkert, V.D.; Carman, D.S.; Gothe, R.W.; Hicks, K.; Ishkhanov, B.S.; Mokeev, V.I. Measurements of $ep \to e'\pi^+\pi^- p'$ Cross Sections with CLAS at 1.40 GeV $< W <$ 2.0 GeV and 2.0 GeV$^2 < Q^2 <$ 5.0 GeV^2. *Phys. Rev. C* **2017**, *96*, 025209.
146. Trivedi, A. Measurement of New Observables from the $\pi^+\pi^-$ p Electroproduction Off the Proton. *Few Body Syst.* **2019**, *60*, 5. [CrossRef]
147. Mokeev, V.I.; Burkert, V.D.; Carman, D.S.; Elouadrhiri, L.; Golovatch, E.; Gothe, R.W.; Hicks, K.; Ishkhanov, B.S.; Isupov, E.L.; Joo, K.; et al. Evidence for the $N'(1720)3/2^+$ Nucleon Resonance from Combined Studies of CLAS $\pi^+\pi^- p$ Photo- and Electroproduction Data. *Phys. Lett. B* **2020**, *805*, 135457. [CrossRef]
148. Mokeev, V.I.; Carman, D.S. New Baryon States in Exclusive Meson Photo-/Electroproduction with CLAS. *Rev. Mex. Fis. Suppl.* **2022**, *3*, 0308024. [CrossRef]
149. Burkert, V.D.; Elouadhriri, L.; Adhikari, K.P.; Adhikari, S.; Amaryan, M.J.; Anderson, D.; Angelini, G.; Antonioli, M.; Atac, H.; Aune, S.; et al. The CLAS12 Spectrometer at Jefferson Laboratory. *Nucl. Instrum. Meth. A* **2020**, *959*, 163419. [CrossRef]
150. Gothe, R.W.; Mokeev, V.; Burkert, V.D.; Cole, P.L.; Joo, K.; Stoler, P. Nucleon Resonance Studies with CLAS12. JLab Experiment E12-09-003. Available online: https://www.jlab.org/exp_prog/proposals/09/PR12-09-003.pdf (accessed on 1 March 2023).
151. Carman, D.S.; Mokeev, V.I.; Burkert, V.D. Exclusive $N^* \to KY$ Studies with CLAS12. JLab Experiment E12-06-108A. Available online: https://www.jlab.org/exp_prog/proposals/14/E12-06-108A.pdf (accessed on 1 March 2023).
152. D'Angelo, A.; Burkert, V.D.; Carman, D.S.; Golovatch, E.; Gothe, R.; Mokeev, V. A Search for Hybrid Baryons in Hall B with CLAS12. JLab Experiment E12-16-010. Available online: https://www.jlab.org/exp_prog/proposals/16/PR12-16-010.pdf (accessed on 1 March 2023).
153. Carman, D.S.; d'Angelo, A.; Lanza, L.; Mokeev, V.I.; Adhikari, K.P.; Amaryan, M.J.; Armstrong, W.R.; Atac, H.; Avakian, H.; Gayoso, C.A.; et al. Beam-Recoil Transferred Polarization in K^+Y Electroproduction in the Nucleon Resonance Region with CLAS12. *Phys. Rev. C* **2022**, *105*, 065201. [CrossRef]

154. Burkert, V.D. N* Experiments and Their Impact on Strong QCD Physics. *Few Body Syst.* **2018**, *59*, 57. [CrossRef]
155. Aznauryan, I.; Burkert, V.; Lee, T.S.; Mokeev, V. Results from the N* Program at JLab. *J. Phys. Conf. Ser.* **2011**, *299*, 012008. [CrossRef]

Disclaimer/Publisher's Note: The statements, opinions and data contained in all publications are solely those of the individual author(s) and contributor(s) and not of MDPI and/or the editor(s). MDPI and/or the editor(s) disclaim responsibility for any injury to people or property resulting from any ideas, methods, instructions or products referred to in the content.

Review

Several Topics on Transverse Momentum-Dependent Fragmentation Functions

Kai-Bao Chen [1], Tianbo Liu [2], Yu-Kun Song [3] and Shu-Yi Wei [2,*]

[1] School of Science, Shandong Jianzhu University, Jinan 250101, China; chenkaibao19@sdjzu.edu.cn
[2] Key Laboratory of Particle Physics and Particle Irradiation (MOE), Institute of Frontier and Interdisciplinary Science, Shandong University, Qingdao 266237, China; liutb@sdu.edu.cn
[3] School of Physics and Technology, University of Jinan, Jinan 250022, China; sps_songyk@ujn.edu.cn
* Correspondence: shuyi@sdu.edu.cn

Abstract: The hadronization of a high-energy parton is described by fragmentation functions which are introduced through QCD factorizations. While the hadronization mechanism per se remains uknown, fragmentation functions can still be investigated qualitatively and quantitatively. The qualitative study mainly concentrates on extracting genuine features based on the operator definition in quantum field theory. The quantitative research focuses on describing a variety of experimental data employing the fragmentation function given by the parameterizations or model calculations. With the foundation of the transverse-momentum-dependent factorization, the QCD evolution of leading twist transverse-momentum-dependent fragmentation functions has also been established. In addition, the universality of fragmentation functions has been proven, albeit model-dependently, so that it is possible to perform a global analysis of experimental data in different high-energy reactions. The collective efforts may eventually reveal important information hidden in the shadow of nonperturbative physics. This review covers the following topics: transverse-momentum-dependent factorization and the corresponding QCD evolution, spin-dependent fragmentation functions at leading and higher twists, several experimental measurements and corresponding phenomenological studies, and some model calculations.

Keywords: fragmentation function; transverse-momentum-dependent factorization; QCD evolution; spin-related effects

1. Introduction

Quantum chromodynamics (QCD) [1] is known as the fundamental theory of strong interaction in the framework of Yang-Mills gauge field theory [2]. As a key property of QCD, the color confinement prohibits direct detection of quarks and gluons, the fundamental degrees of freedom, with any modern detectors. The emergence of color neutral hadrons from colored quarks and gluons is still an unresolved problem and has received particular interest in recent years [3]. With the progress of QCD into the precision era, unraveling the hadronization mechanism in the high-energy scattering processes has become one of the most active frontiers in nuclear and particle physics.

Due to the nonperturbative nature of QCD, it is still challenging to directly calculate the hadronization process from first principles. Similar to the parton distribution functions (PDFs) [4,5], which were originally defined as the probability density of finding a parton inside the parent hadron, the concept of fragmentation functions (FFs) was introduced by Berman, Bjorken, and Kogut [6] right after the parton model to describe the emergence of a system of the hadron from a high-energy parton isolated in the phase space. An alternative name, the parton decay function, has also frequently been used in early literature.

The modern concept of FFs in QCD was first introduced to describe the inclusive production of a desired hadron in the e^+e^- annihilation [7,8], which is still the cleanest reaction currently available to investigate the fragmentation process. Within the QCD-improved parton model, the FF has its foundation in the factorization theorem [9,10], in which the differential cross section is approximated as a convolution of short-distance hard scattering and long-distance matrix elements with corrections formally suppressed by inverse powers of a hard scale, e.g., the center-of-mass (c.m.) energy $Q = \sqrt{s}$ in the e^+e^- annihilation. The predictive power of this theoretical framework relies on the control of the hard probe, which can be achieved by our ability to calculate the partonic cross section order by order in the perturbation theory, and the universality of the long-distance functions, such as the FFs, to be tested in multiple high-energy scattering processes.

For a single-scale process, e.g., $e^+e^- \to hX$, where h represents the identified hadron in the final state and X denotes the undetected particles, the process is not sensitive to the confined motion of quarks and gluons in the hadronization process, and one can apply the colinear factorization with the emergence of the detected hadron described by a colinear FF $D_{f \to h}(z)$, where the subscript f stands for the parton flavor and z is the longitudinal momentum fraction carried by the hadron h with respect to the fragmenting parton. If two hadrons are identified in a process, e.g., $e^+e^- \to h_A h_B X$, where h_A and h_B are detected hadrons in the final state, the reaction becomes a double-scale problem with one scale Q given by the hard probe and the other scale provided by the transverse momentum imbalance, $|p_{A\perp} + p_{B\perp}|$. When the second scale is much smaller than Q, i.e., the two hadrons are nearly back to back, one needs to use the transverse-momentum-dependent (TMD) factorization. The emergence of each of the hadrons is described by a TMD FF $D_{f \to h}(z, k_\perp)$, where $k_\perp$ is the transverse momentum of the fragmenting parton with respect to the observed hadron [8,11]. When the two scales are compatible, the reaction effectively becomes a single-scale process, and one can again use the colinear factorization. The matching between the two regions has been developed. The TMD FFs defined in the e^+e^- annihilation also play an important role in the study of nucleon three-dimensional structures via the semi-inclusive deep inelastic scattering (SIDIS) process [12]. Instead of identifying two hadrons in a reaction, one can also access TMD FFs in the single-hadron production process by reconstructing the thrust axis, which provides the sensitivity to the transverse momentum of the observed hadron, as proposed in recent years [13–16].

Taking the parton spin degree of freedom into account, one can define polarized or spin-dependent TMD FFs. They essentially reflect the correlation between parton transverse momentum and its spin during the hadronization process and result in rich phenomena in high-energy scattering processes. For example, the Collins fragmentation function $H_1^\perp(z, k_\perp)$ [17], naively interpreted as the probability density of a transversely polarized quark fragmenting into an unpolarized hadron, can lead to a single spin asymmetry (SSA) in the SIDIS process with a transversely polarized target [18]. This asymmetry is a key observable for the determination of the quark transversity distribution, the net density of a transversely polarized quark in a transversely polarized nucleon. It also leads to azimuthal asymmetries in e^+e^- annihilation as measured by Belle, BaBar, and BESIII. The progress of experimental techniques to determine the spin state of produced hyperons, such as Λ and Ω, and vector mesons, such as ρ and K^*, offer us the opportunity to extract additional information from FFs. This is far beyond a trivial extension since the spin has been proven to be a powerful quantity to test theories and models, especially in hadron physics. The recent measurement of the spontaneous polarization of Λ from unpolarized e^+e^- annihilation is such an instance [19]. This observation can be explained by a naively time-reversal odd (T-odd) TMD FF $D_{1T}^\perp(z, k_\perp)$ and has received interests from various groups [15,20–32].

In addition to the leading-twist FFs, which usually have probability interpretations, the high-twist FFs have been found o be much more important than expected in recent years for understanding precise experimental data [33–66]. Although the colinear factorization at subleading power was demonstrated some time ago, the TMD factorization beyond the leading power is still under exploration, and some approaches have been proposed [67–75]. Although high-twist contributions are formally power suppressed, their contributions to the cross section might not be negligible and may have significant effects in certain kinematics or observables. The inclusion of high-twist FFs will also modify the evolution equation and consequently affect the leading-twist FFs. The TMD factorization at subleading power was recently explored with different approaches. Overall, many efforts, both theoretical and experimental, are still required to understand the hadronization process and the upcoming data from future electron-ion colliders.

The remainder of this review is organized as follows. In Section 2 we use $e^+e^- \to h_A h_B X$ as an example to present the flow of deriving the TMD factorization and the QCD evolution equation of TMD FFs. In Section 3, we present the FFs up to the twist-4 level for spin-0, -1/2, and 1 hadron productions. In Section 4, we summarize the experimental measurements towards understanding the spin-dependent FFs. In Section 5, we briefly lay out some model calculations. A summary is given in Section 6.

2. Factorization and Evolution

The modern concept of FFs has established on the QCD factorization theorems, which can be derived either from calculating traditional Feynman diagrams in perturbative field theory [10,11,76–81] or in effective theories [82–86]. In the former approach, one first identifies a collection of Feynman diagrams that offers the leading contribution through the Libby–Sterman analysis [87,88]. In this method, the leading contribution is represented by the reduced diagrams.

Taking $e^+e^- \to h_A h_B X$ process with h_A, h_B traveling along almost back-to-back directions as an example [11], the leading regions are presented in Figure 1. The cross section is the product of various ingredients, such as the hard part H, the soft part S, and the colinear parts J_A, J_B. We work in the light-cone coordinate, so that a four momentum p can be written as follows: $p^\mu = (p^+, p^-, \boldsymbol{p}_\perp)$ with $p^\pm = \frac{1}{\sqrt{2}}(p^0 \pm p^3)$. In the kinematic region where TMD factorization applies, the transverse momentum is considerably small compared with that along the longitudinal direction. Therefore, the momenta of the almost back-to-back hadrons A and B scale as $p_A \sim Q(1, \lambda, \sqrt{\lambda})$ and $p_B \sim Q(\lambda, 1, \sqrt{\lambda})$, where Q is the large momentum scale and $\lambda \ll 1$ is a small parameter. The hard part H computes the cross section of interaction among hard partons whose momenta scale as $Q(1,1,1)$ in perturbative field theory. The contribution from colinear partons whose momenta are colinear with the final state hadrons A and B are evaluated in the colinear function $J_{A/B}$. This process results in the gauge invariant bare FFs. The soft part calculates the contribution from soft gluons whose momenta typically take the form of $Q(\lambda, \lambda, \lambda)$. They will be absorbed into the definition of TMD FFs eventually and convert the bare FFs into the renormalized ones.

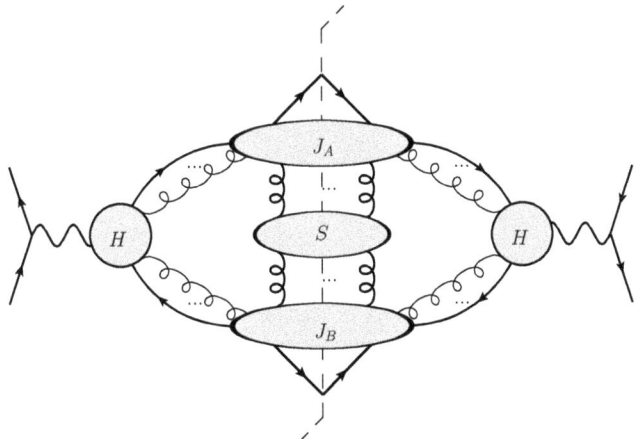

Figure 1. Leading regions for $e^+e^- \to h_A h_B X$.

The interactions between different parts can be eliminated via applying appropriate kinematic approximations and the Ward identity. Finally, the cross section is given by a convolution of those well-separated parts, and we arrive at the factorization theorem of this process.

Depending on the physics of interest, we may derive either colinear factorization or transverse-momentum-dependent (TMD) factorization theorems. For the differential cross section of $e^+e^- \to h_A h_B X$ as a function of the relative transverse momentum between h_A and h_B, the TMD factorization theorem applies.

In the single-photon-exchange approximation, the differential cross section of this process can be written as the production of a leptonic tensor and a hadronic tensor. It reads as follows [43]:

$$\frac{d\sigma}{dy dz_A dz_B d^2 \boldsymbol{P}_{A\perp}} = \frac{2\pi N_c \alpha^2}{Q^4} L_{\mu\nu} W^{\mu\nu}, \tag{1}$$

where, α is the coupling constant, $N_c = 3$ is the color factor, Q is the center-of-mass energy of the colliding leptons, $y = (1 + \cos\theta)/2$ with θ the angle between incoming electron and the outgoing hadron h_A, z_A and z_B are light-cone momentum fractions of h_A and h_B, and $\boldsymbol{P}_{A\perp}$ is the transverse momentum of h_A with respect to the direction of the h_B momentum. For the unpolarized lepton beams, the leptonic tensor $L_{\mu\nu}$ is given by the following:

$$L_{\mu\nu} = l_{1\mu} l_{2\nu} + l_{1\nu} l_{2\mu} - g_{\mu\nu} l_1 \cdot l_2, \tag{2}$$

with l_1 and l_2 being the momenta of colliding leptons. The hadronic tensor $W_{\mu\nu}$ contains nonperturbative quantities and is laid out as follows:

$$W^{\mu\nu} = \sum_f |H_f(Q,\mu)^2|^{\mu\nu} \int d^2 \boldsymbol{k}_{A\perp} d^2 \boldsymbol{k}_{B\perp} \delta^{(2)}(\boldsymbol{k}_{A\perp} + \boldsymbol{k}_{B\perp} - \boldsymbol{q}_\perp)$$
$$\times \left[D_{1q}^{h_A}(z_A, \boldsymbol{p}_{A\perp}; \mu, \zeta_A) D_{1\bar{q}}^{h_B}(z_B, \boldsymbol{p}_{B\perp}; \mu, \zeta_B) + \ldots \right], \tag{3}$$

where $\boldsymbol{q}_\perp = -\boldsymbol{P}_{A\perp}/z_A$, and $H_f(Q,\mu)$ is the hard scattering factor that can be evaluated in the perturbative QCD. Here, $D_{1q}^{h_A}(z_A, \boldsymbol{p}_{A\perp}; \mu, \zeta_A)$ is the TMD FF with $\boldsymbol{p}_{A\perp}$ the transverse momentum of hadron with respect to the fragmenting quark direction, μ is a renormalization scale, and ζ_A is a variable to regularize the rapidity divergence. Notice that $k_{i,\perp}$ is the relative transverse momentum of the fragmenting parton with respect to the hadron momentum. Therefore, we have $p_{i,\perp}$ by $p_{i,\perp} = -z_i k_{i,\perp}$. Please also notice the difference between $\boldsymbol{P}_{A\perp}$ and $\boldsymbol{p}_{A\perp}$. The three-dot symbol stands for various spin-dependent terms which are not explicitly shown.

It is more convenient to perform the TMD evolution in the coordinate space than in the momentum space. Therefore, we need the Fourier transform,

$$D_{1q}^h(z, \boldsymbol{p}_\perp; \mu, \zeta) = \frac{1}{z^2} \int d^2 p_\perp e^{i\frac{1}{z}\boldsymbol{b}_T \cdot \boldsymbol{p}_\perp} \tilde{D}_{1q}^h(z, \boldsymbol{b}_T; \mu, \zeta), \tag{4}$$

to translate the TMD FF into coordinate space one. The hadronic tensor then becomes the following:

$$W^{\mu\nu} = \sum_f |H_f(Q,\mu)^2|^{\mu\nu} \int \frac{d^2 \boldsymbol{b}_T}{(2\pi)^2} e^{-i\boldsymbol{q}_\perp \cdot \boldsymbol{b}_T} \left[\tilde{D}_{1q}^{h_A}(z_A, \boldsymbol{b}_T; \mu, \zeta_A) \tilde{D}_{1\bar{q}}^{h_B}(z_B, \boldsymbol{b}_T; \mu, \zeta_B) + \ldots \right]. \tag{5}$$

The TMD FF in the coordinate space is defined as the product of transition matrix elements between the vacuum and the hadronic final states.

Before presenting the final definition for the TMD FF in the coordinate space, we first show the unsubtracted version, which appears in the LO calculation. For the production of hadron A, it reads as follows:

$$\tilde{D}_{1q}^{h_A,\text{unsub}}(z, \boldsymbol{b}_T; y_{p_A} - y_B) = \frac{1}{4N_c} \text{Tr}_C \text{Tr}_D \frac{1}{z} \sum_X \int \frac{dx^-}{2\pi} e^{ik^+ x^-} \langle 0 | \gamma^+ \mathcal{L}(\frac{x}{2}; +\infty, n_B)$$
$$\times \psi_q(\frac{x}{2}) | h_1, X \rangle \langle h_1, X | \bar{\psi}_q(-\frac{x}{2}) \mathcal{L}(-\frac{x}{2}; +\infty, n_B)^\dagger | 0 \rangle, \quad (6)$$

where the position vector $x = (0, x^-, \boldsymbol{b}_T)$ contains only minus and transverse components, $y_{p_A} = \frac{1}{2} \ln \frac{2(p_A^+)^2}{m_A^2}$ is the rapidity of hadron A, Tr_C is a trace in the color space, and Tr_D is a trace in the Dirac space. The direction of the Wilson lines in the FF of hadron A is specified by the direction of hadron B which is denoted as n_B and vice versa. Notice that the rapidity parameters $y_A \to +\infty$ and $y_B \to -\infty$ are introduce, so that $n_A = (1, -e^{-2y_A}, \boldsymbol{0}_T)$ and $n_B = (-e^{2y_B}, 1, \boldsymbol{0}_T)$ are slightly space-like. Please also notice the difference between y_{p_A} and y_A. The Wilson line starting from the position x is defined as follows:

$$\mathcal{L}(x; +\infty, n)_{ab} = \mathcal{P} \left\{ e^{-ig_0 \int_0^{+\infty} d\lambda n \cdot A_{(0)}^\alpha (x + \lambda n) t^\alpha} \right\}_{ab}. \quad (7)$$

with a and b being the color indices, and g_0 and $A_{(0)}^\alpha$ the bare coupling and the bare gluon field.

Taking the $y_A \to +\infty$ and $y_B \to -\infty$ limit and absorbing the soft factors into the unsubtracted TMD FF, we arrive at the final definition of the TMD FF:

$$\tilde{D}_{1q}^{h_A}(z, \boldsymbol{b}_T; \mu, \zeta_A) = \tilde{D}_{1q}^{h_A,\text{unsub}}(z, \boldsymbol{b}_T; y_{p_A} - (-\infty))$$
$$\times \sqrt{\frac{\tilde{S}_{(0)}(\boldsymbol{b}_T; +\infty, y_n)}{\tilde{S}_{(0)}(\boldsymbol{b}_T; +\infty, -\infty)\tilde{S}_{(0)}(\boldsymbol{b}_T; y_n, -\infty)}} \times Z_D Z_2, \quad (8)$$

where y_n is an arbitrary rapidity introduced to separate $\zeta_A \equiv \frac{m_A^2}{z_A^2} e^{2(y_{p_A} - y_n)}$ from $\zeta_B \equiv \frac{m_B^2}{z_B^2} e^{2(y_n - y_{p_B})}$, and Z_D, Z_2 are renormalization factors. The bare soft factor $\tilde{S}_{(0)}$ is defined as the expectation values of Wilson lines on the vacuum, reading as follows:

$$\tilde{S}_{(0)}(\boldsymbol{b}_T; y_A, y_B) = \frac{1}{N_c} \langle 0 | \mathcal{L}(\frac{\boldsymbol{b}_T}{2}; +\infty, n_B)_{ca}^\dagger \mathcal{L}(\frac{\boldsymbol{b}_T}{2}; +\infty, n_A)_{ad}$$
$$\times \mathcal{L}(-\frac{\boldsymbol{b}_T}{2}; +\infty, n_B)_{bc} \mathcal{L}(-\frac{\boldsymbol{b}_T}{2}; +\infty, n_A)_{db}^\dagger | 0 \rangle. \quad (9)$$

2.1. Evolution Equations for TMD FFs

To regularize the ultraviolet (UV) and rapidity divergences, the energy scale μ and $\sqrt{\zeta}$ are introduced. As a consequence, the TMD FFs differ at different energy scales. The evolution effects are important for phenomenological studies. The QCD evolution for TMD FFs with respect to ζ is controlled by the Collins–Soper (CS) equation [10,11], which is given as follows:

$$\frac{\partial \ln \tilde{D}_{1q}^h(z, \boldsymbol{b}_T; \mu, \zeta)}{\partial \ln \sqrt{\zeta}} = \tilde{K}(\boldsymbol{b}_T; \mu), \quad (10)$$

with $\tilde{K}(\boldsymbol{b}_T; \mu)$ being the CS evolution kernel. The scale dependence of the evolution kernel is governed by

$$\frac{d\tilde{K}(\boldsymbol{b}_T; \mu)}{d \ln \mu} = -\gamma_K(\mu), \quad (11)$$

where $\gamma_K(\mu)$ is the anomalous dimension. It is given by $\gamma_K(\mu) = \frac{2C_F}{\pi}\alpha_s(\mu)$ with $C_F = 4/3$ being the color factor and α_s being the running coupling at the LO accuracy [12].

The μ dependence of the TMD FF is then given by

$$\frac{d \ln \tilde{D}_{1q}^h(z, \bm{b}_T; \mu, \zeta)}{\partial \ln \mu} = \gamma_D(\mu; \frac{\zeta}{\mu^2}), \tag{12}$$

where γ_D is another anomalous dimension. At the LO accuracy [12], it is given as follows: $\gamma_D(\mu, \zeta/\mu^2) = \alpha_s(\mu)\frac{C_F}{\pi}(\frac{3}{2} - \ln\frac{\zeta}{\mu^2})$.

The TMD FF defined by Equation (8) is actually calculable in the colinear factorization approach in the small-b_T regime. However, at the large b_T region, the discrepancy of these two approaches grows in terms of $\Lambda_{\text{QCD}} b_T$. This region is usually referred to as the nonperturbative regime since a large coordinate corresponds to a small energy scale. The perturbative treatment of the QCD evolution in this region is no longer reliable. To have a consistent formula, the b_*-prescription is usually adopted in phenomenology. By introducing $b_* = |\bm{b}_T|/\sqrt{1 + b_T^2/b_{\max}^2}$ and $\mu_b = 2e^{-\gamma_E}/b_*$, we can separate the perturbative part from the nonperturbative part in the QCD evolution. Here, γ_E is the Euler constant, and $b_{\max}$ is an infrared cutoff which is properly chosen to guarantee that $\mu_b \gg \Lambda_{\text{QCD}}$. Employing the b_* prescription, the QCD evolution is always performed in the realm of the perturbative QCD. Therefore, this approach underestimates the contribution from the nonperturbative regime. This part of the contribution can be reintegrated into the final prescription by the introduction of a nonperturbative factor.

Ultimately, we arrive at [12]

$$D_{1q}^h(z, b_T; \mu, \zeta) = D_{1q}^h(z, b_T^*; \mu_0 = \mu_b, \zeta_0 = \mu_b^2)$$
$$\times \exp\left\{ \ln \frac{\sqrt{\zeta}}{\mu_b} \tilde{K}(b_*; \mu_b) + \int_{\mu_b}^{\mu} \frac{d\mu'}{\mu'} \gamma_D(\mu'; \frac{\zeta}{\mu'^2}) \right\}$$
$$\times \exp\left\{ -S_{\text{np}}(z, |\bm{b}_T|, \zeta) \right\}, \tag{13}$$

where the last line is the nonperturbative function that returns the nonperturbative effect that has been deliberately removed from the QCD evolution in the b_*-prescription. There is no theoretical approach that can evaluate this nonperturbative function other than the one that extracts it from experimental data [89–98]. Notice that [99] present a different method to address the nonperturbative physics. Here, $D_{1q}^h(z, b_T^*; \mu_0 = \mu_b, \zeta_0 = \mu_b^2)$ is the FF at the initial scale. In the phenomenology, it is usually chosen to coincide with the colinear FF $D_{1q}^h(z, \mu_f)$ with the factorization scale specified by $\mu_f = \mu_b$.

Similar to the PDF case, QCD evolution tends to broaden the k_T distribution width at higher energy scales. Both unpolarized and spin-dependent FFs show such a behavior [100].

2.2. TMD Factorization at the Higher Twist

In a semi-inclusive process, normally we can find two energy scales: the typical transverse momentum $q_\perp$ and the hardest energy scale Q. In the region of $Q \gg q_\perp \gg \Lambda_{\text{QCD}}$, the TMD factorization framework at the leading twist usually works very well. When $q_\perp \sim Q \gg \Lambda_{\text{QCD}}$, we should fall back to the colinear factorization. However, in between, there is still a large phase space where $q_\perp$ is smaller than Q but not much smaller. This is the kinematic region where both the TMD factorization and the colinear factorization can approximately apply. However, the prediction from the TMD factorization deviates from the experimental measurements when $q_\perp/Q$ becomes not very small, calling for the inclusion of higher twist corrections. The higher twist corrections are also usually referred to as power corrections since they provide contributions in terms of $(q_\perp/Q)^n$. In addition, twist-3 contributions usually introduce new asymmetries that do not appear at the leading twist level. A comprehensive study on the higher twist contributions is thus vital in phenomenology. Contributions from higher twist

TMD PDFs and FFs were studied some time ago [42], but few advances have been made in the systematic derivations of the TMD factorization formula at the higher twist gain. Various theoretical methods have been applied to derive the TMD factorization scheme at the twist-3 level, such as the TMD operator expansion technique [67–69], the soft-collinear effective theory approach [70], factorization from functional integral [71–74], and a very recent work from [75], etc. The TMD factorization at the higher twist level is far from being completed, which requires further theoretical efforts.

3. Spin-Dependent TMD FFs

In semi-inclusive reactions, the experimental observables are usually different azimuthal asymmetries. In the kinematic region of TMD factorization, they are directly linked to TMD PDFs or FFs. The transverse momenta of partons and hadrons are often entangled with their polarizations. As a consequence, there are abundant polarization-dependent azimuthal asymmetries that can be measured in the experiment. This is particularly true for the transverse polarization. It is thought to provide only subleading power contributions compared to the longitudinal polarization at high energy; however, it often generates leading power contributions when correlated with the transverse momenta. In this section, we summarize the definition of the spin-dependent TMD FFs for hadrons with different spins. The following discussion only applies at the LO level since the TMD factorization for higher twist contributions is still far from being concluded. Therefore, we remove the scale dependence from TMD FFs.

3.1. The Intuitive Definition of TMD FFs

FFs represent the momentum distribution of a hadron inside of a hadronic jet produced by the fragmenting high-energy parton. We use $\mathcal{D}^{q \to h}(k;p)$ to denote the probability density of producing a hadron h with momentum p from a quark with momentum k.

In the high-energy limit, we can safely neglect the quark and hadron mass. Therefore, we have $k^2 = p^2 = 0$. In the naive parton model picture, the hadrons move colinearly with the parent quark. We thus have $p = zk$ where p is the hadron momentum, k is the quark momentum, and z is the momentum fraction. In this case, the FF is only a scalar function of z. We have

$$\mathcal{D}^{q \to h}(k;p) = D_1^{q \to h}(z), \tag{14}$$

where $D_1^{q \to h}(z)$ is simply the unpolarized FF.

With the spin degree of freedom being taken into account, the FFs will also depend on additional parameters which characterize the polarization of the final state hadron or the fragmenting quark. For example, for the production of spin-1/2 hadrons, we need to introduce λ_q and λ to describe the helicities and introduce $\vec{s}_{Tq}$ and $\vec{S}_T$ to describe the transverse polarizations of the quark and the hadron. With more available parameters, we can construct two additional scalar structures, $\lambda_q \lambda_h$ and $\vec{s}_{Tq} \cdot \vec{S}_T$, according to the parity conservation. Therefore, the complete decomposition of the FF is given by

$$\mathcal{D}^{q \to h}(k, S_q; p, S) = D_1^{q \to h}(z) + \lambda_q \lambda G_{1L}^{q \to h}(z) + \vec{s}_{Tq} \cdot \vec{S}_T H_{1T}^{q \to h}(z), \tag{15}$$

where $G_{1L}(z)$ and $H_{1T}(z)$ are the longitudinal and transverse spin transfers from the quark to the hadron, respectively. The physical interpretations of these probability densities coincide with those of the leading twist FFs in the colinear factorization approach.

In some cases, the transverse momentum of the final state hadron with respect to the quark momentum becomes relevant to the observable of interest. The interplay between the transverse momentum $p_\perp$ and the polarization parameters induces considerably intriguing phenomena. Again, we use the spin-1/2 hadron production as an example. From the parton model, we obtain the following eight TMD probability densities:

$$\begin{aligned}\mathcal{D}(k,S_q;p,S) =& D_1(z,p_\perp) + \lambda_q \lambda G_{1L}(z,p_\perp) + \vec{s}_{Tq} \cdot \vec{S}_T H_{1T}(z,p_\perp) \\ &+ \frac{1}{M}\vec{S}_T \cdot (\hat{\vec{k}} \times \vec{p}_\perp) D_{1T}^\perp(z,p_\perp) + \frac{1}{M}\lambda_q (\vec{S}_T \cdot \vec{p}_\perp) G_{1T}^\perp(z,p_\perp) \\ &+ \frac{1}{M}\vec{s}_{Tq} \cdot (\hat{\vec{k}} \times \vec{p}_\perp) H_1^\perp(z,p_\perp) + \frac{1}{M}\lambda (\vec{s}_{Tq} \cdot \vec{p}_\perp) H_{1L}^\perp(z,p_\perp) \\ &+ \frac{1}{M^2}(\vec{s}_{Tq} \cdot \vec{p}_\perp)(\vec{S}_T \cdot \vec{p}_\perp) H_{1T}^\perp(z,p_\perp). \end{aligned} \quad (16)$$

Here, we have dropped the $q \to h$ superscript for simplicity. These TMD FFs correspond to the eight leading twist TMD FFs defined in the TMD factorization approach. Among them, we notice in particular the famous Collins function $H_1^\perp$ [17] and the Sivers-type FF $D_{1T}^\perp$ [101,102]. They are usually referred to as the naive-T-odd FFs. In neglecting the interaction among the final state hadrons and the gauge link (which will be explained below), the time-reversal invariance demands that these two functions disappear. However, the time-reversal operation converts the "out" state to the "in" state. The interaction among hadrons suggests that one cannot find a simple relation between the "in" and "out" states any longer. Therefore, the time-reversal invariance actually poses no constraints on FFs. This feature can be fully appreciated in the context of parton correlators in the next subsection. Furthermore, we use H to denote FFs accompanied with the transverse polarization of the fragmenting quark $\vec{s}_{Tq}$. They are chiral-odd FFs. The reason for this will also be explained later.

3.2. The Definition of TMD FFs from the Parton Correlators

In the language of quantum field theory, the quark FFs are defined via the decomposition of parton correlators, such as the quark–quark correlator and the quark–gluon correlator. Usually, we need to define the gauge-invariant quark–quark correlators in the very beginning. From [8,42,43,50,64], we have the following:

$$\hat{\Xi}_{ij}^{(0)}(k;p,S) = \frac{1}{2\pi} \sum_X \int d^4\xi\, e^{-ik\xi} \langle 0|\mathcal{L}^\dagger(0;\infty)\psi_i(0)|p,S;X\rangle \langle p,S;X|\bar{\psi}_j(\xi)\mathcal{L}(\xi;\infty)|0\rangle, \quad (17)$$

where ξ is the coordinate of the quark field, k and p denote the 4-momenta of the fragmenting quark and the produced hadron, respectively; S denotes the hadron spin; and $\mathcal{L}(\xi;\infty)$ is the gauge link that ensures the gauge invariance of the definition of the correlator. We use i and j to represent one component of the corresponding spinor. Therefore, $\hat{\Xi}_{ij}^{(0)}(k;p,S)$ is actually one element in a 4×4 matrix which is denoted by $\hat{\Xi}^{(0)}(k;p,S)$.

As for the TMD FFs, we can integrate the above master correlator over the k^- component and obtain the following TMD quark–quark correlator:

$$\hat{\Xi}_{ij}^{(0)}(z,k_\perp;p,S) = \sum_X \int \frac{p^+ d\xi^-}{2\pi} d^2\xi_\perp e^{-i(p^+\xi^-/z - \vec{k}_\perp \cdot \vec{\xi}_\perp)}$$
$$\times \langle 0|\mathcal{L}^\dagger(0;\infty)\psi_i(0)|p,S;X\rangle \langle p,S;X|\bar{\psi}_j(\xi)\mathcal{L}(\xi;\infty)|0\rangle, \quad (18)$$

where $z = p^+/k^+$ is the longitudinal momentum fraction of the hadron, and $k_\perp$ is the transverse momentum of the fragmenting quark with respect to the hadron momentum. Unlike the discussion in the previous sections, it is more convenient to express the parton correlators as a function of $k_\perp$ instead of $p_\perp$. Nonetheless, since we have the approximation $k_\perp = -p_\perp/z$, these two methods are equivalent.

Although the TMD quark–quark correlator is a nonperturbative object, we can still discuss some general features from the definition. For instance, it possesses hermiticity, parity invariance, and charge-conjugation symmetry. As will be shown below, these properties will constrain the structures of the correlator. However, unlike the case for PDFs, the time-reversal invariance does not mean much for FFs.

Furthermore, the quark–quark correlator is a 4×4 matrix in the Dirac space. Therefore, it can always be decomposed in terms of 16 Γ-matrices, i.e.,

$$\hat{\Xi}^{(0)}(z,k_\perp;p,S) = \Xi^{(0)}(z,k_\perp;p,S) + i\gamma_5 \tilde{\Xi}^{(0)}(z,k_\perp;p,S) + \gamma^\alpha \Xi_\alpha^{(0)}(z,k_\perp;p,S) \\ + \gamma_5 \gamma^\alpha \tilde{\Xi}_\alpha^{(0)}(z,k_\perp;p,S) + i\sigma^{\alpha\beta}\gamma_5 \Xi_{\alpha\beta}^{(0)}(z,k_\perp;p,S). \tag{19}$$

The coefficient functions $\Xi^{(0)}$, $\tilde{\Xi}^{(0)}$, $\Xi_\alpha^{(0)}$, $\tilde{\Xi}_\alpha^{(0)}$ and $\Xi_{\alpha\beta}^{(0)}$ are given by the trace of the corresponding Γ-matrix with the correlator. These coefficient functions can further be decomposed into the products of scalar functions with basic Lorentz covariants according to their Lorentz transformation properties. The basic Lorentz covariants are constructed in terms of the available kinematic variables used in the reaction process. The scalar functions are the corresponding TMD FFs. We will present the detailed decomposition in the following subsections. Notice that the TMD quark–quark correlator given by Equation (18) satisfies the constraints of hermiticity and parity conservation. This will limit the allowed Lorentz structures of the parton correlator.

Higher twist TMD FFs also receive contributions from quark–gluon correlators [42,43,50,64] in addition to the quark–quark correlator mentioned above. For example, the complete decomposition of twist-3 TMD FFs also involves contributions from the following correlator:

$$\hat{\Xi}_{\rho,ij}^{(1)}(k;p,S) = \sum_X \int \frac{d^4\xi}{2\pi} e^{-ik\cdot\xi} \langle 0|\mathcal{L}^\dagger(0;\infty)D_\rho(0)\psi_i(0)|p,S;X\rangle\langle p,S;X|\bar{\psi}_j(\xi)\mathcal{L}(\xi;\infty)|0\rangle, \tag{20}$$

where $D_\rho(y) \equiv -i\partial_\rho + gA_\rho(y)$ and $A_\rho(y)$ denote the gluon field. However, the twist-3 TMD FFs defined via these quark–gluon correlators are not independent from those defined via the quark–quark correlator [64,65]. They are related to each other by a set of equations derived using the QCD equation of motion $\gamma \cdot D(y)\psi(y) = 0$. Therefore, we will only show the explicit decomposition of the TMD FFs from the quark–quark correlator in the following subsections.

3.3. The Spin Dependence

With the spin degree of freedom being taken into account, the basic Lorentz covariants in the decompositions of the coefficient functions in Equation (19) depend on not only momenta but also parameters describing the hadron polarization. The hadron polarization is defined in the rest frame of the hadron and is described by the spin density matrix.

For spin-1/2 hadrons, the spin density matrix is given by

$$\rho = \frac{1}{2}\left(1 + \vec{S}\cdot\vec{\sigma}\right), \tag{21}$$

where $\vec{\sigma}$ is the Pauli matrix, and $\vec{S}$ is the polarization vector in the rest frame of the hadron. The covariant form of the polarization vector reads as follows:

$$S^\mu = \lambda \frac{p^+}{M}\bar{n}^\mu + S_T^\mu - \lambda \frac{M}{2p^+}n^\mu. \tag{22}$$

Here, M is the hadron mass, λ is the helicity, and S_T^μ is the transverse polarization vector of the hadron. We have employed $\bar{n}^\mu$ to represent the light-cone plus direction and n^μ to denote the minus direction. For spin-1/2 hadrons, an additional pseudo-scalar λ and an axial-vector S_T^μ are at our disposal for constructing the basic Lorentz tensors.

For spin-1 hadrons, such as the vector mesons, the polarization is described by a 3×3 density matrix, which is usually given as [47]

$$\rho = \frac{1}{3}(1 + \frac{3}{2}S^i\Sigma^i + 3T^{ij}\Sigma^{ij}). \tag{23}$$

Here, Σ^i is the spin operator of spin-1 particle. The rank-2 tensor polarization basis Σ^{ij} is defined by

$$\Sigma^{ij} \equiv \frac{1}{2}(\Sigma^i\Sigma^j + \Sigma^j\Sigma^i) - \frac{2}{3}\mathbf{1}\delta^{ij}. \tag{24}$$

where the second term subtracts the diagonal elements from the product in the first term to give the relation

$$\Sigma^{xx} + \Sigma^{yy} + \Sigma^{zz} = 0. \tag{25}$$

This can be easily seen from the square of the spin-1 operator, i.e., $\Sigma^2 \equiv \Sigma^x\Sigma^x + \Sigma^y\Sigma^y + \Sigma^z\Sigma^z = s(s+1)\mathbf{1}$ with $s = 1$ for spin-1. From Equation (23), we find that a polarization tensor T is required to fully describe the polarization of a vector meson besides the polarization vector S. The polarization vector S is similar to that of spin-1/2 hadrons. It takes the same covariant form as laid out in Equation (22). The polarization tensor $T^{ij} = \text{Tr}(\rho\Sigma^{ij})$ has five independent components that consist of a Lorentz scalar S_{LL}, a Lorentz vector $S_{LT}^\mu = (0, S_{LT}^x, S_{LT}^y, 0)$ and a Lorentz tensor $S_{TT}^{\mu\nu}$ that has two nonzero independent components ($S_{TT}^{xx} = -S_{TT}^{yy}$ and $S_{TT}^{xy} = S_{TT}^{yx}$). It is parameterized as follows:

$$T = \frac{1}{2}\begin{pmatrix} -\frac{2}{3}S_{LL} + S_{TT}^{xx} & S_{TT}^{xy} & S_{LT}^x \\ S_{TT}^{xy} & -\frac{2}{3}S_{LL} - S_{TT}^{xx} & S_{LT}^y \\ S_{LT}^x & S_{LT}^y & \frac{4}{3}S_{LL} \end{pmatrix}. \tag{26}$$

The Lorentz covariant form for the polarization tensor is expressed as [47]

$$T^{\mu\nu} = \frac{1}{2}\left[\frac{4}{3}S_{LL}\left(\frac{p^+}{M}\right)^2\bar{n}^\mu\bar{n}^\nu + \frac{p^+}{M}\bar{n}^{\{\mu}S_{LT}^{\nu\}} - \frac{2}{3}S_{LL}(\bar{n}^{\{\mu}n^{\nu\}} - g_T^{\mu\nu}) + S_{TT}^{\mu\nu} - \frac{M}{2p^+}\bar{n}^{\{\mu}S_{LT}^{\nu\}} + \frac{1}{3}S_{LL}\left(\frac{M}{p^+}\right)^2 n^\mu n^\nu\right], \tag{27}$$

where we have used the shorthand notation $A^{\{\mu}B^{\nu\}} \equiv A^\mu B^\nu + A^\nu B^\mu$.

For spin-3/2 hadrons, such as the decuplet baryons, the polarization is described by a 4×4 density matrix which is given by [103,104]

$$\rho = \frac{1}{4}\left(1 + \frac{4}{5}S^i\Sigma^i + \frac{2}{3}T^{ij}\Sigma^{ij} + \frac{8}{9}R^{ijk}\Sigma^{ijk}\right). \tag{28}$$

Here, Σ^i is the spin operator of the spin-3/2 particle, and S^i is the corresponding polarization vector. Similar with that for spin-1 case, $\langle\Sigma^{ij}\rangle$ is the polarization tensor basis which has five independent components. It can be constructed from Σ^i and is given by

$$\Sigma^{ij} = \frac{1}{2}\left(\Sigma^i\Sigma^j + \Sigma^j\Sigma^i\right) - \frac{5}{4}\delta^{ij}\mathbf{1}, \tag{29}$$

Notice that the square of the spin-3/2 operator is given by $\sum_i(\Sigma^i)^2 = \frac{3}{2}(\frac{3}{2}+1)\mathbf{1} = \frac{15}{4}\mathbf{1}$. The rank-2 tensor polarization basis for spin-3/2, Σ^{ij}, is also chosen to be traceless as laid out by Equation (25). Therefore, the second term in Equation (29) is different from that in Equation (24) for spin-1 hadrons. The corresponding polarization tensor T^{ij} also has five independent components which are the same as those for spin-1 hadrons. The rank-3 tensor polarization basis Σ^{ijk} is unique for spin-3/2 hadrons. It has seven independent components which can be constructed as follows:

$$\Sigma^{ijk} = \frac{1}{6}\Sigma^{\{i}\Sigma^{j}\Sigma^{k\}} - \frac{41}{60}\left(\delta^{ij}\Sigma^{k} + \delta^{jk}\Sigma^{i} + \delta^{ki}\Sigma^{j}\right)$$
$$= \frac{1}{3}\left(\Sigma^{ij}\Sigma^{k} + \Sigma^{jk}\Sigma^{i} + \Sigma^{ki}\Sigma^{j}\right) - \frac{4}{15}\left(\delta^{ij}\Sigma^{k} + \delta^{jk}\Sigma^{i} + \delta^{ki}\Sigma^{j}\right), \quad (30)$$

where the symbol $\{\cdots\}$ stands for the sum of all possible permutations. The corresponding rank-3 spin tensor R^{ijk} is defined as follows:

$$R^{ijk} = \frac{1}{4}\begin{bmatrix} \begin{pmatrix} -3S^x_{LLT} + S^{xxx}_{TTT} & -S^y_{LLT} + S^{yxx}_{TTT} & -2S_{LLL} + S^{xx}_{LTT} \\ -S^y_{LLT} + S^{yxx}_{TTT} & -S^x_{LLT} - S^{xxx}_{TTT} & S^{xy}_{LTT} \\ -2S_{LLL} + S^{xx}_{LTT} & S^{xy}_{LTT} & 4S^x_{LLT} \end{pmatrix} \\ \begin{pmatrix} -S^y_{LLT} + S^{yxx}_{TTT} & -S^x_{LLT} - S^{xxx}_{TTT} & S^{xy}_{LTT} \\ -S^x_{LLT} - S^{xxx}_{TTT} & -3S^y_{LLT} - S^{yxx}_{TTT} & -2S_{LLL} - S^{xx}_{LTT} \\ S^{xy}_{LTT} & -2S_{LLL} - S^{xx}_{LTT} & 4S^y_{LLT} \end{pmatrix} \\ \begin{pmatrix} -2S_{LLL} + S^{xx}_{LTT} & S^{xy}_{LTT} & 4S^x_{LLT} \\ S^{xy}_{LTT} & -2S_{LLL} - S^{xx}_{LTT} & 4S^y_{LLT} \\ 4S^x_{LLT} & 4S^y_{LLT} & 4S_{LLL} \end{pmatrix} \end{bmatrix}, \quad (31)$$

Meanwhile, the Lorentz covariant form is given as follows:

$$R^{\mu\nu\rho} = \frac{1}{4}\left\{ S_{LLL}\left[\frac{1}{2}\left(\frac{M}{P\cdot\bar{n}}\right)^3 \bar{n}^{\mu}\bar{n}^{\nu}\bar{n}^{\rho} - \frac{1}{2}\left(\frac{M}{P\cdot\bar{n}}\right)\left(\bar{n}^{\{\mu}\bar{n}^{\nu}n^{\rho\}} - \bar{n}^{\{\mu}g^{\nu\rho\}}_T\right)\right.\right.$$
$$\left.+ \left(\frac{P\cdot\bar{n}}{M}\right)\left(\bar{n}^{\{\mu}n^{\nu}n^{\rho\}} - n^{\{\mu}g^{\nu\rho\}}_T\right) - 4\left(\frac{P\cdot\bar{n}}{M}\right)^3 n^{\mu}n^{\nu}n^{\rho}\right]$$
$$+ \frac{1}{2}\left(\frac{M}{P\cdot\bar{n}}\right)^2 \bar{n}^{\{\mu}\bar{n}^{\nu}S^{\rho\}}_{LLT} + 2\left(\frac{P\cdot\bar{n}}{M}\right)^2 n^{\{\mu}n^{\nu}S^{\rho\}}_{LLT} - 2\bar{n}^{\{\mu}n^{\nu}S^{\rho\}}_{LLT} + \frac{1}{2}S^{\{\mu}_{LLT}g^{\nu\rho\}}_T$$
$$\left.+ \frac{1}{4}\left(\frac{M}{P\cdot\bar{n}}\right)\bar{n}^{\{\mu}S^{\nu\rho\}}_{LTT} - \frac{1}{2}\left(\frac{P\cdot\bar{n}}{M}\right)n^{\{\mu}S^{\nu\rho\}}_{LTT} + S^{\mu\nu\rho}_{TTT}\right\}. \quad (32)$$

3.4. Decomposition Result for Spin-Dependent TMD FFs

The results for TMD FFs of spin-1 hadrons defined via quark–quark correlator exist up to twist-4 level in the literature [65]. The leading twist TMD FFs for spin-3/2 hadrons have also been presented in [104]. In this section, we summarize the general decomposition of the quark–quark correlator in terms of TMD FFs for the unpolarized part, polarization-vector-dependent part, rank-2-polarization-tensor-dependent part, and rank-3-polarization-tensor-dependent parts. To describe the production of pseudoscalar mesons, we only need the unpolarized part. To describe the production of baryons, we need to combine the unpolarized and the polarization-vector-dependent parts. The description of the spin-3/2 hadron production requires all four parts. However, it should be noted that different conventions are employed in different works.

The notation system for TMD FFs in this review are laid out here. We use D, G, and H to denote FFs of unpolarized, longitudinally polarized, and transversely polarized quarks, respectively. They are obtained from the decomposition of the γ_μ, $\gamma_5\gamma_\mu$ and $\gamma_5\sigma_{\mu\nu}$ terms of the quark–quark correlator. Those FFs defined from the decomposition of the $\mathbf{1}$ and γ_5 terms are denoted as E. We use the numbers 1 and 3 in the subscripts to denote the leading twist and twist-4 FFs, respectively. Other FFs without numbers in the subscripts are at the twist-3 level. The polarization of the produced hadron will be specified in the subscripts, where L and T represent longitudinal and transverse polarizations, and LL, LT, and TT

stand for the rank-2-tensor polarizations. The symbol $\perp$ in the superscript implies that the corresponding basic Lorentz structure depends on the transverse momentum $k_\perp$.

The decomposition for the unpolarized part is given by the following:

$$z\Xi^{U(0)}(z,k_\perp;p) = ME(z,k_\perp), \tag{33}$$

$$z\tilde{\Xi}^{U(0)}(z,k_\perp;p) = 0, \tag{34}$$

$$z\Xi_\alpha^{U(0)}(z,k_\perp;p) = p^+\bar{n}_\alpha D_1(z,k_\perp) + k_{\perp\alpha}D^\perp(z,k_\perp) + \frac{M^2}{p^+}n_\alpha D_3(z,k_\perp), \tag{35}$$

$$z\tilde{\Xi}_\alpha^{U(0)}(z,k_\perp;p) = -\tilde{k}_{\perp\alpha}G^\perp(z,k_\perp), \tag{36}$$

$$z\Xi_{\rho\alpha}^{U(0)}(z,k_\perp;p) = -\frac{p^+}{M}\bar{n}_{[\rho}\tilde{k}_{\perp\alpha]}H_1^\perp(z,k_\perp) + M\varepsilon_{\perp\rho\alpha}H(z,k_\perp) - \frac{M}{p^+}n_{[\rho}\tilde{k}_{\perp\alpha]}H_3^\perp(z,k_\perp). \tag{37}$$

Here, $\tilde{k}_{\perp\alpha} \equiv \varepsilon_{\perp\mu\alpha}k_\perp^\mu$ denotes the transverse vector orthogonal to $k_{\perp\alpha}$, with $\varepsilon_{\perp\mu\nu}$ being defined as $\varepsilon_{\perp\mu\nu} \equiv \varepsilon_{\mu\nu\alpha\beta}\bar{n}^\alpha n^\beta$. There are eight TMD FFs for the unpolarized part. Among them, the number density D_1 and the Collins function $H_1^\perp$ are at the leading twist. They both have twist-4 companions i.e., D_3 and $H_3^\perp$, respectively. The other four are twist-3 FFs. The TMD FFs $D_{1T}^\perp, G^\perp, H_1^\perp, H$, and $H_3^\perp$ are usually referred to as the naive T-odd FFs. The reader may have already discerned that the T-odd FFs are always associated with the Levi-Civita tensor, $\varepsilon_{\mu\nu\alpha\beta}$. It should be noted that T-odd PDFs can only survive thanks to the gauge link. However, for the FFs, the final state interactions between the produced hadrons in the hadronization process can also contribute to the T-oddness. This difference has a more important impact on the polarization-vector-dependent T-odd PDFs and FFs, which are discussed below.

The decomposition for the vector polarized part is given by the following:

$$z\Xi^{V(0)}(z,k_\perp;p,S) = (\tilde{k}_\perp \cdot S_T)E_T^\perp(z,k_\perp), \tag{38}$$

$$z\tilde{\Xi}^{V(0)}(z,k_\perp;p,S) = M\left[\lambda E_L(z,k_\perp) + \frac{k_\perp \cdot S_T}{M}E_T'^\perp(z,k_\perp)\right], \tag{39}$$

$$z\Xi_\alpha^{V(0)}(z,k_\perp;p,S) = p^+\bar{n}_\alpha \frac{\tilde{k}_\perp \cdot S_T}{M}D_{1T}^\perp(z,k_\perp) - M\tilde{S}_{T\alpha}D_T(z,k_\perp)$$
$$- \tilde{k}_{\perp\alpha}\left[\lambda D_L^\perp(z,k_\perp) + \frac{k_\perp \cdot S_T}{M}D_T^\perp(z,k_\perp)\right] + \frac{M}{p^+}n_\alpha(\tilde{k}_\perp \cdot S_T)D_{3T}^\perp(z,k_\perp), \tag{40}$$

$$z\tilde{\Xi}_\alpha^{V(0)}(z,k_\perp;p,S) = p^+\bar{n}_\alpha\left[\lambda G_{1L}(z,k_\perp) + \frac{k_\perp \cdot S_T}{M}G_{1T}^\perp(z,k_\perp)\right]$$
$$- MS_{T\alpha}G_T(z,k_\perp) - k_{\perp\alpha}\left[\lambda G_L^\perp(z,k_\perp) + \frac{k_\perp \cdot S_T}{M}G_T^\perp(z,k_\perp)\right]$$
$$+ \frac{M^2}{p^+}n_\alpha\left[\lambda G_{3L}(z,k_\perp) + \frac{k_\perp \cdot S_T}{M}G_{3T}^\perp(z,k_\perp)\right], \tag{41}$$

$$z\Xi_{\rho\alpha}^{V(0)}(z,k_\perp;p,S) = p^+\bar{n}_{[\rho}S_{T\alpha]}H_{1T}(z,k_\perp) + \frac{p^+}{M}\bar{n}_{[\rho}k_{\perp\alpha]}\left[\lambda H_{1L}^\perp(z,k_\perp) + \frac{k_\perp \cdot S_T}{M}H_{1T}^\perp(z,k_\perp)\right]$$
$$+ k_{\perp[\rho}S_{T\alpha]}H_T^\perp(z,k_\perp) + M\bar{n}_{[\rho}n_{\alpha]}\left[\lambda H_L(z,k_\perp) + \frac{k_\perp \cdot S_T}{M}H_T'^\perp(z,k_\perp)\right]$$
$$+ \frac{M^2}{p^+}n_{[\rho}S_{T\alpha]}H_{3T}(z,k_\perp) + \frac{M}{p^+}n_{[\rho}k_{\perp\alpha]}\left[\lambda H_{3L}^\perp(z,k_\perp) + \frac{k_\perp \cdot S_T}{M}H_{3T}^\perp(z,k_\perp)\right]. \tag{42}$$

There are in total 24 polarization-vector-dependent TMD FFs. Of these, 6 contribute at the leading twist, 12 at twist-3, and remaining 6 at twist-4. Among the six leading twist FFs, G_{1L} is the longitudinal spin transfer, H_{1T} and $H_{1T}^\perp$ are transverse spin transfers, $G_{1T}^\perp$ is the longitudinal to transverse spin transfer, $H_{1L}^\perp$ is the transverse to longitudinal spin transfer, and $D_{1T}^\perp$ induces the transverse polarization of hadrons in the fragmentation of an unpolarized quark. We note in particular that the $D_{1T}^\perp$ FF resembles the Sivers function in PDFs [101]. It is responsible for the hadron transverse polarization along the normal direction of the production plane in high-energy collisions. It is also a naive T-odd FF. However, as mentioned above, the T-oddness has little meaning in the context of hadronization. The T-odd PDFs arise solely from the gauge link. Therefore, it has been proven theoretically that there is a sign-flip between the Sivers functions in SIDIS and Drell-Yan [105–107]. However, the T-oddness of FFs can also be generated from the interaction among final state hadrons. Therefore, there is no such similar relation for the $D_{1T}^\perp$ FF between different processes. Besides $D_{1T}^\perp$, there are seven other T-odd FFs, namely, $E_T^\perp$, E_L, $E_T^{\prime\perp}$, $D_L^\perp$, D_T, $D_T^\perp$, and $D_{3T}^\perp$. The rest are T-even. All of T-odd FFs are accompanied by the Levi-Civita tensor except for E_L and $E_T^{\prime\perp}$.

The decomposition for the rank-2-polarization-tensor-dependent part is given as follows:

$$z\Xi^{T(0)}(z,k_\perp;p,S) = M\left[S_{LL}E_{LL}(z,k_\perp) + \frac{k_\perp \cdot S_{LT}}{M}E_{LT}^\perp(z,k_\perp) + \frac{S_{TT}^{kk}}{M^2}E_{TT}^\perp(z,k_\perp)\right], \quad (43)$$

$$z\tilde{\Xi}^{T(0)}(z,k_\perp;p,S) = M\left[\frac{\tilde{k}_\perp \cdot S_{LT}}{M}E_{LT}^{\prime\perp}(z,k_\perp) + \frac{S_{TT}^{\tilde{k}k}}{M^2}E_{TT}^{\prime\perp}(z,k_\perp)\right], \quad (44)$$

$$z\Xi_\alpha^{T(0)}(z,k_\perp;p,S) = p^+\bar{n}_\alpha\left[S_{LL}D_{1LL}(z,k_\perp) + \frac{k_\perp \cdot S_{LT}}{M}D_{1LT}^\perp(z,k_\perp) + \frac{S_{TT}^{kk}}{M^2}D_{1TT}^\perp(z,k_\perp)\right]$$
$$+ MS_{LT\alpha}D_{LT}(z,k_\perp) + S_{TT\alpha}^k D_{TT}^{\prime\perp}(z,k_\perp)$$
$$+ k_{\perp\alpha}\left[S_{LL}D_{LL}^\perp(z,k_\perp) + \frac{k_\perp \cdot S_{LT}}{M}D_{LT}^\perp(z,k_\perp) + \frac{S_{TT}^{kk}}{M^2}D_{TT}^\perp(z,k_\perp)\right]$$
$$+ \frac{M^2}{p^+}n_\alpha\left[S_{LL}D_{3LL}(z,k_\perp) + \frac{k_\perp \cdot S_{LT}}{M}D_{3LT}^\perp(z,k_\perp) + \frac{S_{TT}^{kk}}{M^2}D_{3TT}^\perp(z,k_\perp)\right], \quad (45)$$

$$z\tilde{\Xi}_\alpha^{T(0)}(z,k_\perp;p,S) = p^+\bar{n}_\alpha\left[\frac{\tilde{k}_\perp \cdot S_{LT}}{M}G_{1LT}^\perp(z,k_\perp) + \frac{S_{TT}^{\tilde{k}k}}{M^2}G_{1TT}^\perp(z,k_\perp)\right]$$
$$- M\tilde{S}_{LT\alpha}G_{LT}(z,k_\perp) - \tilde{S}_{TT\alpha}^k G_{TT}^{\prime\perp}(z,k_\perp)$$
$$- \tilde{k}_{\perp\alpha}\left[S_{LL}G_{LL}^\perp(z,k_\perp) + \frac{k_\perp \cdot S_{LT}}{M}G_{LT}^\perp(z,k_\perp) + \frac{S_{TT}^{kk}}{M^2}G_{TT}^\perp(z,k_\perp)\right]$$
$$+ \frac{M^2}{p^+}n_\alpha\left[\frac{\tilde{k}_\perp \cdot S_{LT}}{M}G_{3LT}^\perp(z,k_\perp) + \frac{S_{TT}^{\tilde{k}k}}{M^2}G_{3TT}^\perp(z,k_\perp)\right], \quad (46)$$

$$z\Xi_{\rho\alpha}^{T(0)}(z,k_\perp;p,S) = -p^+\bar{n}_{[\rho}\tilde{S}_{LT\alpha]}H_{1LT}(z,k_\perp) - \frac{p^+}{M}\bar{n}_{[\rho}\tilde{S}_{TT\alpha]}^k H_{1TT}^{\prime\perp}(z,k_\perp)$$
$$- \frac{p^+}{M}\bar{n}_{[\rho}\tilde{k}_{\perp\alpha]}\left[S_{LL}H_{1LL}^\perp(z,k_\perp) + \frac{k_\perp \cdot S_{LT}}{M}H_{1LT}^\perp(z,k_\perp) + \frac{S_{TT}^{kk}}{M^2}H_{1TT}^\perp(z,k_\perp)\right]$$
$$+ M\varepsilon_{\perp\rho\alpha}\left[S_{LL}H_{LL}(z,k_\perp) + \frac{k_\perp \cdot S_{LT}}{M}H_{LT}^\perp(z,k_\perp) + \frac{S_{TT}^{kk}}{M^2}H_{TT}^\perp(z,k_\perp)\right]$$
$$+ \bar{n}_{[\rho}n_{\alpha]}\left[(\tilde{k}_\perp \cdot S_{LT})H_{LT}^{\prime\perp}(z,k_\perp) + \frac{S_{TT}^{\tilde{k}k}}{M}H_{TT}^{\prime\perp}(z,k_\perp)\right]$$
$$- \frac{M}{p^+}n_{[\rho}\tilde{k}_{\perp\alpha]}\left[S_{LL}H_{3LL}^\perp(z,k_\perp) + \frac{k_\perp \cdot S_{LT}}{M}H_{3LT}^\perp(z,k_\perp) + \frac{S_{TT}^{kk}}{M^2}H_{3TT}^\perp(z,k_\perp)\right]$$
$$- \frac{M}{p^+}n_{[\rho}M\tilde{S}_{LT\alpha]}\left[H_{3LT}(z,k_\perp) + \tilde{S}_{TT\alpha]}^k H_{3TT}^{\prime\perp}(z,k_\perp)\right]. \quad (47)$$

We have used the shorthanded notations such as $S_{TT}^{kk} \equiv S_{TT}^{\alpha\beta}k_{\perp\alpha}k_{\perp\beta}$. There are in total 40 tensor polarization-dependent TMD FFs; of these. 10 contribute at the leading twist, 20 contribute at twist-3, and the remaining 10 contribute at twist-4. The 24 TMD FFs defined from the decomposition of $\tilde{\Xi}_\alpha^{T(0)}$ and $\Xi_{\rho\alpha}^{T(0)}$ are naive T-odd. Among these TMD FFs, we notice in particularly that the S_{LL} dependent TMD FF D_{1LL}, which is responsible for the spin alignment of the produced vector meson, is decoupled from the quark polarization. This suggests that the vector meson spin alignment can also be observed in the unpolarized high-energy collisions [108–110]. Besides, D_{1LL} also survives the $k_\perp$-integral. Therefore, it also appears in the colinear factorization.

The rank-3-polarization-tensor-dependent TMD FFs are unique for spin-3/2 (or higher) hadrons. A complete set of leading twist quark TMD FFs for spin-3/2 hadrons has been given in [104]. There are in total 14 rank-3-polarization-tensor-dependent TMD FFs that can be defined at the leading twist level. We refer interested readers to [104] for a detailed discussion.

3.5. TMD FFs of Antiquarks and Gluons

One can define antiquark TMD FFs by replacing the fermion fields in the correlator of quark TMD FFs with the charge-conjugated fields. Therefore, it is easy to find that the traces of the correlator with Dirac matrices I, $i\gamma_5$ and $\gamma^\mu\gamma_5$ will have an opposite sign between quark and antiquark cases, while the traces with γ^μ and $i\sigma^\mu\gamma_5$ are the same [42,43,56]. The definition and parameterization of the antiquark TMD FFs are then full analogous to those of quark TMD FFs.

The gluon FFs are defined through the gluon correlator given by [8,111]

$$\hat{\Gamma}^{\mu\nu;\rho\sigma}(k;p,S) = \sum_X \int \frac{d^4\xi}{(2\pi)^4} e^{ik\cdot\xi} \langle 0|F^{\rho\sigma}(\xi)|p,S;X\rangle\langle p,S;X|\mathcal{U}(\xi,0)F^{\mu\nu}(0)|0\rangle, \quad (48)$$

where $F^{\rho\sigma}(\xi) \equiv F^{\rho\sigma,a}T^a$ is the gluon field field strength tensor, and $\mathcal{U}(\xi,0)$ is the Wilson line in the adjoint representation that renders the correlator gauge invariant. Under the assumption that the fragmenting parton moves in the plus direction, an integration over the k^- component is carried out to give the TMD gluon correlator.

At the leading twist, we need to consider

$$M\hat{\Gamma}^{ij}(z,k_\perp;p,S) = \int dk^- \, \Gamma^{+j;+i}(k;p,S), \quad (49)$$

where i and j are transverse Lorentz indices in the transverse directions.

For the spin-1/2 hadron production, there are eight leading twist gluon TMD FFs which are given by the decomposition of the TMD gluon correlator [111]. We have the following:

$$\hat{\Gamma}^{ij}_U(z,k_\perp;p,S) = \frac{p^+}{M}\left[-g_T^{ij}D_{1g}(z,k_\perp) + \left(\frac{k_\perp^i k_\perp^j}{M^2} + g_T^{ij}\frac{k_\perp^2}{2M^2}\right)H^\perp_{1g}(z,k_\perp)\right],$$

$$\hat{\Gamma}^{ij}_L(z,k_\perp;p,S) = -\lambda\frac{p^+}{M}\left[i\varepsilon_\perp^{ij}G_{1Lg}(z,k_\perp) - \frac{\varepsilon_\perp^{k_\perp\{i}k_\perp^{j\}}}{2M^2}H^\perp_{1Lg}(z,k_\perp)\right],$$

$$\hat{\Gamma}^{ij}_T(z,k_\perp;p,S) = -\frac{p^+}{M}\left[g_T^{ij}\frac{\varepsilon_\perp^{k_\perp S_T}}{M}D^\perp_{1Tg}(z,k_\perp) + i\varepsilon_\perp^{ij}\frac{k_\perp\cdot S_T}{M}G^\perp_{1Tg}(z,k_\perp)\right.$$

$$\left. - \frac{\varepsilon_\perp^{k_\perp\{i}S_T^{j\}} + \varepsilon_\perp^{S_T\{i}k_\perp^{j\}}}{4M}H_{1Tg}(z,k_\perp) - \frac{\varepsilon_\perp^{k_\perp\{i}k_\perp^{j\}}}{2M^2}\frac{k_\perp\cdot S_T}{M}H^\perp_{1Tg}(z,k_\perp)\right]. \quad (50)$$

$\hat{\Gamma}_U$, $\hat{\Gamma}_L$, and $\hat{\Gamma}_T$ stand for the unpolarized, the longitudinal, and transverse polarized parts for the hadron production, respectively. Analogously to the quark FFs, we have used D to represent FFs of the unpolarized gluons, G to represent the FFs of the circularly polarized gluons, and H to represent the FFs of the linearly polarized gluons. Higher twist gluon TMD FFs are also discussed in [111], who further detail the parameterizations.

4. Experiment and Phenomenology

In high-energy experiments, the polarization of final state hadrons is usually measured from the angular distribution of their decay products. It is very challenging to acquire accurate experimental data. In light of a considerably large amount of free parameters, the spin-dependent FFs are not well-constrained experimentally. Compared with the case for unpolarized PDFs or FFs, the quantitative study of spin-dependent FFs is still immature. That said, there are already quite a few phenomenological studies making full use of the available experimental data. In this section, we summarize the available experimental data and the corresponding phenomenological studies.

4.1. Λ Hyperons

The polarization of Λ^0 hyperons is usually measured from the angular distribution of the daughter proton in the parity-violating $\Lambda^0 \to p + \pi^-$ decay channel. In the rest frame of Λ^0, the normalized angular distribution of the daughter proton reads as follows:

$$\frac{1}{N}\frac{dN}{d\cos\theta^*} = \frac{1}{2}(1 + \alpha \mathcal{P} \cos\theta^*), \tag{51}$$

where $\alpha = 0.732 \pm 0.014$ is the decay parameter of Λ [112], $\mathcal{P}$ is the polarization of Λ along a specified direction, and θ^* is the angle between the proton momentum and the specified direction to measure the Λ polarization.

The LEP experiment is an e^+e^- collider at the Z^0-pole. Due to the parity violation in the weak interaction, the produced quark and antiquark are strongly polarized along the longitudinal direction. The longitudinal polarizations of those final state quarks and antiquarks in e^+e^- annihilation at different collisional energies can be easily computed at the LO level and are explicitly shown in [113]. At the Z^0-pole, the longitudinal polarization of the final state down-type quarks can reach 0.9. That of the up-type quarks is a bit smaller but is still about $0.6 \sim 0.7$. Based on the $SU(6)$ spin-flavor symmetry, the polarization of Λ^0 is determined by the polarization of the s quark. It is thus proposed in [114] that the final state Λ^0 hyperons are also strongly polarized at LEP, and the measurement of this polarization can probe interesting information on the hadronization mechanism. In the language of QCD factorization, the LEP experiment is the ideal place to study the longitudinal spin transfer $G_{1L}(z)$, which represents the number density of producing longitudinally polarized Λ^0 hyperons from longitudinally polarized quarks. It is the p_T-integrated version of the TMD FF $G_{1L}(z, p_\perp)$.

At the leading order and leading twist, the longitudinal polarization of Λ^0 reads as follows: [108,113]

$$\mathcal{P}_L(y,z) = \frac{\sum_q \lambda_q(y) \omega_q(y) G_{1L,q}(z) + \{q \leftrightarrow \bar{q}; y \leftrightarrow (1-y)\}}{\sum_q \omega_q(y) D_{1,q}(z) + \{q \leftrightarrow \bar{q}; y \leftrightarrow (1-y)\}}, \tag{52}$$

where $\lambda_q(y) = \Delta\omega_q(y)/\omega_q(y)$ is the helicity of the fragmenting quark with $\Delta\omega_q(y)$ and $\omega_q(y)$ being defined as follows:

$$\Delta\omega_q(y) = \chi T_1^q(y) + \chi_{\text{int}}^q I_1^q(y), \tag{53}$$

$$\omega_q(y) = \chi T_0^q(y) + \chi_{\text{int}}^q I_0^q(y) + e_q^2 A(y), \tag{54}$$

$$T_1^q(y) = -2c_V^q c_A^q [(c_V^e)^2 + (c_A^e)^2] A(y) + 2[(c_V^q)^2 + (c_A^q)^2] c_V^e c_A^e B(y), \tag{55}$$

$$T_0^q(y) = [(c_V^q)^2 + (c_A^q)^2][(c_V^e)^2 + (c_A^e)^2] A(y) - 4c_V^q c_A^q c_V^e c_A^e B(y), \tag{56}$$

$$I_1^q(y) = -c_A^q c_V^e A(y) + c_V^q c_A^e B(y), \tag{57}$$

$$I_0^q(y) = c_V^q c_V^e A(y) - c_A^q c_A^e B(y). \tag{58}$$

Here, $y = (1 + \cos\theta)/2$ with θ is the angle between the outgoing Λ and the incoming electron. The coefficient functions are given as $A(y) = (1-y)^2 + y^2$, $B(y) = 1 - 2y$, $\chi = Q^4/[(Q^2 - M_Z^2)^2 + \Gamma_Z^2 M_Z^2] \sin^4 2\theta_W$, and $\chi_{\text{int}}^q = -2e_q Q^2 (Q^2 - M_Z^2)/[(Q^2 - M_Z^2)^2 + \Gamma_Z^2 M_Z^2] \sin^2 2\theta_W$. $c_V^{q/e}$ and $c_A^{q/e}$ are the coupling constants of the vector current and axis-vector current parts of the quark/electron, with Z^0. M_Z being the mass of Z^0 and Γ_Z being the width. Notice that $\lambda_{\bar{q}}(y) = -\lambda_q(1-y)$. The quark helicity and antiquark helicity have the opposite sign. This is in line with the sign flip in Section 3.5. Therefore, the polarization of $\bar{\Lambda}^0$ is expected to have the opposite sign with that of Λ^0.

Since the quark helicity $\lambda_q(y)$ and the production weight $\omega_q(y)$ are calculable in quantum field theory, the measurement of the longitudinal polarization of final state Λ^0 as a function of z can directly provide information of the longitudinal spin transfer. Such experiments were eventually carried out by ALEPH and OPAL collaborations at LEP in the 1990s [115,116]. As shown in Figure 2, the longitudinal polarization increases monotonically with increasing z, which provides a hint on how to parameterize the longitudinal spin transfer.

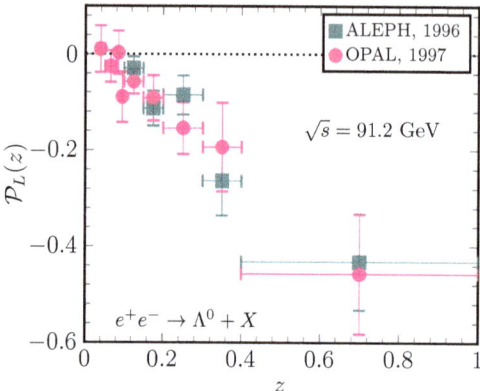

Figure 2. Reproduction of the longitudinal polarization of Λ^0 in e^+e^- annihilation at $\sqrt{s} = 91.2$ GeV measured by the ALEPH [115] and OPAL [116] collaborations at LEP. We have combined the statistical and systematic errors. Neglecting the mass of Λ hyperons in the high-energy limit, the definitions of z in these two experiments are the same as those of the momentum fraction in the light-cone coordinate currently used in the QCD factorizations.

Following the release of these experimental data, many phenomenological studies [108,117–124] were carried out to understand the longitudinal spin transfer $G_{1L}(z)$. Among them, the de Florian–Stratmann–Vogelsang (DSV) parameterization [118] offers three scenarios. The first scenario is based on the naive parton model, which assumes that only the s quark contributes to the longitudinal spin transfer at the initial scale. The second scenario assumes that the u and d quarks contribute to negative $G_{1L}(z)$ at the initial scale. The third scenario assumes that u, d, and s contribute equally. All three can describe the experimental data reasonably well. A more recent Chen–Yang–Zhou–Liang (CYZL) analysis [108] also obtained a good description of the experimental data utilizing the LO formula. The ambiguity again highlights the difficulties in the quantitative study of FFs. It can only be removed through a global analysis of the experimental data in various highenergy reactions. Therefore, many works have also made predictions for the longitudinal polarization of Λ produced in polarized SIDIS [117,125–127] and pp collisions [128–131].

The inclusive DIS process with the polarized lepton beam has been used to probe the spin structure of the nucleon [132–134]. In this process, only the momentum of the final state lepton was measured. Therefore, we can gain information on the nucleon structure but lose those on the hadronization. To restore the access to (spin-dependent) FFs, we have to rely on the semi-inclusive process and measure (the polarization of) one final state hadron (There are two fragmentation regimes in SIDIS, namely the current fragmentation and the target fragmentation. Although the target fragmentation function is also currently a hot topic, it is beyond the scope of this review. We only focus on the study of the current fragmentation function). However, it is not a simple task to do so in the real world. Despite the difficulties, early attempts from the E665 [135] and HERMES [136] collaborations were still successfully performed. Recent measurements from HERMES [137] and COMPASS [138] collaborations have also elevated the quality of experimental data to a level that sheds light on phenomenological studies. These experiments measure the spin transfer coefficient $\mathcal{D}_{LL}(z)$ (it is important to not get confused with the spin-alignment-dependent FF $D_{1LL}(z)$ of vector mesons) which, at the leading order and leading twist approximation, is given by [42]

$$\mathcal{D}_{LL}(x_B, z) = \frac{\sum_q e_q^2 x_B f_{1,q}(x_B) G_{1,q}(z)}{\sum_q e_q^2 x_B f_{1,q}(x_B) D_{1,q}(z)}, \qquad (59)$$

with $f_{1,q}(x_B)$ being the unpolarized PDF. Due to the presence of the unpolarized PDF of proton/nucleus, the polarized SIDIS experiment favors more contributions from the u and d quarks at large x_B than from the e^+e^- collider. We show the HERMES data set as a function of z (integrating over x_B) and the COMPASS data set as a function of x_B (integrating over z) in Figure 3.

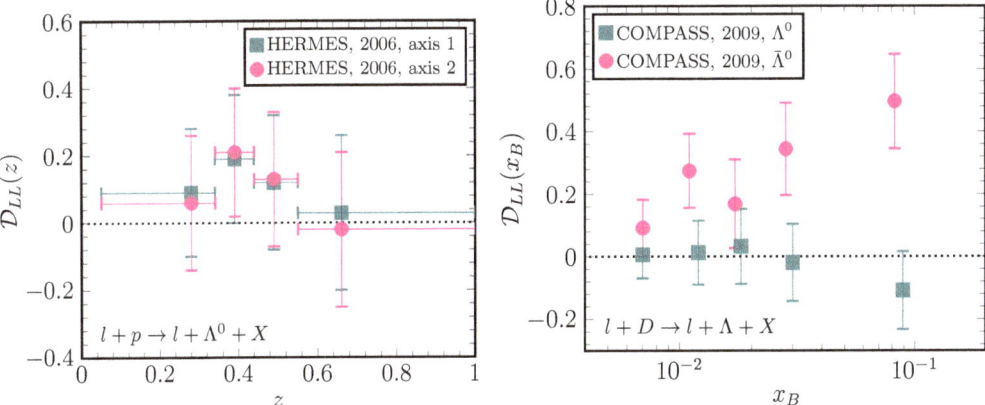

Figure 3. Spin transfer coefficient $\mathcal{D}_{LL}$ as a function of z and x_B in polarized SIDIS measured by the HERMES [136] and COMPASS [138] collaborations. Only the statistic errors are shown in both plots. Axes 1 and 2 refer to two different definitions of the longitudinal direction of Λ in the experiment. They are approximately the same at the high-energy limit. However, at the HERMES energy, they are not parallel to each other.

The experimental data from E665 [135] suggested a difference between $\mathcal{D}_{LL}$ for the Λ^0 production and that for the $\bar{\Lambda}^0$ production. This was later confirmed by the COMPASS [138] experiment. In [139–141], it was shown that such a difference serves as a flavor tag in the study of the G_{1L} FF. More studies on the flavor dependence of PDFs/FFs have been performed [122,142–149]. The NOMAD collaboration also carried out similar measurements in the neutrino SIDIS experiment [150,151]. Because of the flavor-changing feature of the charged weak interaction, this experiment opens more opportunities for quantitative research on the flavor dependence of spin-dependent FFs. A sophisticated investigation was presented in [152].

RHIC is the first and, so far, the only polarized proton–proton collider. The helicity of the incident protons can be transferred to that of the partons through the longitudinal spin transfer $g_{1L}(x)$ of PDFs. Therefore, it also has the capability of probing $G_{1L}(z)$ of the fragmentation. The first measurement was performed in 2009 [153], while an improved analysis was presented in 2018 [154]. These experiments measure the spin transfer coefficient $\mathcal{D}_{LL}$ which is defined as follows:

$$\mathcal{D}_{LL} \equiv \frac{\sigma^{p^+p \to \Lambda^+ + X} - \sigma^{p^+p \to \Lambda^- + X}}{\sigma^{p^+p \to \Lambda^+ + X} + \sigma^{p^+p \to \Lambda^- + X}}. \qquad (60)$$

The + symbol in the superscript denotes the helicity of the corresponding proton or Λ hyperon. The updated experimental data from the STAR collaboration [154] at RHIC are shown in Figure 4. This experimental data tend to favor the first and second scenarios in the DSV parameterization [118]. However, it cannot concretely rule out any scenario yet due to the large uncertainties. Moreover, the Xu–Liang–Sichtermann approach [131] based on the $SU(6)$ spin-flavor symmetry can also describe this data well. Moreover, RHIC also measured the transverse spin transfer coefficient $\mathcal{D}_{TT}$, which is sensitive to the convolution of the transversity PDF and the transversity FF [155].

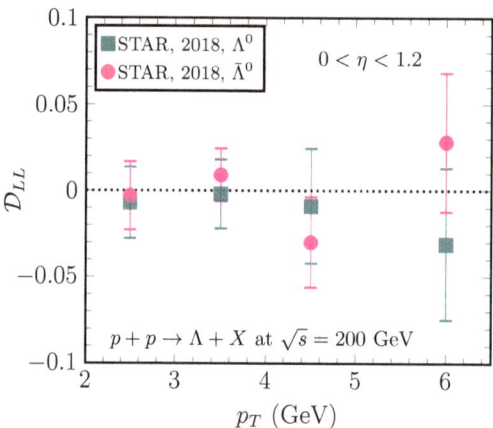

Figure 4. The spin transfer coefficient $\mathcal{D}_{LL}$ in polarized pp collisions at $\sqrt{s} = 200$ GeV measured by the STAR collaboration at RHIC [154]. Data points are taken from [154]. The systematic error and the statistical error have been combined.

The polarizations of partons participating the same hard scattering are strongly correlated. The helicity amplitudes of different partonic processes have been evaluated and summarized in [156]. Thus, [157] proposes the dihadron polarization correlation as a probe to the longitudinal spin transfer G_{1L} in e^+e^- annihilations at low energy where the fragmenting quarks are not polarized. Recently, this idea was further investigated and applied to the unpolarized pp collisions in [158]. By measuring the longitudinal polarization correlation of two almost back-to-back hadrons, we also gain access to the longitudinal spin transfer in unpolarized pp collisions. Since this observable avoids the contamination from the longitudinal spin transfer g_{1L} in PDFs. which is also poorly known, [158] innovated a means to investigating the longitudinal spin transfer G_{1L} in FFs at RHIC, Tevatron, and the LHC. Furthermore, this work can also be used to constrain the FF of circularly polarized gluons.

Recently, the Belle collaboration measured the transverse polarization of Λ hyperons in e^+e^- annihilations [19], sparking considerable theoretical interest [15,20–32]. In this experiment, one first defines the hadron production plane and then measures the transverse polarization along its normal direction. Since there are two transverse directions, we refer to the polarization along one as $\mathcal{P}_N$ and the other one as $\mathcal{P}_T$. The hadron production plane can be defined in two ways. The first one is defined by the thrust axis and the Λ momentum. In the second, the thrust axis is replaced by the momentum of a reference hadron (in the back-to-back side). Therefore, this experiment is dedicated to probing the $D_{1T}^\perp(z, p_\perp)$ FF. While the p_T-differential experimental data of $\mathcal{P}_N$ contain sizable uncertainties, the p_T-integrated version is quite precise, as shown in Figure 5. Employing the Trento convention [159] for the definition of $D_{1T}^{\perp}(z, p_\perp)$, the $p_\perp$ integrated transverse polarization is given by [24,25,28,160]

$$\mathcal{P}_N(z_\Lambda) = \left. \frac{\sum_q e_q^2 \int d^2 \boldsymbol{p}_\perp d^2 \boldsymbol{p}_{h\perp} \frac{-\hat{P}_{\perp\Lambda} \cdot \boldsymbol{p}_\perp}{z_\Lambda M_\Lambda} D_{1,q}^h(z_h, p_{h\perp}) D_{1T,q}^{\perp \Lambda}(z_\Lambda, p_\perp)}{\sum_q e_q^2 \int d^2 \boldsymbol{p}_\perp d^2 \boldsymbol{p}_{h\perp} D_{1,q}^h(z_h, p_{h\perp}) D_{1,q}^\Lambda(z_\Lambda, p_\perp)} \right|_{\boldsymbol{P}_{\perp\Lambda} = \frac{z_\Lambda}{z_h} \boldsymbol{p}_{h\perp} + \boldsymbol{p}_\perp}, \quad (61)$$

where $\hat{P}_{\perp\Lambda}$ is the unit vector along the direction of $P_{\perp\Lambda}$. The integral in the denominator simply reduces to the product of two colinear FFs. However, to evaluate the numerator, we need to first parameterize the $p_\perp$ and $p_{h\perp}$ dependence at the initial scale, which then evolves to the TMD factorization scale through use of the Collins–Soper–Sterman evolution equation. Nonetheless, since the collisional energy at Belle is not very high, a Gaussian ansatz is already a good approximation. More sophisticated approaches incorporating the $p_\perp$ dependence can be found in [29,31,32].

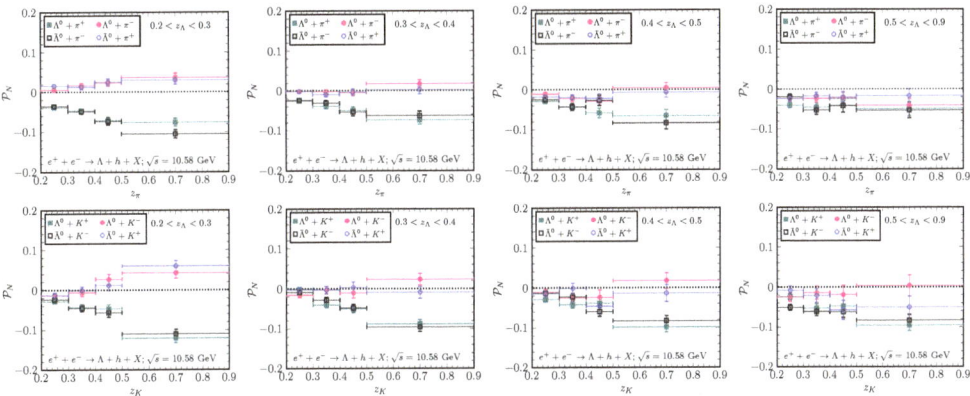

Figure 5. Transverse polarization of Λ in e^+e^- annihilation measured by the Belle collaboration [19]. Data points are taken from [19]. Statistical and systematic errors are combined in quadrature.

As shown in Figure 5, the distinct difference between $\mathcal{P}_N$ measured in the $\Lambda + \pi^+$ (or K^+) and $\Lambda + \pi^-$ (or K^-) processes offers an opportunity to explore the flavor dependence. Early attempts, such as the D'Alesio–Murgia–Zaccheddu (DMZ) [24] and Callos-Kang-Terry (CKT) [25] parameterizations, adopted the strategy that valence parton FFs differ from each other and that parton FFs are the same, i.e., $D_{1T,u}^{\perp \Lambda} \neq D_{1T,d}^{\perp \Lambda} \neq D_{1T,s}^{\perp \Lambda} \neq D_{1T,\text{sea}}^{\perp \Lambda}$. However, this approach violates the isospin symmetry, which is one of the most important features of strong interaction. Furthermore, a model calculation [27] based on the strict $SU(6)$ spin-flavor symmetry failed to describe the experimental data. However, it was first shown in [28] that the isospin symmetric Chen–Liang–Pan–Song–Wei (CLPSW) parameterization can still describe the experimental data well as long as the artificial constraint on sea parton FFs is released. This perspective was further investigated in Ref. [31] recently, which concluded that one can obtain good fit to the Belle data with and without implementing the isospin symmetry constraint after taking into account the charm contribution. This confirms that the current Belle dataset does not represent an isospin symmetry violation in the hadronization. Furthermore, [160] proposed to test the isospin symmetry at the future EIC experiment. By comparing the transverse polarizations in ep and eA scatterings at large x, we can ultimately check the difference between $D_{1T,u}^{\perp}$ and $D_{1T,d}^{\perp}$.

The future EIC is a polarized electron–proton/ion collider with unprecedentedly high luminosity. It will open a new window for the quantitative study of spin-dependent FFs. Several works [161–164] have proposed and made predictions for different observables at the future EIC with polarized proton beams. These observables are sensitive to various combinations of spin-dependent PDFs and FFs. Therefore, the future measurement will reveal information on both hadron structure and hadronization. A recent work [160] also proposed a method to study spin-dependent PDFs/FFs in unpolarized experiments. The key idea is that the polarizations of the final state quark and initial state parton are correlated. Thanks to the Boer–Mulders function in the PDFs, the initial state quark are transversely polarized although the polarization depends on the azimuthal angle. This transverse polarization can further propagate into final state observables through chiral-odd FFs. By measuring the azimuthal-angle-dependent longitudinal and transverse polarizations of final state Λ, we can probe $H_{1L}^\perp$ and $H_{1T}^\perp$ even in the unpolarized SIDIS process. Moreover, we can also measure the azimuthal-angle-dependent polarizations in e^+e^- annihilations to probe combinations of the Collins function and spin-dependent chiral-odd FFs [160]. This idea is akin to those explored in [30,165].

4.2. Vector Mesons

Most vector mesons decay through parity-conserving strong interactions. Their polarization vector does not enter the angular distribution of the daughter hadrons. Therefore, it is not possible to measure their polarization vector. In contrast, the tensor polarization does play a role in the angular distribution and therefore can be measured. Among them, spin alignment, which quantifies the deviation from $1/3$ of ρ_{00} in the spin-density matrix, has received the most attention.

Several collaborations [166–169] at LEP have measured the spin alignment of different vector mesons produced in the e^+e^- annihilation at the Z^0-pole. We show the spin alignment of K^{*0} and ρ^0 measured by the OPAL [166] and DELPHI [167] collaborations in Figure 6. The off-diagonal matrix elements were also measured in some of the experiments. Thereafter, the NOMAD collaboration measured the vector meson spin alignment for the first time in the neutrino DIS experiment [170]. These measurements offer more information on the hadronization mechanism and have led to several phenomenological studies [171–179].

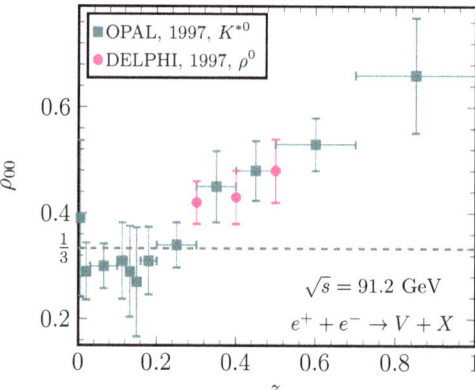

Figure 6. Spin alignment of K^{*0} and ρ^0 measured by the OPAL [166] and DELPHI [167] collaborations at LEP. Data points are taken from [166,167].

Figure 6 shows that ρ_{00} is consistent with $1/3$ (i.e., no spin alignment) at the small-z region. However, at large z, a clear spin alignment is observed. This pattern is similar to that for the longitudinal polarization of Λ also measured at LEP [115,116] (shown in Figure 2).

As mentioned above, the quarks produced at LEP and also those at NOMAD are strongly polarized. Therefore, it is tempting to attribute the tensor polarization of final state vector mesons to the longitudinal polarization of the fragmenting quarks. However, a simple tensor structure analysis [47,63,64,175] shows that this is not the case. The spin alignment of the final state mesons is not coupled with the quark polarization. Instead, it is coupled with the quark-polarization-summed cross section. The vector meson spin alignment in e^+e^- collisions is given by

$$\rho_{00} = \frac{1}{3} - \frac{1}{3}\frac{\sum_q \omega_q(y) D_{1LL,q}(z)}{\sum_q \omega_q(y) D_{1,q}(z)}, \quad (62)$$

where, ω_q is defined to be the same as that for the Λ production in the previous section, and $D_{1LL}(z)$ is the corresponding FF that is responsible for the vector meson spin alignment. As shown in the above equation, the longitudinal polarization of the fragmenting quark does not play a role here. It was thus first proposed in [63] that the vector meson spin alignment can also be observed in other high-energy collisions with unpolarized quarks fragmenting. Fitting to the experimental data from LEP, other work [108,109] extracted $D_{1LL}(z)$ and made predictions for the spin alignment of high p_T vector mesons in unpolarized pp collisions at RHIC and the LHC [109]. Furthermore, from the same mechanism, there will be a significant spin alignment for vector mesons produced in the unpolarized SIDIS. Measuring vector meson spin alignment at the future EIC will cast new light on the quantitative study of the $D_{1LL}(z)$ FF.

Notice that the spin alignment of low-p_T vector mesons in AA collisions has also been measured at RHIC [110,180] and LHC [181] recently. These low-p_T hadrons in relativistic heavy-ion collisions are produced through a different hadronization mechanism than those of fragmentation. Their tensor polarization originates from a different source.

5. Model Calculation

The PDFs and FFs are defined in terms of quark–gluon correlators as laid out in Section 3. Owing to the nonperturbative nature of the hadron state, we cannot directly evaluate them theoretically. Thus far, several proposals for computing quantities that can be related to the PDFs in the lattice QCD approach have been put forward [182–186]. However, it is not possible to study FFs in the lattice QCD yet. In the current stage, the quantitative information is mainly extracted from the experimental data.

However, due to the limited amount of experimental data, the TMD PDFs and FFs are not yet well constrained. As a complementary tool, model calculations have usually been employed to compute different PDFs over the past decades [106,187–219]. These investigations offer quantitative insight into the hadron structure and therefore are indispensable for phenomenological study. The same also goes for the FFs. Most of the models can be used to evaluate both PDFs and FFs. We make a nonexclusive brief summary on FF calculations.

There are quite a few models that can be categorized as a spectator model [220,221]. Among them, the quark–diquark model is a simple one which provides the quark–baryon–diantiquark vertex, so that the baryon FFs can be easily evaluated. In [222], the colinear baryon FFs were calculated at the leading twist using the quark–diquark model, while in [189,223–226], the TMD FFs were further computed at the leading and subleading twists. To compute the meson FFs or the gluon FFs, we need an improved version which offers the vertex among the fragmenting parton, hadron, and spectator. In [227], the Collins function was calculated for pions and kaons in this method. Recently, in [228], approach to calculate leading twist gluon TMD FFs was presented. The chiral invariant model [229] investigated the chiral symmetry and the spontaneous breaking with an effective Lagrangian of quarks, gluons, and goldstone bosons. It can also be classified into the spectator model category. Utilizing this model, several authors [230–235] calculated the pion and kaon FFs. In [236], an extended version was also developed to compute the vector meson FFs. Furthermore, several works [237–240] have evaluated the FFs of different hadrons using a parameterized quark–hadron coupling.

The Nambu–Jona–Lasinio (NJL) model originates from [241,242] who developed an effective theory describing the quark–hadron interaction. It has been employed to evaluate PDFs of different hadrons [191,243–248]. Incorporating with the Feynman–Field model (also known as the quark–jet model) established in [249,250], the NJL=-jet model has been employed to calculate both colinear and TMD FFs of different hadrons [235,251–257]. Recent works have also computed FFs of gluon [258] and charm quark [259] with this approach. The Feynman–Field model relates the total FF to the first rank FF. However, it does not specify how to compute the first rank FF. Therefore, in principle, it can be hybridized with another model which provides with the first rank FF to prolong the applicability of the corresponding model.

We make a final remark on the model calculation to conclude this section. All the above-mentioned models compute FFs employing the effective Lagrangian of partons and hadrons of interest. While these calculations offer quantitative insight into the hadronization scheme, we should draw a line between conclusions that are model-dependent and those that are model-independent.

6. Summary

There are a multitude of topics within the subject of FFs. In this review, we constrain ourselves in a very limited scope that we are familiar with. First, we briefly summarized the derivation of the TMD factorization and the establishment of the QCD evolution equation at the leading twist level. The TMD factorization and the corresponding evolution at the higher twist level are still ongoing topics. Second, we are particularly interested in the spin-related effects. With the spin degree of freedom being taken into account, the interplay between the transverse momentum and the hadron/quark polarization presents a highly intriguing phenomena that can be investigated in experiments. As a result, we need to define more TMD FFs to fully describe the fragmentation process. In quantum field theory, TMD FFs are introduced in the decomposition of parton correlators. We summarized the final results up to the twist-4 level for spin-0, spin-1/2, and spin-1 hadron productions. Finally, although all the TMD FFs have clear definitions in terms of parton fields and hadron states, they are nonperturbative quantities that cannot be directly evaluated from quantum field theory. In contrast to TMD PDFs, FFs cannot be computed even in the lattice QCD approach. The quantitative investigation thus mainly concentrates on the extraction from experimental measurements and model calculations. We summarized several spin-related experiments conducted over the past decades and the corresponding phenomenological studies. In the last section, we also briefly presented several model calculations.

The study of TMD FFs is still a very active field, and many mysteries remain to be explored. The Electron-Ion Collider (EIC) and the Electron-Ion Collider in China (EicC) have been proposed to be built as the new high-energy colliders in the next generation. They will provide new experimental data for the quantitative study of TMD FFs and can significantly boost our understanding of the hadronization mechanism.

Author Contributions: All authors contribute to the writting and proofreading of this review. All authors have read and agreed to the published version of the manuscript.

Funding: K.-B.C. is supported by the National Natural Science Foundation of China under grants no. 12005122 and no. 11947055, as well as by the Shandong Province Natural Science Foundation under grant no. ZR2020QA082. T.L. is supported in part by the National Natural Science Foundation of China under grants no. 12175117 and no. 20221017-1. Y.-K.S. is supported in part by the National Natural Science Foundation of China under grant no. 11505080, and by the Shandong Province Natural Science Foundation under grant no. ZR2018JL006. S.-Y.W. is supported by the Taishan Fellowship of Shandong Province for junior scientists.

Data Availability Statement: Data available on request from the authors.

Conflicts of Interest: The authors declare no conflict of interest.

References

1. Fritzsch, H.; Gell-Mann, M.; Leutwyler, H. Advantages of the Color Octet Gluon Picture. *Phys. Lett. B* **1973**, *47*, 365–368. [CrossRef]
2. Yang, C.N.; Mills, R.L. Conservation of Isotopic Spin and Isotopic Gauge Invariance. *Phys. Rev.* **1954**, *96*, 191–195. [CrossRef]
3. Metz, A.; Vossen, A. Parton Fragmentation Functions. *Prog. Part. Nucl. Phys.* **2016**, *91*, 136–202. [CrossRef]

4. Bjorken, J.D.; Paschos, E.A. Inelastic Electron Proton and gamma Proton Scattering, and the Structure of the Nucleon. *Phys. Rev.* **1969**, *185*, 1975–1982. [CrossRef]
5. Feynman, R.P. Photon-Hadron Interactions. Basic Books, 1973.
6. Berman, S.M.; Bjorken, J.D.; Kogut, J.B. Inclusive Processes at High Transverse Momentum. *Phys. Rev. D* **1971**, *4*, 3388. [CrossRef]
7. Mueller, A.H. Cut Vertices and their Renormalization: A Generalization of the Wilson Expansion. *Phys. Rev. D* **1978**, *18*, 3705. [CrossRef]
8. Collins, J.C.; Soper, D.E. Parton Distribution and Decay Functions. *Nucl. Phys. B* **1982**, *194*, 445–492. [CrossRef]
9. Collins, J.C.; Soper, D.E.; Sterman, G.F. Factorization of Hard Processes in QCD. *Adv. Ser. Direct. High Energy Phys.* **1989**, *5*, 1–91. [CrossRef]
10. Collins, J. *Foundations of Perturbative QCD*; Cambridge University Press: Cambridge, UK, 2013; Volume 32.
11. Collins, J.C.; Soper, D.E. Back-To-Back Jets in QCD. *Nucl. Phys. B* **1981**, *193*, 381; Erratum in *Nucl. Phys. B* **1983**, *213*, 545. [CrossRef]
12. Aybat, S.M.; Rogers, T.C. TMD Parton Distribution and Fragmentation Functions with QCD Evolution. *Phys. Rev. D* **2011**, *83*, 114042. [CrossRef]
13. Kang, Z.B.; Shao, D.Y.; Zhao, F. QCD resummation on single hadron transverse momentum distribution with the thrust axis. *JHEP* **2020**, *12*, 127. [CrossRef]
14. Boglione, M.; Simonelli, A. Factorization of $e^+e^- \to HX$ cross section, differential in z_h, P_T and thrust, in the 2-jet limit. *JHEP* **2021**, *02*, 076. [CrossRef]
15. Boglione, M.; Simonelli, A. Universality-breaking effects in e^+e^- hadronic production processes. *Eur. Phys. J. C* **2021**, *81*, 96. [CrossRef]
16. Makris, Y.; Ringer, F.; Waalewijn, W.J. Joint thrust and TMD resummation in electron-positron and electron-proton collisions. *JHEP* **2021**, *02*, 070. [CrossRef]
17. Collins, J.C. Fragmentation of transversely polarized quarks probed in transverse momentum distributions. *Nucl. Phys. B* **1993**, *396*, 161–182. [CrossRef]
18. Boer, D.; Mulders, P.J.; Pijlman, F. Universality of T odd effects in single spin and azimuthal asymmetries. *Nucl. Phys. B* **2003**, *667*, 201–241. [CrossRef]
19. Guan, Y.; Vossen, A.; Adachi, I.; Adamczyk, K.; Ahn, J.K.; Aihara, H.; Sanuki, T.; Chilikin, K.; Cho, K.; Choi, S.-K.; et al. Observation of Transverse $\Lambda/\bar{\Lambda}$ Hyperon Polarization in e^+e^- Annihilation at Belle. *Phys. Rev. Lett.* **2019**, *122*, 042001. [CrossRef]
20. Matevosyan, H.H.; Kotzinian, A.; Thomas, A.W. Semi-inclusive back-to-back production of a hadron pair and a single hadron in e^+e^- annihilation. *JHEP* **2018**, *10*, 008. [CrossRef]
21. Gamberg, L.; Kang, Z.B.; Pitonyak, D.; Schlegel, M.; Yoshida, S. Polarized hyperon production in single-inclusive electron-positron annihilation at next-to-leading order. *JHEP* **2019**, *01*, 111. [CrossRef]
22. Anselmino, M.; Kishore, R.; Mukherjee, A. Polarizing fragmentation function and the Λ polarization in e^+e^- processes. *Phys. Rev. D* **2019**, *100*, 014029. [CrossRef]
23. Anselmino, M.; Mukherjee, A.; Vossen, A. Transverse spin effects in hard semi-inclusive collisions. *Prog. Part. Nucl. Phys.* **2020**, *114*, 103806. [CrossRef]
24. D'Alesio, U.; Murgia, F.; Zaccheddu, M. First extraction of the Λ polarizing fragmentation function from Belle e^+e^- data. *Phys. Rev. D* **2020**, *102*, 054001. [CrossRef]
25. Callos, D.; Kang, Z.B.; Terry, J. Extracting the transverse momentum dependent polarizing fragmentation functions. *Phys. Rev. D* **2020**, *102*, 096007. [CrossRef]
26. Kang, Z.B.; Lee, K.; Zhao, F. Polarized jet fragmentation functions. *Phys. Lett. B* **2020**, *809*, 135756. [CrossRef]
27. Li, H.; Wang, X.; Yang, Y.; Lu, Z. The transverse polarization of Λ hyperons in $e^+e^- \to \Lambda^\uparrow hX$ processes within TMD factorization. *Eur. Phys. J. C* **2021**, *81*, 289. [CrossRef]
28. Chen, K.B.; Liang, Z.T.; Pan, Y.L.; Song, Y.K.; Wei, S.Y. Isospin Symmetry of Fragmentation Functions. *Phys. Lett. B* **2021**, *816*, 136217. [CrossRef]
29. Gamberg, L.; Kang, Z.B.; Shao, D.Y.; Terry, J.; Zhao, F. Transverse Λ polarization in e^+e^- collisions. *Phys. Lett. B* **2021**, *818*, 136371. [CrossRef]
30. D'Alesio, U.; Murgia, F.; Zaccheddu, M. General helicity formalism for two-hadron production in e^+e^- annihilation within a TMD approach. *JHEP* **2021**, *10*, 078. [CrossRef]
31. D'Alesio, U.; Gamberg, L.; Murgia, F.; Zaccheddu, M. Transverse Λ polarization in e^+e^- processes within a TMD factorization approach and the polarizing fragmentation function. *JHEP* **2022**, *12*, 074. [CrossRef]
32. Boglione, M.; Gonzalez-Hernandez, J.O.; Simonelli, A. Transverse momentum dependent fragmentation functions from recent BELLE data. *Phys. Rev. D* **2022**, *106*, 074024. [CrossRef]
33. Ellis, R.K.; Furmanski, W.; Petronzio, R. Power Corrections to the Parton Model in QCD. *Nucl. Phys. B* **1982**, *207*, 1–14. [CrossRef]
34. Ellis, R.K.; Furmanski, W.; Petronzio, R. Unraveling Higher Twists. *Nucl. Phys. B* **1983**, *212*, 29. [CrossRef]
35. Qiu, J.W. Twist Four Contributions to the Parton Structure Functions. *Phys. Rev. D* **1990**, *42*, 30–44. [CrossRef]
36. Qiu, J.W.; Sterman, G.F. Power corrections in hadronic scattering. 1. Leading $1/Q^2$ corrections to the Drell-Yan cross-section. *Nucl. Phys. B* **1991**, *353*, 105–136. [CrossRef]
37. Qiu, J.W.; Sterman, G.F. Power corrections to hadronic scattering. 2. Factorization. *Nucl. Phys. B* **1991**, *353*, 137–164. [CrossRef]

38. Balitsky, I.I.; Braun, V.M. The Nonlocal operator expansion for inclusive particle production in e^+e^- annihilation. *Nucl. Phys. B* **1991**, *361*, 93–140. [CrossRef]
39. Levelt, J.; Mulders, P.J. Time reversal odd fragmentation functions in semiinclusive scattering of polarized leptons from unpolarized hadrons. *Phys. Lett. B* **1994**, *338*, 357–362. [CrossRef]
40. Levelt, J.; Mulders, P.J. Quark correlation functions in deep inelastic semiinclusive processes. *Phys. Rev. D* **1994**, *49*, 96–113. [CrossRef]
41. Kotzinian, A. New quark distributions and semiinclusive electroproduction on the polarized nucleons. *Nucl. Phys. B* **1995**, *441*, 234–248. [CrossRef]
42. Mulders, P.J.; Tangerman, R.D. The Complete tree level result up to order 1/Q for polarized deep inelastic leptoproduction. *Nucl. Phys. B* **1996**, *461*, 197–237; Erratum in *Nucl. Phys. B* **1997**, *484*, 538–540. [CrossRef]
43. Boer, D.; Jakob, R.; Mulders, P.J. Asymmetries in polarized hadron production in e^+e^- annihilation up to order 1/Q. *Nucl. Phys. B* **1997**, *504*, 345–380. [CrossRef]
44. Kotzinian, A.M.; Mulders, P.J. Probing transverse quark polarization via azimuthal asymmetries in leptoproduction. *Phys. Lett. B* **1997**, *406*, 373–380. [CrossRef]
45. Boer, D.; Mulders, P.J. Time reversal odd distribution functions in leptoproduction. *Phys. Rev. D* **1998**, *57*, 5780–5786. [CrossRef]
46. Boer, D.; Jakob, R.; Mulders, P.J. Leading asymmetries in two hadron production in e^+e^- annihilation at the Z pole. *Phys. Lett. B* **1998**, *424*, 143–151. [CrossRef]
47. Bacchetta, A.; Mulders, P.J. Deep inelastic leptoproduction of spin-one hadrons. *Phys. Rev. D* **2000**, *62*, 114004. [CrossRef]
48. Boer, D.; Jakob, R.; Mulders, P.J. Angular dependences in electroweak semiinclusive leptoproduction. *Nucl. Phys. B* **2000**, *564*, 471–485. [CrossRef]
49. Bacchetta, A.; Mulders, P.J.; Pijlman, F. New observables in longitudinal single-spin asymmetries in semi-inclusive DIS. *Phys. Lett. B* **2004**, *595*, 309–317. [CrossRef]
50. Bacchetta, A.; Diehl, M.; Goeke, K.; Metz, A.; Mulders, P.J.; Schlegel, M. Semi-inclusive deep inelastic scattering at small transverse momentum. *JHEP* **2007**, *02*, 093. [CrossRef]
51. Boer, D. Angular dependences in inclusive two-hadron production at BELLE. *Nucl. Phys. B* **2009**, *806*, 23–67. [CrossRef]
52. Eguchi, H.; Koike, Y.; Tanaka, K. Single Transverse Spin Asymmetry for Large-p(T) Pion Production in Semi-Inclusive Deep Inelastic Scattering. *Nucl. Phys. B* **2006**, *752*, 1–17. [CrossRef]
53. Eguchi, H.; Koike, Y.; Tanaka, K. Twist-3 Formalism for Single Transverse Spin Asymmetry Reexamined: Semi-Inclusive Deep Inelastic Scattering. *Nucl. Phys. B* **2007**, *763*, 198–227. [CrossRef]
54. Koike, Y.; Tanaka, K. Master Formula for Twist-3 Soft-Gluon-Pole Mechanism to Single Transverse-Spin Asymmetry. *Phys. Lett. B* **2007**, *646*, 232–241; Erratum in *Phys. Lett. B* **2008**, *668*, 458–459; . [CrossRef]
55. Kanazawa, K.; Koike, Y. Contribution of twist-3 fragmentation function to single transverse-spin asymmetry in semi-inclusive deep inelastic scattering. *Phys. Rev. D* **2013**, *88*, 074022. [CrossRef]
56. Pitonyak, D.; Schlegel, M.; Metz, A. Polarized hadron pair production from electron-positron annihilation. *Phys. Rev. D* **2014**, *89*, 054032. [CrossRef]
57. Yang, Y.; Lu, Z. Polarized Λ hyperon production in semi-inclusive deep inelastic scattering off an unpolarized nucleon target. *Phys. Rev. D* **2017**, *95*, 074026. [CrossRef]
58. Liang, Z.T.; Wang, X.N. Azimuthal and single spin asymmetry in deep-inelastic lepton-nucleon scattering. *Phys. Rev. D* **2007**, *75*, 094002. [CrossRef]
59. Liang, Z.T.; Wang, X.N.; Zhou, J. The Transverse-momentum-dependent Parton Distribution Function and Jet Transport in Medium. *Phys. Rev. D* **2008**, *77*, 125010. [CrossRef]
60. Gao, J.H.; Liang, Z.T.; Wang, X.N. Nuclear dependence of azimuthal asymmetry in semi-inclusive deep inelastic scattering. *Phys. Rev. C* **2010**, *81*, 065211. [CrossRef]
61. Song, Y.k.; Gao, J.h.; Liang, Z.t.; Wang, X.N. Twist-4 contributions to the azimuthal asymmetry in SIDIS. *Phys. Rev. D* **2011**, *83*, 054010. [CrossRef]
62. Song, Y.k.; Gao, J.h.; Liang, Z.t.; Wang, X.N. Azimuthal asymmetries in semi-inclusive DIS with polarized beam and/or target and their nuclear dependences. *Phys. Rev. D* **2014**, *89*, 014005. [CrossRef]
63. Wei, S.y.; Song, Y.k.; Liang, Z.t. Higher twist contribution to fragmentation function in inclusive hadron production in e^+e^- annihilation. *Phys. Rev. D* **2014**, *89*, 014024. [CrossRef]
64. Wei, S.Y.; Chen, K.b.; Song, Y.k.; Liang, Z.t. Leading and higher twist contributions in semi-inclusive e^+e^- annihilation at high energies. *Phys. Rev. D* **2015**, *91*, 034015. [CrossRef]
65. Chen, K.b.; Yang, W.h.; Wei, S.y.; Liang, Z.t. Tensor polarization dependent fragmentation functions and $e^+e^- \to V\pi X$ at high energies. *Phys. Rev. D* **2016**, *94*, 034003. [CrossRef]
66. Wei, S.y.; Song, Y.k.; Chen, K.b.; Liang, Z.t. Twist-4 contributions to semi-inclusive deeply inelastic scatterings with polarized beam and target. *Phys. Rev. D* **2017**, *95*, 074017. [CrossRef]
67. Vladimirov, A.; Moos, V.; Scimemi, I. Transverse momentum dependent operator expansion at next-to-leading power. *JHEP* **2022**, *1*, 110. [CrossRef]
68. Rodini, S.; Vladimirov, A. Definition and evolution of transverse momentum dependent distribution of twist-three. *JHEP* **2022**, *8*, 031; Erratum in *JHEP* **2022**, *12*, 048. [CrossRef]

69. Rodini, S.; Vladimirov, A. Factorization for Quasi-TMD Distributions of Sub-Leading Power. *arXiv* **2022**. arxiv:2211.04494.
70. Ebert, M.A.; Gao, A.; Stewart, I.W. Factorization for azimuthal asymmetries in SIDIS at next-to-leading power. *JHEP* **2022**, *6*, 7. [CrossRef]
71. Balitsky, I.; Tarasov, A. Higher-twist corrections to gluon TMD factorization. *JHEP* **2017**, *7*, 95. [CrossRef]
72. Balitsky, I.; Tarasov, A. Power corrections to TMD factorization for Z-boson production. *JHEP* **2018**, *5*, 150. [CrossRef]
73. Balitsky, I. Gauge-invariant TMD factorization for Drell-Yan hadronic tensor at small x. *JHEP* **2021**, *5*, 46. [CrossRef]
74. Balitsky, I. Drell-Yan angular lepton distributions at small x from TMD factorization. *JHEP* **2021**, *9*, 22. [CrossRef]
75. Gamberg, L.; Kang, Z.B.; Shao, D.Y.; Terry, J.; Zhao, F. Transverse-momentum-dependent factorization at next-to-leading power. *arXiv* **2022**, arxiv:2211.13209.
76. Collins, J.C.; Soper, D.E. Back-To-Back Jets: Fourier Transform from B to K-Transverse. *Nucl. Phys. B* **1982**, *197*, 446–476. [CrossRef]
77. Collins, J.C.; Soper, D.E.; Sterman, G.F. Transverse Momentum Distribution in Drell-Yan Pair and W and Z Boson Production. *Nucl. Phys. B* **1985**, *250*, 199–224. [CrossRef]
78. Ji, X.D.; Ma, J.P.; Yuan, F. QCD factorization for semi-inclusive deep-inelastic scattering at low transverse momentum. *Phys. Rev. D* **2005**, *71*, 034005. [CrossRef]
79. Ji, X.D.; Ma, J.P.; Yuan, F. QCD factorization for spin-dependent cross sections in DIS and Drell-Yan processes at low transverse momentum. *Phys. Lett. B* **2004**, *597*, 299–308. [CrossRef]
80. Sun, P.; Yuan, C.P.; Yuan, F. Soft Gluon Resummations in Dijet Azimuthal Angular Correlations in Hadronic Collisions. *Phys. Rev. Lett.* **2014**, *113*, 232001. [CrossRef]
81. Sun, P.; Yuan, C.P.; Yuan, F. Transverse Momentum Resummation for Dijet Correlation in Hadronic Collisions. *Phys. Rev. D* **2015**, *92*, 094007. [CrossRef]
82. Becher, T.; Neubert, M. Drell-Yan Production at Small q_T, Transverse Parton Distributions and the Collinear Anomaly. *Eur. Phys. J. C* **2011**, *71*, 1665. [CrossRef]
83. Echevarria, M.G.; Idilbi, A.; Scimemi, I. Factorization Theorem For Drell-Yan At Low q_T And Transverse Momentum Distributions On-The-Light-Cone. *JHEP* **2012**, *7*, 002. [CrossRef]
84. Chiu, J.Y.; Jain, A.; Neill, D.; Rothstein, I.Z. A Formalism for the Systematic Treatment of Rapidity Logarithms in Quantum Field Theory. *JHEP* **2012**, *05*, 084. [CrossRef]
85. Li, Y.; Neill, D.; Zhu, H.X. An exponential regulator for rapidity divergences. *Nucl. Phys. B* **2020**, *960*, 115193. [CrossRef]
86. Kang, Z.B.; Liu, X.; Ringer, F.; Xing, H. The transverse momentum distribution of hadrons within jets. *JHEP* **2017**, *11*, 068. [CrossRef]
87. Sterman, G.F. Mass Divergences in Annihilation Processes. 1. Origin and Nature of Divergences in Cut Vacuum Polarization Diagrams. *Phys. Rev. D* **1978**, *17*, 2773. [CrossRef]
88. Libby, S.B.; Sterman, G.F. Mass Divergences in Two Particle Inelastic Scattering. *Phys. Rev. D* **1978**, *18*, 4737. [CrossRef]
89. Davies, C.T.H.; Webber, B.R.; Stirling, W.J. Drell-Yan Cross-Sections at Small Transverse Momentum *Nucl. Phys. B* **1985**, *256*, 413. [CrossRef]
90. Landry, F.; Brock, R.; Nadolsky, P.M.; Yuan, C.P. Tevatron Run-1 Z boson data and Collins-Soper-Sterman resummation formalism. *Phys. Rev. D* **2003**, *67*, 073016. [CrossRef]
91. Guzzi, M.; Nadolsky, P.M.; Wang, B. Nonperturbative contributions to a resummed leptonic angular distribution in inclusive neutral vector boson production. *Phys. Rev. D* **2014**, *90*, 014030. [CrossRef]
92. Sun, P.; Isaacson, J.; Yuan, C.P.; Yuan, F. Nonperturbative functions for SIDIS and Drell–Yan processes. *Int. J. Mod. Phys. A* **2018**, *33*, 1841006. [CrossRef]
93. Ladinsky, G.A.; Yuan, C.P. The Nonperturbative regime in QCD resummation for gauge boson production at hadron colliders. *Phys. Rev. D* **1994**, *50*, R4239. [CrossRef] [PubMed]
94. Sun, P.; Yuan, F. Transverse momentum dependent evolution: Matching semi-inclusive deep inelastic scattering processes to Drell-Yan and W/Z boson production. *Phys. Rev. D* **2013**, *88*, 114012. [CrossRef]
95. Aidala, C.A.; Field, B.; Gamberg, L.P.; Rogers, T.C. Limits on transverse momentum dependent evolution from semi-inclusive deep inelastic scattering at moderate Q. *Phys. Rev. D* **2014**, *89*, 094002. [CrossRef]
96. Echevarria, M.G.; Idilbi, A.; Kang, Z.B.; Vitev, I. QCD Evolution of the Sivers Asymmetry. *Phys. Rev. D* **2014**, *89*, 074013. [CrossRef]
97. Kang, Z.B.; Xiao, B.W.; Yuan, F. QCD Resummation for Single Spin Asymmetries. *Phys. Rev. Lett.* **2011**, *107*, 152002. [CrossRef] [PubMed]
98. Collins, J.; Rogers, T. Understanding the large-distance behavior of transverse-momentum-dependent parton densities and the Collins-Soper evolution kernel. *Phys. Rev. D* **2015**, *91*, 074020. [CrossRef]
99. Qiu, J.W.; Zhang, X.F. Role of the nonperturbative input in QCD resummed Drell-Yan Q_T distributions. *Phys. Rev. D* **2001**, *63*, 114011. [CrossRef]
100. Kang, Z.B.; Prokudin, A.; Sun, P.; Yuan, F. Extraction of Quark Transversity Distribution and Collins Fragmentation Functions with QCD Evolution. *Phys. Rev. D* **2016**, *93*, 014009. [CrossRef]
101. Sivers, D.W. Single Spin Production Asymmetries from the Hard Scattering of Point-Like Constituents. *Phys. Rev. D* **1990**, *41*, 83. [CrossRef]

102. Sivers, D.W. Hard scattering scaling laws for single spin production asymmetries. *Phys. Rev. D* **1991**, *43*, 261–263. [CrossRef]
103. Song, H. Spin-3/2 Polarization in Production of N^* by Neutrinos. *Phys. Rev.* **1967**, *162*, 1615. [CrossRef]
104. Zhao, J.; Zhang, Z.; Liang, Z.T.; Liu, T.; Zhou, Y.J. Inclusive and semi-inclusive production of spin-3/2 hadrons in e^+e^- annihilation. *Phys. Rev. D* **2022**, *106*, 094006. [CrossRef]
105. Collins, J.C. Leading twist single transverse-spin asymmetries: Drell-Yan and deep inelastic scattering. *Phys. Lett. B* **2002**, *536*, 43–48. [CrossRef]
106. Ji, X.d.; Yuan, F. Parton distributions in light cone gauge: Where are the final state interactions? *Phys. Lett. B* **2002**, *543*, 66–72. [CrossRef]
107. Belitsky, A.V.; Ji, X.; Yuan, F. Final state interactions and gauge invariant parton distributions. *Nucl. Phys. B* **2003**, *656*, 165–198. [CrossRef]
108. Chen, K.b.; Yang, W.h.; Zhou, Y.j.; Liang, Z.t. Energy dependence of hadron polarization in $e^+e^- \to hX$ at high energies. *Phys. Rev. D* **2017**, *95*, 034009. [CrossRef]
109. Chen, K.b.; Liang, Z.t.; Song, Y.k.; Wei, S.y. Spin alignment of vector mesons in high energy pp collisions. *Phys. Rev. D* **2020**, *102*, 034001. [CrossRef]
110. Abdallah, M.S.; Aboona, B.E.; Adam, J.; Adamczyk, L.; Adams, J.R.; Adkins, J.K.; Agakishiev, G.; Aggarwal, I.; Aggarwal, M.M.; Ahammed, Z.; et al. Pattern of global spin alignment of ϕ and K^{*0} mesons in heavy-ion collisions. *Nature* **2023**, *614*, 244–248. [CrossRef]
111. Mulders, P.J.; Rodrigues, J. Transverse momentum dependence in gluon distribution and fragmentation functions. *Phys. Rev. D* **2001**, *63*, 094021. [CrossRef]
112. Zyla, P.A.; Barnett, R.M.; Beringer, J.; Dahl, O.; Dwyer, D.A.; Groom, D.E.; Lin, C.-J.; Lugovsky, K.S.; Pianori, E.; Robinson, D.J.; et al. Review of Particle Physics. *PTEP* **2020**, *2020*, 083C01. [CrossRef]
113. Augustin, J.E.; Renard, F.M. How to Measure Quark Helicities in $e^+e^- \to$ Hadrons. *Nucl. Phys. B* **1980**, *162*, 341. [CrossRef]
114. Gustafson, G.; Hakkinen, J. Lambda polarization in e^+e^- annihilation at the Z0 pole. *Phys. Lett. B* **1993**, *303*, 350–354. [CrossRef]
115. Buskulic, D.; De Bonis, I.; Decamp, D.; Ghez, P.; Goy, C.; Lees, J.P.; Lucotte, A.; Minard, M.N.; Odier, P.; Pietrzyk, B.; et al. Measurement of Lambda polarization from Z decays. *Phys. Lett. B* **1996**, *374*, 319–330. [CrossRef]
116. Ackersta, K.; Alexander, G.; Allison, J.; Altekamp, N.; Anderson, K.J.; Anderson, S.; Arcelli, S.; Asai, S.; Axen, D.; Azuelos, G.; et al. Polarization and forward—Backward asymmetry of Lambda baryons in hadronic Z0 decays. *Eur. Phys. J. C* **1998**, *2*, 49–59. [CrossRef]
117. Kotzinian, A.; Bravar, A.; von Harrach, D. Lambda and anti-Lambda polarization in lepton induced processes. *Eur. Phys. J. C* **1998**, *2*, 329–337. [CrossRef]
118. de Florian, D.; Stratmann, M.; Vogelsang, W. QCD analysis of unpolarized and polarized Lambda baryon production in leading and next-to-leading order. *Phys. Rev. D* **1998**, *57*, 5811–5824. [CrossRef]
119. Liang, Z.T.; Boros, C. Hyperon polarization and single spin left-right asymmetry in inclusive production processes at high-energies. *Phys. Rev. Lett.* **1997**, *79*, 3608–3611. [CrossRef]
120. Boros, C.; Liang, Z.t. Spin content of Lambda and its longitudinal polarization in e^+e^- annihilation at high-energies. *Phys. Rev. D* **1998**, *57*, 4491–4494. [CrossRef]
121. Ma, B.Q.; Soffer, J. Quark flavor separation in Lambda Baryon fragmentation. *Phys. Rev. Lett.* **1999**, *82*, 2250–2253. [CrossRef]
122. Ma, B.Q.; Schmidt, I.; Yang, J.J. Flavor and spin structure of lambda baryon at large x. *Phys. Lett. B* **2000**, *477*, 107–113. [CrossRef]
123. Ma, B.Q.; Schmidt, I.; Yang, J.J. Quark structure of Lambda from Lambda polarization in Z decays. *Phys. Rev. D* **2000**, *61*, 034017. [CrossRef]
124. Liu, C.x.; Liang, Z.t. Spin structure and longitudinal polarization of hyperon in e^+e^- annihilation at high-energies. *Phys. Rev. D* **2000**, *62*, 094001. [CrossRef]
125. Jaffe, R.L. Polarized Λ's in the current fragmentation region. *Phys. Rev. D* **1996**, *54*, R6581–R6585. [CrossRef]
126. de Florian, D.; Stratmann, M.; Vogelsang, W. Polarized Lambda production at HERA. In Proceedings of the Workshop on Physics with Polarized Protons at Hera: 1st Meeting; 1997. Available online: https://inspirehep.net/literature/449966 (accessed on 5 March 2023).
127. Ashery, D.; Lipkin, H.J. Expected polarization of Lambda particles produced in deep inelastic polarized lepton scattering. *Phys. Lett. B* **1999**, *469*, 263–269. [CrossRef]
128. de Florian, D.; Stratmann, M.; Vogelsang, W. Polarized Lambda baryon production in p p collisions. *Phys. Rev. Lett.* **1998**, *81*, 530–533. [CrossRef]
129. Jager, B.; Schafer, A.; Stratmann, M.; Vogelsang, W. Next-to-leading order QCD corrections to high p(T) pion production in longitudinally polarized pp collisions. *Phys. Rev. D* **2003**, *67*, 054005. [CrossRef]
130. Xu, Q.h.; Liu, C.x.; Liang, Z.t. Longitudinal polarization of hyperons in high p(T) jets in singly polarized pp collisions at high-energies. *Phys. Rev. D* **2002**, *65*, 114008. [CrossRef]
131. Xu, Q.h.; Liang, Z.t.; Sichtermann, E. Anti-lambda polarization in high energy pp collisions with polarized beam. *Phys. Rev. D* **2006**, *73*, 077503. [CrossRef]
132. Adeva, B.; Akdogan, T.; Arik, E.; Arvidson, A.; Badelek, B.; Bardin, G.; Baum, G.; Berglund, P.; Betev, L.; Bird, I.G.; et al. Spin asymmetries A(1) and structure functions g1 of the proton and the deuteron from polarized high-energy muon scattering. *Phys. Rev. D* **1998**, *58*, 112001. [CrossRef]

133. Abe, K.; Akagi, T.; Anthony, P.L.; Antonov, R.; Arnold, R.G.; Averett, T.; Band, H.R.; Bauer, J.M.; Borel, H.; Bosted, P.E.; et al. Measurements of the proton and deuteron spin structure functions g(1) and g(2). *Phys. Rev. D* **1998**, *58*, 112003. [CrossRef]
134. Airapetian, A.; Akopov, N.; Akushevich, I.; Amarian, M.; Aschenauer, E.C.; Avakian, H.; Avakian, R.; Avetisian, A.; Bains, B.; Baumgarten, C.; et al. Measurement of the proton spin structure function g1(p) with a pure hydrogen target. *Phys. Lett. B* **1998**, *442*, 484–492. [CrossRef]
135. Adams, M.R.; Aderholz, M.; Aid, S.; Anthony, P.L.; Ashery, D.; Averill, D.A.; Baker, M.D.; Baller, B.R.; Banerjee, A.; Bhatti, A.A.; et al. Lambda and anti-lambda polarization from deep inelastic muon scattering. *Eur. Phys. J. C* **2000**, *17*, 263–267. [CrossRef]
136. Airapetian, A.; Akopov, N.; Amarian, M.; Aschenauer, E.C.; Avakian, H.; Avakian, R.; Avetissian, A.; Bains, B.; Baumgarten, C.; Beckmann, M.; et al. Measurement of longitudinal spin transfer to Lambda hyperons in deep inelastic lepton scattering. *Phys. Rev. D* **2001**, *64*, 112005. [CrossRef]
137. Airapetian, A.; Akopov, N.; Akopov, Z.; Amarian, M.; Andrus, A.; Aschenauer, E.C.; Augustyniak, W.; Avakian, R.; Avetissian, A.; Avetissian, E.; et al. Longitudinal Spin Transfer to the Lambda Hyperon in Semi-Inclusive Deep-Inelastic Scattering. *Phys. Rev. D* **2006**, *74*, 072004. [CrossRef]
138. Alekseev, M.; Alexakhin, V.Y.; Alexandrov, Y.; Alexeev, G.D.; Amoroso, A.; Austregesilo, A.; Badełek, B.; Balestra, F.; Ball, J.; Barth, J.; et al. Measurement of the Longitudinal Spin Transfer to Lambda and Anti-Lambda Hyperons in Polarised Muon DIS. *Eur. Phys. J. C* **2009**, *64*, 171–179. [CrossRef]
139. Ma, B.Q.; Schmidt, I.; Soffer, J.; Yang, J.J. Lambda, anti-Lambda polarization and spin transfer in lepton deep inelastic scattering. *Eur. Phys. J. C* **2000**, *16*, 657–664. [CrossRef]
140. Zhou, S.s.; Chen, Y.; Liang, Z.t.; Xu, Q.h. Longitudinal polarization of hyperon and anti-hyperon in semi-inclusive deep-inelastic scattering. *Phys. Rev. D* **2009**, *79*, 094018. [CrossRef]
141. Chi, Y.; Du, X.; Ma, B.Q. Nucleon strange $s\bar{s}$ asymmetry to the $\Lambda/\bar{\Lambda}$ fragmentation. *Phys. Rev. D* **2014**, *90*, 074003. [CrossRef]
142. Gluck, M.; Reya, E.; Stratmann, M.; Vogelsang, W. Models for the polarized parton distributions of the nucleon. *Phys. Rev. D* **2001**, *63*, 094005. [CrossRef]
143. Ma, B.Q.; Schmidt, I.; Yang, J.J. Nucleon transversity distribution from azimuthal spin asymmetry in pion electroproduction. *Phys. Rev. D* **2001**, *63*, 037501. [CrossRef]
144. Ma, B.Q.; Schmidt, I.; Soffer, J.; Yang, J.J. The Flavor and spin structure of hyperons from quark fragmentation. *Phys. Rev. D* **2000**, *62*, 114009. [CrossRef]
145. Yang, J.J. Quark fragmentation functions in a diquark model for Lambda production. *Phys. Rev. D* **2001**, *64*, 074010. [CrossRef]
146. Leader, E.; Sidorov, A.V.; Stamenov, D.B. A New evaluation of polarized parton densities in the nucleon. *Eur. Phys. J. C* **2002**, *23*, 479–485. [CrossRef]
147. Blumlein, J.; Bottcher, H. QCD analysis of polarized deep inelastic data and parton distributions. *Nucl. Phys. B* **2002**, *636*, 225–263. [CrossRef]
148. Leader, E.; Stamenov, D.B. Can the polarization of the strange quarks in the proton be positive? *Phys. Rev. D* **2003**, *67*, 037503. [CrossRef]
149. Leader, E.; Sidorov, A.V.; Stamenov, D.B. Longitudinal polarized parton densities updated. *Phys. Rev. D* **2006**, *73*, 034023. [CrossRef]
150. Astier, P.; Autiero, D.; Baldisseri, A.; Baldo-Ceolin, M.; Banner, M.; Bassompierre, G.; Benslama, K.; Besson, N.; Bird, I.; Blumenfeld, B.; et al. Measurement of the Lambda polarization in nu/mu charged current interactions in the NOMAD experiment. *Nucl. Phys. B* **2000**, *588*, 3–36. [CrossRef]
151. Astier, P.; Autiero, D.; Baldisseri, A.; Baldo-Ceolin, M.; Banner, M.; Bassompierre, G.; Besson, N.; Bird, I.; Blumenfeld, B.; Bobisut, F.; Bouchez, J.; et al. Measurement of the anti-Lambda polarization in muon-neutrino charged current interactions in the NOMAD experiment. *Nucl. Phys. B* **2001**, *605*, 3–14. [CrossRef]
152. Ellis, J.R.; Kotzinian, A.; Naumov, D.V. Intrinsic polarized strangeness and Lambda0 polarization in deep inelastic production. *Eur. Phys. J. C* **2002**, *25*, 603–613. [CrossRef]
153. Abelev, B.I.; Aggarwal, M.M.; Ahammed, Z.; Alakhverdyants, A.V.; Anderson, B.D.; Arkhipkin, D.; Averichev, G.S.; Balewski, J.; Barannikova, O.; Barnby, L.S.; et al. Longitudinal Spin Transfer to Lambda and anti-Lambda Hyperons in Polarized Proton-Proton Collisions at $\sqrt{s}$ = 200 GeV. *Phys. Rev. D* **2009**, *80*, 111102. [CrossRef]
154. Adam, J.; Adamczyk, L.; Adams, J.R.; Adkins, J.K.; Agakishiev, G.; Aggarwal, M.M.; Ahammed, Z.; Alekseev, I.; Anderson, D.M.; Aoyama, R.; et al. Improved measurement of the longitudinal spin transfer to Λ and $\bar{\Lambda}$ hyperons in polarized proton-proton collisions at $\sqrt{s}$ = 200 GeV. *Phys. Rev. D* **2018**, *98*, 112009. [CrossRef]
155. De Florian, D.; Soffer, J.; Stratmann, M.; Vogelsang, W. Bounds on transverse spin asymmetries for Lambda baryon production in pp collisions at BNL RHIC. *Phys. Lett. B* **1998**, *439*, 176–182. [CrossRef]
156. Gastmans, R.; Wu, T.T. The Ubiquitous Photon: Helicity Method for QED and QCD; 1990; Volume 80. Available online: https://inspirehep.net/literature/302550 (accessed on 5 March 2023).
157. Chen, K.; Goldstein, G.R.; Jaffe, R.L.; Ji, X.D. Probing quark fragmentation functions for spin 1/2 baryon production in unpolarized e^+e^- annihilation. *Nucl. Phys. B* **1995**, *445*, 380–398. [CrossRef]
158. Zhang, H.C.; Wei, S.Y. Probing the longitudinal spin transfer via dihadron polarization correlations in unpolarized e^+e^- and pp collisions. 2023. Available online: https://inspirehep.net/literature/2621901 (accessed on 5 March 2023).

159. Bacchetta, A.; D'Alesio, U.; Diehl, M.; Miller, C.A. Single-spin asymmetries: The Trento conventions. *Phys. Rev. D* **2004**, *70*, 117504. [CrossRef]
160. Chen, K.b.; Liang, Z.t.; Song, Y.k.; Wei, S.y. Longitudinal and transverse polarizations of Λ hyperon in unpolarized SIDIS and e^+e^- annihilation. *Phys. Rev. D* **2022**, *105*, 034027. [CrossRef]
161. Yang, Y.; Wang, X.; Lu, Z. Predicting the $\sin\phi_S$ Single-Spin-Asymmetry of Λ Production off transversely polarized nucleon in SIDIS. *Phys. Rev. D* **2021**, *103*, 114011. [CrossRef]
162. Li, H.; Wang, X.; Lu, Z. $\sin(\phi_\Lambda - \phi_S)$ azimuthal asymmetry in transversely polarized Λ production in SIDIS within TMD factorization at the EIC. *Phys. Rev. D* **2021**, *104*, 034020. [CrossRef]
163. Kang, Z.B.; Lee, K.; Shao, D.Y.; Zhao, F. Spin asymmetries in electron-jet production at the future electron ion collider. *JHEP* **2021**, *11*, 005. [CrossRef]
164. Kang, Z.B.; Terry, J.; Vossen, A.; Xu, Q.; Zhang, J. Transverse Lambda production at the future Electron-Ion Collider. *Phys. Rev. D* **2022**, *105*, 094033. [CrossRef]
165. Ellis, J.; Hwang, D.S. Spin Correlations of Lambda anti-Lambda Pairs as a Probe of Quark-Antiquark Pair Production. *Eur. Phys. J. C* **2012**, *72*, 1877. [CrossRef]
166. Ackerstaff, K.; Alexander, G.; Allison, J.; Altekamp, N.; Anderson, K.J.; Anderson, S.; Arcelli, S.; Asai, S.; Axen, D.; Azuelos, G.; et al. Spin alignment of leading K*(892)0 mesons in hadronic Z0 decays. *Phys. Lett. B* **1997**, *412*, 210–224. [CrossRef]
167. Abreu, P.; Adam, W.; Adye, T.; Alekseev, G.D.; Alemany, R.; Allport, P.P.; Almehed, S.; Amaldi, U.; Amato, S.; Andersson, P.; et al. Measurement of the spin density matrix for the rho0, K*0 (892) and phi produced in Z0 decays. *Phys. Lett. B* **1997**, *406*, 271–286. [CrossRef]
168. Ackersta, K.; Alexander, G.; Allison, J.; Altekamp, N.; Ametewee, K.; Anderson, K.J.; Anderson, S.; Arcelli, S.; Asai, S.; Axen, D.; et al. Study of phi (1020), D*+- and B* spin alignment in hadronic Z0 decays. *Z. Phys. C* **1997**, *74*, 437–449. [CrossRef]
169. Abbiendi, G.; Ackerstaff, K.; Alexander, G.; Allison, J.; Altekamp, N.; Anderson, K.J.; Anderson, S.; Arcelli, S.; Ashby, S.F.; et al. A Study of spin alignment of rho(770)+- and omega(782) mesons in hadronic Z0 decays. *Eur. Phys. J. C* **2000**, *16*, 61–70. [CrossRef]
170. Chukanov, A.; Naumov, D.; Popov, B.A.; Astier, P.; Autiero, D.; Baldisseri, A.; Baldisseri, A.; Baldo-Ceolin, M.; Banner, M.; Bassompierre, G.; et al. Production properties of K*(892)+- vector mesons and their spin alignment as measured in the NOMAD experiment. *Eur. Phys. J. C* **2006**, *46*, 69–79. [CrossRef]
171. Anselmino, M.; Bertini, M.; Murgia, F.; Pire, B. Off diagonal helicity density matrix elements for heavy vector mesons inclusively produced in N N, gamma N and lepton N interactions. *Phys. Lett. B* **1998**, *438*, 347–352. [CrossRef]
172. Anselmino, M.; Bertini, M.; Caruso, F.; Murgia, F.; Quintairos, P. Off diagonal helicity density matrix elements for vector mesons produced in polarized e^+e^- processes. *Eur. Phys. J. C* **1999**, *11*, 529–537. [CrossRef]
173. Schafer, A.; Szymanowski, L.; Teryaev, O.V. Tensor polarization of vector mesons from quark and gluon fragmentation. *Phys. Lett. B* **1999**, *464*, 94–100. [CrossRef]
174. Xu, Q.h.; Liu, C.x.; Liang, Z.t. Spin alignment of vector meson in e^+e^- annihilation at Z0 pole. *Phys. Rev. D* **2001**, *63*, 111301. [CrossRef]
175. Bacchetta, A.; Mulders, P.J. Positivity bounds on spin one distribution and fragmentation functions. *Phys. Lett. B* **2001**, *518*, 85–93. [CrossRef]
176. Shlyapnikov, P.V. Comparing fragmentation of strange quark in Z0 decays and K+ p reactions. *Phys. Lett. B* **2001**, *512*, 18–24. [CrossRef]
177. Xu, Q.h.; Liang, Z.t. Spin alignments of vector mesons in deeply inelastic lepton nucleon scattering. *Phys. Rev. D* **2002**, *66*, 017301. [CrossRef]
178. Xu, Q.h.; Liang, Z.t. Spin alignment of the high p(T) vector mesons in polarized pp collisions at high-energies. *Phys. Rev. D* **2003**, *67*, 114013. [CrossRef]
179. Xu, Q.h.; Liang, Z.t. Spin alignment of vector mesons in unpolarized hadron hadron collisions at high-energies. *Phys. Rev. D* **2003**, *68*, 034023. [CrossRef]
180. Abelev, B.I.; Aggarwal, M.M.; Ahammed, Z.; Anderson, B.D.; Arkhipkin, D.; Averichev, G.S.; Bai, Y.; Balewski, J.; Barannikova, O.; Barnby, L.S.; et al. Spin alignment measurements of the K^{*0}(892) and phi (1020) vector mesons in heavy ion collisions at $\sqrt{s_{NN}}$ = 200 GeV. *Phys. Rev. C* **2008**, *77*, 061902. [CrossRef]
181. Kundu, S. Spin alignment measurements of vector mesons with ALICE at the LHC. *Nucl. Phys. A* **2021**, *1005*, 121912. [CrossRef]
182. Ji, X. Parton Physics on a Euclidean Lattice. *Phys. Rev. Lett.* **2013**, *110*, 262002. [CrossRef]
183. Ji, X.; Zhang, J.H.; Zhao, Y. Renormalization in Large Momentum Effective Theory of Parton Physics. *Phys. Rev. Lett.* **2018**, *120*, 112001. [CrossRef]
184. Radyushkin, A.V. Quasi-parton distribution functions, momentum distributions, and pseudo-parton distribution functions. *Phys. Rev. D* **2017**, *96*, 034025. [CrossRef]
185. Orginos, K.; Radyushkin, A.; Karpie, J.; Zafeiropoulos, S. Lattice QCD exploration of parton pseudo-distribution functions. *Phys. Rev. D* **2017**, *96*, 094503. [CrossRef]
186. Khan, T.; Sufian, R.S.; Karpie, J.; Monahan, C.J.; Egerer, C.; Joó, B.; Morris, W.; Orginos, K.; Radyushkin, A.; Richards, D.G.; et al. Unpolarized gluon distribution in the nucleon from lattice quantum chromodynamics. *Phys. Rev. D* **2021**, *104*, 094516. [CrossRef]
187. Chodos, A.; Jaffe, R.L.; Johnson, K.; Thorn, C.B.; Weisskopf, V.F. A New Extended Model of Hadrons. *Phys. Rev. D* **1974**, *9*, 3471–3495. [CrossRef]

188. Chodos, A.; Jaffe, R.L.; Johnson, K.; Thorn, C.B. Baryon Structure in the Bag Theory. *Phys. Rev. D* **1974**, *10*, 2599. [CrossRef]
189. Jakob, R.; Mulders, P.J.; Rodrigues, J. Modeling quark distribution and fragmentation functions. *Nucl. Phys. A* **1997**, *626*, 937–965. [CrossRef]
190. Petrov, V.Y.; Pobylitsa, P.V.; Polyakov, M.V.; Bornig, I.; Goeke, K.; Weiss, C. Off—forward quark distributions of the nucleon in the large N(c) limit. *Phys. Rev. D* **1998**, *57*, 4325–4333. [CrossRef]
191. Bentz, W.; Hama, T.; Matsuki, T.; Yazaki, K. NJL model on the light cone and pion structure function. *Nucl. Phys. A* **1999**, *651*, 143–173. [CrossRef]
192. Penttinen, M.; Polyakov, M.V.; Goeke, K. Helicity skewed quark distributions of the nucleon and chiral symmetry. *Phys. Rev. D* **2000**, *62*, 014024. [CrossRef]
193. Brodsky, S.J.; Hwang, D.S.; Schmidt, I. Final state interactions and single spin asymmetries in semiinclusive deep inelastic scattering. *Phys. Lett. B* **2002**, *530*, 99–107. [CrossRef]
194. Brodsky, S.J.; Hwang, D.S.; Schmidt, I. Initial state interactions and single spin asymmetries in Drell-Yan processes. *Nucl. Phys. B* **2002**, *642*, 344–356. [CrossRef]
195. Boer, D.; Brodsky, S.J.; Hwang, D.S. Initial state interactions in the unpolarized Drell-Yan process. *Phys. Rev. D* **2003**, *67*, 054003. [CrossRef]
196. Brodsky, S.J.; Hwang, D.S.; Schmidt, I. Single hadronic spin asymmetries in weak interaction processes. *Phys. Lett. B* **2003**, *553*, 223–228. [CrossRef]
197. Gamberg, L.P.; Goldstein, G.R.; Oganessyan, K.A. Novel transversity properties in semiinclusive deep inelastic scattering. *Phys. Rev. D* **2003**, *67*, 071504. [CrossRef]
198. Bacchetta, A.; Schaefer, A.; Yang, J.J. Sivers function in a spectator model with axial vector diquarks. *Phys. Lett. B* **2004**, *578*, 109–118. [CrossRef]
199. Belitsky, A.V.; Ji, X.d.; Yuan, F. Quark imaging in the proton via quantum phase space distributions. *Phys. Rev. D* **2004**, *69*, 074014. [CrossRef]
200. Yuan, F. Sivers function in the MIT bag model. *Phys. Lett. B* **2003**, *575*, 45–54. [CrossRef]
201. Lu, Z.; Ma, B.Q. Non-zero transversity distribution of the pion in a quark-spectator-antiquark model. *Phys. Rev. D* **2004**, *70*, 094044. [CrossRef]
202. Goeke, K.; Meissner, S.; Metz, A.; Schlegel, M. Checking the Burkardt sum rule for the Sivers function by model calculations. *Phys. Lett. B* **2006**, *637*, 241–244. [CrossRef]
203. Cloet, I.C.; Bentz, W.; Thomas, A.W. Transversity quark distributions in a covariant quark-diquark model. *Phys. Lett. B* **2008**, *659*, 214–220. [CrossRef]
204. Meissner, S.; Metz, A.; Goeke, K. Relations between generalized and transverse momentum dependent parton distributions. *Phys. Rev. D* **2007**, *76*, 034002. [CrossRef]
205. Bacchetta, A.; Conti, F.; Radici, M. Transverse-momentum distributions in a diquark spectator model. *Phys. Rev. D* **2008**, *78*, 074010. [CrossRef]
206. Pasquini, B.; Cazzaniga, S.; Boffi, S. Transverse momentum dependent parton distributions in a light-cone quark model. *Phys. Rev. D* **2008**, *78*, 034025. [CrossRef]
207. Avakian, H.; Efremov, A.V.; Schweitzer, P.; Yuan, F. Transverse momentum dependent distribution function $h_{1T}^\perp$ and the single spin asymmetry $A_{UT}^{\sin(3\phi-\phi_S)}$. *Phys. Rev. D* **2008**, *78*, 114024. [CrossRef]
208. Avakian, H.; Efremov, A.V.; Schweitzer, P.; Yuan, F. Pretzelosity distribution function $h_{1T}^\perp$. In Proceedings of the 2nd International Workshop on Transverse Polarization Phenomena in Hard Processes, Ferrara, Italy, 28–31 May 2008.
209. Boffi, S.; Efremov, A.V.; Pasquini, B.; Schweitzer, P. Azimuthal spin asymmetries in light-cone constituent quark models. *Phys. Rev. D* **2009**, *79*, 094012. [CrossRef]
210. Efremov, A.V.; Schweitzer, P.; Teryaev, O.V.; Zavada, P. Transverse momentum dependent distribution functions in a covariant parton model approach with quark orbital motion. *Phys. Rev. D* **2009**, *80*, 014021. [CrossRef]
211. She, J.; Zhu, J.; Ma, B.Q. Pretzelosity $h_{1T}^\perp$ and quark orbital angular momentum. *Phys. Rev. D* **2009**, *79*, 054008. [CrossRef]
212. Avakian, H.; Efremov, A.V.; Schweitzer, P.; Yuan, F. The transverse momentum dependent distribution functions in the bag model. *Phys. Rev. D* **2010**, *81*, 074035. [CrossRef]
213. Pasquini, B.; Yuan, F. Sivers and Boer-Mulders functions in Light-Cone Quark Models. *Phys. Rev. D* **2010**, *81*, 114013. [CrossRef]
214. Lu, Z.; Ma, B.Q. Gluon Sivers function in a light-cone spectator model. *Phys. Rev. D* **2016**, *94*, 094022. [CrossRef]
215. Ma, Z.L.; Zhu, J.Q.; Lu, Z. Quasiparton distribution function and quasigeneralized parton distribution of the pion in a spectator model. *Phys. Rev. D* **2020**, *101*, 114005. [CrossRef]
216. Bacchetta, A.; Celiberto, F.G.; Radici, M.; Taels, P. Transverse-momentum-dependent gluon distribution functions in a spectator model. *Eur. Phys. J. C* **2020**, *80*, 733. [CrossRef]
217. Luan, X.; Lu, Z. Sivers function of sea quarks in the light-cone model. *Phys. Lett. B* **2022**, *833*, 137299. [CrossRef]
218. Tan, C.; Lu, Z. Gluon generalized parton distributions and angular momentum in a light-cone spectator model. *arXiv* **2023**, arXiv:2301.09081.
219. Sharma, S.; Dahiya, H. Twist-4 T-even proton TMDs in the light-front quark-diquark model. *arXiv* **2023**, arXiv:2301.09536.
220. Metz, A. Gluon-exchange in spin-dependent fragmentation. *Phys. Lett. B* **2002**, *549*, 139–145. [CrossRef]

221. Gamberg, L.P.; Mukherjee, A.; Mulders, P.J. Spectral analysis of gluonic pole matrix elements for fragmentation. *Phys. Rev. D* **2008**, *77*, 114026. [CrossRef]
222. Nzar, M.; Hoodbhoy, P. Quark fragmentation functions in a diquark model for proton and Lambda hyperon production. *Phys. Rev. D* **1995**, *51*, 32–36. [CrossRef]
223. Yang, Y.; Lu, Z.; Schmidt, I. Transverse polarization of the Λ hyperon from unpolarized quark fragmentation in the diquark model. *Phys. Rev. D* **2017**, *96*, 034010. [CrossRef]
224. Wang, X.; Yang, Y.; Lu, Z. Double Collins effect in $e^+e^- \to \Lambda\bar{\Lambda}X$ process in a diquark spectator model. *Phys. Rev. D* **2018**, *97*, 114015. [CrossRef]
225. Lu, Z.; Schmidt, I. Twist-3 fragmentation functions in a spectator model with gluon rescattering. *Phys. Lett. B* **2015**, *747*, 357–364. [CrossRef]
226. Yang, Y.; Lu, Z.; Schmidt, I. Twist-3 T-odd fragmentation functions $G^\perp$ and $\tilde{G}^\perp$ in a spectator model. *Phys. Lett. B* **2016**, *761*, 333–339. [CrossRef]
227. Bacchetta, A.; Gamberg, L.P.; Goldstein, G.R.; Mukherjee, A. Collins fragmentation function for pions and kaons in a spectator model. *Phys. Lett. B* **2008**, *659*, 234–243. [CrossRef]
228. Xie, X.; Lu, Z. T-Even Transverse Momentum Dependent Gluon Fragmentation Functions in A Spectator Model. *arXiv* **2022**. arxiv:2210.16532.
229. Manohar, A.; Georgi, H. Chiral Quarks and the Nonrelativistic Quark Model. *Nucl. Phys. B* **1984**, *234*, 189–212. [CrossRef]
230. Ji, X.D.; Zhu, Z.K. Quark Fragmentation Functions in Low-Energy Chiral Theory. 1993. Available online: https://inspirehep.net/literature/363352 (accessed on 5 March 2023).
231. Londergan, J.T.; Pang, A.; Thomas, A.W. Probing charge symmetry violating quark distributions in semiinclusive leptoproduction of hadrons. *Phys. Rev. D* **1996**, *54*, 3154–3161. [CrossRef] [PubMed]
232. Bacchetta, A.; Kundu, R.; Metz, A.; Mulders, P.J. Estimate of the Collins fragmentation function in a chiral invariant approach. *Phys. Rev. D* **2002**, *65*, 094021. [CrossRef]
233. Nam, S.i.; Kao, C.W. Fragmentation functions and parton distribution functions for the pion with the nonlocal interactions. *Phys. Rev. D* **2012**, *85*, 034023. [CrossRef]
234. Nam, S.i.; Kao, C.W. Fragmentation and quark distribution functions for the pion and kaon with explicit flavor-SU(3)-symmetry breaking. *Phys. Rev. D* **2012**, *85*, 094023. [CrossRef]
235. Yang, D.J.; Jiang, F.J.; Kao, C.W.; Nam, S.i. Quark-jet contribution to the fragmentation functions for the pion and kaon with the nonlocal interactions. *Phys. Rev. D* **2013**, *87*, 094007. [CrossRef]
236. Andrianov, A.A.; Espriu, D. Vector mesons in the extended chiral quark model. *JHEP* **1999**, *10*, 022. [CrossRef]
237. Bacchetta, A.; Kundu, R.; Metz, A.; Mulders, P.J. The Collins fragmentation function: A Simple model calculation. *Phys. Lett. B* **2001**, *506*, 155–160. [CrossRef]
238. Bacchetta, A.; Metz, A.; Yang, J.J. Collins fragmentation function from gluon rescattering. *Phys. Lett. B* **2003**, *574*, 225–231. [CrossRef]
239. Gamberg, L.P.; Goldstein, G.R.; Oganessyan, K.A. A Mechanism for the T odd pion fragmentation function. *Phys. Rev. D* **2003**, *68*, 051501. [CrossRef]
240. Amrath, D.; Bacchetta, A.; Metz, A. Reviewing model calculations of the Collins fragmentation function. *Phys. Rev. D* **2005**, *71*, 114018. [CrossRef]
241. Nambu, Y.; Jona-Lasinio, G. Dynamical Model of Elementary Particles Based on an Analogy with Superconductivity 1. *Phys. Rev.* **1961**, *122*, 345–358. [CrossRef]
242. Nambu, Y.; Jona-Lasinio, G. Dynamical Model of Elementary Particles Based on an Analogy With Superconductivity II. *Phys. Rev.* **1961**, *124*, 246–254. [CrossRef]
243. Korpa, C.L.; Meissner, U.G. Flavor Mixing Structure Functions in the Nambu-Jona-Lasinio Model. *Phys. Rev. D* **1990**, *41*, 1679. [CrossRef]
244. Shigetani, T.; Suzuki, K.; Toki, H. Pion structure function in the Nambu and Jona-Lasinio model. *Phys. Lett. B* **1993**, *308*, 383–388. [CrossRef]
245. Davidson, R.M.; Ruiz Arriola, E. Structure functions of pseudoscalar mesons in the SU(3) NJL model. *Phys. Lett. B* **1995**, *348*, 163–169. [CrossRef]
246. Davidson, R.M.; Ruiz Arriola, E. Parton distributions functions of pion, kaon and eta pseudoscalar mesons in the NJL model. *Acta Phys. Polon. B* **2002**, *33*, 1791–1808.
247. Nguyen, T.; Bashir, A.; Roberts, C.D.; Tandy, P.C. Pion and kaon valence-quark parton distribution functions. *Phys. Rev. C* **2011**, *83*, 062201. [CrossRef]
248. Hutauruk, P.T.P.; Cloet, I.C.; Thomas, A.W. Flavor dependence of the pion and kaon form factors and parton distribution functions. *Phys. Rev. C* **2016**, *94*, 035201. [CrossRef]
249. Field, R.D.; Feynman, R.P. A Parametrization of the Properties of Quark Jets. *Nucl. Phys. B* **1978**, *136*, 1. [CrossRef]
250. Field, R.D.; Wolfram, S. A QCD Model for e^+e^- Annihilation. *Nucl. Phys. B* **1983**, *213*, 65–84. [CrossRef]
251. Ito, T.; Bentz, W.; Cloet, I.C.; Thomas, A.W.; Yazaki, K. The NJL-jet model for quark fragmentation functions. *Phys. Rev. D* **2009**, *80*, 074008. [CrossRef]

252. Matevosyan, H.H.; Thomas, A.W.; Bentz, W. Calculating Kaon Fragmentation Functions from NJL-Jet Model. *Phys. Rev. D* **2011**, *83*, 074003. [CrossRef]
253. Matevosyan, H.H.; Thomas, A.W.; Bentz, W. Monte Carlo Simulations of Hadronic Fragmentation Functions using NJL-Jet Model. *Phys. Rev. D* **2011**, *83*, 114010; Erratum in *Phys. Rev. D* **2012**, *86*, 059904. [CrossRef]
254. Matevosyan, H.H.; Bentz, W.; Cloet, I.C.; Thomas, A.W. Transverse Momentum Dependent Fragmentation and Quark Distribution Functions from the NJL-jet Model. *Phys. Rev. D* **2012**, *85*, 014021. [CrossRef]
255. Matevosyan, H.H.; Thomas, A.W.; Bentz, W. Collins Fragmentation Function within NJL-jet Model. *Phys. Rev. D* **2012**, *86*, 034025. [CrossRef]
256. Casey, A.; Matevosyan, H.H.; Thomas, A.W. Calculating Dihadron Fragmentation Functions in the NJL-jet model. *Phys. Rev. D* **2012**, *85*, 114049. [CrossRef]
257. Bentz, W.; Kotzinian, A.; Matevosyan, H.H.; Ninomiya, Y.; Thomas, A.W.; Yazaki, K. Quark-Jet model for transverse momentum dependent fragmentation functions. *Phys. Rev. D* **2016**, *94*, 034004. [CrossRef]
258. Yang, D.J.; Li, H.n. Gluon fragmentation functions in the Nambu–Jona-Lasinio model. *Phys. Rev. D* **2016**, *94*, 054041. [CrossRef]
259. Yang, D.J.; Li, H.n. Charm fragmentation functions in the Nambu–Jona-Lasinio model. *Phys. Rev. D* **2020**, *102*, 036023. [CrossRef]

Disclaimer/Publisher's Note: The statements, opinions and data contained in all publications are solely those of the individual author(s) and contributor(s) and not of MDPI and/or the editor(s). MDPI and/or the editor(s) disclaim responsibility for any injury to people or property resulting from any ideas, methods, instructions or products referred to in the content.

Article

Experimental Determination of the QCD Effective Charge $\alpha_{g_1}(Q)$

Alexandre Deur [1,2,*], Volker Burkert [1], Jian-Ping Chen [1] and Wolfgang Korsch [3]

[1] Thomas Jefferson National Accelerator Facility, Physics Division, Newport News, VA 23606, USA; burkert@jlab.org (V.B.); jpchen@jlab.org (J.-P.C.)
[2] Department of Physics, University of Virginia, Charlottesville, VA 22904, USA
[3] Department of Physics & Astronomy, University of Kentucky, Lexington, KY 40506, USA; wkkors0@g.uky.edu
* Correspondence: deurpam@jlab.org

Abstract: The QCD effective charge $\alpha_{g_1}(Q)$ is an observable that characterizes the magnitude of the strong interaction. At high momentum Q, it coincides with the QCD running coupling $\alpha_s(Q)$. At low Q, it offers a nonperturbative definition of the running coupling. We have extracted $\alpha_{g_1}(Q)$ from measurements carried out at Jefferson Lab that span the very low to moderately high Q domain, $0.14 \leq Q \leq 2.18$ GeV. The precision of the new results is much improved over the previous extractions and the reach in Q at the lower end is significantly expanded. The data show that $\alpha_{g_1}(Q)$ becomes Q-independent at very low Q. They compare well with two recent predictions of the QCD effective charge based on Dyson–Schwinger equations and on the AdS/CFT duality.

Keywords: strong interaction; QCD; nonperturbative; running coupling constant; hadrons; nucleon; spin structure

Citation: Deur, A.; Burkert, V.; Chen, J.-P.; Korsch, W. Experimental Determination of the QCD Effective Charge $\alpha_{g_1}(Q)$. *Particles* **2022**, *5*, 171–179. https://doi.org/10.3390/particles5020015

Academic Editors: Sebastian M. Schmidt, Minghui Ding and Craig Roberts

Received: 2 May 2022
Accepted: 27 May 2022
Published: 31 May 2022

Publisher's Note: MDPI stays neutral with regard to jurisdictional claims in published maps and institutional affiliations.

Copyright: © 2022 by the authors. Licensee MDPI, Basel, Switzerland. This article is an open access article distributed under the terms and conditions of the Creative Commons Attribution (CC BY) license (https://creativecommons.org/licenses/by/4.0/).

1. Introduction

The behavior of quantum chromodynamics (QCD), the gauge theory of the strong interaction, is determined by the magnitude of its coupling α_s. It is large at low momentum, characterized here by $Q \equiv \sqrt{-q^2}$ with q^2 the square of momentum transferred in the process of electromagnetically probing a hadron. For $Q \ll 1$ GeV, $\alpha_s(Q) \gtrsim 1$, which is one of the crucial pieces leading to quark confinement. For $Q \gg 1$ GeV, $\alpha_s(Q) \lesssim 0.2$, which enables the use of perturbative computational techniques (perturbative QCD, pQCD) constituting an accurate analytical approximation of QCD. In this domain, α_s^{pQCD} is well defined and known within an accuracy of 1% at $Q = M_{Z^0} = 91$ GeV, the Z^0 mass, and within a few percents at Q values of a few GeV [1]. However, using pQCD at $Q \lesssim 1$ GeV produces a diverging α_s^{pQCD} (Landau pole) that prohibits any perturbative expansion in α_s^{pQCD} and signals the breakdown of pQCD. In contrast, most nonperturbative methods, including lattice QCD [2], the AdS/CFT (Anti-de-Sitter/Conformal Field Theory) duality [3,4] implemented using QCD's light-front (LF) quantization [5] and a soft-wall AdS potential (Holographic LF QCD, HLFQCD [6]) or solving the Dyson–Schwinger equations (DSEs) [7] yield a finite α_s. In fact, many theoretical approaches predict that α_s "freezes" as $Q \to 0$, viz, it loses its Q-dependence [8].

There are several possible definitions of α_s in the nonperturbative domain ($Q \lesssim 1$ GeV) [8]. We use here the *effective charge* approach that defines α_s from the perturbative series of an observable truncated to its first order in α_s [9]. Although this definition can be applied for any Q value, it was initially proposed for the pQCD domain where it makes α_s the equivalent of the Gell-Mann Low coupling of quantum electrodynamics (QED), α [10]. With this definition, α_s can be evaluated at any Q value, has no low Q divergence and is analytic around quark mass thresholds. Furthermore, since the first order in α_s^{pQCD} of a pQCD approximant is independent of the choice of renormalization scheme

(RS), effective charges are independent of RS and gauge choices. This promotes α_s from a parameter depending on chosen conventions to an observable, albeit with the caveat that it becomes process-dependent since two observables produce generally different effective charges. Yet, pQCD predictability is maintained because effective charges are related without renormalization scale ambiguities by commensurate scale relations (CSR) [11]. CSR are known to hold for pQCD and QED since the latter corresponds to the $N_C \to 0$ limit of QCD, with N_C the number of colors. For example, CSR explicitly relate α_{g_1}, α_{F_3}, α_τ and α_R defined using the generalized Bjorken sum rule [12], the Gross–Llewellyn Smith sum rule [13], and the perturbative approximant for the τ-decay rate [14] and $R_{e^+e^-}$ [15], respectively. In fact, the choice of process to define an effective charge is analogous to an RS choice for α_s^{pQCD} [16] and the procedure of extracting an effective charge, e.g., from τ-decay the τ-scheme is denoted. Here, we discuss the effective charge $\alpha_{g_1}(Q)$ (g_1-scheme) extracted using the generalized Bjorken sum rule:

$$\Gamma_1^{p-n}(Q^2) \equiv \int_0^{1^-} g_1^p(x, Q^2) - g_1^n(x, Q^2) dx = \frac{g_A}{6}\left[1 - \frac{\alpha_s^{pQCD}(Q)}{\pi} - 3.58\left(\frac{\alpha_s^{pQCD}(Q)}{\pi}\right)^2 \right. \\ \left. - 20.21\left(\frac{\alpha_s^{pQCD}(Q)}{\pi}\right)^3 - 175.7\left(\frac{\alpha_s^{pQCD}(Q)}{\pi}\right)^4 + \mathcal{O}\left((\alpha_s^{pQCD})^5\right)...\right] + \sum_{n>1}\frac{\mu_{2n}}{Q^{2n-2}}, \quad (1)$$

where x is the Bjorken scaling variable [17], $g_A = 1.2762(5)$ [2] is the nucleon axial charge, $g_1^{p(n)}$ is the longitudinal spin structure function of the proton (neutron) obtained in polarized lepton-nucleon scattering [18] and μ_{2n} are the operator product expansion's (OPE) nonperturbative higher twist (HT) terms. The integral excludes the elastic contribution at $x = 1$. The series coefficients are computed for $n_f = 3$ and in the $\overline{MS}$ RS for the $n > 1$ α_s^n terms [19]. They originate from the pQCD radiative corrections. Although the expansion (1) is only applicable in the perturbative domain, i.e., at distance scales where confinement effects are weak, the HT terms can be related to the latter [20], and one may picture the terms of Equation (1) as coherently merging together at low Q to produce confinement.

The effective charge α_{g_1} is defined from Equation (1) expressed at first order in coupling and twist:

$$\Gamma_1^{p-n}(Q^2) \equiv \frac{g_A}{6}\left(1 - \frac{\alpha_{g_1}(Q)}{\pi}\right) \longrightarrow \alpha_{g_1}(Q) \equiv \pi\left(1 - \frac{6}{g_A}\Gamma_1^{p-n}(Q)\right). \quad (2)$$

Thus, in the domain where Equation (2) applies, α_{g_1} can be interpreted as a running coupling that not only includes short-distance effects such as vertex correction and vacuum polarization, but all other effects, e.g., pQCD radiative corrections and, in the lower-Q domain of pQCD, HT terms and other nonperturbative effects not formalized by OPE and therefore not included in Equation (2). The latter comes from coherent reactions of a hadron (resonances). In the nonperturbative domain where pQCD radiative corrections and HT effects have merged into global confinement effects, α_{g_1} may approximately retain its interpretation as a coupling if the contribution to Γ_1^{p-n} of nonresonant reactions continues to dominate, as they do at large Q [21].

There are several advantages to α_{g_1} [8]. First, rigorous sum rules constrain $\alpha_{g_1}(Q)$ for $Q \to 0$ (the Gerasimov–Drell–Hearn (GDH) sum rule [22]) and $Q \to \infty$ (the Bjorken sum rule). They provide analytical expressions of $\alpha_{g_1}(Q)$ in these limits (blue dashed line and cyan hatched band in Figure 1). Furthermore, contributions from Δ baryons are quenched in Γ_1^{p-n} [23], enhancing the nonresonant reactions contribution to Γ_1^{p-n} relatively to the resonance contribution, which helps toward interpreting α_{g_1} as a coupling. If so, α_{g_1} would remain approximately equivalent to the Gell-Mann Low coupling in the nonperturbative domain, a crucial property that it is not obvious and may be specific to α_{g_1}. Such a property is supported by the agreement between α_{g_1} and calculations of couplings [24,25] using a definition consistent with α_{g_1}.

Former extractions of α_{g_1} [26] were obtained from experimental data on Γ_1^{p-n} from CERN [27], DESY [28], Jefferson Lab (JLab) [29] and SLAC [30]; see Figure 1. Since the results reported in Ref. [26], progress has occurred on both the experimental and theoretical fronts. Firstly, when Ref. [26] was published, the meaning of α_{g_1} in the nonperturbative region was unclear. Thus, the comparison in [26] of α_{g_1} to theoretical predictions of the nonperturbative coupling was tentative. This is now better understood: as just discussed, α_{g_1} essentially retains its meaning of effective charge at low Q [8,21]. Secondly, new data on Γ_1^{p-n} have become available from CERN (COMPASS experiment) [31] and JLab (EG1dvcs experiment) [32] at high Q, and from JLab (E97110, E03006 and E05111 experiments) [33] at very low Q. Finally, new theoretical studies of the nonperturbative behavior of α_s were conducted, including the first use of the AdS/CFT duality to describe the strong coupling in its nonperturbative domain [24] and the identification of a process-independent (PI) effective charge $\hat{\alpha}_{PI}(Q)$ that unifies a large body of research from DSE and lattice QCD to α_s [25,34]. Connections between the nonperturbative and perturbative effective charges were made [8,16,35], which permitted a prediction of α_s at the Z_0 pole, $\alpha_s^{\overline{MS}}(M_Z^2) = 0.1190 \pm 0.0006$ at N³LO [36] that agrees well with the 2021 Particle Data Group compilation, $\alpha_s(M_Z) = 0.1179 \pm 0.0009$ [2]. In addition to predicting quantities characterizing hadronic structures [3,25,37], the effective charge helps establish conformal behavior at low Q. Through AdS/CFT, this helps the investigation of the physics beyond the standard model [4] or of the quark-gluon plasma [38] in heavy ion collisions [39] and nuclear hydrodynamics [40] for the latter and neutron stars [41].

Here, we report on new experimental data on α_{g_1} extracted from [31–33] and how they compare with the latest theory predictions.

2. Experimental Extraction of α_{g_1}

The new JLab data on $\Gamma_1^{p-n}(Q)$ were taken by four experiments. The first experiment, E97110 [42], occurred in Hall A [43] of JLab. The three others used the CLAS spectrometer [44] in JLab's Hall B and were experiments EG1dvcs [45], E03006 [46] and E05111 [47] (the two latter being referred to as Experimental Group EG4). The four experiments occurred during the 6 GeV era of JLab, before its 12 GeV upgrade. The experiments used a polarized electron beam with energies ranging from 0.8 to 6 GeV. E97110 studied the spin structures of the neutron and ³He using the Hall A polarized ³He target with longitudinal and transverse polarization directions [48]. EG1dvcs, E03006, and E05111 studied the proton, neutron and deuteron spin structures using the Hall B longitudinally polarized ammonia (NH_3 or ND_3) target [49]. The main purpose of EG1dvcs was high Q, up to 2.65 GeV ($Q^2 = 7$ GeV²), exclusive measurements of deep virtual Compton scattering. Therefore, it provided highly precise inclusive Γ_1^{p-n} data compared to the older data in the same domain [27–30]. E97110, E03006 and E05111 were dedicated to test chiral effective field theory predictions by covering very low Q domains: $0.19 \leq Q \leq 0.49, 0.11 \leq Q \leq 0.92$ and $0.14 \leq Q \leq 0.70$ GeV, respectively. To reach low Q while covering the large x range necessary for the Γ_1 integral, high beam energy (up to 4.4 GeV) was needed, and the scattered electrons had to be detected at small angles (down to about 5°). In Hall A, the low angles were reached via a supplementary dipole magnet installed in front of the spectrometer [50]. In Hall B, a Cherenkov counter designed for high efficiency at small angles was installed in one of the six sectors of CLAS [47] for which magnetic field was set to bent outward the scattered electrons. In addition, both the Hall A and B targets were placed about 1 m upstream of their usual positions.

The EG1dvcs data on protons and deuterons were combined to form Γ_1^{p-n} over the range $0.78 \leq Q \leq 2.18$ GeV [32]. The Γ_1^{p-n} formed with the E97110 and EG4 data covers the $0.14 \leq Q \leq 0.70$ GeV range [33]. The α_{g_1} data, obtained following Equation (2), are shown in Figure 1 and given in Table 1. Also shown in the figure are the older data presented in Ref. [29], including α_{F_3} extracted from the data [51] and $\alpha_{g_1(\tau)}$ from the OPAL data on τ-decay [14]. The effective charge α_{F_3} is nearly identical to α_{g_1} [26], and $\alpha_{g_1(\tau)}$ was transformed from the τ-scheme to the g_1-scheme using the CSR [11]. Consequently, α_{F_3} and $\alpha_{g_1(\tau)}$ are directly comparable to α_{g_1}. We also show in Figure 1 the theory predictions from AdS/CFT [24] and DSE [25]. Remarkably, both predictions are parameter-free and gauge-invariant.

Table 1. Data on $\alpha_{g_1}(Q)$ from JLab experiments EG4 (top, from $Q = 0.143$ GeV to 0.704 GeV), EG4/E97110 (middle, from $Q = 0.187$ GeV to 0.490 GeV) and EG1dvcs (bottom, from $Q = 0.775$ GeV to 2.177 GeV).

Q (GeV)	$\alpha_{g_1} \pm$ stat. $\pm$ syst.
0.143	3.064 ± 0.043 ± 0.018
0.156	3.129 ± 0.046 ± 0.019
0.171	2.955 ± 0.046 ± 0.023
0.187	3.083 ± 0.044 ± 0.024
0.204	3.022 ± 0.049 ± 0.024
0.223	3.002 ± 0.052 ± 0.027
0.243	2.988 ± 0.055 ± 0.031
0.266	2.947 ± 0.060 ± 0.035
0.291	2.983 ± 0.065 ± 0.035
Q (GeV)	**$\alpha_{g_1} \pm$ stat. $\pm$ syst.**
0.317	2.961 ± 0.062 ± 0.038
0.347	2.730 ± 0.070 ± 0.044
0.379	2.853 ± 0.077 ± 0.040
0.414	2.745 ± 0.076 ± 0.041
0.452	2.779 ± 0.090 ± 0.043
0.494	2.451 ± 0.094 ± 0.044
0.540	2.397 ± 0.092 ± 0.039
0.590	2.349 ± 0.101 ± 0.040
0.645	2.431 ± 0.109 ± 0.043
0.704	1.996 ± 0.131 ± 0.104
Q (GeV)	**$\alpha_{g_1} \pm$ stat. $\pm$ syst.**
0.187	3.016 ± 0.009 ± 0.027
0.239	2.973 ± 0.015 ± 0.035
0.281	2.952 ± 0.021 ± 0.041
0.316	2.929 ± 0.017 ± 0.048
0.387	2.815 ± 0.021 ± 0.076
0.447	2.704 ± 0.025 ± 0.086
0.490	2.575 ± 0.031 ± 0.053
0.775	1.743 ± 0.007 ± 0.071
0.835	1.571 ± 0.007 ± 0.101
0.917	1.419 ± 0.009 ± 0.132
0.986	1.341 ± 0.010 ± 0.147
1.088	1.272 ± 0.010 ± 0.156
1.167	1.121 ± 0.013 ± 0.153
1.261	0.955 ± 0.016 ± 0.146
1.384	0.874 ± 0.016 ± 0.269
1.522	0.730 ± 0.012 ± 0.280
1.645	0.708 ± 0.009 ± 0.257
1.795	0.617 ± 0.007 ± 0.254
1.967	0.581 ± 0.006 ± 0.223
2.177	0.636 ± 0.003 ± 0.187

The AdS/CFT coupling $\alpha_{g_1}^{HLF}$ is obtained in the HLFQCD approach where QCD is quantized using LF coordinates [5]. The use of the HLFQCD approach incorporates the underlying conformal (i.e., scale-invariant) character of QCD at low and large Q. The deformation of the AdS_5 space is dual to a semiclassical potential that models quark confinement. This potential can be determined with various methods that all lead to the same harmonic oscillator form [3,52,53]. The effective charge $\alpha_{g_1}^{HLF}$ is dual to the product of the AdS_5 coupling *constant* by the AdS_5 space deformation term. Since the latter is dual to the CFT confinement force, the meaning of $\alpha_{g_1}^{HLF}$ is analogous to that of α_{g_1}, which, at low Q, incorporates in α_s confinement effects. The Q-dependence of $\alpha_{g_1}^{HLF}$ is controlled by a single scale, e.g., the proton mass. The coupling is normalized to $\alpha_{g_1}^{HLF}(0) = \pi$ to obey the kinematic constraint that $\Gamma_1^{p-n}(0) = 0$, i.e., $\alpha_{g_1}(0) = \pi$, see Equation (2). This normalization amounts to the RS choice of pQCD [16]. Thus, the $\alpha_{g_1}^{HLF}(Q)$ prediction is parameter-free. Above $Q \simeq 1$ GeV, HLFQCD ceases to be valid because its *semiclassical* potential does not include, by definition, the short distance quantum effects responsible for the running of a coupling. This is palliated by matching HLFQCD and pQCD near $Q \simeq 1$ GeV where both formalisms apply, thereby providing $\alpha_{g_1}^{HLF}(Q)$ at all Q [16].

The DSE effective charge $\hat{\alpha}_{PI}$ [25] is obtained starting with the pinch technique [54] and background field method [55]. They allow us to define a process-independent QCD coupling in terms of a mathematically reconstructed gluon two-point function analogous to the Gell-Mann Low effective charge of QED. The $\hat{\alpha}_{PI}$ is then computed by combining the solution of DSE compatible with lattice QCD results. The definition of $\hat{\alpha}_{PI}$ explicitly factors in a renormalization group invariant interaction, thus causing it, like $\alpha_{g_1}(Q)$ and $\alpha_{g_1}^{HLF}(Q)$, to incorporate confinement [56]. Like them, $\hat{\alpha}_{PI}(Q)$ freezes at low Q with a predicted infrared fixed-point of $\hat{\alpha}_{PI}(0) = (0.97 \pm 0.04)\pi$. The mechanism at the origin of the freezing in the DSE framework is the emergence of a dynamical gluon mass $m_g(Q)$ [54,57] that (A) regulates the Landau pole and (B) decouples the dynamics at scales $Q \lesssim m_g(0)$, thereby causing the coupling to lose its Q-dependence [58]. Like $\alpha_{g_1}^{HLF}$, $\hat{\alpha}_{PI}$ is parameter-free and gauge-invariant but, in contrast to the former and α_{g_1}, $\hat{\alpha}_{PI}$ is also process-independent. No parameter is varied to predict the infrared fixed-point $\hat{\alpha}_{PI}(0)$ since it is largely fixed by the value of $m_g(0)$, nor is a matching necessary to ensure agreement with the perturbative determination of $\alpha_{g_1}^{pQCD}$ from the renormalization group equations and the Bjorken sum rule. Crucially, the practical determination of $\hat{\alpha}_{PI}(Q)$ consistently incorporates the extensive information from Lattice QCD on the gluon and ghost propagators, thereby connecting this technique to α_{g_1}.

The new data on α_{g_1} agree well with the older data and display a much improved precision over the whole Q range covered. In addition, the data now reach clearly the freezing domain of QCD at very low Q. That α_{g_1} freezes could be already inferred with the old data but only by complementing them with the GDH sum rule or/and the $\alpha_{g_1}(0) = \pi$ constraint. For the first time, the onset of freezing is now visible with data only. One notes that only three of the lowest Q points agree with the GDH expectation. This may signal a fast arising Q-dependence beyond the leading behavior given by GDH. The data agree well with the $\alpha_{g_1}^{HLF}$ and $\hat{\alpha}_{PI}$ predictions. That such agreements would occur was not obvious and is a significant finding. The possible tension between the data and $\hat{\alpha}_{PI}$ for the range $0.3 \lesssim Q \lesssim 0.5$ GeV may be because α_{g_1} and $\hat{\alpha}_{PI}$ are not exactly the same effective charges (e.g., at high Q, $\alpha_{g_1}/\hat{\alpha}_{PI} \simeq 1 + 0.05\alpha_s^{pQCD} \neq 1$), but it is noteworthy that it occurs only in the moderately low Q domain where the ghost-gluon vacuum effect as computed in the Landau gauge contributes the most to $\hat{\alpha}_{PI}$.

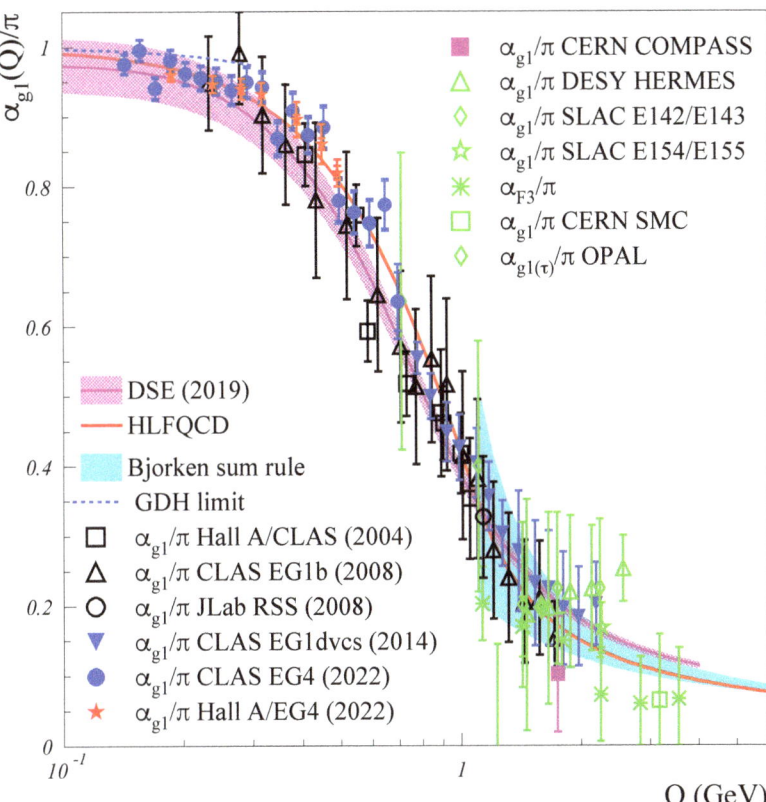

Figure 1. Effective charge $\alpha_{g_1}(Q)/\pi$ obtained from JLab experiments E03006/E97110 [33] (solid stars), E03006/E05111 [33] (solid circles) and EG1dvcs [32] (solid triangles) and from COMPASS [31] (solid square). Inner error bars represent the statistical uncertainties, and outer ones represent the systematic and statistical uncertainties added quadratically. The open symbols show the older world data [27–30] with the error bars the quadratic sum of the systematic and statistical uncertainties. Also shown are the HLFQCD [24] (red line, using the HLFQCD scale $\kappa = 0.534$ GeV [59]) and DSE [25] (magenta line and hatched band) parameter-free predictions of effective charges. The dashed line and hatched cyan band are $\alpha_{g_1}(Q)/\pi$ obtained from the GDH and Bjorken sum rules, respectively.

3. Summary and Conclusions

We used the new JLab data and COMPASS datum on the Bjorken sum to extract the QCD effective charge $\alpha_{g_1}(Q)$ in the Q-range $0.14 \leq Q \leq 2.18$ GeV. The new result displays a significantly higher precision compared to the older extractions of $\alpha_{g_1}(Q)$, and improve the low Q reach by about a factor of 2.

The new data show that $\alpha_{g_1}(Q)$ "freezes", viz, loses its Q-dependence, at small Q, saturating at an infrared fixed-point $\alpha_{g_1}(Q \simeq 0) \simeq \pi$. This was already apparent with the older data when combined with the GDH sum rule expectation, but the new data explicitly display the behavior without needing the sum rule and with significantly higher precision. The freezing of $\alpha_{g_1}(Q)$ together with the smallness of the light quark masses makes QCD approximately conformal at low Q. The conformal behavior vanishes when transiting from the low-Q effective degrees of freedom of QCD (hadrons) to the large-Q fundamental ones (partons) where conformality is then restored (the long-known Bjorken scaling [17]). This transition is revealed by the drastic change of value of the effective charge. It occurs at a Q value indicative of the chiral symmetry breaking parameter, $\Lambda_B \simeq 1$ GeV. The breaking at low Q of chiral symmetry, one of the crucial properties of QCD, is believed to cause the emergence of the global properties of hadrons.

The new data agree well with sum rule predictions and with the latest predictions from DSE and from a AdS/CFT-based approach. They show that a strong coupling can be consistently defined in the nonperturbative domain of QCD, namely as an effective charge analogous to the definition used in QED, and that it can then be used to compute a large variety of hadronic quantities and other phenomena in which the strong interaction plays a role.

Author Contributions: Conceptualization, J.-P.C., A.D. and W.K.; methodology, A.D.; software, A.D.; validation, V.B., J.-P.C., A.D. and W.K.; formal analysis, A.D.; investigation, V.B., J.-P.C., A.D. and W.K.; resources, V.B.; data curation, A.D.; writing—original draft preparation, A.D.; writing—review and editing, V.B., J.-P.C., A.D. and W.K.; visualization, V.B., J.-P.C., A.D. and W.K.; supervision, V.B.; project administration, A.D.; funding acquisition, V.B. and W.K. All authors have read and agreed to the published version of the manuscript.

Funding: This research was funded by the U.S. Department of Energy, Office of Science, Office of Nuclear Physics, contracts DE-AC05-06OR23177 and DE-FG02-99ER41101.

Data Availability Statement: The results reported in this article are available in Table 1 included in this article. They are based on data available in Refs. [32,33].

Acknowledgments: The authors thank D. Binosi, S. J. Brodsky, Z.-F. Cui, G. F. de Téramond, J. Papavassiliou, C. D. Roberts and J. Rodríguez-Quintero for their valuable comments on the manuscript. This work is supported by the U.S. Department of Energy, Office of Science, Office of Nuclear Physics, contracts DE-AC05-06OR23177 and DE-FG02-99ER41101.

Conflicts of Interest: The authors declare no conflict of interest.

References

1. D'Enterria, D.; Kluth, S.; Zanderighi, G.; Ayala, C.; Benitez-Rathgeb, M.A.; Bluemlein, J.; Xie, K. The strong coupling constant: State of the art and the decade ahead. *arXiv* **2022**, arXiv:2203.08271.
2. Yao, W.M.; Amsler, C.; Asner, D.; Barnett, R.M.; Beringer, J.; Burchat, P.R.; Staney, T. Review of Particle Physics. *J. Phys. G Nucl. Part. Phys.* **2006**, *33*, 1–1232. [CrossRef]
3. Brodsky, S.J.; de Téramond, G.F.; Dosch, H.G.; Erlich, J. Light-front holographic QCD and emerging confinement. *Phys. Rep.* **2015**, *584*, 1–105. [CrossRef]
4. Dobado, A.; Espriu, D. Strongly coupled theories beyond the Standard Model. *Prog. Part. Nucl. Phys.* **2020**, *115*, 103813. [CrossRef]
5. Dirac, P.A.M. Forms of relativistic dynamics. *Rev. Mod. Phys.* **1949**, *21*, 392. [CrossRef]
6. Brodsky, S.J.; de Téramond, G.F. Light-front hadron dynamics and AdS/CFT correspondence. *Phys. Lett. B* **2004**, *582*, 211–221. [CrossRef]
7. Maris, P.; Roberts, C.D. Dyson-Schwinger equations: A Tool for hadron physics. *Int. J. Mod. Phys. E* **2003**, *12*, 297–365. [CrossRef]
8. Deur, A.; Brodsky, S.J.; de Téramond, G.F. The QCD Running Coupling. *Prog. Part. Nucl. Phys.* **2016**, *90*, 1. [CrossRef]
9. Grunberg, G. Renormalization Group Improved Perturbative QCD. *Phys. Lett.* **1980**, *95B*, 70; Erratum in *Phys. Lett.* **1982**, *110B*, 501. [CrossRef]
10. Gell-Mann, M.; Low, F.E. Quantum electrodynamics at small distances. *Phys. Rev.* **1954**, *95*, 1300. [CrossRef]
11. Brodsky, S.J.; Lu, H.J. Commensurate scale relations in quantum chromodynamics. *Phys. Rev. D* **1995**, *51*, 3652. [CrossRef] [PubMed]
12. Bjorken, J.D. Applications of the Chiral $U(6) \times U(6)$ Algebra of Current Densities. *Phys. Rev.* **1966**, *148*, 1467. [CrossRef]
13. Gross, D.J.; Smith, C.H.L. High-energy neutrino-nucleon scattering, current algebra and partons. *Nucl. Phys. B* **1969**, *14*, 337. [CrossRef]
14. Brodsky, S.J.; Menke, S.; Merino, C.; Rathsman, J. On the behavior of the effective QCD coupling alpha(tau)(s) at low scales. *Phys. Rev. D* **2003**, *67*, 055008. [CrossRef]
15. Gorishnii, S.G.; Kataev, A.L.; Larin, S.A. The $O(\alpha_s^3)$-corrections to $\sigma_{tot}(e^+e^- \to hadrons)$ and $\Gamma(\tau^- \to \nu_\tau + hadrons)$ in QCD. *Phys. Lett. B* **1991**, *259*, 144–150 [CrossRef]
16. Deur, A.; Brodsky, S.J.; de Téramond, G.F. Connecting the hadron mass scale to the fundamental mass scale of quantum chromodynamics. *Phys. Lett. B* **2015**, *750*, 528. [CrossRef]
17. Bjorken, J.D. Asymptotic Sum Rules at Infinite Momentum. *Phys. Rev.* **1969**, *179*, 1547–1553. [CrossRef]
18. Deur, A.; Brodsky, S.J.; de Téramond, G.F. The Spin Structure of the Nucleon. *Rep. Prog. Phys.* **2019**, *82*, 7. [CrossRef]
19. Kataev, A.L. The Ellis-Jaffe sum rule: The estimates of the next-to-next-to-leading order QCD corrections. *Phys. Rev. D* **1994**, *50*, 5469. [CrossRef]

20. Burkardt, M. Transverse force on quarks in deep-inelastic scattering. *Phys. Rev. D* **2013**, *88*, 114502. [CrossRef]
21. Deur, A. Spin Sum Rules and the Strong Coupling Constant at large distance. *AIP Conf. Proc.* **2009**, *1155*, 112–121.
22. Gerasimov, S.B. A Sum rule for magnetic moments and the damping of the nucleon magnetic moment in nuclei. *Sov. J. Nucl. Phys.* **1966**, *2*, 430.
23. Burkert, V.D. Comment on the generalized Gerasimov-Drell-Hearn sum rule in chiral perturbation theory. *Phys. Rev. D* **2001**, *63*, 097904. [CrossRef]
24. Brodsky, S.J.; de Téramond, G.F.; Deur, A. Nonperturbative QCD Coupling and its β-function from Light-Front Holography. *Phys. Rev. D* **2010**, *81*, 096010. [CrossRef]
25. Binosi, D.; Mezrag, C.; Papavassiliou, J.; Roberts, C.D.; Rodriguez-Quintero, J. Process-independent strong running coupling. *Phys. Rev. D* **2017**, *96*, 054026. [CrossRef]
26. Deur, A.; Burkert, V.; Chen, J.P.; Korsch, W. Experimental determination of the effective strong coupling constant. *Phys. Lett. B* **2007**, *650*, 244. [CrossRef]
27. Adeva, B.; Ahmad, S.; Arvidson, A.; Badelek, B.; Ballintijn, M.K.; Bardin, G.; Voss, R. Measurement of the spin dependent structure function g1(x) of the deuteron. *Phys. Lett. B* **1993**, *302*, 533. [CrossRef]
28. Airapetian, A.; Akopov, N.; Akushevich, I.; Amarian, M.; Arrington, J.; Aschenauer, E.C.; Schüler, K.P. The Q^2 dependence of the generalized Gerasimov-Drell-Hearn integral for the proton. *Phys. Lett. B* **2000**, *494*, 1–8. [CrossRef]
29. Deur, A.; Bosted, P.; Burkert, V.; Cates, G.; Chen, J.P.; Choi, S.; Yun, J. Experimental determination of the evolution of the Bjorken integral at low Q^2. *Phys. Rev. Lett.* **2004**, *93*, 212001. [CrossRef]
30. Anthony, P.L.; Arnold, R.G.; Band, H.R.; Borel, H.; Bosted, P.E.; Breton, V.; Zapalac, G. Deep inelastic scattering of polarized electrons by polarized He-3 and the study of the neutron spin structure. *Phys. Rev. D* **1996**, *54*, 6620. [CrossRef]
31. Alekseev, M.G.; Alexakhin, V.Y.; Alexandrov, Y.; Alexeev, G.D.; Amoroso, A.; Austregesilo, A.; Padee, A. The Spin-dependent Structure Function of the Proton g_1^p and a Test of the Bjorken Sum Rule. *Phys. Lett. B* **2010**, *690*, 466–472. [CrossRef]
32. Deur, A.; Prok, Y.; Burkert, V.; Crabb, D.; Girod, F.X.; Griffioen, K.A.; Kvaltine, N. High precision determination of the Q^2 evolution of the Bjorken Sum. *Phys. Rev. D* **2014**, *90*, 012009. [CrossRef]
33. Deur, A.; Chen, J.P.; Kuhn, S.E.; Peng, C.; Ripani, M.; Sulkosky, V.; Zheng, X. Experimental study of the behavior of the Bjorken sum at very low Q2. *Phys. Lett. B* **2022**, *825*, 136878. [CrossRef]
34. Rodríguez-Quintero, J.; Binosi, D.; Mezrag, C.; Papavassiliou, J.; Roberts, C.D. Process-independent effective coupling. From QCD Green's functions to phenomenology. *Few Body Syst.* **2018**, *59*, 121. [CrossRef]
35. Deur, A.; Shen, J.M.; Wu, X.G.; Brodsky, S.J.; de Téramond, G.F. Implications of the Principle of Maximum Conformality for the QCD Strong Coupling. *Phys. Lett. B* **2017**, *773*, 98. [CrossRef]
36. Deur, A.; Brodsky, S.J.; de Téramond, G.F. Determination of $\Lambda_{\overline{MS}}$ at five loops from holographic QCD. *J. Phys. G* **2017**, *44*, 105005. [CrossRef]
37. Cui, Z.F.; Ding, M.; Gao, F.; Raya, K.; Binosi, D.; Chang, L.; Roberts, C.D.; Rodriguez-Quintero, J.; Schmidt, S.M. Kaon and pion parton distributions. *Eur. Phys. J. C* **2020**, *80*, 1064. [CrossRef]
38. Janik, R.A. The Dynamics of Quark-Gluon Plasma and AdS/CFT. *Lect. Notes Phys.* **2011**, *828*, 147–181.
39. Busza, W.; Rajagopal, K.; van der Schee, W. Heavy Ion Collisions: The Big Picture, and the Big Questions. *Ann. Rev. Nucl. Part. Sci.* **2018**, *68*, 339–376. [CrossRef]
40. Florkowski, W.; Heller, M.P.; Spalinski, M. New theories of relativistic hydrodynamics in the LHC era. *Rept. Prog. Phys.* **2018**, *81*, 046001. [CrossRef]
41. Jokela, N.; Järvinen, M.; Remes, J. Holographic QCD in the Veneziano limit and neutron stars. *JHEP* **2019**, *3*, 41. [CrossRef]
42. Sulkosky, V.; Singh, J.T.; Peng, C.; Chen, J.P.; Deur, A.; Abrahamyan, S.; Zhu, L. Measurement of the ^3He spin-structure functions and of neutron (^3He) spin-dependent sum rules at $0.035 \leq Q^2 \leq 0.24$ GeV^2. *Phys. Lett. B* **2020**, *805*, 135428. [CrossRef]
43. Alcorn, J.; Anderson, B.D.; Aniol, K.A.; Annand, J.R.M.; Auerbach, L.; Arrington, J.; McKeown, R.D. Basic Instrumentation for Hall A at Jefferson Lab. *Nucl. Instrum. Meth. A* **2004**, *522*, 294–346. [CrossRef]
44. Mecking, B.A.; Adams, G.; Ahmad, S.; Anciant, E.; Anghinolfi, M.; Asavapibhop, B.; Vlassov, A.V. The CEBAF Large Acceptance Spectrometer (CLAS). *Nucl. Instrum. Meth. A* **2003**, *503*, 513–553. [CrossRef]
45. Prok, Y.; Bosted, P.; Kvaltine, N.; Adhikari, K. Precision measurements of g_1 of the proton and the deuteron with 6 GeV electrons. *Phys. Rev. C* **2014**, *90*, 025212. [CrossRef]
46. Zheng, X.; Deur, A.; Kang, H.; Kuhn, S.E.; Ripani, M.; Zhang, J.; Zachariou, N. Measurement of the proton spin structure at long distances. *Nat. Phys.* **2021**, *17*, 736. [CrossRef]
47. Adhikari, K.P.; Deur, A.; El Fassi, L.; Kang, H.; Kuhn, S.E.; Ripani, M.; CLAS Collaboration. Measurement of the Q2-dependence of the deuteron spin structure function g1 and its moments at low Q2 with CLAS. *Phys. Rev. Lett.* **2018**, *120*, 062501. [CrossRef]
48. Sulkosky, V. The Spin Structure of ^3He and the Neutron at Low Q^2: A Measurement of the Generalized Gdh Integrand. Ph.D. Thesis, College of William & Mary, Williamsburg, VA, USA, 2007.
49. Keith, C.D.; Anghinolfi, M.; Battaglieri, M.; Bosted, P.; Branford, D.; Bültmann, S.; Witherspoon, S. A polarized target for the CLAS detector. *Nucl. Instrum. Meth. A* **2003**, *501*, 327–339. [CrossRef]
50. Garibaldi, F.; Acha, A.; Ambrozewicz, P.; Aniol, K.A.; Baturin, P.; Benaoum, H.; Jefferson Lab Hall A Collaboration. High-resolution hypernuclear spectroscopy at Jefferson Lab, Hall A. *Phys. Rev. C* **2019**, *99*, 054309. [CrossRef]

51. Kim, J.H.; Harris, D.A.; Arroyo, C.G.; de Barbaro, L.; de Barbaro, P.; Bazarko, A.O.; Bernstein, R.H.; Bodek, A.; Bolton, T.; Budd, H.; et al. A measurement of $\alpha_s(Q^2)$ from the Gross–Llewellyn Smith sum rule. *Phys. Rev. Lett.* **1998**, *81*, 3595. [CrossRef]
52. De Alfaro, V.; Fubini, S.; Furlan, G. Conformal invariance in quantum mechanics. *Nuovo Cim. A* **1976**, *34*, 569. [CrossRef]
53. Trawinski, A.P.; Glazek, S.D.; Brodsky, S.J.; de Téramond, G.F.; Dosch, H.G. Effective confining potentials for QCD. *Phys. Rev. D* **2014**, *90*, 074017. [CrossRef]
54. Cornwall, J.M. Dynamical Mass Generation in Continuum QCD. *Phys. Rev. D* **1982**, *26*, 1453. [CrossRef]
55. Abbott, L.F. The Background Field Method Beyond One Loop. *Nucl. Phys. B* **1981**, *185*, 189–203. [CrossRef]
56. Binosi, D.; Chang, L.; Papavassiliou, J.; Roberts, C.D. Bridging a gap between continuum-QCD and ab initio predictions of hadron observables. *Phys. Lett. B* **2015**, *742*, 183–188. [CrossRef]
57. Aguilar, A.C.; Binosi, D.; Papavassiliou, J. Gluon and ghost propagators in the Landau gauge: Deriving lattice results from Schwinger-Dyson equations. *Phys. Rev. D* **2008**, *78*, 025010. [CrossRef]
58. Brodsky, S.J.; Shrock, R. Maximum wavelength of confined quarks and gluons and properties of quantum chromodynamics. *Phys. Lett. B* **2008**, *666*, 95. [CrossRef]
59. Sufian, R.S.; Liu, T.; de Téramond, G.F.; Dosch, H.G.; Brodsky, S.J.; Deur, A.; Hlfhs Collaboration. Nonperturbative strange-quark sea from lattice QCD, light-front holography, and meson-baryon fluctuation models. *Phys. Rev. D* **2018**, *98*, 114004. [CrossRef]

Review

Emergence of Hadron Mass and Structure

Minghui Ding [1,*,†], Craig D. Roberts [2,3,*,†] and Sebastian M. Schmidt [1,4,*,†]

[1] Helmholtz-Zentrum Dresden-Rossendorf, Bautzner Landstraße 400, D-01328 Dresden, Germany
[2] School of Physics, Nanjing University, Nanjing 210093, China
[3] Institute for Nonperturbative Physics, Nanjing University, Nanjing 210093, China
[4] III. Physikalisches Institut B, RWTH Aachen University, D-52074 Aachen, Germany
* Correspondence: m.ding@hzdr.de (M.D.); cdroberts@nju.edu.cn (C.D.R.); s.schmidt@hzdr.de (S.M.S.)
† All authors contributed equally to this work.

Abstract: Visible matter is characterised by a single mass scale; namely, the proton mass. The proton's existence and structure are supposed to be described by quantum chromodynamics (QCD); yet, absent Higgs boson couplings, chromodynamics is scale-invariant. Thus, if the Standard Model is truly a part of the theory of Nature, then the proton mass is an emergent feature of QCD; and emergent hadron mass (EHM) must provide the basic link between theory and observation. Nonperturbative tools are necessary if such connections are to be made; and in this context, we sketch recent progress in the application of continuum Schwinger function methods to an array of related problems in hadron and particle physics. Special emphasis is given to the three pillars of EHM—namely, the running gluon mass, process-independent effective charge, and running quark mass; their role in stabilising QCD; and their measurable expressions in a diverse array of observables.

Keywords: confinement of gluons and quarks; continuum Schwinger function methods; Dyson–Schwinger equations; emergence of hadron mass; parton distribution functions; hadron form factors; hadron spectra; hadron structure and interactions; nonperturbative quantum field theory; quantum chromodynamics

Contents

1. Introduction 58
2. Hadron Mass Budgets 60
3. Gluons and the Emergence of Mass 62
4. Process-Independent Effective Charge 64
5. Confinement 67
6. Spectroscopy 72
7. Baryon Wave Functions 78
8. Meson Form Factors 82
9. Baryon Form Factors 85
10. Transition Form Factors of Heavy+Light Mesons 92
11. Distribution Functions 98
12. Conclusions 105
References 107

Citation: Ding, M.; Roberts, C.D.; Schmidt, S.M. Emergence of Hadron Mass and Structure. *Particles* **2023**, *6*, 57–120. https://doi.org/10.3390/particles6010004

Academic Editor: Armen Sedrakian

Received: 21 November 2022
Revised: 18 December 2022
Accepted: 19 December 2022
Published: 11 January 2023

Copyright: © 2023 by the authors. Licensee MDPI, Basel, Switzerland. This article is an open access article distributed under the terms and conditions of the Creative Commons Attribution (CC BY) license (https://creativecommons.org/licenses/by/4.0/).

1. Introduction

Our Universe exists; and even the small part that we occupy contains much which might be considered miraculous. Nevertheless, science typically assumes that the Universe's evolution can be explained by some collection of equations—even a single equation, perhaps, which replaces distinct theories of many things with a single theory of everything. Choosing not to approach that frontier, then, within the current paradigm, the Standard Model of particle physics (SM) is given a central role; and it must account for a huge array of observable phenomena. Herein, we focus on one especially important aspect, viz. the fact that the mass of the vast bulk of visible material in the Universe is explained as soon as one understands why the proton is absolutely stable and how it comes to possess a mass $m_p \approx 1$ GeV. In elucidating this connection, we will argue that the theory of strong interactions may deliver far more than was originally asked of it.

We have evidently supposed that quantum gauge field theory is the correct paradigm for understanding Nature. In this connection, it is important to note that, in our tangible Universe, time and space give us four noncompact dimensions. Consider, therefore, that quantum gauge field theories in $D \neq 4$ dimensions are characterised by an explicit, intrinsic mass scale: the basic couplings generated by minimal substitution are mass-dimensioned and set the scale for all calculated quantities. For $D > 4$, such theories manifest uncontrollable ultraviolet divergences, making them of little physical use. In contrast, for $D < 4$, they are super-convergent, but are afflicted with a hierarchy problem, viz. dynamical mass-generation effects are typically very small when compared with the theory's explicit scale [1–5]. Hence, perhaps unsurprisingly, $D = 4$ is a critical point. Removing Higgs boson couplings, the classical gauge theory elements of the SM are scale-invariant. Taking the step to quantum theories, they are all (at least perturbatively) renormalisable; and that procedure introduces a mass scale. As we have noted, the scale for visible matter is $m_{\text{Nature}} \approx m_p \approx 1$ GeV. However, the size of this scale is not determined by the theory; so, whence does it come? Further, how much tolerance does Nature give us? Is the Universe habitable when $m_{\text{Nature}} \to (1 \pm \delta) m_{\text{Nature}}$, with $\delta = 0.1$ or 0.2, etc.? It is comforting to imagine that our (ultimate?) theory of Nature will answer these questions, but the existence of such a theory is not certain.

Returning to concrete issues, strong interactions within the SM are described by quantum chromodynamics (QCD). Therefore, consider the classical Lagrangian density that serves as the starting point on the road to QCD:

$$\mathcal{L}_{\text{QCD}} = \sum_{f=u,d,s,\ldots} \bar{q}_f [\gamma \cdot \partial + ig \tfrac{1}{2} \lambda^a \gamma \cdot A^a + m_f] q_f + \tfrac{1}{4} G^a_{\mu\nu} G^a_{\mu\nu}, \quad (1a)$$

$$G^a_{\mu\nu} = \partial_\mu A^a_\nu + \partial_\nu A^a_\mu - g f^{abc} A^b_\mu A^c_\nu, \quad (1b)$$

where $\{q_f \,|\, f = u, d, s, c, b, t\}$ are fields associated with the six known flavours of quarks; $\{m_f\}$ are their current-masses, generated by the Higgs boson; $\{A^a_\mu \,|\, a = 1, \ldots, 8\}$ represent the gluon fields, whose matrix structure is encoded in $\{\tfrac{1}{2}\lambda^a\}$, the generators of SU(3) in the fundamental representation; and g is the *unique* QCD coupling, using which one conventionally defines $\alpha = g^2/[4\pi]$. As remarked above, if one removes Higgs boson couplings into QCD, so that $\{m_f \equiv 0\}$ in Equation (1), then the classical action associated with this Lagrangian is scale-invariant. A scale-invariant theory cannot produce compact bound states; indeed, scale-invariant theories do not support dynamics, only kinematics [6]. So, if Equation (1) is really capable of explaining, amongst other things, the proton's mass, size, and stability, then remarkable features must emerge via the process of defining quantum chromodynamics.

This point is placed in stark relief when one appreciates that the gluon and quark fields used to express the one-line Lagrangian of QCD are not the degrees-of-freedom measured in detectors. This is an empirical manifestation of confinement. Amongst other things, a solution of QCD will reveal the meaning of confinement, predict the observable states, and explain how they are built from the Lagrangian's gluon and quark partons. However, the search for a solution presumes that QCD is actually a theory. Effective theories are tools for use in obtaining a realistic description of phenomena perceived at a given scale. A true theory must be rigorously defined at all scales and unify phenomena perceived at vastly different energies. If QCD really is a well-defined quantum field theory, then it may serve as a paradigm for physics far beyond the SM.

Having raised this possibility, then it is appropriate to provide a working definition of "well-defined" in relation to quantum field theory. Aspects of the mathematical problem are discussed elsewhere [7,8]. Herein, we consider that a quantum (gauge) field theory is well-defined if its ultraviolet renormalisation can be accomplished with a finite number of renormalisation constants, $\{Z_j | j = 1, \ldots, N\}$, $N \lesssim 10$,[1] all of which can (*a*) be computed nonperturbatively and (*b*) remain bounded real numbers as any regularisation scale is removed. Further, that the renormalisation of ultraviolet divergences is sufficient to ensure that any/all infrared divergences are eliminated, i.e., the theory is infrared-complete.

Quantum electrodynamics (QED) is not well-defined owing to the existence of a Landau pole in the far ultraviolet (see, e.g., Reference [9] (Ch. 13) and References [10–13]). Furthermore, weak interactions are essentially perturbative because the inclusion of the Higgs scalar boson introduces an enormous infrared scale that suppresses all nonperturbative effects; moreover, the Higgs boson mass is quadratically divergent, making the theory non-renormalisable.

On the other hand, as we will explain herein, it is beginning to seem increasingly likely that QCD satisfies the tests listed above; hence, is the first well-defined quantum field theory that humanity has developed. QCD may thus stand alone as an internally consistent theory, so that after quantisation of Equation (1), with nothing further added, it is a genuinely predictive mathematical framework for the explanation of natural phenomena.

We have used a Euclidean metric and consistent Dirac matrices in writing Equation (1) because if there is any hope of arriving at a rigorous definition of QCD, then it is by formulating the theory in Euclidean space. There are many reasons for adopting this perspective. Amongst the most significant being the fact that a lattice regularisation of the theory is only possible in Euclidean space, where one can use the action associated with Equation (1) to define a probability measure [14] (Section 2.1). Notably, a choice must be made because any "Wick rotation" between Minkowski space and Euclidean space is a purely formal exercise, whose validity is only guaranteed for perturbative calculations [15,16]. If QCD really does (somehow) explain the emergence of hadron mass and structure, then nonlinear, nonperturbative dynamics must be crucial. Consequently, one cannot assume that any of the requirements necessary to mathematically justify a Wick rotation are satisfied when calculating and summing the necessarily infinite collection of processes associated with a given experimental observable.

One concrete example may serve to illustrate the point. Both continuum and lattice analyses of the gluon two-point Schwinger function (often called the Euclidean space gluon propagator) yield a result whose analytic properties are very different from those one would obtain in perturbation theory at any finite order [17]. As a consequence, the Minkowski space gluon gap equation that is obtained from the Euclidean form via the standard transcriptions used to implement the Wick rotation [16] (Section 2.3), whilst being similar in appearance, cannot possess the same solutions. Thus, to avoid confusion, one should begin with all such equations formulated in Euclidean space, where the solutions determined have a direct and unambiguous connection with results obtained using numerical simulations of the lattice-regularised theory. Anything else is an unnecessary and potentially misleading pretence. Furthermore, only those Schwinger functions corresponding to observable quantities need have a continuation to Minkowski space, and that can be accomplished following standard notions from constructive field theory ([15] (Sections 3 and 4), [16] (Section 2.3)).

---

[1] Here, the value "10" is arbitrary. More generally, the number should be small enough to ensure that predictive power is not lost through a need to fit too many renormalised observables to measured quantities.

We proceed then by supposing that QCD is defined by the Euclidean-space generating functional built using the Lagrangian density in Equation (1). Here, a new choice presents itself. One might attempt to solve the thus-quantised theory using a lattice regularisation [18,19]. Lattice-regularised QCD (lQCD) is a popular framework, which, owing to growth in computer power and algorithm improvements, is becoming more effective—see, e.g., Reference [20]. On the other hand, continuum Schwinger function methods (CSMs) are also available [14,16,21–25]. Much has been achieved using this approach, especially during the past decade [26–31] and particularly in connection with elucidating the origins and wide-ranging expressions of emergent hadron mass (EHM) [32–37]. It is upon those advances that we focus herein.

2. Hadron Mass Budgets

There is one generally recognised mass-generating mechanism in the SM; namely, that associated with Higgs boson couplings [38,39]. Insofar as QCD is concerned, there are six distinct such couplings, each of which generates the current–mass of a different quark flavour. Those current–quark masses exhibit a remarkable hierarchy of scales, ranging from an electron-like size for the u and d quarks up to a value five-orders-of-magnitude larger for the t quark ([40] (p. 32)). Faced with such discordance, we choose to begin our discussion of mass by considering the proton and its closest relatives, viz. the π- and ρ-mesons.

The proton is defined as the lightest state constituted from the valence quark combination $u + u + d$. π^+ is a pseudoscalar meson built from $u + \bar{d}$ valence quarks, and the ρ^+ is its kindred vector meson partner: in quark models, the π and ρ are identified as $^1S_0$ and $^3S_1$ states, respectively ([40] (Section 63)). Table 1 presents a breakdown of the masses of these states into three contributions: the simplest to count is that associated with the Higgs-generated current–masses of the valence quarks (HB); the least well understood is that part which has no connection with the Higgs boson (EHM); and the remainder is that arising from constructive interference between these two sources of mass (EHM+HB).

Table 1. Mass budgets of a collection of hadrons, with each panel ordered according to the contribution from Higgs boson couplings into QCD (HB) and including the component that is entirely unrelated to the Higgs (EHM) and that arising from constructive interference between these two mass sources (EHM+HB) (separation at $\zeta = 2\,GeV$, produced using information from References [35,40–43]).

Hadron (Mass/GeV)	HB	Mass Fraction (%) EHM+HB	EHM
$p\,(0.938)$	1	6	93
$\rho\,(0.775)$	1	2	97
$D^*\,(2.010)$	63	30	7
$B^*\,(5.325)$	78	21	1
$\pi\,(0.140)$	5	95	0
$K\,(0.494)$	20	80	0
$D\,(1.870)$	68	32	0
$B\,(5.279)$	79	21	0

The information listed in Rows 1, 2, 5, and 6 of Table 1 is represented pictorially in Figure 1: plainly, there are significant differences between the upper and lower panels. Regarding the proton and ρ-meson, the HB-alone component of their masses is just 1% in each case. Notwithstanding that, their masses are large and remain so even in the absence of Higgs boson couplings into QCD, i.e., in the chiral limit. This overwhelmingly dominant component is a manifestation of EHM in the SM. It produces roughly 95% of the measured mass. Evidently, baryons and vector mesons are similar in these respects.

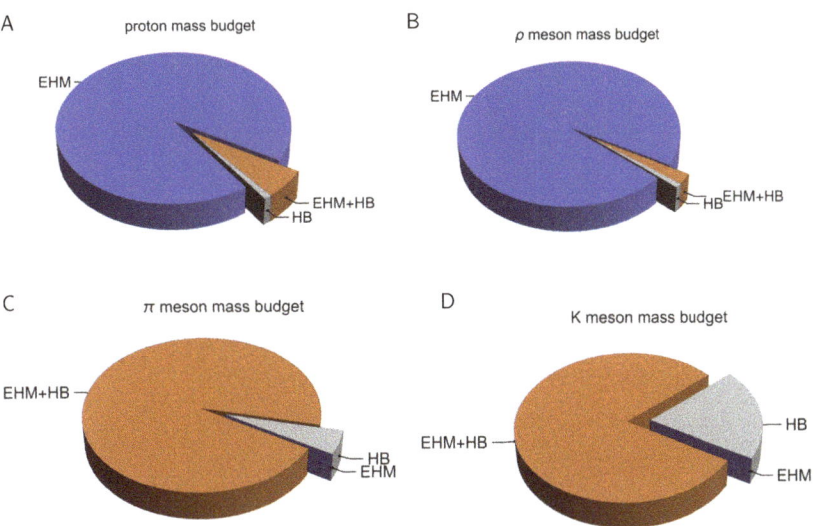

Figure 1. Mass budgets: (**A**) proton; (**B**) ρ-meson; (**C**) pion; (**D**) kaon. Each is drawn using a Poincaré-invariant decomposition and the numerical values listed in Table 1 (separation at $\zeta = 2\,GeV$, calculated using information from References [40–43]).

Conversely and yet still owing to EHM via its dynamical chiral symmetry breaking (DCSB) corollary, the pion is massless in the chiral limit—it is the SM's Nambu–Goldstone (NG) mode [27,44–52]. Returning to the quark model picture, the only difference between ρ- and π-mesons is a spin-flip: in the ρ, the constituent quark spins are aligned, whereas they are antialigned in π. Yet, their mass budgets are fundamentally different: Figure 1B; cf. Figure 1C. An inability to explain this difference is a conspicuous failure of quark models: whilst it is easy to obtain a satisfactory mass for the ρ, a low-mass pion can only be obtained by fine-tuning the quark model's potential. Nature, however, does not fine-tune the pion: in the absence of Higgs boson couplings, it is massless irrespective of the size of m_ρ and, in fact, the mass of any other hadron.

The kaon mass budget is also drawn—see Figure 1D. In the chiral limit, then, like the π, the K-meson is an NG boson. However, with realistic values of Higgs boson couplings into QCD, the s quark current–mass is approximately 27-times the average of the u and d current–masses [40]: $2m_s \approx 27(m_u + m_d)$. Consequently, the HB wedge in Figure 1D accounts for 20% of m_K. The remaining 80% is generated by constructive EHM+HB interference. It follows that comparisons between π and K properties present good opportunities for studying Higgs boson modulation of EHM, because the HB mass fraction is four-times larger in kaons than in pions. Moreover, the array of images in Figure 1 highlights that additional, complementary information can be obtained from comparisons between baryons/vector mesons and the set of kindred pseudoscalar mesons. For instance, studies of spectra (Section 6), transitions between vector mesons and pseudoscalar mesons (Section 10), and comparative analyses of proton and pion parton distribution functions (DFs) (see Section 11). In all cases, predominantly EHM systems, on the one hand, are contrasted/overlapped with final states that possess varying degrees of EHM+HB interference.

These observations highlight that EHM—whatever it is—can be accessed via experiment. The task for theory is to identify and explain its source, then elucidate a broad range of observable consequences so that the origins and explanations can be validated.

3. Gluons and the Emergence of Mass

The requirement of gauge invariance ensures that the Higgs boson does not couple to gluons and precludes any other means of generating an explicit mass term for the gluon fields in Equation (1). Consequently, it is widely believed that gluons are massless; and this is recorded by the Particle Data Group (PDG) [40] (p. 25). (We stress that gluon partons are massless).

In QCD, this "gauge invariance" statement is properly translated into a property of the two-point gluon Schwinger function. Namely, using the class of covariant gauges as an illustrative tool, characterised by a gauge-fixing parameter ξ, the inverse of the gluon two-point function can be expressed in terms of a gluon vacuum polarisation (or self-energy):

$$D_{\mu\nu}^{-1}(k) = \delta_{\mu\nu} k^2 - k_\mu k_\nu (1-\xi) + \Pi_{\mu\nu}(k) =: {}_0 D_{\mu\nu}^{-1}(k) + \Pi_{\mu\nu}(k) + \xi k_\mu k_\nu, \quad (2)$$

where k is the gluon momentum. (Regarding ξ, common choices in perturbation theory are $\xi = 0, 1$, viz. Landau and Feynman gauges, respectively.) Gauge invariance (BRST symmetry of the quantised theory [53] (Ch. II)) is expressed in the following Slavnov–Taylor identity [54,55]:

$$k_\mu \Pi_{\mu\nu}(k) = 0 = \Pi_{\mu\nu}(k) k_\nu. \quad (3)$$

This restrictive, yet generous, constraint states that interactions cannot affect the four-longitudinal component of the gluon two-point function, but leaves room for modifications of the propagation characteristics of the three four-transverse degrees-of-freedom.

Equation (3) means

$$\Pi_{\mu\nu}(k) = [\delta_{\mu\nu} k^2 - k_\mu k_\nu] \Pi(k^2) =: T_{\mu\nu}^k k^2 \Pi(k^2), \quad (4)$$

where $\Pi(k^2)$ is the dimensionless gluon self-energy; hence, the gauge invariance constraint entails

$$D_{\mu\nu}(k) = T_{\mu\nu}^k \frac{1}{k^2[1+\Pi(k^2)]} + \xi \frac{k_\mu k_\nu}{k^4} =: T_{\mu\nu}^k \overline{D}(k^2) + \xi \frac{k_\mu k_\nu}{k^4}. \quad (5)$$

This is the propagator of a massless vector boson, *unless*

$$\Pi(k^2) \stackrel{k^2 \simeq 0}{=} \frac{m_J^2}{k^2}, \quad (6)$$

in the event of which both the dressed gluon acquires a mass and all symmetry constraints are preserved. That Equation (6) is possible in an interacting quantum gauge field theory was first shown in a study of two-dimensional QED [56,57], and the phenomenon is now known as the Schwinger mechanism of gauge boson mass generation. Three-dimensional QED supports a similar outcome [3–5,58,59], as does $D = 3$ QCD [60,61]; but, as already noted above, there is a difference between both these examples and QCD. Namely, whereas the Lagrangian couplings in $D < 4$ theories carry a mass dimension, which explicitly breaks scale invariance, this is not the case for $D = 4$ chromodynamics.

The existence of a Schwinger mechanism in QCD was first conjectured forty years ago [62]. The idea has subsequently been explored and refined [63–67], so that, today, a detailed picture is emerging, which unifies both the gauge and matter sectors [68]. The dynamical origin of the QCD Schwinger mechanism and its intimate connection with non-perturbative dynamics in the three-gluon vertex are elucidated elsewhere [36,37]. This is an area of continuing research, where synergies between continuum and lattice QCD are being exploited [69,70]. For our purposes, it is sufficient to know that Equation (6) is realised in QCD. Indeed, owing to their self-interactions, gluon partons transmogrify into gluon quasi-particles, whose propagation characteristics are determined by a momentum-dependent mass function. That mass function is power-law suppressed in the ultraviolet—hence, invisible in perturbation theory, yet large at infrared momenta, being characterised by a renormalisation-point-independent value [71]:

$$m_0 = 0.43(1)\,\text{GeV}. \tag{7}$$

The renormalisation-group-invariant (RGI) gluon mass function is drawn in Figure 2.

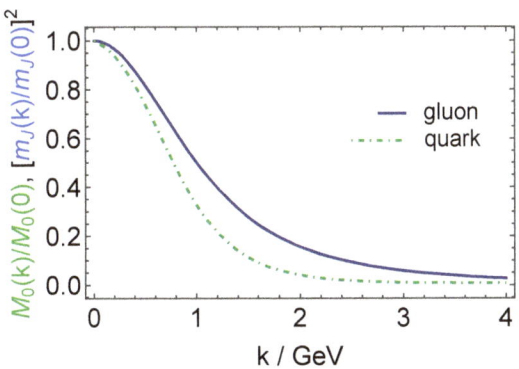

Figure 2. Renormalisation-group-invariant dressed gluon mass function (solid blue curve) calculated, following the method in Reference [72], from a gluon two-point function obtained using the lQCD configurations in References [73–75]. The mass-squared curve is plotted, normalised by its $k=0$ value, and compared with the kindred chiral-limit dressed quark mass function drawn from Reference [76] (dotted-dashed green curve). It is this pair of curves that is $1/k^2$-suppressed in the ultraviolet, each with additional logarithmic corrections.

Before closing this section, it is worth stressing the importance of Poincaré covariance in modern physics.[2] If one chooses to formulate a problem in quantum field theory using a scheme that does not ensure Poincaré invariance of physical quantities, then artificial or "pseudodynamical" effects are typically encountered [77]. In connection with gauge theory Schwinger functions, Poincaré covariance very effectively limits the nature and number of independent amplitudes that are required for a complete representation. In contrast, analyses and quantisation procedures that violate Poincaré covariance engender a rapid proliferation in the number of such functions. For instance, the covariant-gauge gluon two-point function in Equation (5) is fully specified by one scalar function, whereas, in the class of axial gauges, two unconnected functions are required, and unphysical, kinematic singularities appear in the associated tensors [78,79]. This is why covariant gauges are normally employed for concrete calculations in both continuum and lattice-regularised QCD. In fact, Landau gauge, i.e., $\xi = 0$ in Equation (5), is often used because, amongst other things, it is a fixed point of the renormalisation group [53] (Ch. IV) and implemented readily in lQCD [80]. We typically refer to Landau gauge results herein. Naturally, gauge covariance of Schwinger functions ensures that expressions of EHM in physical observables are independent of the gauge used for their elucidation.

Equation (7) is the cleanest expression of EHM in Nature, being truly a manifestation of mass emerging from *nothing*: infinitely many massless gluon partons fuse together so that, for all intents and purposes, they behave as coherent quasiparticle fields with a long-wavelength mass, which is almost half that of the proton. The implications of this result are enormous and far-reaching, including, e.g., key steps toward the elimination of

---

[2] When working with a Euclidean formulation, as we do, Poincaré covariance maps straightforwardly into Euclidean covariance, *viz.* valid Schwinger functions must transform covariantly under O(4) rotations and linear translations in $\mathbb{R}^4$. Owing to the simplicity of this connection, we avoid transliteration and speak of Poincaré covariance and invariance throughout.

the problem of Gribov ambiguities [81], which were long thought to prevent a rigorous definition of QCD.

4. Process-Independent Effective Charge

In classical field theories, couplings and masses are constants. Typically, this is also true in quantum mechanics models of strong interaction phenomena. However, this is not the case in renormalisable quantum gauge field theories, as highlighted by the Gell-Mann–Low effective charge/running coupling in QED [82], which is a textbook case [9] (Ch. 13.1).

A highlight of Twentieth Century physics was the realisation that QCD, in particular, and non-Abelian gauge theories, in general, express asymptotic freedom ([83–85], [86] (Ch. 7.1)), i.e., the feature that the interaction between charge carriers in the theory becomes weaker as k^2, the momentum-squared characterising the scattering process, becomes larger. Analysed perturbatively at one-loop order in the modified minimal subtraction renormalisation scheme, $\overline{MS}$, the QCD running coupling takes the form

$$\alpha_{\overline{MS}}(k^2) = \frac{\gamma_m \pi}{\ln k^2/\Lambda_{QCD}^2}, \tag{8}$$

where $\gamma_m = 12/[33 - 2n_f]$, with n_f the number of quark flavours whose mass does not exceed k^2, and $\Lambda_{QCD} \sim 0.2$ GeV is the RGI mass parameter that sets the scale for perturbative analyses.

Asymptotic freedom comes with a "flip side", which came to be known as *infrared slavery* ([87] (Section 3.1.2)). Namely, beginning with some $k^2 \gg \Lambda_{QCD}^2$, then the interaction strength grows as k^2 is reduced, with the coupling diverging at $k^2 = \Lambda_{QCD}^2$. (This is the Landau pole.) Qualitatively, this statement is true at any finite order in perturbation theory: whilst the value of Λ_{QCD} changes somewhat, the divergence remains. In concert with the area law demonstrated in Reference [18], which entails that the potential between any two infinitely massive colour sources grows linearly with their separation, many practitioners were persuaded that the complex dynamical phenomenon of confinement could simply be explained by an unbounded potential that grows with parton separation. As we shall see, that is not the case, but the notion is persistent.

Given the character of QCD's perturbative running coupling, two big questions arise:

(a) Does QCD possess a unique, nonperturbatively well-defined and calculable effective charge, viz. a veritable analogue of QED's Gell-Mann–Low running coupling; and

(b) Does Equation (8) express the large-k^2 behaviour of that charge?

If both questions can be answered in the affirmative, then great strides have been made toward verifying that QCD is truly a theory.

Following roughly forty years of two practically disjoint research efforts, one focused on QCD's gauge sector [66,67,88] and another on its matter sector [21,23–25], a key step on the path to answering these questions was taken in Reference [68]. The two distinct efforts were designated therein as the top-down approach—ab initio computation of the interaction via direct analyses of gauge sector gap equations; and the bottom-up scheme—inferring the interaction by describing data within a well-defined truncation of those matter sector equations that are relevant to bound state properties. Reference [68] showed that the top-down and bottom-up approaches are unified when the RGI running interaction predicted by then-contemporary analyses of QCD's gauge sector is used to explain ground state hadron observables using nonperturbatively improved truncations of the matter sector bound state equations. The first such truncation was introduced in Reference [89].

It was a short walk from this point to a realisation [90] that in QCD, by means of the pinch technique [65,91,92] and background field method [93], one can define and calculate a unique, process-independent (PI) and RGI analogue of the Gell-Mann–Low effective charge, now denoted $\hat{\alpha}(k^2)$. The analysis was refined in Reference [71], which combined modern results from continuum analyses of QCD's gauge sector and lQCD configurations generated with three domain-wall fermions at the physical pion mass [73–75] to obtain a parameter-free prediction of $\hat{\alpha}(k^2)$. The resulting charge is drawn in Figure 3. It is reliably interpolated by writing

$$\hat{\alpha}(k^2) = \frac{\gamma_m \pi}{\ln\left[\mathcal{K}^2(k^2)/\Lambda_{QCD}^2\right]}, \quad \mathcal{K}^2(y = k^2) = \frac{a_0^2 + a_1 y + y^2}{b_0 + y}, \quad (9)$$

with (in GeV^2): $a_0 = 0.104(1)$, $a_1 = 0.0975$, $b_0 = 0.121(1)$. The curve was obtained using a momentum subtraction renormalisation scheme: $\Lambda_{QCD} = 0.52$ GeV when $n_f = 4$.

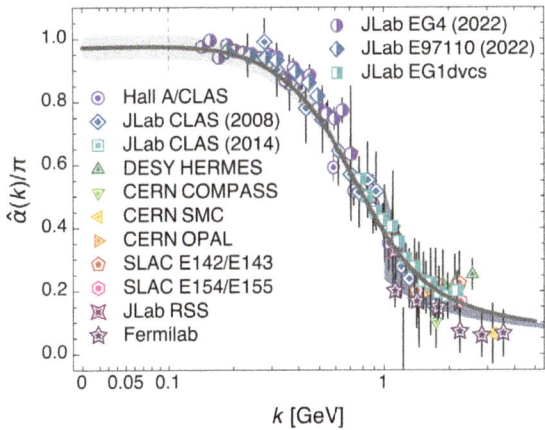

Figure 3. Process-independent effective charge, $\hat{\alpha}(k)/\pi$, obtained by combining modern results from continuum and lattice analyses of QCD's gauge sector [71]. Existing data on the process-dependent charge α_{g_1} [94,95], defined via the Bjorken sum rule, is shown for comparison—see References [95–121]. (Image courtesy of D. Binosi).

Notably, $\hat{\alpha}(k^2)$ is PI and RGI in any gauge; but, it is sufficient to know $\hat{\alpha}(k^2)$ in Landau gauge, $\xi = 0$ in Equation (5), which is the choice both for easiest calculation and the result in Equation (9). This is because $\hat{\alpha}(k^2)$ is form-invariant under gauge transformations, as may be shown using identities discussed elsewhere [122], and gauge covariance ensures that any such transformations can be absorbed into the Schwinger functions of the quasiparticles whose interactions are described by $\hat{\alpha}(k^2)$ [123].

The following physical features of $\hat{\alpha}$ deserve to be highlighted because they expose a great deal about QCD.

Absence of a Landau pole. Whereas the perturbative running coupling, e.g., Equation (8), diverges at $k^2 = \Lambda_{QCD}^2$, revealing the Landau pole, the PI charge is a smooth function on $k^2 \geq 0$: the Landau pole is eliminated owing to the appearance of a gluon mass scale, Equation (7).

Implicit in the "screening function", $\mathcal{K}(k^2)$, is a screening mass:

$$\zeta_\mathcal{H} = \mathcal{K}(k^2 = \Lambda_{QCD}^2) \approx 1.4\,\Lambda_{QCD} < m_p, \quad (10)$$

at which point the perturbative coupling would diverge but the PI coupling passes through an inflection point on its way to saturation. On $\sqrt{k^2} \lesssim \zeta_\mathcal{H}$, the PI charge enters a new domain, upon which the running slows, practically ceasing on $\sqrt{k^2} \leq m_0/2$, so that QCD is once again effectively a conformal theory and the charge saturates to a constant infrared value $\hat{\alpha}(k^2 = 0) = \pi \times 0.97(4)$. This value is a prediction: within 3(4)%, the coupling saturates to a value of π at $k^2 = 0$. It is not yet known whether this proximity to π has any deeper significance in Nature, but a potential explanation is provided in the next bullet.

These features emphasise the role of EHM as expressed in Equation (7): the existence of $m_0 \approx m_p/2$ guarantees that long-wavelength gluons are screened, so play no dynamical role. Consequently, ζ_H marks the boundary between soft/nonperturbative and hard/perturbative physics. It is therefore a natural choice for the "hadron scale", viz. the renormalisation scale at which valence quasiparticle degrees-of-freedom should be used to formulate and solve hadron bound state problems [71]. Implementing that notion, then those quasiparticles carry all hadron properties at $\zeta = \zeta_H$. This approach is today being used to good effect in the prediction of hadron parton distribution functions (DFs)—see Section 11 and References [124–136].

Match with the Bjorken process-dependent charge. The theory of process-*dependent* (PD) charges was introduced in References [137,138]: *"...to each physical quantity depending on a single scale variable is associated an effective charge, whose corresponding Stückelberg–Peterman–Gell-Mann–Low function is identified as the proper object on which perturbation theory applies."* PD charges have since been widely canvassed [94,139,140]. One of the most fascinating things about the PI running coupling is highlighted by its comparison with the data in Figure 3, which express measurements of the PD effective charge, $\alpha_{g_1}(k^2)$, defined via the Bjorken sum rule [141,142]. The charge calculated in Reference [71] is an essentially PI charge. There are no parameters; and, prima facie, no reason to expect that it should match $\alpha_{g_1}(k^2)$. The almost precise agreement is a discovery, given more weight by new results on $\alpha_{g_1}(k^2)$ [95], which now reach into the conformal window at infrared momenta.

Mathematically, at least part of the explanation lies in the fact that the Bjorken sum rule is an isospin non-singlet relation, which eliminates many dynamical contributions that might distinguish between the two charges. It is known that the two charges are not identical; yet, equally, on any domain for which perturbation theory is valid, the charges are nevertheless very much alike:

$$\frac{\alpha_{g_1}(k^2)}{\hat{\alpha}(k^2)} \stackrel{k^2 \gg m_0^2}{=} 1 + \frac{1}{20}\alpha_{\overline{\text{MS}}}(k^2), \qquad (11)$$

where $\alpha_{\overline{\text{MS}}}$ is given in Equation (8). At the c quark current–mass, the ratio is 1.007, i.e., indistinguishable from unity insofar as currently achievable precision is concerned. At the other extreme, in the far infrared, the Bjorken charge saturates to $\alpha_{g_1}(k^2 = 0) = \pi$; hence,

$$\frac{\alpha_{g_1}(k^2)}{\hat{\alpha}(k^2)} \stackrel{k^2 \ll m_0^2}{=} 1.03(4). \qquad (12)$$

Evidently, the PD charge determined from the Bjorken sum rule is, for practical intents and purposes, indistinguishable from the PI charge generated by QCD's gauge sector dynamics [71,90].

Infrared completion. Being process-independent, $\hat{\alpha}(k^2)$ serves numerous purposes and unifies many observables. It is therefore a good candidate for that long-sought running coupling which describes QCD's effective charge at all accessible momentum scales [139], from the deep infrared to the far ultraviolet, and at all scales in between, without any modification.

Significantly, the properties of $\hat{\alpha}(k^2)$ support the conclusion that QCD is actually a theory, viz. a well-defined $D = 4$ quantum gauge field theory. QCD therefore emerges as a viable tool for use in moving beyond the SM by giving substructure to particles that today seem elementary. A good example was suggested long ago; namely, perhaps all spin-$J = 0$ bosons may be [57] *"...secondary dynamical manifestations of strongly coupled primary fermion fields and vector gauge fields ..."*. Adopting this position, the SM's Higgs boson might also be composite, in which case, inter alia, the quadratic divergence of Higgs boson mass corrections would be eliminated.

Qualitatively equivalent remarks have been developed using light-front holographic models of QCD based on anti-de Sitter/conformal field theory (AdS/CFT) duality [143,144].

Returning to the two questions posed following Equation (8) in Items (a) and (b), it is now apparent that they are answerable in the affirmative: QCD does possess a unique, nonperturbatively well-defined and calculable effective charge whose large-k^2 behaviour connects smoothly with that in Equation (8). These facts provide strong support for the view that QCD is a well-defined 4D quantum gauge field theory.

5. Confinement

Confinement is much discussed, but little understood. In large part, both these things stem from the absence of a clear, agreed definition of confinement. With certainty, it is only known that nothing with quantum numbers matching those of the gluon or quark fields in Equation (1) has ever reached a detector.

An interpretation of confinement is included in the official description of the Yang–Mills Millennium Problem [145]. The simpler background statement is worth repeating:

"Quantum Yang–Mills theory is now the foundation of most of elementary particle theory, and its predictions have been tested at many experimental laboratories, but its mathematical foundation is still unclear. The successful use of Yang–Mills theory to describe the strong interactions of elementary particles depends on a subtle quantum mechanical property called the 'mass gap': the quantum particles have positive masses, even though the classical waves travel at the speed of light. This property has been discovered by physicists from experiment and confirmed by computer simulations, but it still has not been understood from a theoretical point of view. Progress in establishing the existence of the Yang–Mills theory and a mass gap will require the introduction of fundamental new ideas both in physics and in mathematics."

The formulation of this problem focuses entirely on quenched-QCD, i.e., QCD without quarks; so, its solution is not directly relevant to our Universe. Confinement in pure quantum SU(3) gauge theory and in QCD proper are probably very different because the pion exists and is unnaturally light on the hadron scale [27]. On the other hand, the remarks concerning the emergence of a "mass gap" relate directly to Figure 2 and Equation (7) herein. Whilst these properties of QCD may be considered proven by the canons of theoretical physics, such arguments do not meet the standards of mathematical physics and constructive field theory because they involve input from numerical analyses of QCD Schwinger functions. Hereafter, therefore, we will continue within the theoretical physics perspective.

As noted above, a mechanism for the total confinement of infinitely massive charge sources has been identified in the lattice-regularised treatment of quantum field theories using compact representations of Abelian or non-Abelian gauge fields [18], viz. the area law ≡ linear source–antisource potential. However, no treatment of the continuum meson bound state problem has yet been able to demonstrate how such an area law emerges as the masses of the meson's valence degrees-of-freedom grow to infinity.

In the era of infrared slavery, it was widely assumed that some sort of nonperturbatively improved one-gluon exchange could simultaneously produce asymptotic freedom and a linearly rising potential between quarks; and many models were developed with just such features [146–150]. However, as highlighted by our discussion of QCD's effective charge, ongoing developments in the study of mesons, using rigorous treatments of the Schwinger functions involved, do not support this picture of confinement via dressed one-gluon exchange. The path to an area law is far more complex.

One direction that deserves exploration is connected with the gluon "H-diagrams" drawn in Reference [151] (Figure 8) and reproduced in Figure 4A. Imagine a valence quark and antiquark scattering via such a process, as drawn in Figure 4B; then keep adding H-diagrams within H-diagrams, exploiting both gluon–quark and gluon self-couplings. Such H-diagram scattering processes produce an infrared divergence in the perturbative computation of a static quark potential [152], viz. a contribution that exhibits unbounded growth as the source–antisource separation increases. Nonperturbatively, that divergence is tamed because the effective charge saturates—Figure 3. On the other hand, there are infinitely many such contributions; and in the limit of static valence degrees-of-freedom, the entire unbounded sum of planar H-diagrams is contracted to a point connection of infinitely dense fisherman's net/spider's web diagrams on both the source and antisource. It is conceivable that the confluence of these effects could yield the long-sought area law via the Bethe–Salpeter equation [151,153].

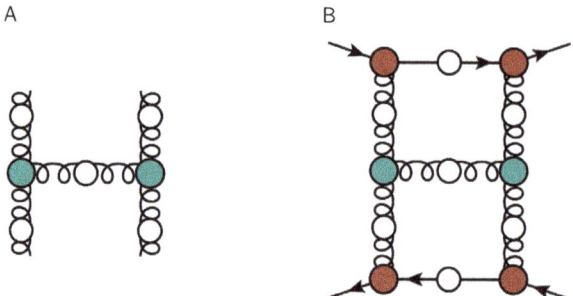

Figure 4. Panel (**A**): primitive gluon H-diagram. Panel (**B**): one-H-diagram contribution to quark+antiquark scattering. Legend: Dressed gluon two-point function—spring with open circle insertion; dressed quark two-point function—straight line with open circle insertion; blue-filled circle at 3-gluon junction—dressed 3-gluon vertex; red-filled circle at gluon–quark junction—dressed gluon–quark vertex. Repeated insertion of H-diagrams within H-diagrams, exploiting both gluon–quark and gluon self-couplings, leads to a plaquette-like area-filling structure, reminiscent of the planar summation of elementary squares in Reference [18].

Real-world QCD, however, is characterised by light degrees-of-freedom: u and d quarks with electron-size current–masses; s quarks with a mass roughly one order-of-magnitude larger, so still much less than m_p. Pions and kaons are constituted from such valence degrees-of-freedom, and these mesons are light. In fact, the pion has a lepton-like mass [40]: $m_\pi \approx m_\mu$, where m_μ is the mass of the μ lepton. Owing to the presence of such degrees-of-freedom, light particle annihilation and creation effects are essentially nonperturbative in QCD. Consequently, despite continuing dedicated efforts [154–156], it has thus far proven impossible to either define or calculate a static quantum mechanical potential between two light quarks.

This may be illustrated by apprehending that a potential which increases with separation can be described by a flux tube extending between the source and antisource. As the source–antisource separation increases, so does the potential energy stored in the flux tube. However, it can only increase until the stored energy matches that required to produce a particle+antiparticle pair of the theory's lightest asymptotic states—in QCD, a $\pi^+\pi^-$ pair. Numerical simulations of lQCD reveal [157,158] that once the energy exceeds this critical value, the flux tube then dissolves along its entire length, leaving two isolated colour singlet systems. Given that $m_\pi = 0.14\,\text{GeV}$, then this disintegration must occur at source+antisource centre-of-mass separation $r \approx (1/3)\,\text{fm}$ [159], which is well within the interior of any hadron. This example assumes that the source and antisource are static. The situation is even more complex for real, dynamical quarks. Thus, at least in the u, d, s quark sector, confinement is manifested in features of Schwinger functions that are far more subtle than can be captured in typical potential models.

One non-static, i.e., dynamical, picture of confinement has emerged from studies of the analytic properties of the two-point Schwinger functions associated with the propagation of coloured gluon and quark quasiparticles—see, e.g., Figure 2. The development of this perspective may be traced back to its beginning almost forty years ago [15,160–164]. It has subsequently been carefully explored [17,22,33,51,81,163,165–170]; and in this connection, one may profitably observe that only Schwinger functions which satisfy the axiom of reflection positivity [7,171,172] can be connected with states that appear in the Hilbert space of observables.

The axioms referred to here are those first presented in References [171,172] and subsequently modified in [7], which identify the properties of Schwinger functions that are necessary and sufficient to ensure equivalence between the formulation of a given quantum field theory in Euclidean and Minkowski space. (A contemporary literature compilation is presented elsewhere [173].) In effect, this means that all and only those Schwinger functions which satisfy the five Osterwalder–Schrader axioms possess connections with elements in the Hilbert space of physical states. Regarding strong interactions, all physical states are colour singlets. Consequently, for QCD to be the theory of strong interactions, all its colour singlet Schwinger functions must satisfy the Osterwalder–Schrader axioms; equally, all its colour-nonsinglet functions must violate at least one.

Reflection positivity is a severe constraint. It requires that the Fourier transform of the momentum space Schwinger function, treated as a function of analytic, Poincaré-invariant arguments, is a positive-definite function. To illustrate, consider the gluon Schwinger function in Equation (5). A massless partonic gluon is described by $\overline{D}(k^2) = 1/k^2$; and the 4D Fourier transform of this function is

$$\int \frac{d^4k}{(2\pi)^4} e^{ik\cdot x} \frac{1}{k^2} = \frac{1}{4\pi^2 x^2} > 0 \,\forall x^2 > 0. \tag{13}$$

More generally, regarding two-point functions, viz. those connected with the propagation of elementary excitations in QCD, reflection positivity is satisfied if, and only if, the Schwinger function has a Källén–Lehmann representation. Returning to the gluon Schwinger function in Equation (5), this means one must be able to write

$$\overline{D}(k^2) = \int_0^\infty d\varsigma \frac{\rho(\varsigma)}{k^2 + \varsigma^2}, \quad \rho(\varsigma) > 0 \,\forall \varsigma > 0. \tag{14}$$

Plainly, $\rho(\varsigma) = \delta(\varsigma)$ yields $\overline{D}(k^2) = 1/k^2$, i.e., the two-point function for a bare gluon parton. Hence, according to Equations (13) and (14), absent dressing, the gluon parton could appear in the Hilbert space of physical states.

It is important to observe that any function which satisfies Equation (14) is positive-definite itself. Moreover, given Equation (14),

$$sgn\left([\frac{d}{dk^2}]^n \overline{D}(k^2)\right) = (-1)^n; \tag{15}$$

consequently, inter alia, treated as a function of the analytic, Poincaré-invariant variable k^2, no function with a Källén–Lehmann representation of the form written in Equation (14) can possess an inflection point. Conversely, any function that exhibits an inflection point or, more generally, has a second derivative that changes sign must violate the axiom of reflection positivity [22]; hence, the associated excitation cannot appear in the Hilbert space of observables.

Take another step, and consider the following configuration space Schwinger function ($\tau = x_4$, $\ell = k_4$):

$$\Delta(\tau) = \int d^3x \int \frac{d^4k}{(2\pi)^4} e^{ik\cdot x} \overline{D}(k^2) = \frac{1}{\pi} \int_0^\infty d\ell \cos(\ell\tau) \overline{D}(\ell^2). \tag{16}$$

Suppose that interactions generate a constant mass for the gluon parton, so that $\overline{D}(k^2) = 1/(k^2 + \mu^2)$. Does that trigger confinement? The answer is "no" because this Schwinger function has a spectral representation with $\rho(\varsigma) = \delta(\varsigma^2 - \mu^2)$; Equation (15) is satisfied; and so is positivity:

$$\Delta(\tau) = \frac{1}{2\mu} e^{-\mu\tau}. \tag{17}$$

Suppose instead that interactions produce a momentum-dependent mass-squared function like that in Figure 2, which is $1/k^2$ suppressed in the ultraviolet:

$$m_J^2(k^2) = \frac{\mu_0^4}{k^2 + \mu_0^2} \Rightarrow \overline{D}(k^2) = \frac{k^2 + \mu_0^2}{k^2(k^2 + \mu_0^2) + \mu_0^4}. \tag{18}$$

The mass function itself is a monotonically decreasing, concave-up function; yet, in this case, the Schwinger function has an inflection point at $k^2 = 0.53\mu_0^2$. Hence, it does not have a Källén–Lehmann representation; so, the associated excitation cannot appear in the Hilbert space of observables. Furthermore, evaluation of the configuration space Schwinger function defined by Equation (16) yields [81]

$$\Delta(\tau) = \frac{1}{\mu_0} e^{-\tau\mu_0 \frac{\sqrt{3}}{2}} \cos\frac{\mu_0\tau}{2} =: \Delta_p(\tau)\cos\frac{\mu_0\tau}{2}, \tag{19}$$

using which the curve in Figure 5 is drawn: plainly, the configuration space Schwinger function violates reflection positivity. (Notably, the algebraic calculation of $\Delta(\tau)$ is often difficult and not always possible; so, uniform positivity of the second-derivative, Equation (15), is a much quicker means of testing for reflection positivity. Nevertheless, when it can be obtained, an explicit form of $\Delta(\tau)$ does provide additional insights.)

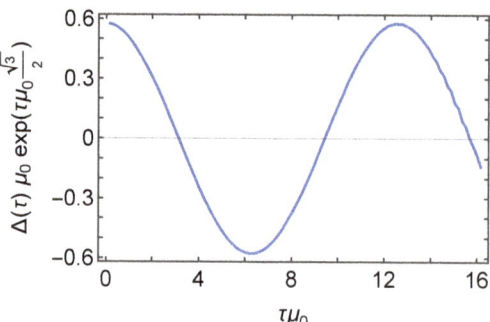

Figure 5. $\Delta(\tau)\mu_0 \exp(\mu_0\tau \frac{\sqrt{3}}{2})$ computed from the Schwinger function in Equation (18) using Equation (16).

It is interesting to extend Equation (18) using $m_J^2(k^2) = \alpha\mu_0^4/[k^2 + \mu_0^2]$, in which case

$$\overline{D}(k^2 = \mu_0^2 y) = \frac{1}{\mu_0^2} \frac{1+y}{y^2 + y + \alpha} =: \frac{1}{\mu_0^2} d(y). \tag{20}$$

This type of Schwinger function lies within the so-called "refined Gribov–Zwanziger" class [166]. For $\alpha > \frac{1}{2}$, as in the example above, the function in Equation (20) exhibits an inflection point at some $y > 0$; and when $\alpha \in (\frac{1}{4}, \frac{1}{2})$, the inflection point is found at a location $y \in (-\frac{1}{2}, 0)$. With $\alpha \in (0, \frac{1}{4})$, on the other hand, the function in Equation (20) separates into a sum of two terms:

$$d(y) = \frac{n_1}{y + p_1} - \frac{n_2}{y + p_2}, \tag{21}$$

where $p_{1,2} \in \mathbb{R}$ and $sgn(p_1/p_2) = +1$. In this case, there is no inflection point; nevertheless, the second derivative does change sign, switching $\pm\infty \leftrightarrow \mp\infty$ as y passes through the pole locations. Consequently, any excitation whose propagation is described by a Schwinger function obtained with $\alpha > 0$ in Equation (20) cannot appear in the Hilbert space of observable states.

Inserting Equation (20) into Equation (16), one finds, after some careful algebra [81] ($a = \alpha^{1/4}$):

$$\Delta(\tau) = \frac{1}{2\mu_0 a s_{\varphi/2}} e^{-\tau\mu_0 a c_{\varphi/2}}$$
$$\times \left[(1 + \tfrac{1}{a^2})s_{\varphi/2}\cos(\tau\mu_0 a s_{\varphi/2}) - (1 - \tfrac{1}{a^2})c_{\varphi/2}\sin(\tau\mu_0 a s_{\varphi/2})\right], \quad (22)$$

where $c_{\varphi/2} = [\tfrac{1}{2} + \tfrac{1}{4a^2}]^{1/2}$, $s_{\varphi/2} = [1 - c_{\varphi/2}^2]^{1/2}$. (Equation (19) is a special case of this result.) Equation (22) reveals that the distance from $\tau = 0$ of the first zero in the configuration space Schwinger function, τ_z, increases with decreasing α, i.e., as the infrared value of the gluon mass-squared function is reduced. Thus, confinement would practically be lost if τ_z were to become much greater than π/μ_0. Considering realistic gluon two-point functions, however, one finds $\mu_0 \approx (3/4)m_p$, $\alpha \approx 1$ [81]; so, $\tau_z \approx 1$ fm, expressing a natural confinement length scale. (It is worth observing that if τ_z were instead measured on the Å scale, then the notion of confinement would be lost because modern detectors are able to directly image targets of this size [174].)

This discussion is readily summarised. Owing to complex nonlinear dynamics in QCD, gluon and quark partons acquire momentum-dependent mass functions, as a consequence of which they emerge as quasiparticles whose propagation characteristics are described by two-point Schwinger functions that are incompatible with reflection positivity. Normally, the dynamical generation of running masses is alone sufficient to ensure this outcome. It follows that the dressed gluons and quarks cannot appear in the Hilbert space of physical states. In this sense, they are confined. The associated confinement length scale is $\tau_z \approx 1$ fm. It is worth stressing that the use of such two-point functions in the calculation of colour singlet matrix elements ensures the absence of coloured particle+antiparticle production thresholds [175], thereby securing the empirical expression of real-QCD confinement.

Considering these quasiparticle Schwinger functions further, one may also define a parton *persistence* or *fragmentation* length, τ_F, as the scale whereat the deviation of the Schwinger function from parton-like behaviour is 50%: $\Delta(\tau_F)/\Delta_p(\tau_F) = 0.5$. Referring to Equation (19), one reads $\tau_F = (2/3)\tau_z$. This result is also found using realistic gluon two-point functions [81]. (The value of 50% is merely a reasonable choice. At this level, 30% would also be acceptable, in which case $\tau_F \to \tau'_F = (1/2)\tau_z$.)

A physical picture of dynamical confinement now becomes apparent [165]. Namely, once a gluon or quark parton is produced, it begins to propagate in spacetime. However, after traversing a spacetime distance characterised by τ_F, interactions occur, causing the parton to lose its identity, sharing it with others. Ultimately, combining the effects on this parton with similar impacts on those produced along with it, a countable infinity of partons (a parton cloud) is produced, from which detectable colour singlet final states coalesce. This train of events is the physics expressed in parton fragmentation functions (FFs) [176]. Such distributions describe how the QCD partons in Equation (1), generated in a high-energy event and almost massless in perturbation theory, transform into a shower of massive hadrons, viz. they describe how hadrons with mass emerge from practically massless partons. It is natural, therefore, to view FFs as the cleanest expression of dynamical confinement in QCD. Furthermore, in the neighbourhood of their common boundary of support, DFs and FFs are related by crossing symmetry [177]: FFs are timelike analogues of DFs. Hence, an understanding of FFs and their deep connection with DFs can deliver fundamental insights into EHM. This picture of parton propagation, hadronisation, and confinement—of DFs and FFs—can be tested in experiments at modern and planned facilities [178–186]. A pressing demand on theory is delivery of predictions for FFs before such experiments are completed so as, for instance, to guide development of facilities and detectors. As yet, however, there are no realistic computations of FFs. In fact, even a formulation of this problem remains uncertain.

Before moving on, it is worth reiterating that confinement means different things to different people. Whilst some see confinement only in an area law for Wilson loops [18], our perspective stresses a dynamical picture, in which dynamically driven changes in the analytic structure of coloured Schwinger functions ensures the absence of colour-carrying objects from the Hilbert space of observable states. In time, perhaps, as strong QCD is better understood, it may be found that these two realisations are connected. The only certain thing is the necessity to keep an open mind on this subject.

6. Spectroscopy

Insofar as the spectrum of hadrons is concerned, results from nonrelativistic or somewhat relativised quark models [187–189] are still often cited as benchmarks. Indeed, a standard reference ([40] (Section 63)) includes the following assertions: *"The spectrum of baryons and mesons exhibits a high degree of regularity. The organizational principle which best categorizes this regularity is encoded in the quark model. All descriptions of strongly interacting states use the language of the quark model."* This is despite the facts that neither the "quarks" nor the potentials in quark models have been shown to possess any mathematical link with Equation (1)—rigorous or otherwise; and, furthermore, the orbital angular momentum and spin used to label quark model states are not Poincaré-invariant (observable) quantum numbers.

In step with improvements in computer performance, lQCD is delivering interesting results for hadron spectra [190,191], amongst which one may highlight indications for the existence of hybrid and exotic hadrons [192–195]. Continuum studies in quantum field theory are lagging behind owing in part to impediments placed by the character of the Bethe–Salpeter equation; primarily the fact that it is impossible to write the complete Bethe–Salpeter kernel in a closed form.

A systematic approach to truncating the integral equations associated with bound state problems in QCD was introduced almost thirty years ago [196,197]. Amongst other things, the scheme highlighted the importance of preserving continuous and discrete symmetries when formulating bound state problems; enabled proof of Goldberger–Treiman identities and the Gell-Mann–Oakes–Renner relation in QCD [198,199]; and opened the door to symmetry-preserving, Poincaré-invariant predictions of hadron observables, including elastic and transition form factors and DFs [21,23–25,28,125,200–207]. Some of the more recent developments are sketched below.

An issue connected with the leading-order (rainbow ladder (RL)) term in the truncation scheme of References [196,197] is that it only serves well for those ground state hadrons which possess little rest-frame orbital angular momentum, L, between the dressed valence constituents [208–218]. This limitation can be traced to its inability to realistically express the impacts of EHM on hadron observables, a weakness that is not overcome at any finite order of elaboration [210]. Improved schemes, which express EHM in the kernels, have been identified [68,89,151,219–221]. They have shown promise in applications to ground state mesons constituted from u and d valence quarks and/or antiquarks. However, that is a small subset of the hadron spectrum; so, a recent extension to the spectrum and decay constants of u, d, and s meson ground and first-excited states is welcome [222].

Returning to quark models, it was long ago claimed [149] *"... that all mesons—from the pion to the upsilon—can be described in a unified framework."* The context for this assertion was a model potential built using one-gluon-like exchange combined with an infrared slavery "confinement" term that increases linearly with colour source separation. The basic mass scales in such potential models are set by the constituent quark masses; and one might draw a qualitative link between those scales and the far-infrared values of the momentum-dependent dressed quark running masses ([35] (Figure 2.5)): $M_{u,d}(0) \simeq 0.41$ GeV, $M_s(0) \simeq 0.53$ GeV. Thereafter, mass splittings and level orderings are arranged by tuning details of the potential. Such a procedure can be quantitatively efficacious; however, it is qualitatively incorrect. This is readily seen by recalling the Gell-Mann–Oakes–Renner relation [46,198,223]: $m_\pi^2 \propto \hat{m}$, where $\hat{m}$ is Nature's explicit source of chiral symmetry breaking, generated by Higgs boson couplings to quarks in the SM. Such behaviour is impossible in a potential model [25,27], but natural in the CSM treatment of bound states—see, e.g., References [198,208], [21] (Figure 3.3), [224] (Figure 7), [89] (Figure 1A). Thus, whilst potential models might deliver a fit to hadron spectra, they do not provide an explanation.

That such challenges are surmounted when using CSMs to treat hadron bound state problems is further exemplified in Reference [222], which adapts a novel scheme for including EHM effects in the Bethe–Salpeter kernel [221] to simultaneously treat ground and first-excited states of u, d, and s quarks. As revealed in Figure 6, the empirical spectrum displays some curious features, e.g.: consistent with quark mass–scale counting, $m_\rho < m_{K^*}$, but this ordering is reversed for the first excitations of these states; the first excited state of the π is lighter than that of the ρ, but the ordering is switched for K, K^*; and all axial vector mesons are nearly degenerate, with the larger mass of the s quarks appearing to have little or no impact. In delivering the first symmetry-preserving analysis of this collection of states to employ an EHM-improved kernel, Reference [222] supplies fresh insights into the dynamical foundations of the properties of lighter quark mesons.

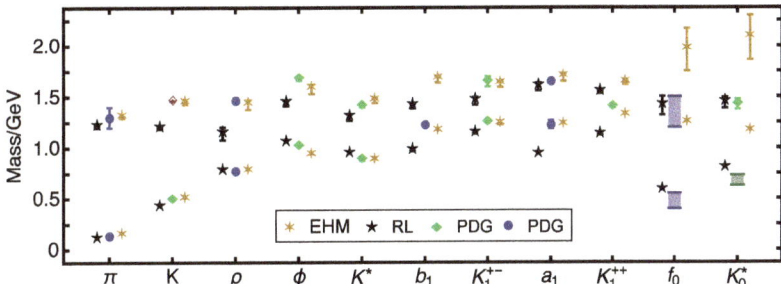

Figure 6. Empirical spectrum [40] (PDG summary tables): blue circles (bars)—u, d states; green diamonds (bar)—systems with s and/or $\bar{s}$ quarks. Little is known about $K(1460)$, which is therefore drawn as an open red diamond. Gold six-pointed stars—spectrum of low-lying u, d, and s mesons predicted by the EHM-improved Bethe–Salpeter kernel developed in Reference [222]; black five-pointed stars—same spectrum computed using RL truncation.

In order to sketch that effort, we note that, using CSMs, the dressed propagator (two-point Schwinger function) for a quark with flavour g is obtained as the solution of the following gap equation:

$$S_g^{-1}(k) = i\gamma \cdot k\, A_g(k^2) + B_g(k^2) = [i\gamma \cdot k + M_g(k^2)]/Z_g(k^2)\,, \qquad (23a)$$

$$= Z_2\, (i\gamma \cdot k + m_g) + \Sigma_g(k)\,, \qquad (23b)$$

$$\Sigma(k) = Z_1 \int_{dq}^{\Lambda} g^2 D_{\mu\nu}(k-q) \frac{\lambda^a}{2} \gamma_\mu S_g(q) \frac{\lambda^a}{2} \Gamma_\nu^g(k,p)\,, \qquad (23c)$$

where $M_g(k^2)$ is RGI;

$$D_{\mu\nu}(k) = \Delta(k^2)\, _0D_{\mu\nu}(k)\,; \qquad (24)$$

Γ_ν^g is the quark-gluon vertex; $\int_{dq}^{\Lambda}$ denotes a Poincaré-invariant regularisation of the four-dimensional integral, with Λ the regularisation mass scale; and $Z_{1,2}(\zeta^2, \Lambda^2)$ are, respectively, the vertex and quark wave function renormalisation constants, with ζ the renormalisation point. (In such applications, when renormalisation is necessary, a mass-independent scheme is important, as discussed elsewhere [225].)

It was anticipated almost forty years ago [226,227], and confirmed more recently [76,228–230], that EHM engenders a large anomalous chromomagnetic moment (ACM) for the lighter quarks; and with development of the first EHM-improved kernels, it was shown that such an ACM has a big impact on the u, d meson spectrum [219–221].

The aim in Reference [222] was to extend References [219–221] and highlight additional impacts of an ACM on the spectrum of mesons constituted from u, d, and s quarks. An ACM emerges as a feature of the dressed gluon–quark vertex, a three-point Schwinger function. Its character and impacts can be exposed by writing ($l = k - q$)

$$\Gamma^g_\nu(q,k) = \gamma_\nu + \tau_\nu(l), \quad \tau_\nu(l) = \eta \kappa(l^2) \sigma_{l\nu}, \qquad (25)$$

$\sigma_{l\nu} = \sigma_{\rho\nu} l_\rho$, $\kappa(l^2) = (1/\omega) \exp(-l^2/\omega^2)$. Here, $\eta > 0$ is the strength of the ACM term, $\tau_\nu(l)$; and it is assumed that the vertex is flavour-independent, which is a sound approximation for the lighter quarks [224,231]. When considering Equation (25), one might remark that the complete gluon–quark vertex is far more complicated—potentially containing twelve distinct terms—and, in QCD, $\kappa(l^2)$ is power-law suppressed in the ultraviolet. Notwithstanding these things, illustrative purposes are well served by Equation (25).

ACM effects are most immediately felt by the dressed quark propagator. The presence of an ACM in the kernel of Equation (23) increases positive EHM-induced feedback on dynamical mass generation. Consequently, as shown elsewhere [76], realistic values of the dressed quark mass at infrared momenta are achieved using the PI effective charge in Figure 3. Such an outcome requires tuning when using the PI charge in a rainbow truncation of the gap equation; in fact, DCSB cannot be guaranteed in that case [68].

Following References [89,221,228] in continuing to emphasise clarity over numerical complexity, Reference [222] also simplified the kernel in Equation (23), writing

$$g^2 \Delta(k^2) = 4\pi \hat{\alpha}(0) \frac{D(\eta)}{\omega^4} e^{-k^2/\omega^2}, \qquad (26)$$

where $\omega = 0.8\,\text{GeV}$, a value matching that suggested by analyses of QCD's gauge sector [68,71], and $D(\eta) = D_{\text{RL}}(1 + 0.27\eta)/(1 + 1.47\eta)$, with $\omega D_{\text{RL}} = (1.286\,\text{GeV})^3$ chosen to achieve $m_\rho = 0.77\,\text{GeV}$ in RL truncation. The η-dependence of $D(\eta)$ was fixed a posteriori by requiring that m_ρ remain unchanged as η is increased. Since $\eta > 0$ adds EHM strength to the gap equation's kernel, then D must become smaller as η grows in order to maintain a fixed value of m_ρ. Following this procedure, m_ρ becomes the benchmark against which all ACM-induced changes are measured.

It is worth noting that when one identifies $(g^2/[4\pi])\Delta(k^2 = 0) = 1/\mu_0^2$, then $\mu_0 = 0.39\,\text{GeV}$ in RL truncation and $\mu_0 = 0.57\,\text{GeV}$ at $\eta = 1.2$. Therefore, the interaction specified by Equation (26) is consistent with gluon mass generation, as it is described in Section 3. On the other hand, the large-k^2 behaviour of Equation (26) does not respect the renormalisation group flow of QCD. This would be an issue if one were using it, e.g., to calculate hadron form factors at large-Q^2 [200,201,203,204], where Q^2 is momentum transfer squared, or parton distribution functions and amplitudes near the endpoints of their support domains [124–127].[3] However, it is far less important when calculating global, integrated properties, like hadron masses. In such applications, Equation (26) is satisfactory. Indeed, good results can even be obtained using a symmetry-preserving treatment of a momentum-independent interaction [232–234]. A key merit of Equation (26) lies in its elimination of the need for renormalisation, which simplifies analyses without materially affecting relevant results.

---

[3] This is well known and explains why the truncated interaction in Equation (26) was not used for any of the calculations described in Sections 7–11 below. All those studies are based on interactions that at least preserve QCD's ultraviolet power-law behaviour, where more has not yet been achieved—Sections 7 and 9—and also the one-loop logarithmic improvement, when the necessary algorithms are already available—Sections 8, 10, and 11.

The scheme introduced in Reference [221] provides a direct route from any reasonable set of gap equation elements to closed-form Bethe–Salpeter kernels for meson bound state problems. Thus having specified physically reasonable gap equations via Equations (25) and (26), Reference [222] adapted that scheme to arrive at Bethe–Salpeter equations for each of the mesons identified in Figure 6, obtaining solutions in all cases. The image compares the experiment with the RL truncation results, also calculated in Reference [222], and predictions obtained using the EHM-improved kernel.[4]

It is first worth mentioning the RL truncation mass predictions in Figure 6. On the whole, the mean absolute relative difference, $\overline{\text{ard}}$, between RL results and central experimental values is 13(8)%. This is tolerable. However, there is substantial scatter and there are many qualitative discrepancies.

In contrast, compared with central experimental values, the EHM-improved masses in Figure 6 agree at the level of $\overline{\text{ard}} = 2.9(2.7)\%$. This is a factor of 4.6 of improvement over the RL spectrum. Moreover, correcting RL truncation flaws and reproducing empirical results: $m_{K'} > m_{\pi'}, m_{\rho'} > m_{\pi'}, m_{\rho'} \approx m_{K^{*\prime}}$; the mass splittings a_1-ρ and b_1-ρ match the empirical values because including the ACM in the kernel has markedly increased the masses of the a_1 and b_1 mesons, whilst m_ρ was deliberately kept unchanged; $m_{\phi'} - m_\phi$ agrees with experiment to within 2%; the K_1^{+-}, K_1^{++}-level order is correct; and quark+antiquark scalar mesons are heavy, providing room for the addition of strongly attractive resonant contributions to the bound state kernels [235,236].

Using the Bethe–Salpeter amplitudes obtained in solving for the meson spectrum, canonically normalised in the standard fashion ([237] (Section 3)), Reference [222] also delivered predictions for the entire array of associated leptonic decay constants, f_H, including many that have not yet been measured. The predicted values are depicted in Figure 7, which also includes the few available empirical results. The ground state leptonic decay constants in Figure 7 were calculated directly on-shell, but extrapolation was necessary to obtain on-shell values for those of the excited states. For these observables, two extrapolation schemes were used and they yielded consistent results in all cases.

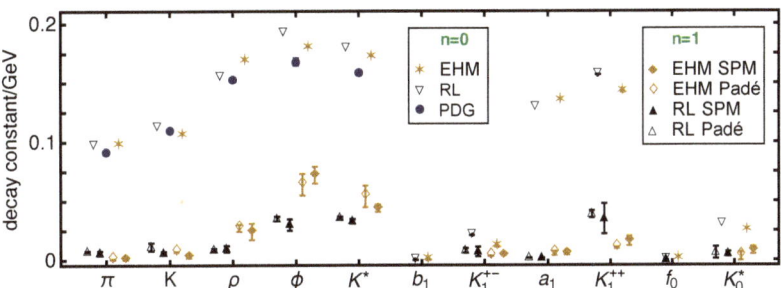

Figure 7. Leptonic decay constants for all states whose masses are reported in Figure 6: ground states, $n = 0$; lowest lying radial excitations, $n = 1$. For the excited states, two extrapolation results are presented for each state, viz. one obtained with Padé approximants and the other employing the Schlessinger point method (SPM) [206,207,238,239]—the distinct approaches yield consistent outcomes. Results inferred from data are also plotted, where available [40] (PDG).

Given that Reference [222] used a simplified interaction, viz. Equation (26), then the Figure 7 comparison between predicted ground state decay constants and the few

---

[4] Dressed-quark propagators form an important part of the kernels of all bound state equations. As on-shell meson masses increase, poles in those propagators enter the complex plane integration domain sampled by the Bethe–Salpeter equation [199]. For such cases—here, meson excited states—a direct on-shell solution cannot be obtained using simple algorithms. Therefore, to obtain the masses of those mesons, Reference [222] used an extrapolation procedure based on Padé approximants. This is the origin of the uncertainty bar on the CSM predictions.

known empirical values is favourable, particularly because decay constants are sensitive to ultraviolet physics, which was omitted. There are also indications that the EHM-improved kernels deliver better agreement.

The decay constants of radially excited states are especially interesting. Quantum mechanics models of positronium-like systems produce a single zero in the radial wave function of $n = 1$ states. The decay constant of a first radial excitation is thus $(1/8)$-times that of the ground state. The predictions in Reference [222] are generally consistent with this pattern, except for $J^P = 0^-$ mesons. In the pseudoscalar channel, as a corollary of EHM, QCD predicts $f_H^{n=1} \equiv 0$ in the chiral limit [208,209,240,241]. The results in Reference [222] meet this requirement, whereas such outcomes cannot be achieved in quark models without tuning parameters. For this reason alone, the decay constant predictions of [222] warrant testing.

Notwithstanding the simplifications used in formulating the problem, Reference [222] delivered the first Poincaré-invariant analysis of the spectrum and decay constants of the u, d, and s meson ground and first-excited states. The results include predictions for masses of as-yet unseen mesons and many unmeasured decay constants. One may look forward to extensions of the approach to heavy+light mesons [239,242,243], hybrid/exotic mesons [213,244–246], and glueballs [247–250]. These directions are especially important owing to worldwide investments in studies of the former and searches for the latter [179,183,186,251–253].

Such progress with meson properties should not obscure the need to calculate the spectrum of baryons. Indeed, baryons are the most fundamental three-body systems in Nature; if we do not understand how QCD, a Poincaré-invariant quantum field theory, structures the spectrum of baryons, then we do not understand Nature. Within the context of the truncation scheme introduced in References [196,197], baryon masses and bound state amplitudes have been calculated using a Poincaré-covariant Faddeev equation that describes a six-point Schwinger function for three-quark → three-quark scattering. The first solution of this problem for the nucleon (N) was presented in Reference [254]; continuing studies are reviewed elsewhere [28,31,181,255]; efforts are now under way to adapt the methods in References [221,222] to the formulation and solution of the baryon Faddeev equations.

Meanwhile, the quark + dynamical diquark approach to baryon properties, introduced in References [256–259], is also being pursued vigorously. This treatment begins with solutions of the equation illustrated in Figure 8. As sketched, e.g., in Reference [21] (Section 5.1), this is an approximation to the three-body Faddeev equation, whose kernel is constructed using dressed quark and nonpointlike diquark degrees-of-freedom. Binding energy is lodged within the diquark correlation and also produced by the exchange of a dressed quark, which, as drawn in Figure 8, emerges in the break-up of one diquark and propagates to be absorbed into formation of another. In the general case, five distinct diquark correlations are possible: isoscalar–scalar, $(I, J^P = 0, 0^+)$; isovector–axial vector; isoscalar–pseudoscalar; isoscalar–vector; isovector–vector. Channel dynamics within a given baryon determines the relative strengths of these correlations therein.

Given the extensive coverage of the role of diquark correlations in hadron structure presented in Reference [255], herein, we will only draw some recent highlights from analyses of the baryon spectrum using the Faddeev equation in Figure 8, drawing largely from Reference [232]. That study was built upon a symmetry-preserving treatment of a vector×vector contact interaction (SCI), which was introduced a little over a decade ago [260] and has since been employed with success in numerous applications, some of which are reviewed in this volume [261]. Amongst the merits of the SCI are its algebraic simplicity; limited number of parameters; simultaneous applicability to many systems and processes; and potential for generating insights that connect and explain numerous phenomena.

Reference [232] used the SCI to calculate the ground state masses of $J^P = 0^\pm, 1^\pm$ ($f\bar{g}$) mesons and $J^P = 1/2^\pm, 3/2^\pm$ (fgh) baryons, where $f, g, h \in \{u, d, s, c, b\}$. Using $J^P = \frac{1}{2}^\pm$ states as exemplars, Figure 9 highlights the level of quantitative accuracy.

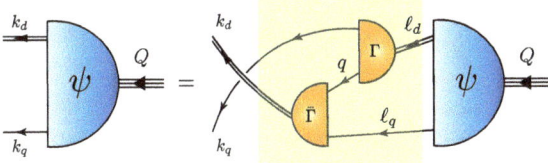

Figure 8. Figurative representation of the integral equation satisfied by the Poincaré-covariant matrix-valued function Ψ, viz. the Faddeev amplitude for a baryon with total momentum $Q = \ell_q + \ell_d = k_q + k_d$ built from three valence quarks, two of which are always participants in a nonpointlike, interacting diquark correlation. Ψ describes the sharing of relative momentum between the dressed quarks and diquarks. *Shaded rectangle*—Faddeev kernel. Legend: *single line*—dressed quark propagator, $S(q)$; $\Gamma^{J^P}(k; K)$—diquark correlation amplitude; *double line*—diquark propagator, $\Delta^{J^P}(K)$.

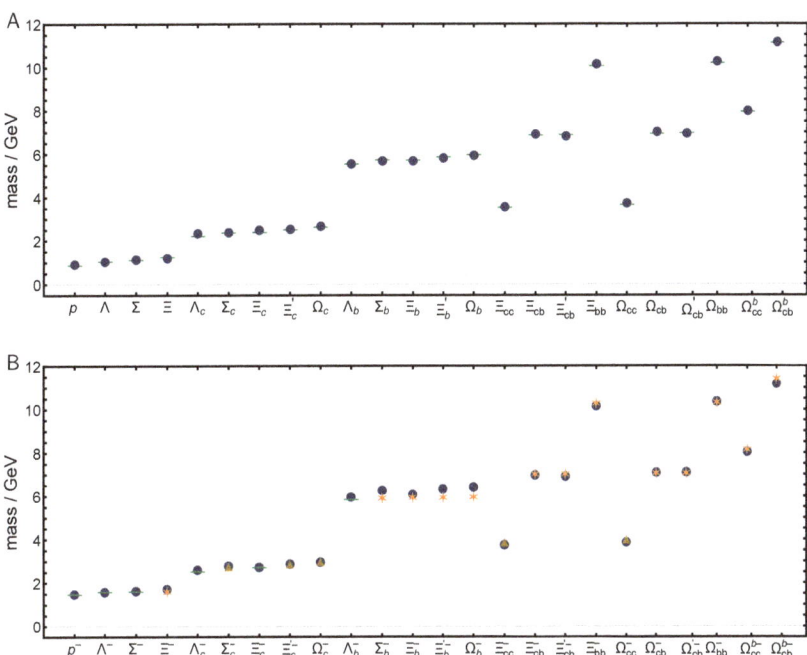

Figure 9. *Upper panel* (**A**): SCI-calculated masses of ground state flavour-SU(5) $J^P = 1/2^+$ baryons [232] compared with either experiment (first 15) [40] or lQCD (last 9) [262,263]. *Lower panel* (**B**): SCI masses of ground state flavour-SU(5) $J^P = 1/2^-$ baryons in [232] compared with experiment [40] (green bars), lQCD [264] (gold triangles), or three-body Faddeev equation results [218] (orange asterisks). Analogous plots for $J^P = 3/2^\pm$ baryons are presented elsewhere ([232] (Figures 4B and 5B)).

Regarding the 33 mesons, then $\overline{\mathrm{ard}} = 5(6)\%$ when comparing SCI predictions with empirical masses. In achieving this outcome, it was found that sound expressions of EHM were crucial. Turning to baryons, the SCI generated 88 distinct bound states, namely every possible three-quark $1/2^\pm, 3/2^\pm$ ground state. In this collection, 34 states have already been identified empirically, and lQCD results are available for another 30: for these 64 states, comparing SCI prediction with experiment, where available, or lQCD mass otherwise, $\overline{\mathrm{ard}} = 1.4(1.2)\%$. This level of agreement was only achieved through the implementation of EHM-induced effects associated with spin–orbit repulsion in $1/2^-$ baryons. Notably, the same 88 ground states are also produced by a three-body Faddeev equation [218]: in comparison with those results, $\overline{\text{ard}} = 3.4(3.0)\%$.

Overall, Reference [232] delivered SCI predictions for 164 distinct quantities, 114 of which have either been measured or calculated using lQCD: performing a comparison on this subset yields $\overline{\text{ard}} = 4.5(7.1)\%$. Such quantitative success means that credibility should be given to the qualitative conclusions that follow from the SCI analysis. We list them here: (*i*) Nonpointlike, dynamical diquarks play a significant role in all baryons. Usually, the lightest allowed diquark is the most important part of a baryon's Faddeev amplitude. This remains true, even if the lightest correlation is a (sometimes called bad) axial vector diquark, and also for baryons containing one or more heavy quarks. In the latter connection, this means one cannot safely assume that singly heavy baryons may realistically be described as two-body light diquark+heavy quark ($qq' + Q$) bound states or that doubly heavy baryons (qQQ') can be treated as two-body light quark+heavy diquark bound states, $q + QQ'$. Corresponding statements apply to the treatment of tetra- and penta-quark problems. (*ii*) Positive-parity diquarks dominate in positive-parity baryons. Axial vector diquarks are prominent in all states. (*iii*) Negative-parity diquarks play a minor role in positive-parity baryons. On the other hand, owing to EHM, they are significant and sometimes dominant in $J = 1/2^-$ baryons. (*iv*) Curiously, however, $J = 3/2^-$ baryons are built (almost) exclusively from $J = 1^+$ diquark correlations. These conclusions are being checked using Faddeev equations with momentum-dependent exchange interactions; hence, a closer link to QCD. Where results are already available, the SCI conclusions have been confirmed [265–268].

Following more than fifty years of hadron spectroscopy based on quark models, we are beginning to see real progress with the use of bound state equations in quantum field theory. Poincaré-invariant, symmetry-preserving analyses that reveal the expressions of EHM in hadron masses and level orderings are becoming available. This increases the value of experimental hadron spectra measurements, making them a clearer window onto strong QCD.

7. Baryon Wave Functions

Concerning baryon structure, as noted when opening Section 6, quark models are still considered to provide a useful picture [187–189,269]. In such models, baryons built from combinations of u, d, and s valence quarks are grouped into multiplets of $SU(6) \otimes O(3)$. The multiplets are labelled by their flavour content—$SU(3)$, spin—$SU(2)$, and orbital angular momentum—$O(3)$. However, as has been emphasised, quark potential models do not have an explicit link with QCD, a Poincaré-invariant quantum gauge field theory.

For the lightest four $(I, J^P = \frac{1}{2}, \frac{1}{2}^{\pm})$ baryons, with I denoting isospin, a comparison between quark model expectations and insights drawn from solutions of the Poincaré-covariant Faddeev equation is presented elsewhere [265]. Herein, we will illustrate the qualitative character of such comparisons by considering more recent studies of $(\frac{3}{2}, \frac{3}{2}^{\pm})$, $(\frac{1}{2}, \frac{3}{2}^{\mp})$ baryons [267,268].

These systems were studied in Reference [217] using RL truncation and direct calculations of all primary Schwinger functions. With current algorithms, owing to singularities that enter the integration domains sampled by the Faddeev equations [199], this approach limits the ability to compute wave functions because the on-shell point for many systems is inaccessible. (Similar issues are encountered with meson structure studies—Section 6.) To circumvent this issue, References [267,268] employed the QCD-kindred framework introduced elsewhere [270], in which, instead of calculating all primary Schwinger functions, one uses physics-constrained algebraic representations of the Faddeev kernel elements. This weakens the connection with QCD, but that loss is well compensated because, with reliably informed choices for the representation functions, the expedient enables access to on-shell baryon wave functions. The QCD-kindred framework has widely been used with success—see, e.g., References [271–276].

Within the quark model framework and using standard spectroscopic notation, $n\,{}^{2S+1}L_J$, where n is the radial quantum number with "0" labelling the ground state, the lightest four $(\frac{3}{2}, \frac{3}{2}^{\pm})$ Δ baryons, constructed from isospin $I = \frac{3}{2}$ combinations of three u and/or d quarks, are understood as follows [277]:

1. $\Delta(1232)\frac{3}{2}^{+} \ldots 0\,{}^4S_{\frac{3}{2}}$ = S-wave ground state;
2. $\Delta(1600)\frac{3}{2}^{+} \ldots 1\,{}^4S_{\frac{3}{2}}$ = S-wave radial excitation of $\Delta(1232)\frac{3}{2}^{+}$;
3. $\Delta(1700)\frac{3}{2}^{-} \ldots 0\,{}^2P_{\frac{3}{2}}$ = P-wave orbital angular momentum excitation of $\Delta(1232)\frac{3}{2}^{+}$;
4. $\Delta(1940)\frac{3}{2}^{-} \ldots 1\,{}^4P_{\frac{3}{2}}$ = P-wave excitation of $\Delta(1600)\frac{3}{2}^{+}$.

Analogously, the $(\frac{1}{2}, \frac{3}{2}^{\mp})$ states are interpreted thus [277]:

1. $N(1520)\frac{3}{2}^{-} \ldots 0\,{}^2P_{\frac{1}{2}}$ = P-wave ground state in this channel and an angular momentum coupling partner of $N(1535)\frac{1}{2}^{-}$;
2. $N(1700)\frac{3}{2}^{-} \ldots 0\,{}^4P_{\frac{3}{2}}$ = P-wave angular momentum coupling partner of $N(1520)\frac{3}{2}^{-}$;
3. $N(1720)\frac{3}{2}^{+} \ldots 0\,{}^2D_{\frac{3}{2}}$ = D-wave orbital angular momentum excitation of $N(1520)\frac{3}{2}^{-}$;
4. $N(1900)\frac{3}{2}^{+} \ldots 0\,{}^4D_{\frac{3}{2}}$ = D-wave orbital angular momentum excitation of $N(1700)\frac{3}{2}^{-}$.

On the other hand, Poincaré-invariant quantum field theory does not readily admit such assignments. Instead, the states appear as poles in the six-point Schwinger functions associated with the given (I, J^P) channels. Here, "(1 ↔ 3)" and "(2 ↔ 4)" in each block above are related as parity partners. All differences between positive- and negative-parity states can be attributed to chiral symmetry breaking in quantum field theory. This is highlighted by the ρ-a_1 meson complex [219–222,278]. Regarding light quark hadrons, such symmetry breaking is almost exclusively dynamical [24,279–284]. As noted above, DCSB is a corollary of EHM [32–37]; hence, quite probably linked tightly with confinement—Section 5. These features imbue quantum field theory analyses of $(\frac{3}{2}, \frac{3}{2}^{\pm})$, $(\frac{1}{2}, \frac{3}{2}^{\mp})$ baryons with particular interest; consequently, experiments that test predictions made for structural differences between parity partners are highly desirable.

Working with the Faddeev equation sketched in Figure 8, then, a priori, the $(\frac{3}{2}, \frac{3}{2}^{\pm})$ baryons are the simpler systems because they can only contain isovector–axial vector and isovector–vector diquarks, whereas the $(\frac{1}{2}, \frac{3}{2}^{\mp})$ systems may involve all five distinct types of diquarks: $(0,0^+)$, $(1,1^+)$, $(0,0^-)$, $(1,1^-)$, $(0,1^-)$. Nonetheless, the formulation of the bound state problems in both channels is practically identical, using the same dressed quark and diquark propagators; diquark correlation amplitudes; etc. This way, one guarantees a unified description of all states in the spectrum. The propagators are parametrised using entire functions [163,285,286]; hence, satisfy the confinement constraints described in Section 5. It is this feature that enables on-shell calculations for all baryons.

The calculated spectrum of states is displayed in Figure 10. As highlighted elsewhere [29,265,267,268,287,288], the kernel in Figure 8 does not include contributions that may be understood as meson–baryon final state interactions. These are the interactions that transform a bare baryon into the observed state, e.g., via dynamical coupled channels calculations [289–293]. The Faddeev amplitudes and masses calculated in References [265,267,268] should therefore be seen as describing the *dressed quark core* of the bound state, not the fully dressed, observable object [294–296]. That explains why the masses are uniformly too large. Evidently and importantly, in each sector, a single subtraction constant is sufficient to realign the masses and produce a good description of the spectrum.

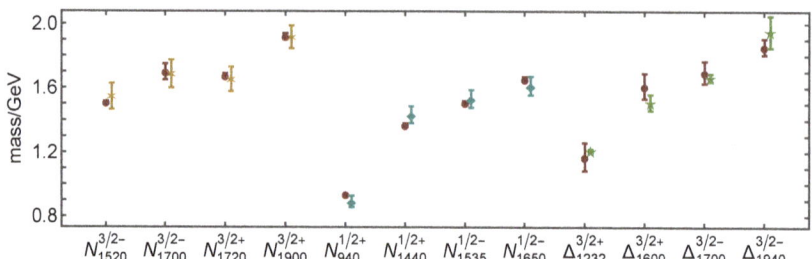

Figure 10. Real part of empirical pole position for each designated baryon [40] (red circle) compared with: calculated masses in Reference [268] (gold asterisks) after subtracting $\delta_{\mathrm{MB}}^{N_{3/2}} = 0.13\,\mathrm{GeV}$ from each; Reference [265] (teal diamonds) after subtracting $\delta_{\mathrm{MB}}^{N_{1/2}} = 0.30\,\mathrm{GeV}$; and Reference [267] (green five-pointed stars) after subtracting $\delta_{\mathrm{MB}}^{\Delta_{3/2}} = 0.17\,\mathrm{GeV}$. All Faddeev equation predictions are drawn with an uncertainty that reflects a $\pm 5\%$ change in diquark masses.

Regarding $(\frac{3}{2}, \frac{3}{2}^{\pm})$ baryons, Reference [267] found that although these states may contain both $(1,1^+)$ and $(1,1^-)$ quark+quark correlations, one can neglect the $(1,1^-)$ diquarks because they have practically no impact on the masses or wave functions. After this simplification, the Poincaré-covariant wave functions involve eight independent matrix-valued terms, each multiplied by a scalar function of two variables: $(k^2, k \cdot Q)$, with k the quark+diquark relative momentum. Studying the properties of these functions, one may conclude that $\Delta(1600)\frac{3}{2}^+$ is fairly interpreted as a radial excitation of the $\Delta(1232)\frac{3}{2}^+$, as suggested by the quark model. However, the wave functions of the $\Delta(1700)\frac{3}{2}^-$ and $\Delta(1940)\frac{3}{2}^-$ states are complicated and do not readily admit direct analogies with quark model pictures.

Projecting the Poincaré-covariant Faddeev wave functions of $(\frac{3}{2}, \frac{3}{2}^{\pm})$ baryons into their respective rest-frames, one arrives at a $J = L + S$ separation which is comparable to that associated with quark models. (Here, L is quark–diquark orbital angular momentum.) Following this procedure, Reference [267] found that the angular momentum structure of all these states is much more complicated than is typically generated in quark models—see Figure 11.

Evidently, making a link to quark models, the $\Delta(1232)\frac{3}{2}^+$ and $\Delta(1600)\frac{3}{2}^+$ are characterised by a dominant S-wave component and the $\Delta(1700)\frac{3}{2}^-$ by a prominent P-wave. The $\Delta(1940)\frac{3}{2}^-$, however, does not fit this picture: contrary to quark model expectations, indicated on page 79, this state is S-wave-dominated. Moreover, each state contains every admissible partial wave.

Combining all gathered information, Reference [267] furthermore concluded that the negative-parity Δ baryons are not merely orbital angular momentum excitations of positive parity ground states. In this observation, the results match those obtained earlier for $(\frac{1}{2}, \frac{1}{2}^{\pm})$ baryons [265].

Recalling now that the interpolating fields for positive- and negative-parity hadrons are related by chiral rotation of the quark spinors used in their construction, then the highlighted structural differences are largely generated by DCSB. Regarding the $\Delta(1940)\frac{3}{2}^-$ in particular, these novel structural predictions may be expected to encourage new experimental efforts aimed at extracting reliable information about this little-understood state from exclusive $\pi^+\pi^- p$ electroproduction data [297,298].

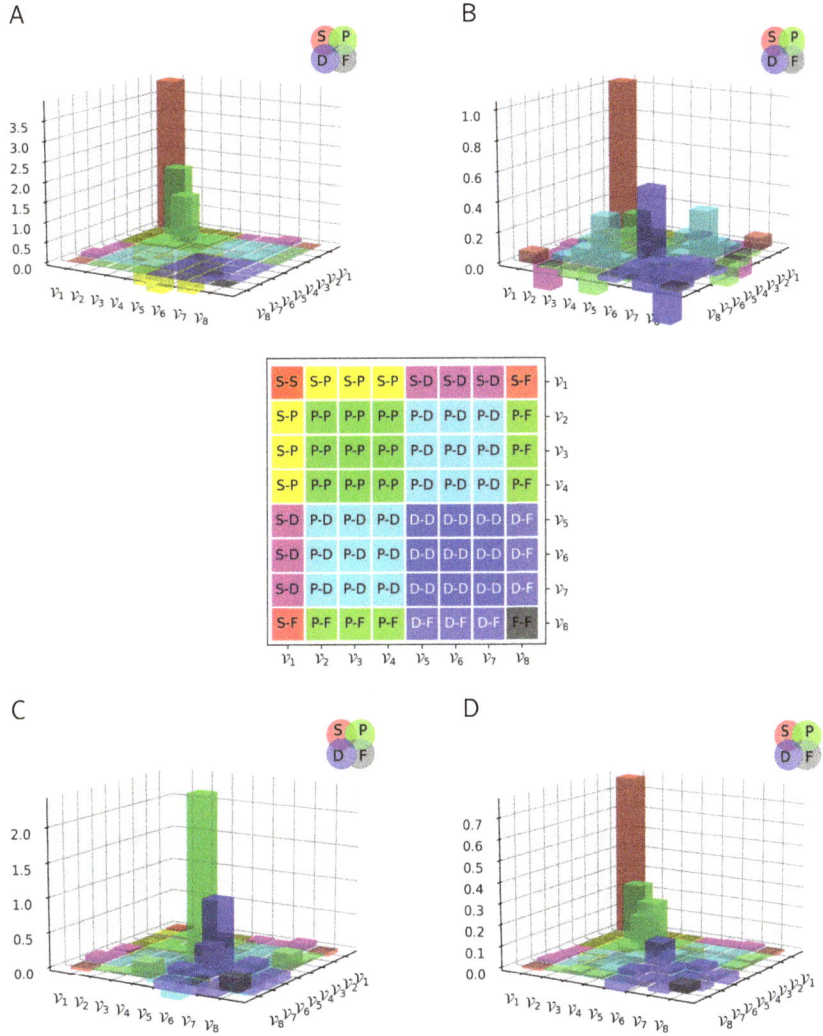

Figure 11. Rest-frame quark+axial vector–diquark orbital angular momentum content of $(\frac{3}{2}, \frac{3}{2}^{\pm})$ baryons considered in Reference [267], as measured by the contribution from the various components to the associated canonical normalisation constant: **A** $\Delta(1232)\frac{3}{2}^{+}$; **B** $\Delta(1600)\frac{3}{2}^{+}$; **C** $\Delta(1700)\frac{3}{2}^{-}$; **D** $\Delta(1940)\frac{3}{2}^{-}$. The overall positive normalisations receive both positive (above plane) and negative (below plane) contributions. The central image provides the legend for the interpretation of the other panels, identifying interference between the various identified orbital angular momentum basis components in the baryon rest-frame.

Observations upon similar features were made about $(\frac{1}{2}, \frac{3}{2}^{\pm})$ baryons in Reference [268]. To begin, despite the fact that such states may contain all possible diquark correlations, the analyses showed that a good approximation is obtained by keeping only $(0, 0^{+})$, $(1, 1^{+})$ correlations. This runs counter to the nature of $(\frac{1}{2}, \frac{1}{2}^{-})$ systems, in which $(0, 0^{-})$, $(0, 1^{-})$ diquarks are also important [265,299]. Projecting the Poincaré-covariant Faddeev wave functions into the baryon rest-frames and considering the baryon mass fraction contributed by each partial wave, this collection of states form a set related via orbital angular momentum excitation: as in quark models, the parity-negative states are primarily P-wave in nature, whereas the parity-positive states are D wave—see Reference [268] (Figure 4). However, looking with finer resolution, using charts of the canonical normalisation constant contributions from the various partial waves in the Poincaré-covariant wave functions, like Figure 11 herein, far greater L-complexity was observed than is usually found in quark models—see Reference [268] (Figure 7). Here, too, one may anticipate that these structural predictions can be tested using data from measurements of resonance electroexcitation at large momentum transfers. For the $N(1520)\frac{3}{2}^-$, data are already available [300–305], and calculations of the electroproduction form factors are underway. Large-Q^2 data on the other states is not available; so, the predictions in Reference [268] will also encourage new experimental efforts in this area.

Parity partner channels are identical when chiral symmetry is restored [14,30]. It is therefore interesting to note that the mass splitting between partner states does not exhibit a simple pattern, viz. empirically [40]:

$$
\begin{array}{ll}
\text{states} & \text{mass splitting/GeV} \\
N(1535)\tfrac{1}{2}^- - N(940)\tfrac{1}{2}^+ & 0.57, \\
N(1650)\tfrac{1}{2}^- - N(1440)\tfrac{1}{2}^+ & 0.29, \\
\Delta(1700)\tfrac{3}{2}^- - \Delta(1232)\tfrac{3}{2}^+ & 0.46, \\
\Delta(1940)\tfrac{3}{2}^- - \Delta(1600)\tfrac{3}{2}^+ & 0.44, \\
N(1720)\tfrac{3}{2}^+ - N(1520)\tfrac{3}{2}^- & 0.17, \\
N(1900)\tfrac{3}{2}^+ - N(1700)\tfrac{3}{2}^- & 0.22.
\end{array}
\tag{27}
$$

This system dependence of the mass splitting is also linked to the deeper structural differences between these states that are expressed in their complex wave functions.

Using a familiar quantum mechanics framework, quark models produce baryon wave functions that have an appealing simplicity. However, far richer structures are found when quantum field theory is used to solve baryon bound state problems. The growing body of quantum field theory predictions can be tested, e.g., in modern and future large-Q^2 measurements of baryon elastic and transition form factors. In fact, the large-Q^2 character of such experiments is alone sufficient to demand the sort of Poincaré-invariant, symmetry-preserving treatment that only analyses in quantum field theory can deliver. One may, therefore, expect studies using the Faddeev equation approach outlined in this section to become steadily more widespread.

8. Meson Form Factors

The truncation scheme explained in References [196,197] has been used to calculate many meson elastic and transition form factors [21]; and modern algorithms have enabled predictions to be delivered on the entire domain of spacelike Q^2 [200,201,306–309], making it possible to draw connections with hard scattering formulae derived using QCD perturbation theory [310–312]. These new predictions, which unify the infrared and ultraviolet Q^2 domains, are providing the impetus for measurements at new-generation high-energy and high-luminosity facilities [35,180,182,184]. Recalling that QCD is not found in form factor scaling, but in scaling violations, then the goal of these new experiments is, of course, to discover the breakaway from scaling in a hard exclusive process and thus reveal the hand of QCD.

In connection with these new facilities, it has been argued that the interaction of a heavy vector meson, $V = J/\psi, Y$, with a proton, p, may provide access to a QCD van der Waals interaction, produced by multiple gluon exchange [313,314] and/or the QCD trace anomaly [6,315–317]. The van der Waals interaction is of interest because it might relate to, amongst other things, the observation of hidden-charm pentaquark states [318]; and the trace anomaly is topical because of its connection with EHM.

Lacking vector meson beams, ongoing and anticipated experiments at electron (e) accelerators are based on an expectation that the desired $V + p$ interactions can be accessed through the electromagnetic production of vector mesons from the proton, in reactions like $e + p \to e' + V + p$ [183,252,319]. This is because some practitioners imagine that single-pole vector meson dominance (VMD) [320–322] can reliably be employed to draw a clean link between $e + p \to e' + V + p$ and the desired $Vp \to Vp$ cross-sections. In this picture (see Figure 12), the interaction is supposed to proceed via the following sequence of steps: (i) $e \to e' + \gamma^{(*)}(Q)$; (ii) $\gamma^{(*)}(Q) \to V$; and (iii) $V + p \to V + p$. $\gamma^{(*)}(Q)$ is a virtual photon; and step (ii) expresses the VMD hypothesis. As commonly used, VMD assumes: (a) that a photon, which is generally spacelike, so that $Q^2 > 0$, transforms into an on-shell vector meson, with timelike momentum $Q^2 = -m_V^2$; and (b) that the $Q^2 > 0$ strength and form of the transition in (ii) is the same as that measured in the real vector meson decay process, $V \to \gamma^*(Q^2 = -m_V^2) \to e^+ + e^-$. Property (b) means that $\gamma_{\gamma V}$, the associated decay constant, is fixed at its meson on-shell value and acquires no momentum dependence:

$$\gamma_{\gamma V}^2 = 4\pi \alpha_{\text{em}} m_V^2 f_V^2, \quad (28)$$

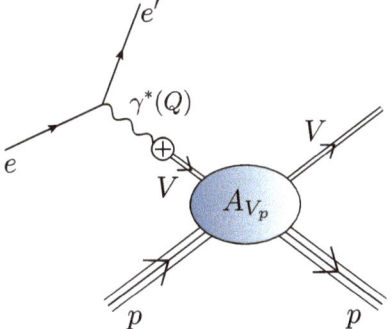

Figure 12. Electroproduction of a vector meson from the proton: $e + p \to e' + V + p$, often interpreted as providing access to $V + p \to V + p$ using a vector meson dominance model. The $\gamma^{(*)}(Q) \to V$ VMD transition is indicated by the crossed circle. It is usually assumed to occur with a momentum-independent strength $\gamma_{\gamma V}$ [320–322]. However, regarding heavy mesons, at least, the VMD hypothesis is unsound, as discussed below and in, e.g., References [204,323].

where α_{em} is the QED fine-structure constant, m_V is the vector meson mass, and f_V measures the strength of the meson's Bethe–Salpeter wave function at the origin in configuration space [324] (Section IIB).

The fidelity of these VMD assumptions was recently subjected to scrutiny via analyses of the photon vacuum polarisation and photon–quark vertex [204]. Regarding the photon vacuum polarisation, it was shown that there is no vector meson contribution to this polarisation at the photoproduction point, $Q^2 = 0$. Consequently, massless real photons cannot readily be linked with massive vector bosons, and the current field identity, Equation (28), typical of VMD implementations, should not be used literally because it entails violations of Ward–Green–Takahashi identities (breaking of symmetries) in QED.

Turning to the dressed photon–quark vertex, $\Gamma_\nu^\gamma(k; Q)$, this three-point Schwinger function exhibits a pole at the mass of any vector meson bound state. A physical property, it expresses the fact that the decay $V \to e^+ e^-$ proceeds via a timelike virtual photon. The VMD hypothesis may thus be seen as a claim that $\Gamma_\nu^\gamma(k; Q)|_{Q^2 \simeq 0}$ maintains a rigorous link, in both magnitude and momentum dependence, with the Bethe–Salpeter amplitude of an on-shell vector meson. However, as shown in Figure 13, that is not the case. The panels in Figure 13 depict the ratios:

$$R_V(k^2;Q^2=0) = \frac{2f_V}{m_V}\frac{F_1^0(k^2;-m_V^2)}{G_1^0(k^2;Q^2=0)}, \quad R_V^B(k^2;Q^2=0) = \frac{f_V}{m_V}\frac{F_8^0(k^2;-m_V^2)}{G_3^0(k^2;Q^2=0)}, \quad (29)$$

where G_1^0 is the zeroth Chebyshev moment[5] of the dominant amplitude in the photon–quark vertex, associated with the matrix structure $\gamma \cdot \epsilon(Q)$, where $\epsilon(Q)$ is the photon polarisation vector—in fact, using the vector Ward–Green–Takahashi identity and Equation (23), $G_1^0(k^2;Q^2=0) = A_g(k^2)$, where g is the flavour of the meson's valence quark; F_1^0 is its analogue in the vector meson bound state amplitude; G_3^0 is the zeroth moment of that term in the photon–quark vertex, which is directly linked to the scalar piece of the dressed quark self-energy via the vector Ward–Green–Takahashi identity, i.e., $G_3^0(k^2;Q^2=0) = -2B_g'(k^2)$; and F_8^0 is its analogue in the vector meson bound state amplitude. Were VMD to be a sound assumption, then all these curves would lie near the thin horizontal line drawn at unity in both panels of Figure 13. However, whilst one might discuss the case for lighter vector mesons, the VMD hypothesis is plainly false for heavy vector mesons: the momentum-dependence of the $Q^2 = 0$ photon–quark vertex is completely different from that of the vector meson Bethe–Salpeter amplitude.

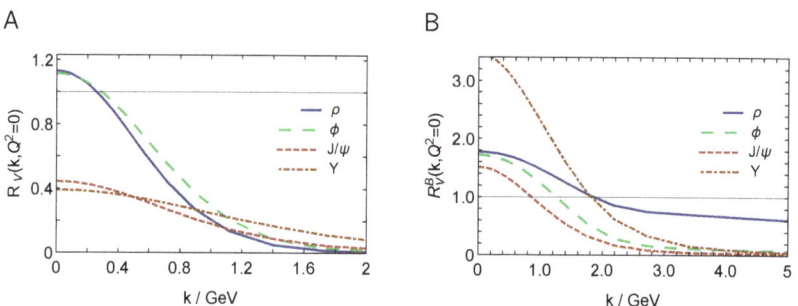

Figure 13. *Left panel* (**A**): First ratio in Equation (29) computed using matched solutions of the gap and Bethe–Salpeter equations for $V = \rho, \phi, J/\psi, \Upsilon$. *Right panel* (**B**): Second ratio in Equation (29), computed similarly. In cases where the VMD hypothesis were sound, all these curves would lie close to the thin horizontal line drawn at unity.

One is consequently led to conclude that, insofar as heavy mesons are concerned, no extant attempt to link $e + p \to e' + V + p$ reactions with $V + p \to V + p$ via VMD is reliable. A similar conclusion is drawn in Reference [323] using arguments within perturbative QCD. It is strengthened by Reference [325], which demonstrates that even if VMD were valid, then contributing relevant coupled channels processes would obscure connections between $e + p \to e' + V + p$ and $V + p \to V + p$ reactions. These analyses make tenuous any interpretation of $e + p \to e' + V + p$ reactions as a path to hidden-charm pentaquark production or as a route to understanding the origin of the proton mass. In fact, the crudity of existing models prevents sound conclusions being drawn about the capacity of heavy meson production to reveal something about the SM [326].

On the other hand, numerous applications [28,31–33,327], including $\gamma^*\gamma \to \pi^0, \eta, \eta', \eta_c$, and η_b [201,308], show that a viable alternative to the VMD hypothesis exists in adapting the CSMs discussed herein to a direct analysis of processes like $\gamma^{(*)} + p \to V + p$. Considering vector meson photoproduction/electroproduction from the proton,

---

[5] The Chebyshev (or hyperspherical) expansion of Poincaré-invariant functions of two scalar variables is discussed, e.g., in Reference [199] (IV.B).

References [326,328,329] illustrate how one might proceed; and given developments in the past vicennium, improvements of such studies are now possible.

Following the first studies almost thirty years ago [330], the CSM treatment of meson elastic and transition form factors has now reached a mature level. Today, as sketched above, sound predictions with a traceable connection to QCD are being delivered. This enables new opportunities to be exploited, such as informative comparison with results from lQCD [200], weak transitions of heavy mesons (see References [205–207] and Section 10), calculations of form factors describing light-by-light scattering contributions to the muon anomalous magnetic moment [331], and a beginning to the analysis of meson gravitational form factors [332].

9. Baryon Form Factors

Advances in the calculation of meson form factors are complemented by progress with the parameter-free prediction of baryon elastic and transition form factors. The first direct RL truncation study of nucleon elastic form factors was presented in Reference [333]. It was extended to nucleon axial and pseudoscalar form factors in Reference [334], the elastic form factors of Δ and Ω baryons in Reference [335], and nucleon tensor charges in Reference [31].

Many other direct RL truncation analyses are in train. However, in all such studies, the challenge of quark propagator singularities moving into the complex integration domain must also be overcome [199]. Using existing algorithms, the singularities limit the Q^2 reach of baryon form factor calculations. Here, again, the QCD-kindred quark plus fully interacting, nonpointlike diquark picture of baryon structure—outlined in Section 7—can profitably be exploited. Some of the successes are summarised elsewhere [181,255]. Recently, the predictions made for $\gamma + p \to \Delta(1600)$ transition form factors [273] have been tested in analyses of $\pi^+\pi^- p$ electroproduction data collected at Jefferson Lab (JLab), with preliminary results supporting the quark+diquark picture [336].

Scanning the sources indicated above, it becomes apparent that, hitherto, nucleon properties have largely been probed in $e + N$ scattering. Aspects of this field are reviewed elsewhere [337]. An entirely new window onto baryon structure is opened when one uses neutrino scattering. In fact, reliable predictions of nucleon and $N \to \Delta(1232)$ electroweak form factors are crucial for understanding new-generation long-baseline neutrino oscillation experiments [338–344]. Importantly, in this connection, recent developments within the framework of CSMs have enabled practitioners to deliver the first Poincaré-invariant parameter-free predictions for such form factors on a momentum transfer domain that extends to $Q^2 = 10\,\text{GeV}^2$ [274–276]. Extensions to even larger Q^2 are feasible. Where data are available, the predictions confirm the measurements. More significantly, the results are serving as motivation for new experiments at high-luminosity facilities.

The key step in the use of CSMs was the construction of a symmetry-preserving current that describes the coupling between axial vector and pseudoscalar probes and baryons whose structure is determined by the Faddeev equation in Figure 8. This current is illustrated in Figure 14 and explained by the legend in Table 2. The origins and characters of Diagrams 1–3 are obvious: the probe must interact with every "constituent" that carries a weak charge. Diagram 4 is a two-loop diagram, made necessary by the quark exchange nature of the kernel in Figure 8: the object exchanged in binding also carries a weak charge. Given the presence of Diagram 4, so-called seagull diagrams—Diagrams 5 and 6—are necessary to ensure symmetry preservation. The analogous contributions for baryon electromagnetic currents were derived in Reference [345], but it took more than twenty years before the seagull terms were derived for axial vector and pseudoscalar currents [275]. These contributions are both two-loop diagrams.

Axial vector interactions of a nucleon are described by two form factors—$G_A(Q^2)$ (axial) and $G_P(Q^2)$ (induced pseudoscalar)—associated with the following matrix element:

$$\hat{\mathcal{J}}_{5\mu}^j(K,Q) := \langle N(P_f)|\mathcal{A}_{5\mu}^j(0)|N(P_i)\rangle \tag{30a}$$

$$= \bar{u}(P_f)\frac{\tau^j}{2}\gamma_5\left[\gamma_\mu G_A(Q^2) + i\frac{Q_\mu}{2m_N}G_P(Q^2)\right]u(P_i), \tag{30b}$$

where $P_{i,f}$ are, respectively, the initial and final nucleon momenta, with $P_{i,f}^2 = -m_N^2$, m_N is the nucleon mass, and $u(P)$ is the associated Euclidean spinor. (Associated conventions are specified elsewhere, e.g., Reference [271] (Appendix B).) The average momentum of the system is $K = (P_i + P_f)/2$, and $Q = P_f - P_i$ is the momentum transferred between the initial and final states. It is usual to consider the $SU(2)_F$ isospin limit $m_u = m_d =: m_q$, with the flavour structure described using Pauli matrices $\{\tau^j | j = 1,2,3\}$: $\tau^{1\pm i2} := (\tau^1 \pm i\tau^2)/2$ correspond to the weak charged currents and τ^3 is the neutral current. Moreover, the isovector axial current operator is

$$\mathcal{A}_{5\mu}^j(x) = \bar{q}(x)\frac{\tau^j}{2}\gamma_5\gamma_\mu q(x), \quad q = \begin{pmatrix} u \\ d \end{pmatrix}. \tag{31}$$

A third form factor is defined via the kindred pseudoscalar current, a matrix element of $\mathcal{P}_5^j(x) = \bar{q}(x)\frac{\tau^j}{2}\gamma_5 q(x)$:

$$\hat{\mathcal{J}}_5^j(K,Q) := \langle N(P_f)|\mathcal{P}_5^j(0)|N(P_i)\rangle = \bar{u}(P_f)\frac{\tau^j}{2}\gamma_5 G_5(Q^2) u(P_i). \tag{32}$$

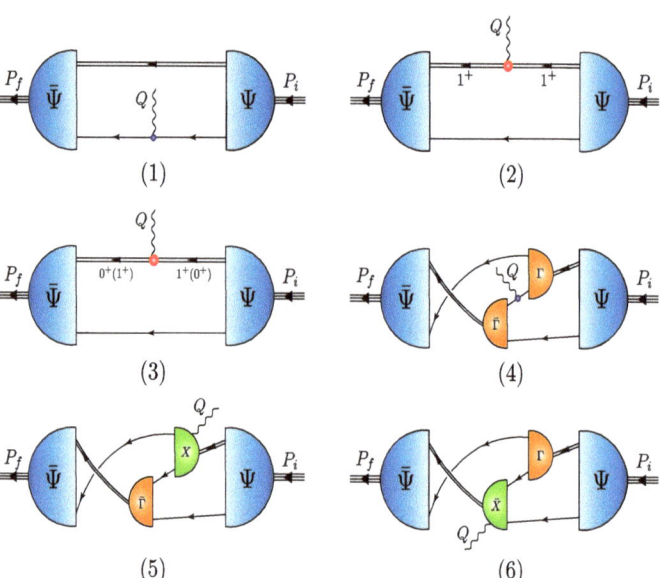

Figure 14. Axial/pseudoscalar current that ensures a symmetry-preserving interaction with an on-shell baryon described by a Faddeev amplitude obtained from the equation depicted in Figure 8: *single line*, dressed quark propagator; *undulating line*, axial/pseudoscalar current; Γ, diquark correlation amplitude; *double line*, diquark propagator; χ, seagull terms. A detailed legend is provided in Table 2.

Using the framework deployed to calculate the baryon wave functions discussed in Section 7, References [274,275] delivered the parameter-free prediction for $G_A(Q^2)$ displayed in Figure 15: the lighter blue band expresses the impact of $\pm 5\%$ variations in the diquark masses: $m_{0^+} = 0.80(1 \pm 0.05)$; $m_{1^+} = 0.89(1 \pm 0.05)$. The calculated values for the nucleon axial charge $g_A = G_A(Q^2 = 0)$, associated axial charge radius, and axial mass are, respectively,

$$g_A = 1.25(3), \quad \langle r_A^2 \rangle^{1/2} m_N = 3.25(4), \quad m_A/m_N = 1.23(3). \tag{33}$$

For comparison, empirically, $g_A = 1.2756(13)$ [40] (PDG) and [346] $\langle r_A^2 \rangle^{1/2} m_N = 3.23(72)$, $m_A/m_N = 1.15(08)$. Evidently, the CSM predictions agree with extant data. Regarding the axial mass, we note that it is sometimes convenient, when comparing with other analyses, to use a dipole *Ansatz* as an approximation for the axial form factor:

$$G_A(Q^2) = \frac{g_A}{\left(1 + Q^2/m_A^2\right)^2}; \tag{34}$$

therefore References [274,275] extracted m_A using Equation (34) to interpolate the *global* Q^2 behaviour of G_A on $x \in [0, 1.6]$, in which case m_A is not simply related to the axial radius. It is worth remarking that scalar and axial vector diquark mass variations interfere destructively, e.g., reducing m_{0^+} increases g_A, whereas g_A decreases with the same change in the axial vector mass.

Table 2. Enumeration of terms in the baryon axial vector/pseudoscalar current, drawn in Figure 14.

1. Diagram 1, two terms: $\langle J \rangle_q^S$—probe strikes dressed quark with scalar diquark spectator; $\langle J \rangle_q^A$—probe strikes dressed quark with axial vector diquark spectator.
2. Diagram 2: $\langle J \rangle_{qq}^{AA}$—probe strikes axial vector diquark with dressed quark spectator.
3. Diagram 3: $\langle J \rangle_{qq}^{\{SA\}}$—probe mediates transition between scalar and axial vector diquarks, with dressed quark spectator.
4. Diagram 4, three terms: $\langle J \rangle_{ex}^{SS}$—probe strikes dressed quark "in-flight" between one scalar diquark correlation and another; $\langle J \rangle_{ex}^{\{SA\}}$—dressed quark "in-flight" between a scalar diquark correlation and an axial vector correlation; $\langle J \rangle_{ex}^{AA}$—"in-flight" between one axial vector correlation and another.
5. Diagrams 5 and 6, seagull diagrams describing the probe coupling into the diquark correlation amplitudes: $\langle J \rangle_{\text{sg}}$. There is one contribution from each diagram to match every term in Diagram (4).

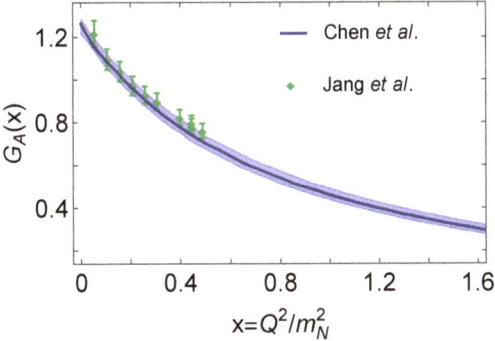

Figure 15. Predicted result for $G_A(x)$ in Reference [275] (Chen et al.) (blue curve within lighter blue uncertainty band), compared with lQCD results from Reference [347] (Jang et al.) (green) diamonds. With respect to the CSM central results, this comparison may be quantified by reporting the mean-χ^2 value, which is 0.27.

Turning to the induced pseudoscalar form factor, muon capture experiments ($\mu + p \to \nu_\mu + n$) may be used to determine the induced pseudoscalar charge:

$$g_p^* = \frac{m_\mu}{2m_N} G_P(Q^2 = 0.88\, m_\mu^2). \tag{35}$$

The CSM prediction is $g_p^* = 8.80(23)$. Compared with the recent MuCap Collaboration result, $g_p^* = 8.06(55)$ [348,349], it agrees within uncertainties, but is slightly larger. The CSM value is nicely aligned with the world average [350]: $g_p^* = 8.79(1.92)$.

The pseudoscalar form factor, $G_5(Q^2)$, is of interest because, inter alia, it is used to define the pion–nucleon form factor $G_{\pi NN}(Q^2)$ via

$$G_5(Q^2) =: \frac{m_\pi^2}{Q^2 + m_\pi^2} \frac{f_\pi}{m_q} G_{\pi NN}(Q^2), \tag{36}$$

where f_π is the pion leptonic decay constant (appearing in Figure 7) and $G_{\pi NN}(-m_\pi^2) = g_{\pi NN}$ is the πNN coupling, a key input to nucleon+nucleon potentials. The CSM prediction is [274,275]: $g_{\pi NN}/m_N = 14.02(33)/\text{GeV}$. This value overlaps with that inferred from pion–nucleon scattering [351] ($g_{\pi NN}/m_N = 13.97(10)/\text{GeV}$) and compares favourably with a determination based on the Granada 2013 np and pp scattering database [352] ($g_{\pi NN}/m_N = 14.11(3)/\text{GeV}$) and a recent analysis of nucleon–nucleon scattering using effective field theory and related tools [353] ($g_{\pi NN}/m_N = 14.09(4)/\text{GeV}$). All these results are compared in Figure 16, which also highlights their error-weighted average:

$$g_{\pi NN}/m_N = 14.10(2)/\text{GeV}. \tag{37}$$

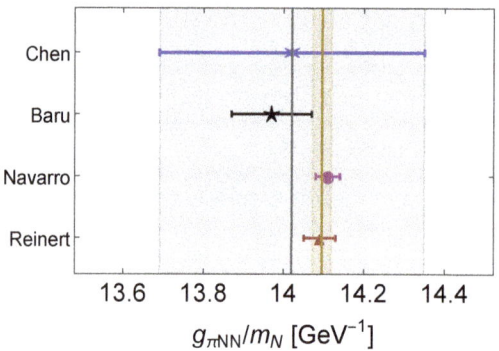

Figure 16. Comparison of the CSM prediction for $g_{\pi NN}/m_N$ (blue asterisk) [275] with values extracted from pion–nucleon scattering [351] (Baru) (black star), the Granada 2013 np and pp scattering database [352] (Navarro) (purple circle), and nucleon–nucleon scattering [353] (Reinert) (red triangle). The vertical grey band marks the estimated uncertainty in the CSM prediction. The error-weighted average of the depicted results, Equation (37), is drawn as the gold line within the like-coloured band.

Continuing with an effort to inform nuclear physics potentials using hadron physics results, it is worth noting that on $-m_\pi^2 < Q^2 < 2\,m_N^2$, a fair approximation to the CSM prediction for the pion–nucleon form factor is provided by ($x = Q^2/m_N^2$):

$$G_{\pi NN}^d(x) = \frac{13.47\, m_N}{(1 + x/0.845^2)^2}. \tag{38}$$

This corresponds to a πNN dipole mass $\Lambda_{\pi NN} = 0.845\, m_N = 0.79\,\text{GeV}$, viz. a soft form factor. (A commensurate value was obtained previously in a simpler quark+scalar-diquark model [354].) Being just $\sim 20\%$ greater, the CSM prediction is qualitatively equivalent to the πNN dipole mass inferred in a dynamical coupled channels analysis of πN, γN

interactions [292]. Future such coupled channels studies may profit by implementing couplings and range parameters determined in CSM analyses.

An extension of this effort to the more general case of hyperon+nucleon potentials can be found in Reference [355]. Using an SCI, predictions were made therein for an array of meson+octet-baryon couplings. Comparing the results with extant phenomenological interactions [292,356,357], one finds a mean absolute relative deviation (P is the pseudoscalar meson absorbed in the baryon transition $B \to B'$, and $g_{PB'B}$ is the associated coupling):

$$\delta_r^g := \{|g_{PB'B}^{\text{SCI}}/g_{PB'B}^{\text{phen.}} - 1|\} = 0.18(14). \tag{39}$$

This result strengthens the case in favour of using CSM predictions for the couplings as new constraints in the development of baryon+baryon potentials.

A little algebra reveals that the proton axial charge can be interpreted as a measure of the valence quark contributions to the proton light-front helicity, e.g., Reference [358] (Equations (6)–(8)):

$$g_A = \int_0^1 dx \left[\Delta u^p(x; \zeta_\mathcal{H}) - \Delta d^p(x; \zeta_\mathcal{H}) \right] =: g_A^u - g_A^d, \tag{40}$$

where $\Delta q^p(x; \zeta_\mathcal{H}) = q_\uparrow^p(x; \zeta_\mathcal{H}) - q_\downarrow^p(x; \zeta_\mathcal{H})$ is the light-front helicity DF for a quark q. Plainly, Δq^p is the difference between the light-front number density of quarks with helicity parallel and antiparallel to that of the proton. It is scale-dependent.

Equation (40) conveys additional significance to a flavour separation of the axial charge form factor:

$$G_A(Q^2) = G_A^u(Q^2) - G_A^d(Q^2). \tag{41}$$

A detailed analysis is presented in Reference [276] (Section 4), which reveals the following diagram contributions to the separate u, d axial form factors:

$$G_A^u = \langle J \rangle_q^S - \langle J \rangle_q^A + \langle J \rangle_{qq}^{AA} + \tfrac{1}{2}\langle J \rangle_{qq}^{\{SA\}} + 2\langle J \rangle_{\text{ex}}^{\{SA\}} + \tfrac{4}{5}\langle J \rangle_{\text{ex}}^{AA}, \tag{42a}$$

$$-G_A^d = 2\langle J \rangle_q^A + \tfrac{1}{2}\langle J \rangle_{qq}^{\{SA\}} + \langle J \rangle_{\text{ex}}^{SS} - \langle J \rangle_{\text{ex}}^{\{SA\}} + \tfrac{1}{5}\langle J \rangle_{\text{ex}}^{AA}, \tag{42b}$$

where the nomenclature of Table 2 is used. Identified according to Equations (42), the calculated $Q^2 = 0$ contributions are listed in Table 3. It is worth stressing that Equations (42) express the fact that since a 0^+ diquark cannot couple to an axial vector current, then Diagram 1 in Figure 14 only supplies a u quark contribution to the proton $G_A(Q^2)$, viz. $\langle J \rangle_q^S$. It follows that in a scalar-diquark-only proton, a d quark contribution to $G_A(Q^2)$ can only arise from Figure 14, Diagram 4, i.e., $\langle J \rangle_{\text{ex}}^{SS}$; and $|\langle J \rangle_{\text{ex}}^{SS}/\langle J \rangle_q^S| \approx 0.06$. Notably, many scalar-diquark-only models omit Diagram 4, in which case $G_A^d(Q^2) \equiv 0$. An extension of these observations to the complete array of octet baryons is described elsewhere [355] (Section IV).

Table 3. Flavour and diagram—Figure 14—separation of the proton axial charge: $g_A^u = G_A^u(0)$, $g_A^d = G_A^d(0)$; $g_A^u - g_A^d = 1.25(3)$. The listed uncertainties express the effect of ±5% variations in the diquark masses, e.g., $0.886_\mp \Rightarrow 0.88 \mp 0.06$.

	$\langle J \rangle_q^S$	$\langle J \rangle_q^A$	$\langle J \rangle_{qq}^{AA}$	$\langle J \rangle_{qq}^{\{SA\}}$	$\langle J \rangle_{\text{ex}}^{SS}$	$\langle J \rangle_{\text{ex}}^{\{SA\}}$	$\langle J \rangle_{\text{ex}}^{AA}$
g_A^u	$0.886_\mp$	$-0.080_\pm$	$0.030_\mp$	$0.080_\mp$	0	≈ 0	$0.031_\pm$
$-g_A^d$	0	$0.160_\pm$	0	$0.080_\pm$	$0.051_\pm$	≈ 0	$0.010_\pm$

Using the solution of the Faddeev equation (Figure 8), Reference [276] reports

$$g_A^u/g_A = 0.76 \pm 0.01, \quad g_A^d/g_A = -0.24 \pm 0.01, \quad g_A^d/g_A^u = -0.32 \pm 0.02. \tag{43}$$

In nonrelativistic quark models with uncorrelated wave functions, $g_A^d/g_A^u = -1/4$. Hence, the relevant comparison reveals that the highly correlated wave function obtained by solving the Faddeev equation gives the valence d quark a markedly larger fraction of the proton's light-front helicity than is found in simple quark models. Reviewing the discussion after Equation (42), it becomes apparent that this feature owes to the presence of axial vector diquarks in the proton: the current contribution arising from the $\{uu\}$ correlation—underlined term in Equation (42b)—measuring the probe striking the valence d quark, is twice as strong as that from the $\{ud\}$ correlation—underlined term in Equation (42a)—in which the probe strikes the valence u quark.

It is natural to enquire after the robustness of the results in Equation (43). Consider, therefore, that assuming SU(3) flavour symmetry in analyses of octet baryon axial charges, these charges are expressed in terms of two low-energy constants ([359] (Table 1)): D, F, with $g_A^u = 2F$, $g_A^d = F - D$. (This assumption is accurate to roughly 4%—see, e.g., Reference [355].) In this case, the values in Equation (43) correspond to

$$D = 0.78(2), \quad F = 0.48(1), \quad F/D = 0.61(2). \tag{44}$$

On the other hand, using available empirical information [40], one obtains $D = 0.774(26)$, $F = 0.503(27)$, and $g_A^u/g_A = 0.79(4)$, $g_A^d/g_A = -0.21(3)$, and $g_A^d/g_A^u = -0.27(4)$, values that are consistent with the results in Equations (43) and (44).[6] In addition, the SCI predicts [355]

$$D = 0.78, \quad F = 0.43, \quad F/D = 0.56, \tag{45}$$

and a covariant baryon chiral perturbation theory analysis yields $D = 0.80(1)$, $F = 0.47(1)$, $F/D = 0.59(1)$ [360].

Given the favourable realistic proton wave function comparisons presented above, the values in Equation (43) can be viewed as reliable. This is important because of the connection between flavour-separated axial charges and the so-called proton "spin crisis" [361,362]. At any given resolving scale, the singlet, triplet, and octet axial charges of the proton are, respectively:

$$a_0 = g_A^u + g_A^d + g_A^s, \quad a_3 = g_A = g_A^u - g_A^d, \quad a_8 = g_A^u + g_A^d - 2g_A^s. \tag{46}$$

If working at the hadron scale, $\zeta_\mathcal{H}$, where dressed valence quasiparticles carry all proton properties [126,127,132–135], then $g_A^s \equiv 0$ $a_0 = a_8$; hence [276],

$$a_0 = 0.65(2). \tag{47}$$

In general, $a_{3,8}$ are conserved charges, viz. they are the same at all resolving scales, ζ. However, that is not true of the individual terms in their definitions: the separate valence quark charges g_A^u, g_A^d, g_A^s evolve with ζ [362]. Consequently, the value of a_0, which is the fraction of the proton's total $J = 1/2$ carried by valence degrees-of-freedom, changes with scale: the result in Equation (47) is the maximum value of a_0, and the fraction falls slowly with increasing ζ.

Returning to expectations based on simple, nonrelativistic quark models, textbook-level algebra yields $a_0 = 1$. Therefore, in such pictures, all the proton spin derives from that of the constituent quarks. On the other hand, the CSM analysis in Reference [276] predicts that proton dressed valence degrees-of-freedom carry only two-thirds of the spin. Since there are no other degrees-of-freedom at $\zeta_\mathcal{H}$ and the Poincaré-covariant proton wave function properly describes a $J = 1/2$ system, then the "missing" part of the total J must be associated with quark+diquark orbital angular momentum. Similar conclusions apply for all ground state octet baryons [355].

The study in Reference [276] delivered Poincaré-invariant parameter-free predictions for the proton axial form factor and its flavour separation out to $Q^2 \approx 10\, m_N^2$. The axial form factor

---

[6] If one eliminates axial vector diquarks from the proton wave function, then $g_A^d/g_A^u = -0.054(13)$, a result disfavoured by experiments at the level of 5.1σ, i.e., the probability that the scalar-diquark-only proton result could be consistent with data is $1/7,100,000$.

itself agrees with available data [363,364], which extends to $Q^2 \approx 5\,m_N^2$—see Reference [276] (Figure 3). More importantly, the results will likely serve as motivation for new experiments aimed at exploring nucleon structure with axial vector probes instead of the photon, opening a new window onto hadron structure. It is worth highlighting here that a dipole fit to data is only a good approximation on the fitting domain. With increasing Q^2, the dipole increasingly overestimates the actual result, being 56(5)% too large at $Q^2 = 10\,m_N^2$—see Reference [276] (Figure 4B). It therefore becomes an unsound tool for developing qualitative insights and quantitative cross-section estimates.

Furthermore, with flavour-separated form factors in hand on such a large-Q^2-domain, Reference [276] was able to calculate and contrast the u and d quark contributions to the associated light-front transverse spatial density profiles:

$$\hat{\rho}_A^f(|\hat{b}|) = \int \frac{d^2\vec{q}_\perp}{(2\pi)^2}\,e^{i\vec{q}_\perp\cdot\hat{b}}\,G_A^f(x)\,, \tag{48}$$

with $G_A^f(x)$ interpreted in a frame defined by $Q^2 = m_N^2 q_\perp^2$, $m_N q_\perp = (\vec{q}_{\perp 1}, \vec{q}_{\perp 2}, 0, 0) = (Q_1, Q_2, 0, 0)$. These profiles are depicted in Figure 17. We note that $|\hat{b}|$ and $\hat{\rho}_A^f$ are dimensionless; so, the images drawn in Figure 17 can be mapped into physical units using:

$$\rho_A^f(|b| = |\hat{b}|/m_N) = m_N^2\,\hat{\rho}_A^f(|\hat{b}|)\,, \tag{49}$$

in which case $|\hat{b}| = 1$ corresponds to $|b| \approx 0.2$ fm and $\hat{\rho}_1^f = 0.1 \Rightarrow \rho_1^f \approx 2.3/\text{fm}^2$.

The top row of Figure 17 provides two-dimensional renderings of the flavour-separated transverse density profiles calculated from a proton wave function which does not include axial vector diquarks, i.e., a scalar-diquark-only proton. In this case, the u quark profile is far more diffuse than that of the d quark, viz. its extent in the light-front transverse spatial plane is much greater. One may quantify this by reporting the associated radii: $r_{A^d}^\perp = 0.24$ fm, $r_{A^u}^\perp = 0.48$ fm, so the d/u ratio of radii is ≈ 0.5.

The bottom row of Figure 17 was obtained using a realistic proton wave function, in which both scalar and axial vector diquarks are present with the strength determined by the Faddeev equation in Figure 8. In this realistic case, the d quark profile is not very different from that of the u quark: relative to the u quark profile, the intensity peak is only somewhat broader for the d quark; and comparing radii,

$$r_{A^d}^\perp = 0.43\text{ fm}, \quad r_{A^u}^\perp = 0.49\text{ fm}, \quad r_{A^d}^\perp/r_{A^u}^\perp \approx 0.9\,. \tag{50}$$

The CSM predictions for nucleon axial and pseudoscalar form factors discussed in this section complement those for the large-Q^2 behaviour of nucleon electromagnetic elastic and transition form factors reported, e.g., in References [272,365]. One may now anticipate that predictions for form factors characterising weak interaction induced $N \to \Delta(1232)$ and $N \to N^*(1535)$ transitions will soon become available. Each will shed new light on nucleon structure; and the former, calculated on a domain that stretches from low-to-large-Q^2, will likely prove valuable in developing a more reliable understanding of neutrino scattering from nucleons and nuclei.

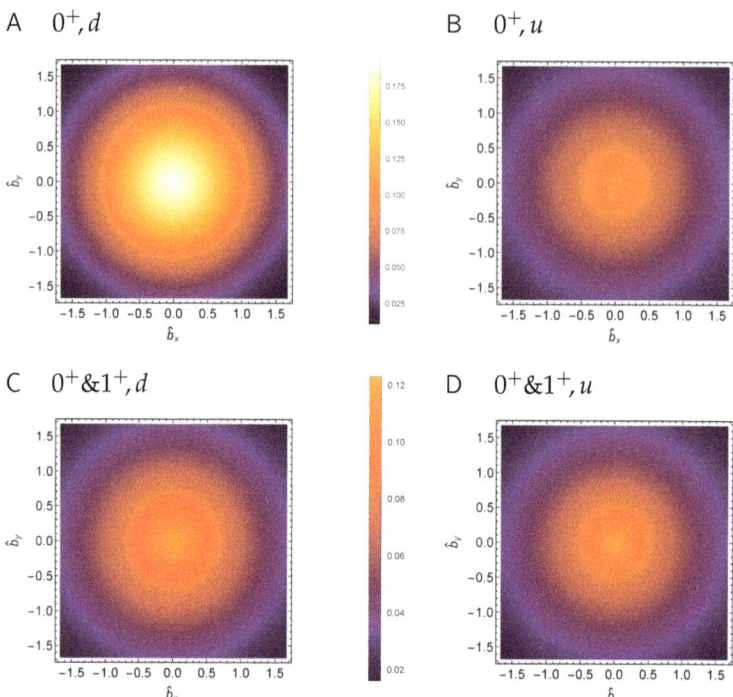

Figure 17. Transverse density profiles, Equation (48), calculated in Reference [276] from flavour-separated proton axial form factors. Top row: scalar-diquark-only proton. Panel **A**: two-dimensional plot of $\hat{\rho}_A^d(|\hat{b}|)/g_A^d$; Panel **B**: similar plot of $\hat{\rho}_A^u(|\hat{b}|)/g_A^u$. Removing the $1/g_A^f$ normalisation, the $b = 0$ values are $\hat{\rho}_A^d(0) = -0.009$, $\hat{\rho}_A^u(0) = 0.097$. Bottom row: realistic proton Faddeev amplitude, including axial vector diquarks as predicted by the Faddeev equation in Figure 8. Panel **C**: $\hat{\rho}_A^d(|\hat{b}|)/g_A^d$; Panel **D**: $\hat{\rho}_A^u(|\hat{b}|)/g_A^u$. Removing the $1/g_A^f$ normalisation, the $b = 0$ values of these profiles are $\hat{\rho}_A^d(0) = -0.038$, $\hat{\rho}_A^u(0) = 0.12$. N.B. $\int d^2\hat{b}\,\hat{\rho}_A^f(|\hat{b}|)/g_A^f = 1, f = u, d$.

10. Transition Form Factors of Heavy+Light Mesons

Heavy mesons ($B_{q=c,s,u}$, D_s, D) are special for many reasons; and their mass budgets and role in exposing constructive interference between Nature's two known sources of mass are of particular interest herein. Consider, therefore, Figure 18 and contrast the images with those in Figure 1. Evidently, for heavy meson masses: (*i*) the HB component is largest in each case and its relative size grows as the current–masses of the valence constituents increase; (*ii*) all receive a significant EHM+HB interference component, but its relative strength diminishes with increasing current–masses; and (*iii*) for vector heavy mesons, but not pseudoscalar mesons, there is an EHM component, but its relative strength drops as the HB component grows.

Next consider semileptonic weak-interaction transitions between heavy and light mesons. Comparing Figures 1 and 18, it is apparent that heavy-pseudoscalar to light-pseudoscalar transitions serve to probe the relative impacts of the strength of EHM+HB interference in the initial and final states, whereas heavy-pseudoscalar to light-vector transitions overlap systems in which HB mass is dominant with those whose mass is owed almost entirely to EHM. Both classes of transitions, therefore, present excellent opportunities for exposing the influence of Nature's two known sources of mass on physical observables. These cases are of special interest, of course, because the transitions have long been used to place constraints on the values of the elements of the Cabibbo–Kobayashi–Maskawa (CKM) matrix, which parametrises quark flavour mixing in the SM.

A unifying analysis of both classes of transitions was recently completed using the SCI [233,366], yielding results that compare favourably with other reliable experimental or independent theory analyses. The SCI branching fraction predictions should therefore be a reasonable guide. This is important because predictions were made for several branching fraction ratios—$R_{D_{(s)}^{(*)}}$, $R_{J/\psi}$, R_{η_c}—whose values are direct tests of lepton universality in weak interactions: the SCI values confirm SM predictions, hence, speak for universality, as we discuss below. The analyses also used $B_{(s)} \to D_{(s)}^*$ transitions to predict the precursor functions which evolve into the universal Isgur–Wise function [367], obtaining results in agreement with empirical inferences—References [368] (Equations (177) and (181)) and [369] (Belle).

The SCI's successes in these applications highlight the need for kindred studies using an interaction with a closer connection to QCD. The impediment has always been the large disparity in masses that typically exists between initial and final states. That mass imbalance requires, inter alia, that any approach to the problem be simultaneously able to deal with both chiral and heavy quark limits in quantum field theory. To date, compared with the SCI, no framework with a better link to QCD can directly surmount this difficulty. Nevertheless, following Reference [205], practicable and effective algorithms for continuum studies do now exist. They exploit the strengths of the statistical SPM as a tool for interpolating data (broadly defined) and therefrom deliver extrapolations with a rigorously defined and calculable uncertainty [238]. Namely, results are calculated on domains of current–quark mass for which transition form factors may straightforwardly be obtained. The SPM is then used to extrapolate those results and arrive at predictions for the physical processes of interest.

At present, CSM RL truncation predictions are available for the following transitions [206,207]: $B_c \to J/\psi, \eta_c$; $B_{(s)} \to \pi(K)$; $D_s \to K$; $D \to \pi, K$; and $K \to \pi$. The last process is something of a test for the approach because such $K_{\ell 3}$ transitions have long been of experimental and theoretical interest [40] (Section 62). The calculated branching fractions are gathered in Table 4, from which it will be seen that CSMs deliver sound results.

Table 4. CSM predictions for pseudoscalar meson semileptonic branching fractions [206,207]—each such fraction is to be multiplied by 10^{-3}. The column labelled "ratio" is the ratio of the preceding two entries in the row, so *no* factor of 10^{-3} is applied in this column. (A 1σ SPM uncertainty is listed for the CSM predictions.) Reference [40] (PDG) lists the following values for the CKM matrix elements: $|V_{us}| = 0.2245(8)$, $|V_{cd}| = 0.221(4)$, $|V_{cs}| = 0.987(11)$ $|V_{ub}| = 0.00382(24)$; and the following lifetimes (in seconds): $\tau_{K^+} = 1.2379(21) \times 10^{-8}$, $\tau_{D^0} = 4.10 \times 10^{-13}$, $\tau_{D_s^\pm} = 5.04 \times 10^{-13}$, $\tau_{B^0} = 1.519 \times 10^{-12}$, $\tau_{B_s^0} = 1.515 \times 10^{-12}$, $\tau_{B_c^\pm} = 0.51 \times 10^{-12}$.

$\mathcal{B}_{I \to F(\ell \nu_\ell)}$	References [206,207]			PDG [40] or Other, If Indicated		
	$e^+ \nu_e$	$\mu^+ \nu_\mu$	ratio	$e^+ \nu_e$	$\mu^+ \nu_\mu$	ratio
$K^+ \to \pi^0$	50.0(9)	33.0(6)	0.665	50.7(6)	33.5(3)	0.661(07)
$D^0 \to \pi^-$	2.70(12)	2.66(12)	0.987(02)	2.91(4)	2.67(12)	0.918(40)
$D_s^+ \to K^0$	2.73(12)	2.68(12)	0.982(01)	3.25(36) [370]		
$D^0 \to K^-$	39.0(1.7)	38.1(1.7)	0.977(01)	35.41(34)	34.1(4)	0.963(10)
	$\mu^- \bar{\nu}_\mu$	$\tau^- \bar{\nu}_\tau$	ratio	$\mu^- \bar{\nu}_\mu$	$\tau^- \bar{\nu}_\tau$	ratio
$\bar{B}^0 \to \pi^+$	0.162(44)	0.120(35)	0.733(02)	0.150(06)		
$\bar{B}_s^0 \to K^+$	0.186(53)	0.125(37)	0.667(09)			
$B_c \to \eta_c$	8.10(45)	2.54(10)	0.31(2)			
$B_c \to J/\psi$	17.2(1.9)	4.17(66)	0.24(5)			

One of the key results in Reference [206] concerns a ratio of $B_c^+ \to J/\psi$ branching fractions measured for the first time by the LHCb Collaboration fairly recently [371]:

$$R_{J/\psi} := \frac{\mathcal{B}_{B_c^+ \to J/\psi \tau \nu}}{\mathcal{B}_{B_c^+ \to J/\psi \mu \nu}} = 0.71 \pm 0.17 \, (\text{stat}) \pm 0.18 \, (\text{syst}) \,. \tag{51}$$

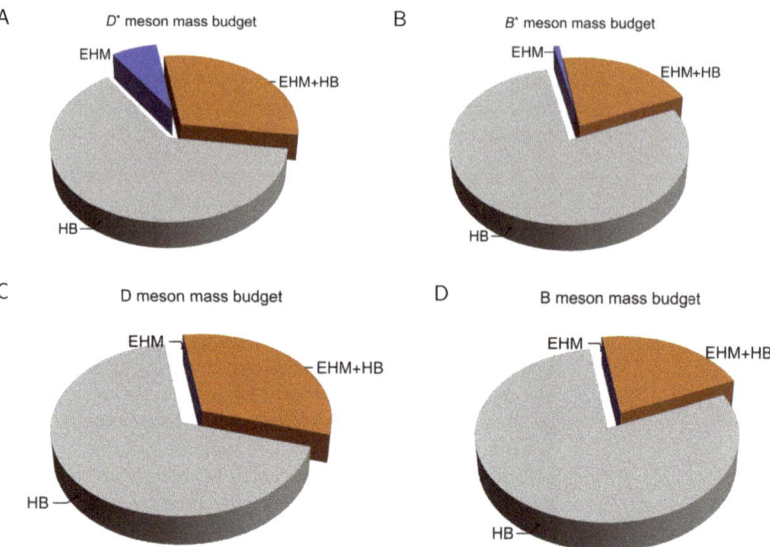

Figure 18. Mass budgets: (**A**): D^* meson; (**B**): B^*-meson; (**C**): D meson; (**D**): B meson. Each is drawn using a Poincaré-invariant decomposition and the numerical values listed in Table 1. (Separation at $\zeta = 2\,GeV$, calculated using information from References [35,40]).

This value is plotted in Figure 19 and compared with the CSM prediction and other Standard Model calculations. Evidently, the LHCb measurement lies approximately 2σ above the values predicted by reliable SM calculations. If future, precision experiments do not deliver a markedly lower central value, then one might begin to judge that lepton flavour universality is violated in $B_c \to J/\psi$ semileptonic decays. As yet, however, the experimental precision is insufficient to support such a claim. Furthermore, a compelling case would need to include information on $B_c \to \eta_c$ semileptonic decays. The CSM prediction is $R_{\eta_c} = 0.313(22)$—see Table 4; the SCI result is $R_{\eta_c} = 0.25$ [233]; and a mean value of $0.31(4)$ is obtained from modern continuum analyses [372–377]. An experimental value is lacking.

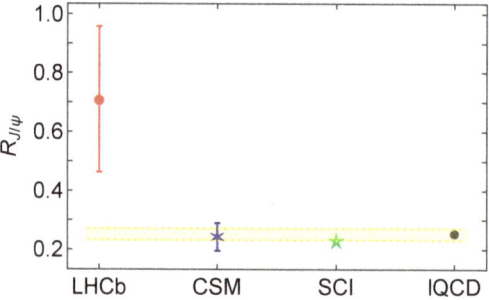

Figure 19. Ratio $R_{J/\psi}$ in Equation (51)—red circle, empirical result [371] (LHCb); blue asterisk—CSM prediction [206]; green star—SCI prediction [366]; grey circle—lQCD result [378,379]; gold band—unweighted mean of central values from several calculations [372–377].

The array of analyses in Reference [207] yields novel results in other areas. Of particular interest are the discussions of $D^0 \to K^-$ transition form factors and the value of $|V_{cs}|$. Two independent form factors characterise $0^+ \to 0^+$ transitions, viz. vector and scalar, $f_{+,0}(t)$, respectively, where t is the Mandelstam variable, whose value expresses the momentum transferred to the final state. The CSM predictions are plotted in Figure 20 and compared with available data [380–383]. The CSM result is largely consistent with this collection, although there may be a hint that it is too high at lower t values. Concerning branching fractions, form factor contributions from this domain are important. It is therefore notable that, within mutual uncertainties, the CSM value for $f_+^{D\to K}(0) = 0.796(9)$ agrees with the $N_f = 2+1+1$ lQCD result in Reference [384]: $0.765(31)$.

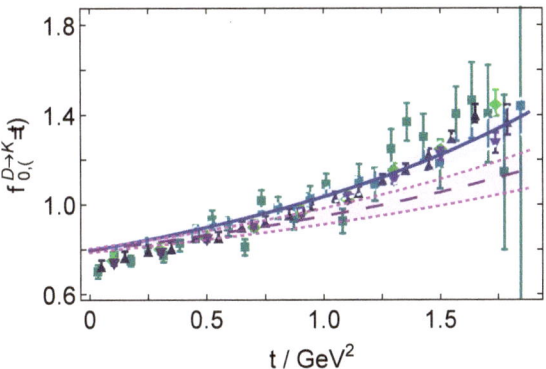

Figure 20. $D \to K$ transition form factors. f_+—solid blue curve; f_0—long-dashed purple curve; like-coloured shaded bands—associated SPM uncertainty in each case. Data: cyan squares [380]; green diamonds [381]; dark-blue up-triangles [382]; indigo down-triangles [383].

With the form factors in Figure 20, Reference [207] obtained the $D^0 \to K^-$ branching fractions listed in Table 4 when using the value of $|V_{cs}|$ listed in the caption: evidently, both the $e^+\nu_e$ and $\mu^+\nu_\mu$ fractions exceed their respective PDG values. On the other hand, the ratio agrees within 1.4σ; so, a common overall factor can remedy the mismatch. Adopting this perspective, then the value $|V_{cs}| = 0.937(17)$ combined with the CSM form factors delivers branching fractions that match the PDG values, viz. $3.52(18)\%$ and $3.44(18)\%$, respectively. Actually, referring to Reference [40] (Section 12.2.4), one sees that the inferred CSM value is both commensurate with and more precise than one of the two used to arrive at the PDG average listed in the caption of Table 4. With $|V_{cs}| = 0.937(17)$ used instead to compute this average, one finds a slightly more precise central value that is 1σ lower:

$$|V_{cs}| = 0.974(10). \tag{52}$$

Predictions for semileptonic $\bar{B}^0 \to \pi^+$, $\bar{B}_s^0 \to K^+$ transition form factors and branching fractions were also delivered in Reference [207]. As emphasised above, such processes present challenges because π, K are Nambu–Goldstone bosons and there is a huge disparity between the masses of the initial and final states. Consequently, comparisons with data serve as a stringent test of the new CSM algorithms.

CSM predictions for the $\bar{B}^0 \to \pi^+$ transition form factors are depicted in Figure 21A. Regarding $f_+^{B\to\pi}$, data have been collected by two collaborations [385–388]: within mutual uncertainties, the CSM predictions agree with these data. The data support a value

$$f_+^{B\to\pi}(t=0) = 0.27(2), \tag{53}$$

which is consistent with the CSM prediction [207]: $0.29(5)$.

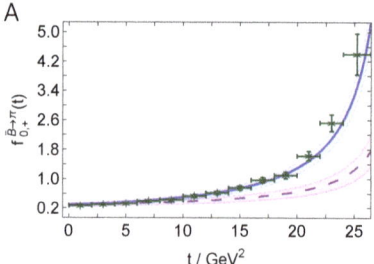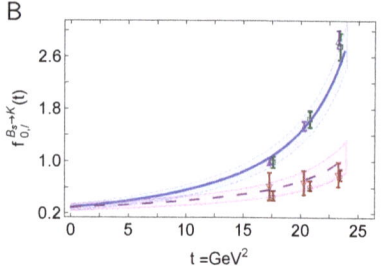

Figure 21. CSM predictions for $B_{(s)}$ semileptonic transition form factors: f_+—solid blue curve; f_0—long-dashed purple curve; SPM uncertainty in each—like-coloured shaded bands. *Left panel* (A): $\bar{B}^0 \to \pi^+$. Data (green stars): reconstructed from the average [368] (Table 81) of data reported in References [385–388]. *Right panel* (B): $\bar{B}_s^0 \to K^+$. lQCD results: f_+—indigo open up-triangles [389] and green open boxes [390]; f_0—brown open down-triangles [389] and red open circles [390].

Using the form factors in Figure 21A, one obtains the $\bar{B}^0 \to \pi^+$ branching fractions in Table 4. The PDG lists a result for the $\mu^- \nu_\mu$ final state, which matches the CSM prediction within mutual uncertainties. Precise agreement is obtained using

$$|V_{ub}| = 0.00374(44). \tag{54}$$

This value is commensurate with other analyses of $\mathcal{B}_{\bar{B}^0 \to \pi^+ \mu^- \bar{\nu}_\mu}$ [40] (Section 76.3), thus increases tension with the higher value inferred from inclusive decays. No data are available on the $\tau^- \nu_\tau$ final state; so, the $\tau{:}\mu$ ratio is empirically unknown. Here, a $N_f = 2 + 1$-flavour lQCD study yields 0.69(19) [390], which, within its uncertainty, matches the CSM result: 0.733(2).

Figure 21B displays CSM predictions for the $\bar{B}_s \to K^+$ form factors. Although the $B_s \to K^-$ transition was recently observed [391], with the measurement yielding the branching fraction:

$$\mathcal{B}_{\bar{B}_s^0 \to K^- \mu^+ \nu_\mu} = [0.106 \pm 0.005_{\text{stat}} \pm 0.008_{\text{syst}}] \times 10^{-3}, \tag{55}$$

no form factor data are yet available. Comparisons are therefore made in Figure 21B with results obtained using $N_f = 2 + 1$-flavour lQCD [389,390]. Owing to difficulties encountered when using lattice methods to calculate form factors of heavy+light mesons, lQCD results are limited to a few points on the domain $t \gtrsim 17 \text{ GeV}^2$—see Figure 21B. Today, lattice analyses typically employ such results to construct a least-squares fit to the form factor points, using some practitioner-favoured functional form. That fit is then employed to define the form factor on the whole kinematically accessible domain: $0 \lesssim t \lesssim 25 \text{ GeV}^2$ in this case. It is worth noting that, at this time, given the small number of points and their limited precision, the SPM cannot gainfully be used to develop function-form unbiased interpolations and extrapolations of the lQCD output.

The CSM form factors in Figure 21B yield the $\bar{B}_s^0 \to K^+$ branching fractions in Table 4. Figure 22A compares the $\mu^- \bar{\nu}_\mu$ value with the measurement in Equation (55) and also results obtained via various other means. Experiment and theory only agree because the theory uncertainty is large. The unweighted theory average is 0.141(44)‰, and the uncertainty-weighted mean is 0.139(08)‰. These values increase when Entries V–VI [390,392] are omitted: unweighted 0.159(38)‰ and uncertainty weighted 0.156(10)‰. The extrapolations employed in V–VI [390,392] lead to values of $f_+^{\bar{B}_s \to K}(0)$ that are $\sim 50\%$ of those obtained in I–IV [389,393–395]: 0.148(53) vs. 0.299(86). This can explain the difference in branching fractions: V–VI vs. I–IV in Figure 22A. Significantly, a different approach to fitting and extrapolating lQCD results, using the LHCb datum, Equation (55), as an additional constraint, produces [396]: $f_+^{\bar{B}_s \to K}(0) = 0.211(3)$.

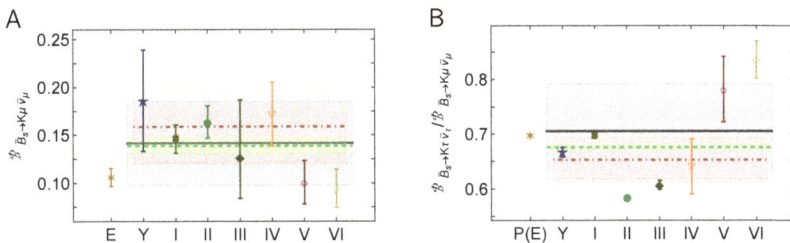

Figure 22. *Left panel* (**A**): Branching fraction $\mathcal{B}_{\bar{B}_s^0 \to K^+ \mu^- \bar{\nu}_\mu}$ computed in Reference [207], "Y", compared with the value in Equation (55), "E", viz. a measurement of $\mathcal{B}_{\bar{B}_s^0 \to K^- \mu^+ \nu_\mu}$ [391], and some results obtained using other approaches: continuum I–III [393–395]; lattice IV–VI [389,390,392]. *Right panel* (**B**): Branching fraction ratio $\mathcal{B}_{\bar{B}_s^0 \to K^+ \tau^- \bar{\nu}_\tau}/\mathcal{B}_{\bar{B}_s^0 \to K^+ \mu^- \bar{\nu}_\mu}$ computed in Reference [207] compared with some results obtained using other approaches. The legend matches Panel A, except "P(E)" is the result from [396], i.e., an estimate constrained by the datum in Reference [391]. Both panels: Grey line: unweighted mean of theory results. Pink dotted-dashed line: unweighted mean of theory results, omitting V–VI. Green dashed line: uncertainty-weighted average of theory results. Like-coloured bands mark associated uncertainties in each case.

Reference [396] also infers $f_+^{\bar{B} \to \pi}(0) = 0.255(5)$, leading to $f_+^{\bar{B}_s \to K}(0)/f_+^{\bar{B} \to \pi}(0) < 1$. This outcome conflicts with the CSM prediction, which has $f_+^{\bar{B}_s \to K}(0)/f_+^{\bar{B} \to \pi}(0) > 1$ at the 85% confidence level [207] (Equation (9)), and the results in a raft of other studies, e.g., References [393,394,397–403]. It is probable, therefore, that the value for $f_+^{\bar{B}_s \to K}(0)$ in Reference [396] is too small. It is worth remarking that the SCI is unclear on the value of this ratio. It produces $f_+^{\bar{B}_s \to K}(0)/f_+^{\bar{B} \to \pi}(0) < 1$, but the individual $t = 0$ values are too large by a factor of two [233] (Table 3A). On the other hand, the $t = 0$ value of the kindred ratio of vector form factors in $\bar{B}_s \to K^*$, $\bar{B} \to \rho$ transitions is greater than unity [366] (Table 1A). Notably, the ratio $f_+^{\bar{B}_s \to K}(0)/f_+^{\bar{B} \to \pi}(0)$ is a marker for SU(3)-flavour symmetry breaking and its modulation by EHM; so, it is worth reaching a sound conclusion on the value of the ratio. It is here relevant to observe that $f_K/f_\pi = 1.2 > 1$.

Regarding the $\tau\nu_\tau$ final state in $\bar{B}_s \to K$ transitions, no empirical information is currently available, hence none on the $|V_{ub}|$-independent ratio that would test lepton flavour universality. In Figure 22B, therefore, we compare the CSM prediction for this ratio, drawn from Table 4, with results obtained via other means. The unweighted average of theory results is $0.705(87)$, and the uncertainty weighted mean is $0.678(03)$. Omitting Entries V–VI [390,392], these values are: unweighted $0.653(41)$ and uncertainty weighted $0.677(03)$. Within sound analyses, many uncertainties cancel in this ratio; so, the results should be more reliable than any calculation of either fraction alone. Nevertheless, the values are widely scattered, indicating that there is ample room for improving the precision of $\bar{B}_s^0 \to K$ theory. Of course, measurements enabling extraction of $\bar{B}_s^0 \to K^-$ form factors would be very useful, too, for refining both (*a*) comparisons with theory and between theory analyses and (*b*) making progress toward a more accurate value of $|V_{ub}|$.

Such predictions for heavy-to-light meson electroweak transition form factors are a new branch of application for CSMs. They are far more sophisticated and robust than the Schwinger function parametrisation-based analyses in, e.g., References [324,403,404] and significantly improve upon earlier RL truncation studies of $\pi_{\ell 3}$ and $K_{\ell 3}$ transitions [405,406]. The keys to these advances are an improved understanding of RL truncation and the capacity to greatly expand its quark mass and mass splitting domains of applicability using the SPM. Fairly soon, one can expect these advances to be exploited in the study of kindred baryon transitions.

11. Distribution Functions

Hadron parton DFs are probability densities: each one describes the light-front fraction, x, of the hadron's total momentum carried by a given parton species within the bound state [407]. They are a much-prized source of hadron structure information; and following the quark discovery experiments fifty years ago [408–411], measurements interpretable in terms of hadron DFs have been awarded a high priority. For much of this time, DFs were inferred from global fits to data, with the results viewed as benchmarks. Such fitting remains crucial, providing input for the conduct of a huge number of experiments worldwide; but the past decade has seen the dawn of a new theory era. Continuum and lattice studies of QCD are beginning to yield robust predictions for the pointwise behaviour of DFs; and these developments are exposing potential conflicts with the fitting results [35,129,132–136].

Notwithstanding the enormous expense of time and effort, much yet remains to be learnt before proton and pion structure may be judged as understood in terms of DFs. For instance and most simply, it is still unclear whether there are differences between the distributions of partons within the proton (Nature's most fundamental bound state) and the pion (Nature's most fundamental (near) Nambu–Goldstone boson). Plainly, if there are differences, then they must be explained. As we have stressed above, answering the question of similarity/difference between proton and pion DFs is particularly important today as science seeks to expose and explain EHM [32–37].

Regarding DFs measured in processes that do not resolve beam or target polarisation, practitioners experienced and involved with solving bound state problems in QCD have learnt that, at the hadron scale, $\zeta_\mathcal{H} < m_p$, valence quark DFs in the proton and pion behave as follows [132–134,412,413]:

$$d^p(x;\zeta_\mathcal{H}), u^p(x;\zeta_\mathcal{H}) \overset{x \simeq 1}{\propto} (1-x)^3, \quad \bar{d}^\pi(x;\zeta_\mathcal{H}), u^\pi(x;\zeta_\mathcal{H}) \overset{x \simeq 1}{\propto} (1-x)^2. \quad (56)$$

It subsequently follows from the DGLAP equations [177,414–416] that the large-x power on the related gluon DF is approximately one unit larger; and that for sea quark DFs is roughly two units larger. Moreover, as the resolving scale increases to $\zeta > \zeta_\mathcal{H}$, all these exponents grow logarithmically. However, fuelling controversy and leading some to question the veracity of QCD [132,133,417], these constraints are typically ignored in fits to the world's data on deep inelastic scattering (DIS) and kindred processes [418–422]. Furthermore, largely because pion data are scarce [35] (Table 9.5), proton and pion data have never been considered simultaneously. Therefore, the unified body of results in Reference [135], which uses a single symmetry-preserving framework to predict the pointwise behaviour of all proton and pion DFs—valence, glue, and four-flavour-separated sea—is a significant advance.

In order to sketch this progress, it is necessary to recall that the modern approach to the CSM prediction of hadron DFs[7] is based on a single proposition [131–135]:

P1 There is an effective charge, $\alpha_{1\ell}(k^2)$, which, when used to integrate the one-loop perturbative-QCD DGLAP equations, defines a DF evolution scheme that is all orders exact.

As noted in connection with Figure 3, charges of this sort are discussed in References [137–139]. They need not be process-independent (PI), hence not unique. Moreover, the results delivered are independent of the explicit form of $\alpha_{1\ell}(k^2)$. Notwithstanding these things, a suitable PI charge is available, viz. the coupling discussed in Section 4, which has proven efficacious. In being defined by an observable—in this instance, structure functions—each such $\alpha_{1\ell}(k^2)$ is [94]: consistent with the renormalisation group and renormalisation scheme independent; everywhere analytic and finite; and, crucially, provides an infrared completion of any standard perturbative running coupling.

---

[7] Contemporary continuum methods for obtaining light-front amplitudes and density distributions from Euclidean space Schwinger functions are detailed, e.g., in References [134,423]; [35] – Sections 3, 5; [125] – Section IV; [127] – Sections 2, 5.

P1 was used in References [126–128,131,136] to deliver meson DFs with a flavour-symmetric sea. A generalisation, which expresses key quark current–mass effects in the evolution kernels, was introduced in Reference [135] and used for the proton and pion. It features a threshold function $\mathcal{P}_{qg}^{\zeta} \sim \theta(\zeta - \delta_f)$, which ensures that a given quark flavour only becomes active in DF evolution when the energy scale exceeds a value determined by the quark's mass [35] (Figure 2.5): $\delta_{u,d} \approx 0$, $\delta_s \approx 0.1$ GeV, $\delta_c \approx 0.9$ GeV. The impact of this modification is readily anticipated. Supposing that all quark flavours are light, then each would be emitted with equal probability on $\zeta > \zeta_\mathcal{H}$; so, evolution would produce a certain gluon momentum fraction in the hadron plus a sea quark fraction shared equally between all quark flavours. Considering mass differences between the quarks, with some flavours being heavier than the light quark threshold, then evolution on $\zeta > \zeta_\mathcal{H}$ will generate a gluon momentum fraction that is practically unchanged from the all-light quark case and a sea quark fraction divided amongst the quarks in roughly inverse proportion to their mass.

It is worth reiterating here that $\zeta_\mathcal{H}$ is the scale at which the valence quasiparticle degrees-of-freedom carry all properties of a given hadron [124–136]. Moreover, the value of this scale is a prediction. Using the PI charge discussed in Section 4 to construct bound state kernels informed by References [68,212], then

$$\zeta_\mathcal{H} = 0.331(2) \text{ GeV}. \qquad (57)$$

The value in Equation (57) is the same for all hadrons.

Furthermore, combined with evolution according to P1, the character of $\zeta_\mathcal{H}$ ensures that all hadron DFs are intertwined at every scale ζ. Hence, this perspective suggests that it is incorrect to choose independent, uncorrelated functions to parametrise the DFs of different parton species when fitting data at any scale $\zeta > \zeta_\mathcal{H}$. If one nevertheless chooses to ignore the innate associations, then DFs with unphysical features may be obtained—see, e.g., Reference [132] [Figure 6].

CSM predictions for the $\zeta = \zeta_\mathcal{H}$ proton and pion valence DFs are drawn in Figure 23A. The following points are significant.

(i) Each DF is consistent with the relevant large-x scaling law in Equation (56). Hence, from the outset, whilst the $\zeta = \zeta_\mathcal{H}$ momentum sum rules for each hadron are necessarily saturated by valence degrees-of-freedom, viz.

$$\langle x \rangle_{u_p}^{\zeta_\mathcal{H}} = 0.687, \ \langle x \rangle_{d_p}^{\zeta_\mathcal{H}} = 0.313, \ \langle x \rangle_{u_\pi}^{\zeta_\mathcal{H}} = 0.5, \qquad (58)$$

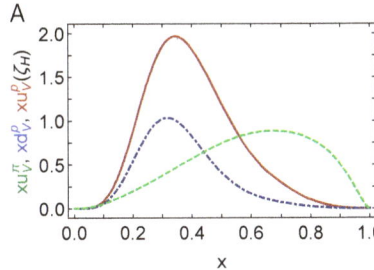
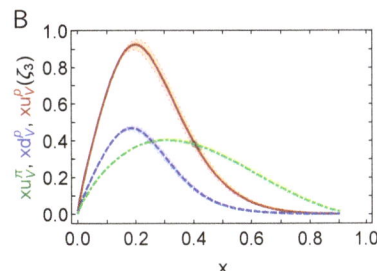

Figure 23. *Left panel* (**A**); Proton and pion hadron scale valence parton DFs: $xu^p(x; \zeta_\mathcal{H})$—solid red curve; $xd^p(x; \zeta_\mathcal{H})$—dotted-dashed blue curve; $xu^\pi(x; \zeta_\mathcal{H})$—dashed green curve. *Right panel* (**B**): Valence DFs in Panel A evolved to $\zeta_3 = m_{J/\psi} = 3.097$ GeV according to P1. The band surrounding each CSM curve expresses the response to a $\pm 5\%$ variation in $\zeta_\mathcal{H}$.

The proton and pion valence DFs nevertheless have markedly different x-dependence. (Nature's approximate $\mathcal{G}$-parity symmetry [424] entails $\bar{d}_\pi(x; \zeta) = u_\pi(x; \zeta)$.) (ii) Owing to DCSB [14,76,279–283], an important corollary of EHM, QCD dynamics simultaneously produce a dressed light quark mass function, $M_{u,d}(k^2)$, that is large at infrared momenta,

$M_D := M_{u,d}(k^2 \simeq 0) \approx 0.4\,\text{GeV}$, and an almost massless pion, $m_\pi^2/M_D^2 \approx 0.1$—see Reference [35] (Section 2). As a result, $u^\pi(x;\zeta_H)$ is Nature's most dilated hadron-scale valence DF. This is highlighted by Figure 23A and Refs. [126,127] and implicit in numerous other symmetry-preserving analyses, e.g., References [201,239,425,426].

Evolving the DFs in Figure 23A according to P1, one obtains the $\zeta = m_{J/\psi} =: \zeta_3$ distributions in Figure 23B. Plainly, although the profiles change, the relative dilation of the DFs is preserved and is therefore a verifiable prediction of the EHM paradigm.

Given that Figure 23B depicts the first CSM predictions for proton valence quark DFs, one might question their reliability. That issue can partly be addressed through a comparison with lQCD results. The calculation of individual valence DFs using lQCD is problematic owing to difficulties in handling so-called disconnected contributions [427]. In the continuum limit, however, disconnected diagrams do not contribute to the isovector DF $[u^p(x;\zeta) - d^p(x;\zeta)]$, so computations of this difference are available [428,429]. Both analyses use the quasidistribution approach [430], but the lattice algorithms and configurations are somewhat different. Their comparison with CSM predictions is depicted in Figure 24. The level of agreement is encouraging, especially because refinements of both continuum and lattice calculations may be anticipated. For instance, the CSM predictions were obtained using a simplified proton Faddeev amplitude, and the lattice studies must address issues with, inter alia, the pion masses used, lattice artefacts and systematic errors, and convergence of the boost expansion in the quasidistribution approach. (The last of these is a particular hindrance to lQCD extractions of DF endpoint behaviour [431].)

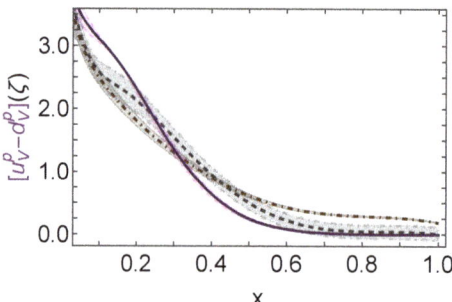

Figure 24. Isovector distribution $[u^p(x;\zeta) - d^p(x;\zeta)]$. CSM prediction—solid purple curve, $\zeta = \zeta_3$; lQCD result from Reference [428]—dashed grey curve, $\zeta = \zeta_3$; lQCD result from Reference [429]—dotted-dashed brown curve, $\zeta = \zeta_2$. Like-coloured band bracketing each curve indicates associated uncertainty.

In typical evolution kernels, gluon splitting yields quark+antiquark pairs of all flavours with equal probability. However, it was long ago argued [432] that, because the proton contains two valence u quarks and one valence d quark, the Pauli exclusion principle should force gluon splitting to prefer $d + \bar{d}$ production over $u + \bar{u}$. Consequently, when implementing evolution of proton singlet and glue DFs, Reference [135] followed Reference [134] and introduced a small Pauli blocking factor into the gluon splitting function. This correction preserves the baryon number, but shifts momentum into $d + \bar{d}$ from $u + \bar{u}$, otherwise leaving the sum of sea quark momentum fractions unchanged. It vanishes with increasing ζ, in order to express the declining influence of valence quarks as the proton's sea and glue content increases.

The resulting CSM predictions for $\zeta = \zeta_3$ proton and pion glue DFs are drawn in Figure 25A. The glue-in-π DF is directly related to the $\zeta = 2\,\text{GeV} =: \zeta_2$ result discussed in Reference [129], which is drawn in Figure 25B: evidently, it agrees with a recent lQCD calculation of the glue-in-π DF [433]. Furthermore, reproducing the pattern seen with valence quark DFs in Figures 23 and 25A reveals that the glue-in-π DF possesses significantly more support on the valence domain, $x \gtrsim 0.1$, than the glue-in-p DF. Once again, this feature is a measurable expression of EHM.

The $\zeta = \zeta_3$ light quark sea DFs for the proton and pion are depicted in Figure 25C. The EHM-induced pattern is also apparent here, viz. the sea-in-π DF possesses greater support on $x \gtrsim 0.1$ than the kindred sea-in-p DFs. DFs of the heavier sea quarks are also generated via evolution, with the results drawn in Figure 25D. Interestingly, the $\zeta = \zeta_3$ s and c quark DFs are similar in size to those of the light quark sea DFs; and for these heavier quarks, as well, the pion DFs have significantly greater support on the valence domain, $x \gtrsim 0.1$, than the related proton DFs.

An analysis of the endpoint exponents of all $\zeta = \zeta_3$ DFs is also contained in Reference [135] along with simple interpolations of each DF that can readily be used by any practitioner—Reference [135] (Table 1). It is worth reiterating the following remarks about the endpoint exponents.

(*i*) The power-laws express *measurable* effective exponents, obtained from separate linear fits to $\ln[xp(x)]$ on the domains $0 < x < 0.005$, $0.85 < x < 1$. (Here, $p(x)$ denotes a generic DF.)

(*ii*) Within mutual uncertainties, proton and pion DFs have the same power-law behaviour on $x \simeq 0$:
$$\alpha^{p,\pi}_{\text{valence}} \approx -0.22, \quad \alpha^{p,\pi}_{\text{glue}} \approx -1.6, \quad \alpha^{p,\pi}_{\text{sea}} \approx -1.5. \tag{59}$$

(*iii*) On $x \simeq 1$, the following relationships exist for and between pion and proton DF exponents:
$$\beta^{\pi}_{\text{valence}} \approx 2.5, \quad \beta^{p}_{\text{valence}} \approx \beta^{\pi}_{\text{valence}} + 1.6, \tag{60a}$$
$$\beta^{p,\pi}_{\text{glue}} \approx \beta^{p,\pi}_{\text{valence}} + 1.4, \quad \beta^{p,\pi}_{\text{sea}} \approx \beta^{p,\pi}_{\text{valence}} + 2.4. \tag{60b}$$

(*iv*) Given (*ii*) and (*iii*), then the CSM predictions are consistent with the QCD expectations discussed in connection with Equation (56).

(*v*) Existing phenomenological fits to relevant scattering data typically arrive at DFs which are inconsistent with (*ii*) and (*iii*); hence, fail to meet many QCD-based expectations, e.g., References [419,421,422,434,435]. This point is also discussed elsewhere [132,133,436].

Owing to the Pauli blocking factor described above and as evident in Figure 25C, the DFs calculated in Reference [135] express an in-proton separation between $\bar{d}$ and $\bar{u}$ distributions. This entails a violation of the Gottfried sum rule [437,438], which has been seen in experiments [439–443]. Using the proton DFs in Figure 25C then, on the domain covered by the measurements in References [439,440], one obtains
$$\int_{0.004}^{0.8} dx \left[\bar{d}(x; \zeta_3) - \bar{u}(x; \zeta_3) \right] = 0.116(12) \tag{61}$$

for the Gottfried sum rule discrepancy. This value matches that inferred from recent fits to a large sample of high-precision data ($\zeta = 2\,\text{GeV}$) [419] (CT18), 0.110(80), and is far more precise.

The result in Equation (61) corresponds to a strength for the Pauli blocking term in the gluon splitting function that shifts just $\approx 25\%$ of the u quark sea momentum fraction into the d quark sea at $\zeta = \zeta_2$. Changing the strength by $\pm 25\%$ leads to the uncertainty indicated in Equation (61). Data from the most recent experiment focused on the asymmetry of antimatter in the proton [443] (E906) are presented in Figure 26A. They may be compared with the CSM result obtained using the proton DFs in Figure 25C. Evidently, a modest Paul blocking effect in the gluon splitting function is sufficient to explain modern data.

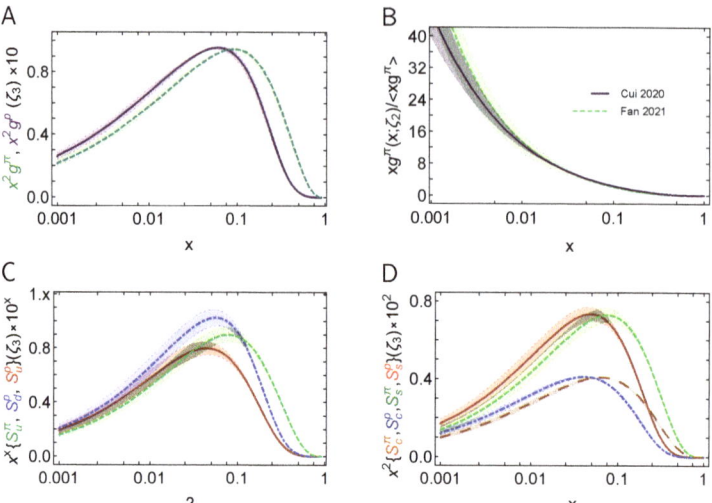

Figure 25. *Left upper panel* (**A**): Glue DFs—$x^2 g$, in the proton (solid purple curve) and pion (dashed green curve) at $\zeta = \zeta_3$. *Right upper panel* (**B**): Comparison between continuum [127] (Cui 2020) and lattice [433] (Fan 2021) results for the glue-in-pion DF at $\zeta = 2$ GeV. *Left lower panel* (**C**): Proton and pion light quark sea DFs: $x^2 S_u^p(x;\zeta_3)$—solid red curve; $x^2 S_d^p(x;\zeta_3)$—dashed blue curve; $x^2 S_u^\pi(x;\zeta_3)$—dotted-dashed green curve. *Right lower panel* (**D**): Proton and pion c- and s quark sea DFs: $x^2 S_s^p(x;\zeta_3)$—solid red curve; $x^2 S_s^\pi(x;\zeta_3)$—dashed green curve; $x^2 S_c^p(x;\zeta_3)$—dotted-dashed blue curve; $x^2 S_c^\pi(x;\zeta_3)$—long-dashed orange curve. The band surrounding each CSM curve expresses the response to a $\pm 5\%$ variation in $\zeta_\mathcal{H}$. The uncertainty in the lQCD result is similarly indicated in Panel B.

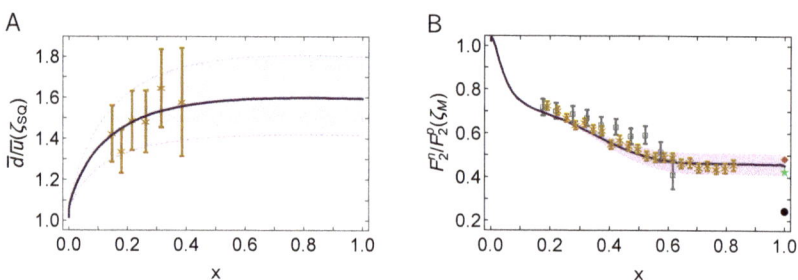

Figure 26. *Left panel* (**A**): Ratio of light antiquark DFs. Data: Reference [443] (E906). CSM result (solid purple curve) obtained from valence quark DFs in Figure 23 after evolution to $\zeta^2 = \zeta_{SQ}^2 = 30$ GeV^2 [135]. The shaded band expresses the impact of a $\pm 25\%$ variation in the strength of Pauli blocking. *Right panel* (**B**): Neutron-to-proton structure function ratio. Data: [444] (BoNuS)—open grey squares; [445] (MARATHON)—gold asterisks. Contemporary CSM results (solid purple curve) obtained from valence quark DFs in Figure 23 after evolution to $\zeta = \zeta_M = 2.7$ GeV [134,135]. Other predictions: green star—helicity conservation in the QCD parton model [412,446,447]; red diamond—large-x estimate based on Faddeev equation solutions [448]; retaining only scalar diquarks in the proton wave function, which produces a large-x value for this ratio that lies in the neighbourhood of the filled circle [449,450]. The band surrounding the CSM curve expresses the response to a $\pm 5\%$ variation in the size of axial vector diquark contributions to the proton charge. It is only noticeable on the valence quark domain.

Proton and, in principle, neutron structure functions—$F_2^{p,n}$—can be measured in the DIS of electrons from nucleons [408–411]. The ratio $F_2^n(x)/F_2^p(x)$ is recognised as a sensitive measure of $d^p(x)/u^p(x)$ on $x \gtrsim 0.4$ [407], and the latter ratio is important because it is a keen discriminator between pictures of proton structure [444,445,448]. The obstacle to an empirical result for $F_2^n(x)/F_2^p(x)$ is the measurement of F_2^n: since isolated neutrons decay rather quickly, a suitable, effective "free neutron target" must be found. Following References [451,452], many experiments have used the deuteron. However, despite this being a weakly bound system, the representation dependence of proton–neutron interactions leads to large theory uncertainties in the extracted ratio on $x \gtrsim 0.7$ [453].

A more favourable approach is provided by DIS measurements on ^{3}H and ^{3}He. In this case, nuclear interaction effects cancel to a very large degree when extracting $F_2^n(x)/F_2^p(x)$ from the ^{3}H:^{3}He ratio of scattering rates [454,455]. Of course, ^{3}H is highly radioactive; so, careful planning and implementation are required to deliver a safe target. Recently, after years of development, all challenges were overcome, and such an experiment was completed [445]: the extracted data are drawn in Figure 26B. Importantly, within mutual uncertainties, the results from Reference [445] match those inferred from an analysis of nuclear DIS reactions, exploiting targets ranging from the deuteron to lead and accounting for the effects of short-range correlations in the nuclei [456]. This speaks in support of the reliability of the analyses in both cases.

As described in Section 7, the Faddeev equation in Figure 8 makes firm statements about proton structure. In particular, well-constrained studies predict that axial vector diquark correlations are responsible for approximately 40% of the proton's charge—see Reference [267] (Figure 2)—e.g., this strength is confirmed in studies of nucleon axial vector and pseudoscalar currents—see Section 9. Consequently, this is the size of the axial vector diquark fraction in the nucleon Faddeev amplitudes used to calculate proton DFs in References [134,135]. Using the results therein, one may readily predict the neutron–proton structure function ratio:

$$\frac{F_2^n(x;\zeta)}{F_2^p(x;\zeta)} = \frac{\mathcal{U}(x;\zeta) + 4\mathcal{D}(x;\zeta) + \Sigma(x;\zeta)}{4\mathcal{U}(x;\zeta) + \mathcal{D}(x;\zeta) + \Sigma(x;\zeta)}, \quad (62)$$

where, in terms of quark and antiquark DFs, $\mathcal{U}(x;\zeta) = u^p(x;\zeta) + \bar{u}^p(x;\zeta)$, $\mathcal{D}(x;\zeta) = d^p(x;\zeta) + \bar{d}^p(x;\zeta)$, and $\Sigma(x;\zeta) = s^p(x;\zeta) + \bar{s}^p(x;\zeta) + c^p(x;\zeta) + \bar{c}^p(x;\zeta)$. Supposing that valence quarks dominate on $x \simeq 1$, then the limiting cases $d^p(x) \equiv 0$ and $u^p(x) \equiv 0$ yield the Nachtmann bounds [457]:

$$1/4 \leq F_2^n(x)/F_2^p(x) \leq 4 \quad \text{on } x \simeq 1. \quad (63)$$

The $\zeta = 2.7$ GeV $=: \zeta_M$ CSM prediction for $F_2^n(x)/F_2^p(x)$ is drawn in Figure 26B. Its comparison with modern data [445] (MARATHON) may be quantified by noting that the central curve yields χ^2/ degree-of-freedom = 1.3. It is worth stressing that the x-dependence of the CSM prediction in Figure 26B was made without reference to any data. Consequently, the agreement with the results published in Reference [445] (MARATHON) is meaningful and should serve to allay any concerns that the associated data analysis omitted some systematic effect deriving from nuclear structure modelling.

Such heightened confidence in the MARATHON data adds impact to the model-independent SPM analysis of that data described in Reference [458]; so, it is worth recapitulating some of the material therein. The final results are highlighted by Figure 27, which compares the MARATHON-based SPM prediction for $F_2^n/F_2^p\big|_{x\to 1}$ with: the nuclear DIS value [456]; theory predictions [271,412,446,450]; and the phenomenological fit result in Reference [434]. The figure also marks the Nachtmann lower bound, Equation (63), which is saturated if valence d quarks play no significant role at $x = 1$; namely, when there are practically no valence d quarks in the proton: $d^p/u^p\big|_{x\to 1} = 0$. This outcome is characteristic of proton wave function models in which the valence d quark is (almost) always paired with one of the valence u quarks inside a scalar diquark [255,449,459]. Even allowing for the quark exchange dynamics in Figure 8, one still finds $d^p/u^p|_{x \to 1} \approx 0$ if only scalar diquarks are retained [450].

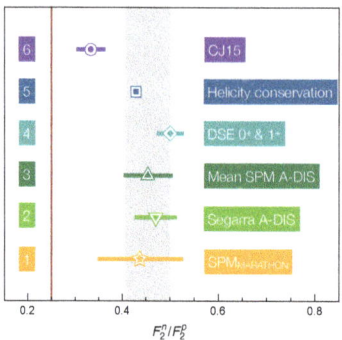

Figure 27. $\lim_{x \to 1} F_2^n(x)/F_2^p(x)$. SPM prediction derived from Reference [445] (MARATHON) compared with results inferred from: nuclear DIS [456]; large-x estimate from CSM studies [271,450]; quark counting (helicity conservation) [446]; and a phenomenological fit (CJ15) [434]. The red vertical line indicates the Nachtmann lower limit, Equation (63), which is saturated if valence d quarks play no material role at $x = 1$; Row 3 is the average in Equation (64).

The following observations serve as a summary of the analyses in Reference [458]:

Observation A ... Applied to MARATHON data, the SPM yields $F_2^n/F_2^p\big|_{x \to 1} = 0.437(85)$ $\Rightarrow d^p/u^p|_{x \to 1} = 0.227(100).$[8] The possibility $d^p/u^p|_{x \to 1} = 0$ is thus excluded with a 98.7% level of confidence; hence, scalar-diquark-only models of proton structure are excluded with equal likelihood. On the other hand, with this same 98.7% level of confidence, the SPM analysis confirms the QCD parton model prediction [412,446]: $d^p(x) \propto u^p(x)$ on $x \simeq 1$.

Observation B ... The value of $F_2^n/F_2^p\big|_{x \to 1}$ inferred from nuclear DIS [456] agrees with the SPM prediction; hence, they may be averaged to yield

$$F_2^n/F_2^p\big|_{x \to 1}^{\text{SPM \& DIS}-A} = 0.454 \pm 0.047. \tag{64}$$

This result is drawn on Row 3 in Figure 27. It corresponds to

$$\lim_{x \to 1} \frac{d^p(x)}{u^p(x)} = 0.230 \pm 0.057 \tag{65}$$

and entails that the probability that scalar-diquark-only models of proton structure might be consistent with available data is $1/141,000$. In fact, with a high level of confidence, one may discard any proton structure model that delivers a result for $F_2^n/F_2^p\big|_{x \to 1}$ that differs significantly from Equation (64). (As reviewed in Section 9, the ratio g_A^d/g_A^u places an even harder exclusion bound on scalar-diquark-only models.)

Observation C ... Within uncertainties, the result in Equation (64) agrees with both: (i) the value obtained by assuming an SU(4)-symmetric spin–flavour wave function for the proton and helicity conservation in high-Q^2 interactions [412,446]; and (ii) the

---

[8] Extrapolations based on [1,1] Padé fits to MARATHON data, obtained using a one-point jackknife procedure, yield $F_2^n/F_2^p = 0.395(3)$ on $x \simeq 1 \Rightarrow d^p/u^d = 0.169(3)$ [182]. Another analysis [460], employing practitioner-chosen polynomials as the basis for extrapolation, obtains $F_2^n/F_2^p = 0.37(7)$ on $x \simeq 1 \Rightarrow d^p/u^d = 0.13(8)$. The latter is less precise, but both results are consistent with the function form unbiased SPM prediction.

prediction developed from proton Faddeev wave functions that contain both scalar and axial vector diquarks, with the axial vector contributing approximately 40% of the proton charge [271,450]. (Recall that this was the axial vector diquark fraction built into the analyses in References [134,135].)

Following common practice, Reference [135] (Table 2) lists low-order $\zeta = \zeta_2, \zeta_3$ Mellin moments of all proton and pion DFs. Given P1 and the character of the hadron scale, then comparable momentum fractions in the proton and pion are necessarily identical and, as anticipated above, the total sea quark momentum fraction is shared between the quark flavours in roughly inverse proportion to their infrared dressed mass, M_f. Significantly, given that they are *predictions*, the calculated values of the proton DF moments in Reference [135] (Table 2) are in fair agreement with those produced by phenomenological fits—see e.g., Reference [419] (Table VI): referring to the CT18 column, the CSM results match at the level of $1.7(1.5)\,\sigma$. Furthermore, within mutual uncertainties, the pion valence quark DF moments agree with recent lQCD results [461,462].

It is worth emphasising that the quantitative similarities also extend to the c quark: Reference [135] predicts $\langle x \rangle_{\bar{c}_p}^{\zeta_2} = 1.32(5)\%$, $\langle x \rangle_{\bar{c}_p}^{\zeta_3} = 1.82(6)\%$, cf. $1.7(4), 2.5(4)\%$ in Reference [435] (Figure 60). Moreover, the CSM study predicts $\langle x \rangle_{\bar{c}}^{\zeta=1.5\,\text{GeV}} = 0.64(3)\%$ in both the pion and proton. Regarding the pion, nothing is known about this momentum fraction; and in the proton, phenomenological estimates are inconclusive, ranging from 0-2% [435] (Figure 59). Plainly, a significant c quark momentum fraction is obtained under P1 evolution without recourse to "intrinsic charm" [463]. This outcome, which is independent of the explicit form of $\alpha_{1\ell}(k^2)$, potentially challenges the findings in Reference [464]. Notwithstanding the size of these calculated fractions, we stress that, as apparent in Figure 25, $S_c^{\pi,p}(x)$ have sea quark profiles.

Given these observations, one is led to re-evaluate what is meant by intrinsic charm in the proton or any other hadron. With $\zeta_\mathcal{H}$ being the scale at which valence quasiparticle degrees-of-freedom carry all measurable properties of a given hadron, then the Fock space components, which might be interpreted as intrinsic charm or intrinsic strangeness, etc., are sublimated into the nonperturbatively computed $\zeta = \zeta_\mathcal{H}$ Schwinger functions that completely express the bound state's structure. In this context, such Fock space components are interpreted as being members of the set of basis eigenvectors representing a free-field light-front Hamiltonian. One may exemplify this by noting that any true QCD solution for the dressed quark Schwinger function (propagator) must contain infinitely many and all possible contributing Fock space vectors. The putative "intrinsic" components within a bound state are then revealed by evolving the hadron-scale Schwinger functions to higher scales, whereat an interpretation of data in terms of a Fock space expansion is relevant and practicable. In one study or another, the actual expressions of the characters of intrinsic charm, strangeness, etc., will depend on the sophistication of the kernels used to calculate the hadron-scale Schwinger functions. Nevertheless, as highlighted by Reference [135], any reasonable kernels will predict that a measurable fraction of the proton's light-front momentum is carried by the charm quark sea at all resolving scales for which data may be interpreted in terms of DFs.

With a symmetry-preserving framework in hand that has the demonstrated ability to provide simultaneous predictions for the entire array of proton and pion DFs, one is potentially in a position to bring a new order to the study of hadron structure functions. Stringent new tests of the approach, including P1—see page 98—and the general character of the hadron scale, Equation (57), will be found in, amongst other things, studies of helicity-dependent DFs. New insights into proton spin structure may then be forthcoming.

12. Conclusions

We sketched some recent advances in the use of continuum Schwinger function methods (CSMs) to link QCD and hadron observables. The connecting bridge from theory to observation is supported by the three pillars of emergent hadrons mass (EHM): (*i*) dynamical generation of a gluon mass scale, whose size is roughly one half the proton mass;

(ii) existence of a unique, process-independent effective charge, $\hat{\alpha}(k^2)$, which runs to a finite value at infrared momenta, $\hat{\alpha}(0)/\pi \approx 1$; and (iii) emergence of a running quark mass in the chiral limit, whose infrared value matches that typically identified as the constituent quark mass. Our subsequent commentary stressed that the single phenomenon of EHM manifests itself differently in the diverse array of measurable quantities that define hadron physics. No single observable is alone sufficient to validate the EHM paradigm for understanding strong interactions within the Standard Model (SM). Instead, theory should identify a broad range of empirical consequences of EHM so that the order brought to a collective body of experimental results—existing and future—can be recognised as the signature of EHM.

Our developing understanding of EHM suggests that QCD is unique amongst known fundamental theories of natural phenomena. It might be the first well-defined four-dimensional quantum field theory ever contemplated. If so, then QCD could provide the archetype for theories that take physics beyond the SM.

Science has delivered theories of many things. The best of them remain a part of the grander theories developed in response to new observations. The basic question yet remains unanswered, viz.: Is there a theory of everything? Hadron physics and QCD might be pointing us toward an answer in exposing the special qualities of strong-coupling non-Abelian quantum gauge field theories.

Funding: Work supported by: National Natural Science Foundation of China (Grant No. 12135007); and Helmholtz-Zentrum Dresden Rossendorf under the High Potential Programme.

Data Availability Statement: Not applicable.

Acknowledgments: This contribution is based on results obtained and insights developed through collaborations with many people, to all of whom we are greatly indebted.

Conflicts of Interest: The authors declare no conflict of interest.

Abbreviations

The following abbreviations are used in this manuscript:

ACM	anomalous chromomagnetic moment
AdS/CFT (duality)	anti-de Sitter/conformal field theory (duality)
$\overline{\text{ard}}$	mean absolute relative difference
CKM	Cabibbo–Kobayashi–Maskawa (matrix)
CSMs	continuum Schwinger function methods
DCSB	dynamical chiral symmetry breaking
DF	(parton) distribution function
DIS	deep inelastic scattering
DSE	Dyson–Schwinger equation
EHM	emergent hadron mass
FF	(parton) fragmentation function
JLab	Thomas Jefferson National Accelerator Facility
lQCD	lattice-regularised quantum chromodynamics
NG (mode/boson)	Nambu–Goldstone (mode/boson)
PD (charge)	process-dependent (charge)
PDG	Particle Data Group and associated publications
PI (charge)	process-independent (charge)
QCD	quantum chromodynamics
QED	quantum electrodynamics
RGI	renormalisation-group-invariant
RL	rainbow ladder (truncation)
SCI	symmetry-preserving treatment of a vector×vector contact interaction
SM	Standard Model of particle physics
SPM	Schlessinger point method
VMD	vector meson dominance

References

1. Appelquist, T.; Pisarski, R.D. High-Temperature Yang–Mills Theories and Three-Dimensional Quantum Chromodynamics. *Phys. Rev. D* **1981**, *23*, 2305. [CrossRef]
2. Appelquist, T.W.; Bowick, M.J.; Karabali, D.; Wijewardhana, L.C.R. Spontaneous Chiral Symmetry Breaking in Three-Dimensional QED. *Phys. Rev. D* **1986**, *33*, 3704. [CrossRef] [PubMed]
3. Bashir, A.; Raya, A.; Cloet, I.C.; Roberts, C.D. Regarding confinement and dynamical chiral symmetry breaking in QED3. *Phys. Rev. C* **2008**, *78*, 055201. [CrossRef]
4. Bashir, A.; Raya, A.; Sánchez-Madrigal, S.; Roberts, C.D. Gauge invariance of a critical number of flavours in QED3. *Few Body Syst.* **2009**, *46*, 229–237. [CrossRef]
5. Braun, J.; Gies, H.; Janssen, L.; Roscher, D. Phase structure of many-flavor QED_3. *Phys. Rev. D* **2014**, *90*, 036002. [CrossRef]
6. Roberts, C.D. Perspective on the origin of hadron masses. *Few Body Syst.* **2017**, *58*, 5. [CrossRef]
7. Glimm, J.; Jaffee, A. *Quantum Physics: A Functional Point of View*; Springer: New York, NY, USA, 1981.
8. Seiler, E. *Gauge Theories as a Problem of Constructive Quantum Theory and Statistical Mechanics*; Springer: New York, NY, USA, 1982.
9. Itzykson, C.; Zuber, J.B. *Quantum Field Theory*; McGraw-Hill Inc.: New York, NY, USA, 1980.
10. Rakow, P.E.L. Renormalization group flow in QED: An investigation of the Schwinger-Dyson equations. *Nucl. Phys. B* **1991**, *356*, 27–45. [CrossRef]
11. Gockeler, M.; Horsley, R.; Linke, V.; Rakow, P.E.L.; Schierholz, G.; Stuben, H. Probing the continuum limit in noncompact QED: New results on large lattices. *Nucl. Phys. B Proc. Suppl.* **1995**, *42*, 660–662. [CrossRef]
12. Reenders, M. On the nontriviality of Abelian gauged Nambu-Jona-Lasinio models in four dimensions. *Phys. Rev. D* **2000**, *62*, 025001. [CrossRef]
13. Kızılersü, A.; Sizer, T.; Pennington, M.R.; Williams, A.G.; Williams, R. Dynamical mass generation in unquenched QED using the Dyson–Schwinger equations. *Phys. Rev. D* **2015**, *91*, 065015. [CrossRef]
14. Roberts, C.D.; Schmidt, S.M. Dyson–Schwinger equations: Density, temperature and continuum strong QCD. *Prog. Part. Nucl. Phys.* **2000**, *45*, S1–S103. [CrossRef]
15. Roberts, C.D.; Williams, A.G.; Krein, G. On the implications of confinement. *Int. J. Mod. Phys. A* **1992**, *7*, 5607–5624. [CrossRef]
16. Roberts, C.D.; Williams, A.G. Dyson–Schwinger equations and their application to hadronic physics. *Prog. Part. Nucl. Phys.* **1994**, *33*, 477–575. [CrossRef]
17. Binosi, D.; Tripolt, R.A. Spectral functions of confined particles. *Phys. Lett. B* **2020**, *801*, 135171. [CrossRef]
18. Wilson, K.G. Confinement of quarks. *Phys. Rev. D* **1974**, *10*, 2445–2459. [CrossRef]
19. Wilson, K.G. The origins of lattice gauge theory. *Nucl. Phys. Proc. Suppl.* **2005**, *140*, 3–19. [CrossRef]
20. Cali, S.; Detmold, W.; Korcyl, G.; Korcyl, P.; Shanahan, P. Implementation of the conjugate gradient algorithm for heterogeneous systems. In Proceedings of the 38th International Symposium on Lattice Field Theory—PoS(LATTICE2021), Cambridge, MA, USA, 26–30 July 2021; Volume 396.
21. Maris, P.; Roberts, C.D. Dyson–Schwinger equations: A tool for hadron physics. *Int. J. Mod. Phys. E* **2003**, *12*, 297–365. [CrossRef]
22. Roberts, C.D. Hadron Properties and Dyson–Schwinger Equations. *Prog. Part. Nucl. Phys.* **2008**, *61*, 50–65. [CrossRef]
23. Chang, L.; Roberts, C.D.; Tandy, P.C. Selected highlights from the study of mesons. *Chin. J. Phys.* **2011**, *49*, 955–1004.
24. Bashir, A.; Chang, L.; Cloet, I.C.; El-Bennich, B.; Liu, Y.X.; Roberts, C.D.; Tandy, P.C. Collective perspective on advances in Dyson–Schwinger Equation QCD. *Commun. Theor. Phys.* **2012**, *58*, 79–134. [CrossRef]
25. Roberts, C.D. Strong QCD and Dyson–Schwinger Equations. *IRMA Lect. Math. Theor. Phys.* **2015**, *21*, 356–458.
26. Roberts, C.D. Three Lectures on Hadron Physics. *J. Phys. Conf. Ser.* **2016**, *706*, 022003. [CrossRef]
27. Horn, T.; Roberts, C.D. The pion: an enigma within the Standard Model. *J. Phys. G* **2016**, *43*, 073001. [CrossRef]
28. Eichmann, G.; Sanchis-Alepuz, H.; Williams, R.; Alkofer, R.; Fischer, C.S. Baryons as relativistic three-quark bound states. *Prog. Part. Nucl. Phys.* **2016**, *91*, 1–100. [CrossRef]
29. Burkert, V.D.; Roberts, C.D. Roper resonance: Toward a solution to the fifty-year puzzle. *Rev. Mod. Phys.* **2019**, *91*, 011003. [CrossRef]
30. Fischer, C.S. QCD at finite temperature and chemical potential from Dyson–Schwinger equations. *Prog. Part. Nucl. Phys.* **2019**, *105*, 1–60. [CrossRef]
31. Qin, S.X.; Roberts, C.D. Impressions of the Continuum Bound State Problem in QCD. *Chin. Phys. Lett.* **2020**, *37*, 121201. [CrossRef]
32. Roberts, C.D.; Schmidt, S.M. Reflections upon the Emergence of Hadronic Mass. *Eur. Phys. J. Spec. Top.* **2020**, *229*, 3319–3340. [CrossRef]
33. Roberts, C.D. Empirical Consequences of Emergent Mass. *Symmetry* **2020**, *12*, 1468. [CrossRef]
34. Roberts, C.D. On Mass and Matter. *AAPPS Bull.* **2021**, *31*, 6. [CrossRef]
35. Roberts, C.D.; Richards, D.G.; Horn, T.; Chang, L. Insights into the emergence of mass from studies of pion and kaon structure. *Prog. Part. Nucl. Phys.* **2021**, *120*, 103883. [CrossRef]
36. Binosi, D. Emergent Hadron Mass in Strong Dynamics. *Few Body Syst.* **2022**, *63*, 42. [CrossRef]
37. Papavassiliou, J. Emergence of mass in the gauge sector of QCD. *Chin. Phys. C* **2022**, *46*, 112001. [CrossRef]
38. Englert, F. Nobel Lecture: The BEH mechanism and its scalar boson. *Rev. Mod. Phys.* **2014**, *86*, 843. [CrossRef]
39. Higgs, P.W. Nobel Lecture: Evading the Goldstone theorem. *Rev. Mod. Phys.* **2014**, *86*, 851. [CrossRef]

40. Workman, R.L.; Burkert, V.D.; Crede, V.; Klempt, E.; Thoma, U.; Tiator, L.; Agashe, K.; Aielli, G.; Allanach, B.C.; Amsler, C.; et al. Review of Particle Physics. *Prog. Theor. Exp. Phys.* **2022**, *2022*, 083C01.
41. Flambaum, V.V.; Höll, A.; Jaikumar, P.; Roberts, C.D.; Wright, S.V. Sigma terms of light quark hadrons. *Few Body Syst.* **2006**, *38*, 31–51. [CrossRef]
42. Ruiz de Elvira, J.; Hoferichter, M.; Kubis, B.; Meißner, U.G. Extracting the σ-term from low-energy pion–nucleon scattering. *J. Phys. G* **2018**, *45*, 024001. [CrossRef]
43. Aoki, S.; Aoki, Y.; Bečirević, D.; Blum, T.; Colangelo, G.; Collins, S.; Della Morte, M.; Dimopoulos, P.; Dürr, S.; Fukaya, H.; et al. FLAG Review 2019. *Eur. Phys. J. C* **2020**, *80*, 113.
44. Nambu, Y. Quasiparticles and Gauge Invariance in the Theory of Superconductivity. *Phys. Rev.* **1960**, *117*, 648–663. [CrossRef]
45. Goldstone, J. Field Theories with Superconductor Solutions. *Nuovo Cim.* **1961**, *19*, 154–164. [CrossRef]
46. Gell-Mann, M.; Oakes, R.J.; Renner, B. Behavior of current divergences under SU(3) x SU(3). *Phys. Rev.* **1968**, *175*, 2195–2199. [CrossRef]
47. Casher, A.; Susskind, L. Chiral magnetism (or magnetohadrochironics). *Phys. Rev.* **1974**, *D9*, 436–460. [CrossRef]
48. Brodsky, S.J.; Shrock, R. Condensates in Quantum Chromodynamics and the Cosmological Constant. *Proc. Natl. Acad. Sci. USA* **2011**, *108*, 45–50. [CrossRef]
49. Brodsky, S.J.; Roberts, C.D.; Shrock, R.; Tandy, P.C. New perspectives on the quark condensate. *Phys. Rev. C* **2010**, *82*, 022201. [CrossRef]
50. Chang, L.; Roberts, C.D.; Tandy, P.C. Expanding the concept of in-hadron condensates. *Phys. Rev. C* **2012**, *85*, 012201. [CrossRef]
51. Brodsky, S.J.; Roberts, C.D.; Shrock, R.; Tandy, P.C. Confinement contains condensates. *Phys. Rev. C* **2012**, *85*, 065202. [CrossRef]
52. Cloet, I.C.; Roberts, C.D. Explanation and Prediction of Observables using Continuum Strong QCD. *Prog. Part. Nucl. Phys.* **2014**, *77*, 1–69. [CrossRef]
53. Pascual, P.; Tarrach, R. *QCD: Renormalization for the Practitioner, Lecture Notes in Physics*; Springer: Berlin/Heidelberg, Germany, 1984; Volume 194.
54. Taylor, J.C. Ward Identities and Charge Renormalization of the Yang–Mills Field. *Nucl. Phys. B* **1971**, *33*, 436–444. [CrossRef]
55. Slavnov, A.A. Ward Identities in Gauge Theories. *Theor. Math. Phys.* **1972**, *10*, 99–107. [CrossRef]
56. Schwinger, J.S. Gauge Invariance and Mass. *Phys. Rev.* **1962**, *125*, 397–398. [CrossRef]
57. Schwinger, J.S. Gauge Invariance and Mass. 2. *Phys. Rev.* **1962**, *128*, 2425–2429. [CrossRef]
58. Appelquist, T.; Nash, D.; Wijewardhana, L. Critical Behavior in (2+1)-Dimensional QED. *Phys. Rev. Lett.* **1988**, *60*, 2575. [CrossRef] [PubMed]
59. Maris, P. The influence of the full vertex and vacuum polarization on the fermion propagator in QED3. *Phys. Rev.* **1996**, *D54*, 4049–4058.
60. Aguilar, A.C.; Binosi, D.; Papavassiliou, J. Nonperturbative gluon and ghost propagators for d=3 Yang–Mills. *Phys. Rev. D* **2010**, *81*, 125025. [CrossRef]
61. Cornwall, J.M. Exploring dynamical gluon mass generation in three dimensions. *Phys. Rev. D* **2016**, *93*, 025021. [CrossRef]
62. Cornwall, J.M. Dynamical Mass Generation in Continuum QCD. *Phys. Rev. D* **1982**, *26*, 1453. [CrossRef]
63. Aguilar, A.C.; Binosi, D.; Papavassiliou, J. Gluon and ghost propagators in the Landau gauge: Deriving lattice results from Schwinger-Dyson equations. *Phys. Rev. D* **2008**, *78*, 025010. [CrossRef]
64. Boucaud, P.; Leroy, J.; Le Yaouanc, A.; Micheli, J.; Pene, O.; Rodríguez-Quintero, J. On the IR behaviour of the Landau-gauge ghost propagator. *J. High Energy Phys.* **2008**, *0806*, 099. [CrossRef]
65. Binosi, D.; Papavassiliou, J. Pinch Technique: Theory and Applications. *Phys. Rept.* **2009**, *479*, 1–152. [CrossRef]
66. Boucaud, P.; Leroy, J.P.; Le-Yaouanc, A.; Micheli, J.; Pene, O.; Rodríguez-Quintero, J. The Infrared Behaviour of the Pure Yang–Mills Green Functions. *Few Body Syst.* **2012**, *53*, 387–436. [CrossRef]
67. Aguilar, A.C.; Binosi, D.; Papavassiliou, J. The Gluon Mass Generation Mechanism: A Concise Primer. *Front. Phys. China* **2016**, *11*, 111203. [CrossRef]
68. Binosi, D.; Chang, L.; Papavassiliou, J.; Roberts, C.D. Bridging a gap between continuum-QCD and ab initio predictions of hadron observables. *Phys. Lett. B* **2015**, *742*, 183–188. [CrossRef]
69. Aguilar, A.C.; Ferreira, M.N.; Papavassiliou, J. Exploring smoking-gun signals of the Schwinger mechanism in QCD. *Phys. Rev. D* **2022**, *105*, 014030. [CrossRef]
70. Pinto-Gómez, F.; De Soto, F.; Ferreira, M.N.; Papavassiliou, J.; Rodríguez-Quintero, J. Lattice three-gluon vertex in extended kinematics: planar degeneracy. *arXiv* **2022**, arXiv:2208.01020.
71. Cui, Z.F.; Zhang, J.L.; Binosi, D.; de Soto, F.; Mezrag, C.; Papavassiliou, J.; Roberts, C.D.; Rodríguez-Quintero, J.; Segovia, J.; Zafeiropoulos, S. Effective charge from lattice QCD. *Chin. Phys. C* **2020**, *44*, 083102. [CrossRef]
72. Aguilar, A.C.; De Soto, F.; Ferreira, M.; Papavassiliou, J.; Rodríguez-Quintero, J.; Zafeiropoulos, S. Gluon propagator and three-gluon vertex with dynamical quarks. *Eur. Phys. J. C* **2020**, *80*, 154. [CrossRef]
73. Blum, T.; Boyle, P.A.; Christ, N.H.; Frison, J.; Garron, N.; Hudspith, R.J.; Izubuchi, T.; Janowski, T.; Jung, C.; Jüttner, A.; et al. Domain wall QCD with physical quark masses. *Phys. Rev. D* **2016**, *93*, 074505. [CrossRef]
74. Boyle, P.A.; Christ, N.H.; Garron, N.; Jung, C.; Jüttner, A.; Kelly, C.; Mawhinney, R.D.; McGlynn, G.; Murphy, D.J.; Ohta, S.; et al. Low energy constants of SU(2) partially quenched chiral perturbation theory from N_f=2+1 domain wall QCD. *Phys. Rev. D* **2016**, *93*, 054502. [CrossRef]

75. Boyle, P.A.; Del Debbio, L.; Jüttner, A.; Khamseh, A.; Sanfilippo, F.; Tsang, J.T. The decay constants f_D and f_{D_s} in the continuum limit of $N_f = 2 + 1$ domain wall lattice QCD. *J. High Energy Phys.* **2017**, *12*, 008. [CrossRef]
76. Binosi, D.; Chang, L.; Papavassiliou, J.; Qin, S.X.; Roberts, C.D. Natural constraints on the gluon–quark vertex. *Phys. Rev. D* **2017**, *95*, 031501. [CrossRef]
77. Brodsky, S.J.; Deur, A.; Roberts, C.D. Artificial dynamical effects in quantum field theory. *Nat. Rev. Phys.* **2022**, *4*, 489–495. [CrossRef]
78. West, G.B. General Infrared and Ultraviolet Properties of the Gluon Propagator in Axial Gauge. *Phys. Rev. D* **1983**, *27*, 1878. [CrossRef]
79. Brown, N.; Pennington, M.R. Studies of Confinement: How the Gluon Propagates. *Phys. Rev. D* **1989**, *39*, 2723. [CrossRef] [PubMed]
80. Cucchieri, A.; Mendes, T.; Santos, E.M.S. Covariant gauge on the lattice: A New implementation. *Phys. Rev. Lett.* **2009**, *103*, 141602. [CrossRef] [PubMed]
81. Gao, F.; Qin, S.X.; Roberts, C.D.; Rodríguez-Quintero, J. Locating the Gribov horizon. *Phys. Rev. D* **2018**, *97*, 034010. [CrossRef]
82. Gell-Mann, M.; Low, F.E. Quantum electrodynamics at small distances. *Phys. Rev.* **1954**, *95*, 1300–1312. [CrossRef]
83. Politzer, H.D. The dilemma of attribution. *Proc. Natl. Acad. Sci. USA* **2005**, *102*, 7789–7793. [CrossRef]
84. Wilczek, F. Asymptotic freedom: From paradox to paradigm. *Proc. Natl. Acad. Sci. USA* **2005**, *102*, 8403–8413. [CrossRef]
85. Gross, D.J. The discovery of asymptotic freedom and the emergence of QCD. *Proc. Natl. Acad. Sci. USA* **2005**, *102*, 9099–9108. [CrossRef]
86. Pickering, A. *Constructing Quarks. A Sociological History of Particle Physics*; University of Chicago Press: Chicago, IL, USA, 1999.
87. Marciano, W.J.; Pagels, H. Quantum Chromodynamics: A Review. *Phys. Rept.* **1978**, *36*, 137. [CrossRef]
88. Fischer, C.S. Infrared properties of QCD from Dyson–Schwinger equations. *J. Phys. G* **2006**, *32*, R253–R291. [CrossRef]
89. Chang, L.; Roberts, C.D. Sketching the Bethe–Salpeter kernel. *Phys. Rev. Lett.* **2009**, *103*, 081601. [CrossRef]
90. Binosi, D.; Mezrag, C.; Papavassiliou, J.; Roberts, C.D.; Rodríguez-Quintero, J. Process-independent strong running coupling. *Phys. Rev. D* **2017**, *96*, 054026. [CrossRef]
91. Pilaftsis, A. Generalized pinch technique and the background field method in general gauges. *Nucl. Phys. B* **1997**, *487*, 467–491. [CrossRef]
92. Cornwall, J.M.; Papavassiliou, J.; Binosi, D. *The Pinch Technique and its Applications to Non-Abelian Gauge Theories*; Cambridge University Press: Cambridge, UK, 2010.
93. Abbott, L.F. Introduction to the Background Field Method. *Acta Phys. Polon. B* **1982**, *13*, 33.
94. Deur, A.; Brodsky, S.J.; de Teramond, G.F. The QCD Running Coupling. *Prog. Part. Nucl. Phys.* **2016**, *90*, 1–74. [CrossRef]
95. Deur, A.; Burkert, V.; Chen, J.P.; Korsch, W. Experimental determination of the QCD effective charge $\alpha_{g_1}(Q)$. *Particles* **2022**, *5*, 171. [CrossRef]
96. Deur, A.; Burkert, V.; Chen, J.P.; Korsch, W. Experimental determination of the effective strong coupling constant. *Phys. Lett. B* **2007**, *650*, 244–248. [CrossRef]
97. Deur, A.; Burkert, V.; Chen, J.P.; Korsch, W. Determination of the effective strong coupling constant $\alpha_{g_1}(s)$ from CLAS spin structure function data. *Phys. Lett. B* **2008**, *665*, 349–351. [CrossRef]
98. Deur, A.; Prok, Y.; Burkert, V.; Crabb, D.; Girod, F.X.; Griffioen, K.A.; Guler, N.; Kuhn, S.E.; Kvaltine, N. High precision determination of the Q^2 evolution of the Bjorken Sum. *Phys. Rev. D* **2014**, *90*, 012009. [CrossRef]
99. The HERMES Collaboration; Ackerstaff, K. Measurement of the neutron spin structure function g_1^n with a polarized ^{3}He internal target. *Phys. Lett. B* **1997**, *404*, 383–389. [CrossRef]
100. Ackerstaff, K.; Airapetian, A.; Akopov, N.; Akushevich, I.; Amarian, M.; Aschenauer, E.; Avakian, H.; Avetissian, A.; Bains, B.; Barrow, S.; et al. Determination of the deep inelastic contribution to the generalized Gerasimov-Drell-Hearn integral for the proton and neutron. *Phys. Lett. B* **1998**, *444*, 531–538. [CrossRef]
101. Airapetian, A.; Akopov, N.; Akushevich, I.; Amarian, M.; Aschenauer, E.; Avakian, H.; Avetissian, A.; Bains, B.; Baumgarten, C.; Beckmann, M.; et al. Measurement of the proton spin structure function g_1^p with a pure hydrogen target. *Phys. Lett. B* **1998**, *442*, 484–492. [CrossRef]
102. Airapetian, A.; Akopov, N.; Akopov, Z.; Amarian, M.; Ammosov, V.V.; Aschenauer, E.C.; Avakian, H.; Avetissian, A.; Bailey, P.; Baturin, V.; et al. Evidence for quark hadron duality in the proton spin asymmetry A_1. *Phys. Rev. Lett.* **2003**, *90*, 092002. [CrossRef]
103. Airapetian, A.; Akopov, N.; Akopov, Z.; Andrus, A.; Aschenauer, E.C.; Augustyniak, W.; Tchuiko, B. Precise determination of the spin structure function g_1 of the proton, deuteron and neutron. *Phys. Rev. D* **2007**, *75*, 012007. [CrossRef]
104. NuTeV Collaboration. A Measurement of $\alpha_s(Q^2)$ from the Gross-Llewellyn Smith sum rule. *Phys. Rev. Lett.* **1998**, *81*, 3595–3598. [CrossRef]
105. Alexakhin, V.Y.; Alexandrov, Y.; Alexeev, G.D.; Alexeev, M.; Amoroso, A.; Badełek, B.; Mikhailov, Y.V. The Deuteron Spin-dependent Structure Function g_1^d and its First Moment. *Phys. Lett. B* **2007**, *647*, 8–17. [CrossRef]
106. Alekseev, M.G.; Alexakhin, V.Y.; Alexandrov, Y.; Alexeev, G.D.; Amoroso, A.; Austregesilo, A.; Padee, A. The Spin-dependent Structure Function of the Proton g_1^p and a Test of the Bjorken Sum Rule. *Phys. Lett. B* **2010**, *690*, 466–472. [CrossRef]
107. COMPASS Collaboration; Bordalo, P.; Franco, C.; Nunes, A.S.; Quaresma, M.; Quintans, C.; Ramos, S.; Silva, L.; Stolarski, M. The spin structure function g_1^p of the proton and a test of the Bjorken sum rule. *Phys. Lett. B* **2016**, *753*, 18–28. [CrossRef]

108. Anthony, P.L.; Arnold, R.G.; Band, H.R.; Borel, H.; Bosted, P.E.; Breton, V.; Cates, G.D.; Chupp, T.E.; Dietrich, F.S.; Dunne, J.; et al. Determination of the neutron spin structure function. *Phys. Rev. Lett.* **1993**, *71*, 959–962. [CrossRef] [PubMed]
109. E143 Collaboration; Abe, K.; Akagi, T.; Anthony, P.L.; Antonov, R.; Arnold, R.G.; Averett, T. Precision measurement of the proton spin structure function g_1^p. *Phys. Rev. Lett.* **1995**, *74*, 346–350. [CrossRef]
110. E143 Collaboration; Abe, K.; Akagi, T.; Anthony, P.L.; Antonov, R.; Arnold, R.G.; Averett, T. Precision measurement of the deuteron spin structure function g_1^d. *Phys. Rev. Lett.* **1995**, *75*, 25–28. [CrossRef] [PubMed]
111. Abe, K.; Akagi, T.; Anthony, P.L.; Antonov, R.; Arnold, R.G.; Averett, T.; Band, H.R.; Bauer, J.M.; Borel, H.; Bosted, P.E.; et al. Measurements of the proton and deuteron spin structure function g_2 and asymmetry A_2. *Phys. Rev. Lett.* **1996**, *76*, 587–591. [CrossRef] [PubMed]
112. Abe, K.; Akagi, T.; Anthony, P.; Antonov, R.; Arnold, R.; Averett, T.; Band, H.; Bauer, J.; Borel, H.; Bosted, P.; et al. Measurements of the Q^2 dependence of the proton and deuteron spin structure functions g_1^p and g_1^d. *Phys. Lett. B* **1995**, *364*, 61–68. [CrossRef]
113. Anthony, P.L.; Arnold, R.G.; Band, H.R.; Borel, H.; Bosted, P.E.; Breton, V.; Cates, G.D.; Chupp, T.E.; Dietrich, F.S.; Dunne, J.; et al. Deep inelastic scattering of polarized electrons by polarized ^{3}He and the study of the neutron spin structure. *Phys. Rev. D* **1996**, *54*, 6620–6650. [CrossRef]
114. Abe, K.; Akagi, T.; Anderson, B.D.; Anthony, P.L.; Arnold, R.G.; Averett, T.; Band, H.R.; Berisso, C.M.; Bogorad, P.; Borel, H.; et al. Precision determination of the neutron spin structure function g_1^n. *Phys. Rev. Lett.* **1997**, *79*, 26–30. [CrossRef]
115. Abe, K.; Akagi, T.; Anderson, B.; Anthony, P.; Arnold, R.; Averett, T.; Band, H.; Berisso, C.; Bogorad, P.; Borel, H.; et al. Measurement of the neutron spin structure function g_2^n and asymmetry A_2^n. *Phys. Lett. B* **1997**, *404*, 377–382. [CrossRef]
116. Abe, K.; Akagi, T.; Anderson, B.; Anthony, P.; Arnold, R.; Averett, T.; Band, H.; Berisso, C.; Bogorad, P.; Borel, H.; et al. Next-to-leading order QCD analysis of polarized deep inelastic scattering data. *Phys. Lett. B* **1997**, *405*, 180–190. [CrossRef]
117. Abe, K.; Akagi, T.; Anthony, P.L.; Antonov, R.; Arnold, R.G.; Averett, T.; Band, H.R.; Bauer, J.M.; Borel, H.; Bosted, P.E.; et al. Measurements of the proton and deuteron spin structure functions g_1 and g_2. *Phys. Rev.* **1998**, *D58*, 112003.
118. E155 Collaboration; Anthony; P.L.; Arnold, R.G.; Averett, T.; Band, H.R.; Berisso, M.C.; Borelg, H.; Bosteda, P.E.; Bültmanns, S.L.; Buenerd, M.; et al. Measurement of the proton and deuteron spin structure functions g_2 and asymmetry A_2. *Phys. Lett. B* **1999**, *B458*, 529–535.
119. Anthony, P.; Arnold, R.; Averett, T.; Band, H.; Berisso, M.; Borel, H.; Bosted, P.; Bültmann, S.; Buenerd, M.; Chupp, T.; et al. Measurement of the deuteron spin structure function $g_1^d(x)$ for $1\,(\text{GeV}/c)^2 < Q^2 < 40\,(\text{GeV}/c)^2$. *Phys. Lett. B* **1999**, *463*, 339–345.
120. Anthony, P.; Arnold, R.; Averett, T.; Band, H.; Berisso, M.; Borel, H.; Bosted, P.; Bültmann, S.; Buenerd, M.; Chupp, T.; et al. Measurements of the Q^2 dependence of the proton and neutron spin structure functions g_1^p and g_1^n. *Phys. Lett. B* **2000**, *493*, 19–28. [CrossRef]
121. Anthony, P.; Arnold, R.; Averett, T.; Band, H.; Benmouna, N.; Boeglin, W.; Borel, H.; Bosted, P.; Bültmann, S.; Court, G.; et al. Precision measurement of the proton and deuteron spin structure functions g_2 and asymmetries A_2. *Phys. Lett. B* **2003**, *553*, 18–24. [CrossRef]
122. Binosi, D.; Quadri, A. Anti-BRST symmetry and background field method. *Phys. Rev. D* **2013**, *88*, 085036. [CrossRef]
123. Aslam, M.J.; Bashir, A.; Gutierrez-Guerrero, L.X. Local Gauge Transformation for the Quark Propagator in an SU(N) Gauge Theory. *Phys. Rev. D* **2016**, *93*, 076001. [CrossRef]
124. Ding, M.; Raya, K.; Binosi, D.; Chang, L.; Roberts, C.D.; Schmidt, S.M. Drawing insights from pion parton distributions. *Chin. Phys. C (Lett.)* **2020**, *44*, 031002. [CrossRef]
125. Ding, M.; Raya, K.; Binosi, D.; Chang, L.; Roberts, C.D.; Schmidt, S.M. Symmetry, symmetry breaking, and pion parton distributions. *Phys. Rev. D* **2020**, *101*, 054014. [CrossRef]
126. Cui, Z.F.; Ding, M.; Gao, F.; Raya, K.; Binosi, D.; Chang, L.; Roberts, C.D.; Rodríguez-Quintero, J.; Schmidt, S.M. Higgs modulation of emergent mass as revealed in kaon and pion parton distributions. *Eur. Phys. J. A (Lett.)* **2021**, *57*, 5. [CrossRef]
127. Cui, Z.F.; Ding, M.; Gao, F.; Raya, K.; Binosi, D.; Chang, L.; Roberts, C.D.; Rodríguez-Quintero, J.; Schmidt, S.M. Kaon and pion parton distributions. *Eur. Phys. J. C* **2020**, *80*, 1064. [CrossRef]
128. Han, C.; Xie, G.; Wang, R.; Chen, X. An Analysis of Parton Distribution Functions of the Pion and the Kaon with the Maximum Entropy Input. *Eur. Phys. J. C* **2021**, *81*, 302. [CrossRef]
129. Chang, L.; Roberts, C.D. Regarding the distribution of glue in the pion. *Chin. Phys. Lett.* **2021**, *38*, 081101. [CrossRef]
130. Xie, G.; Han, C.; Wang, R.; Chen, X. Tackling the kaon structure function at EicC. *Chin. Phys. C* **2022**, *46*, 064107. [CrossRef]
131. Raya, K.; Cui, Z.F.; Chang, L.; Morgado, J.M.; Roberts, C.D.; Rodríguez-Quintero, J. Revealing pion and kaon structure via generalised parton distributions. *Chin. Phys. C* **2022**, *46*, 013105. [CrossRef]
132. Cui, Z.F.; Ding, M.; Morgado, J.M.; Raya, K.; Binosi, D.; Chang, L.; Papavassiliou, J.; Roberts, C.D.; Rodríguez-Quintero, J.; Schmidt, S.M. Concerning pion parton distributions. *Eur. Phys. J. A* **2022**, *58*, 10. [CrossRef]
133. Cui, Z.F.; Ding, M.; Morgado, J.M.; Raya, K.; Binosi, D.; Chang, L.; De Soto, F.; Roberts, C.D.; Rodríguez-Quintero, J.; Schmidt, S.M. Emergence of pion parton distributions. *Phys. Rev. D* **2022**, *105*, L091502. [CrossRef]
134. Chang, L.; Gao, F.; Roberts, C.D. Parton distributions of light quarks and antiquarks in the proton. *Phys. Lett. B* **2022**, *829*, 137078. [CrossRef]
135. Lu, Y.; Chang, L.; Raya, K.; Roberts, C.D.; Rodríguez-Quintero, J. Proton and pion distribution functions in counterpoint. *Phys. Lett. B* **2022**, *830*, 137130. [CrossRef]

136. De Paula, W.; Ydrefors, E.; Nogueira Alvarenga, J.H.; Frederico, T.; Salmè, G. Parton distribution function in a pion with Minkowskian dynamics. *Phys. Rev. D* **2022**, *105*, L071505. [CrossRef]
137. Grunberg, G. Renormalization Scheme Independent QCD and QED: The Method of Effective Charges. *Phys. Rev. D* **1984**, *29*, 2315. [CrossRef]
138. Grunberg, G. On Some Ambiguities in the Method of Effective Charges. *Phys. Rev. D* **1989**, *40*, 680. [CrossRef]
139. Dokshitzer, Y.L. Perturbative QCD theory (includes our knowledge of $\alpha(s)$) - hep-ph/9812252. In Proceedings of the 29th International Conference High-Energy Physics (ICHEP'98), Vancouver, BC, Canada, 23–29 July 1998; Volume 1, pp. 305–324.
140. Prosperi, G.M.; Raciti, M.; Simolo, C. On the running coupling constant in QCD. *Prog. Part. Nucl. Phys.* **2007**, *58*, 387–438. [CrossRef]
141. Bjorken, J.D. Applications of the Chiral U(6) x (6) Algebra of Current Densities. *Phys. Rev.* **1966**, *148*, 1467–1478. [CrossRef]
142. Bjorken, J.D. Inelastic Scattering of Polarized Leptons from Polarized Nucleons. *Phys. Rev. D* **1970**, *1*, 1376–1379. [CrossRef]
143. Brodsky, S.J.; de Téramond, G.F. Light-front hadron dynamics and AdS/CFT correspondence. *Phys. Lett. B* **2004**, *582*, 211–221. [CrossRef]
144. Brodsky, S.J.; de Teramond, G.F.; Deur, A. Nonperturbative QCD Coupling and its β-function from Light-Front Holography. *Phys. Rev. D* **2010**, *81*, 096010. [CrossRef]
145. Carlson, J.; Jaffe, A.; Wiles, A. (Eds.) The Millennium Prize Problems; American Mathematical Society: Providence, RI, USA, 2006.
146. Eichten, E.; Gottfried, K.; Kinoshita, T.; Lane, K.D.; Yan, T.M. Charmonium: The Model. *Phys. Rev. D* **1978**, *17*, 3090; Erratum: *Phys. Rev. D* **1980**, *21*, 313. [CrossRef]
147. Richardson, J.L. The Heavy Quark Potential and the Upsilon, J/psi Systems. *Phys. Lett. B* **1979**, *82*, 272–274. [CrossRef]
148. Buchmuller, W.; Tye, S.H.H. Quarkonia and Quantum Chromodynamics. *Phys. Rev. D* **1981**, *24*, 132. [CrossRef]
149. Godfrey, S.; Isgur, N. Mesons in a Relativized Quark Model with Chromodynamics. *Phys. Rev. D* **1985**, *32*, 189–231. [CrossRef]
150. Lucha, W.; Schöberl, F.F.; Gromes, D. Bound states of quarks. *Phys. Rept.* **1991**, *200*, 127–240. [CrossRef]
151. Binosi, D.; Chang, L.; Qin, S.X.; Papavassiliou, J.; Roberts, C.D. Symmetry preserving truncations of the gap and Bethe–Salpeter equations. *Phys. Rev. D* **2016**, *93*, 096010. [CrossRef]
152. Smirnov, A.V.; Smirnov, V.A.; Steinhauser, M. Three-loop static potential. *Phys. Rev. Lett.* **2010**, *104*, 112002. [CrossRef] [PubMed]
153. Brodsky, S.J.; Deur, A.; de Téramond, G.F.; Dosch, H.G. Light-Front Holography and Superconformal Quantum Mechanics: A New Approach to Hadron Structure and Color Confinement. *Int. J. Mod. Phys. Conf. Ser.* **2015**, *39*, 1560081. [CrossRef]
154. Brodsky, S.J.; de Teramond, G.F.; Dosch, H.G.; Erlich, J. Light-Front Holographic QCD and Emerging Confinement. *Phys. Rept.* **2015**, *584*, 1–105. [CrossRef]
155. Reinhardt, H.; Burgio, G.; Campagnari, D.; Ebadati, E.; Heffner, J.; Quandt, M.; Vastag, P.; Vogt, H. Hamiltonian approach to QCD in Coulomb gauge - a survey of recent results. *Adv. High Energy Phys.* **2018**, *2018*, 2312498. [CrossRef]
156. Hoyer, P. *Journey to the Bound States*; Springer Briefs in Physics; Springer: Berlin/Heidelberg, Germany, 2021.
157. Bali, G.S.; Neff, H.; Duessel, T.; Lippert, T.; Schilling, K. Observation of string breaking in QCD. *Phys. Rev. D* **2005**, *71*, 114513. [CrossRef]
158. Prkacin, Z.; Bali, G.S.; Dussel, T.; Lippert, T.; Neff, H.; Schilling, K. Anatomy of string breaking in QCD. *High Energy Phys. Lattice* **2006**, *LAT2005*, 308.
159. Chang, L.; Cloet, I.C.; El-Bennich, B.; Klähn, T.; Roberts, C.D. Exploring the light quark interaction. *Chin. Phys. C* **2009**, *33*, 1189–1196.
160. Munczek, H.J.; Nemirovsky, A.M. The Ground State $q\bar{q}$ Mass Spectrum in QCD. *Phys. Rev. D* **1983**, *28*, 181–186. [CrossRef]
161. Stingl, M. Propagation Properties and Condensate Formation of the Confined Yang–Mills Field. *Phys. Rev. D* **1986**, *34*, 3863–3881. [Erratum: *Phys. Rev. D* **36**, 651 (1987)]. [CrossRef] [PubMed]
162. Zwanziger, D. Local and Renormalizable Action From the Gribov Horizon. *Nucl. Phys. B* **1989**, *323*, 513–544. [CrossRef]
163. Burden, C.J.; Roberts, C.D.; Williams, A.G. Singularity structure of a model quark propagator. *Phys. Lett. B* **1992**, *285*, 347–353. [CrossRef]
164. Gribov, V.N. The theory of quark confinement. *Eur. Phys. J. C* **1999**, *10*, 91–105. [CrossRef]
165. Stingl, M. A Systematic extended iterative solution for quantum chromodynamics. *Z. Phys. A* **1996**, *353*, 423–445. [CrossRef]
166. Dudal, D.; Gracey, J.A.; Sorella, S.P.; Vandersickel, N.; Verschelde, H. A Refinement of the Gribov–Zwanziger approach in the Landau gauge: Infrared propagators in harmony with the lattice results. *Phys. Rev. D* **2008**, *78*, 065047. [CrossRef]
167. Qin, S.X.; Rischke, D.H. Quark Spectral Function and Deconfinement at Nonzero Temperature. *Phys. Rev. D* **2013**, *88*, 056007. [CrossRef]
168. Lucha, W.; Schöberl, F.F. Analytic Bethe–Salpeter Description of the Lightest Pseudoscalar Mesons. *Phys. Rev. D* **2016**, *93*, 056006. [CrossRef]
169. Dudal, D.; Oliveira, O.; Roelfs, M.; Silva, P. Spectral representation of lattice gluon and ghost propagators at zero temperature. *Nucl. Phys. B* **2020**, *952*, 114912. [CrossRef]
170. Fischer, C.S.; Huber, M.Q. Landau gauge Yang–Mills propagators in the complex momentum plane. *Phys. Rev. D* **2020**, *102*, 094005. [CrossRef]
171. Osterwalder, K.; Schrader, R. Axioms for Euclidean Green's Functions. *Commun. Math. Phys.* **1973**, *31*, 83–112. [CrossRef]
172. Osterwalder, K.; Schrader, R. Axioms for Euclidean Green's Functions. 2. *Commun. Math. Phys.* **1975**, *42*, 281. [CrossRef]
173. Dedushenko, M. Snowmass White Paper: The Quest to Define QFT. *arXiv* **2022**, arXiv:2203.08053.

174. Stodolna, A.S.; Rouzée, A.; Lépine, F.; Cohen, S.; Robicheaux, F.; Gijsbertsen, A.; Jungmann, J.H.; Bordas, C.; Vrakking, M.J.J. Hydrogen Atoms under Magnification: Direct Observation of the Nodal Structure of Stark States. *Phys. Rev. Lett.* **2013**, *110*, 213001. [CrossRef] [PubMed]
175. Bhagwat, M.; Pichowsky, M.; Tandy, P.C. Confinement phenomenology in the Bethe–Salpeter equation. *Phys. Rev. D* **2003**, *67*, 054019. [CrossRef]
176. Field, R.D.; Feynman, R.P. A Parametrization of the Properties of Quark Jets. *Nucl. Phys. B* **1978**, *136*, 1–76. [CrossRef]
177. Gribov, V.N.; Lipatov, L.N. Deep inelastic electron scattering in perturbation theory. *Phys. Lett. B* **1971**, *37*, 78–80. [CrossRef]
178. Brodsky, S.J.; Deshpande, A.L.; Gao, H.; McKeown, R.D.; Meyer, C.A.; Meziani, Z.E.; Milner, R.G.; Qiu, J.; Richards, D.G.; Roberts, C.D. QCD and Hadron Physics. *arXiv* **2015**, arXiv:1502.05728.
179. Denisov, O. Letter of Intent (Draft 2.0): A New QCD facility at the M2 beam line of the CERN SPS *arXiv* **2018**, arXiv:1808.00848.
180. Aguilar, A.C.; Ahmed, Z.; Aidala, C.; Ali, S.; Andrieux, V.; Arrington, J.; Bashir, A.; Berdnikov, V.; Binosi, D.; Chang, L.; et al. Pion and Kaon Structure at the Electron-Ion Collider. *Eur. Phys. J. A* **2019**, *55*, 190. [CrossRef]
181. Brodsky, S.J.; Burkert, V.D.; Carman, D.S.; Chen, J.P.; Cui, Z.-F.; Döring, M.; Dosch, H.G.; Draayer, J.; Elouadrhiri, L.; Glazier, D.I.; et al. Strong QCD from Hadron Structure Experiments. *Int. J. Mod. Phys. E* **2020**, *29*, 2030006. [CrossRef]
182. Chen, X.; Guo, F.K.; Roberts, C.D.; Wang, R. Selected Science Opportunities for the EicC. *Few Body Syst.* **2020**, *61*, 43. [CrossRef]
183. Anderle, D.P.; Bertone, V.; Cao, X.; Chang, L.; Chang, N.; Chen, G.; Chen, X.; Chen, Z.; Cui, Z.; Dai, L.; et al. Electron-ion collider in China. *Front. Phys.* **2021**, *16*, 64701. [CrossRef]
184. Arrington, J.; Gayoso, C.A.; Barry, P.C.; Berdnikov, V.; Binosi, D.; Chang, L.; Diefenthaler, M.; Ding, M.; Ent, R.; Frederico, T.; et al. Revealing the structure of light pseudoscalar mesons at the electron–ion collider. *J. Phys. G* **2021**, *48*, 075106. [CrossRef]
185. Aoki, K.; Fujioka, H.; Gogami, T.; Hidaka, Y.; Hiyama, E.; Honda, R.; Hosaka, A.; Ichikawa, Y.; Ieiri, M.; Isaka, M.; et al. Extension of the J-PARC Hadron Experimental Facility: Third White Paper. *arXiv* **2021**, arXiv:2110.04462.
186. Quintans, C. The New AMBER Experiment at the CERN SPS. *Few Body Syst.* **2022**, *63*, 72. [CrossRef]
187. Capstick, S.; Roberts, W. Quark models of baryon masses and decays. *Prog. Part. Nucl. Phys.* **2000**, *45*, S241–S331. [CrossRef]
188. Giannini, M.M.; Santopinto, E. The hypercentral Constituent Quark Model and its application to baryon properties. *Chin. J. Phys.* **2015**, *53*, 020301.
189. Plessas, W. The constituent-quark model — Nowadays. *Int. J. Mod. Phys. A* **2015**, *30*, 1530013. [CrossRef]
190. Dudek, J.J.; Edwards, R.G.; Peardon, M.J.; Richards, D.G.; Thomas, C.E. Toward the excited meson spectrum of dynamical QCD. *Phys. Rev. D* **2010**, *82*, 034508. [CrossRef]
191. Edwards, R.G.; Dudek, J.J.; Richards, D.G.; Wallace, S.J. Excited state baryon spectroscopy from lattice QCD. *Phys. Rev. D* **2011**, *84*, 074508. [CrossRef]
192. Liu, L.; Moir, G.; Peardon, M.; Ryan, S.M.; Thomas, C.E.; Vilaseca, P.; Dudek, J.J.; Edwards, R.G.; Joo, B.; Richards, D.G. Excited and exotic charmonium spectroscopy from lattice QCD. *J. High Energy Phys.* **2012**, *7*, 126. [CrossRef]
193. Dudek, J.J.; Edwards, R.G. Hybrid Baryons in QCD. *Phys. Rev. D* **2012**, *85*, 054016. [CrossRef]
194. Ryan, S.M.; Wilson, D.J. Excited and exotic bottomonium spectroscopy from lattice QCD. *J. High Energy Phys.* **2021**, *02*, 214. [CrossRef]
195. Woss, A.J.; Dudek, J.J.; Edwards, R.G.; Thomas, C.E.; Wilson, D.J. Decays of an exotic 1−+ hybrid meson resonance in QCD. *Phys. Rev. D* **2021**, *103*, 054502. [CrossRef]
196. Munczek, H.J. Dynamical chiral symmetry breaking, Goldstone's theorem and the consistency of the Schwinger-Dyson and Bethe–Salpeter Equations. *Phys. Rev. D* **1995**, *52*, 4736–4740. [CrossRef] [PubMed]
197. Bender, A.; Roberts, C.D.; von Smekal, L. Goldstone Theorem and Diquark Confinement Beyond Rainbow- Ladder Approximation. *Phys. Lett. B* **1996**, *380*, 7–12. [CrossRef]
198. Maris, P.; Roberts, C.D.; Tandy, P.C. Pion mass and decay constant. *Phys. Lett. B* **1998**, *420*, 267–273. [CrossRef]
199. Maris, P.; Roberts, C.D. π and K meson Bethe–Salpeter amplitudes. *Phys. Rev. C* **1997**, *56*, 3369–3383. [CrossRef]
200. Chen, M.; Ding, M.; Chang, L.; Roberts, C.D. Mass-dependence of pseudoscalar meson elastic form factors. *Phys. Rev. D* **2018**, *98*, 091505. [CrossRef]
201. Ding, M.; Raya, K.; Bashir, A.; Binosi, D.; Chang, L.; Chen, M.; Roberts, C.D. $\gamma^*\gamma \to \eta, \eta'$ transition form factors. *Phys. Rev. D* **2019**, *99*, 014014. [CrossRef]
202. Eichmann, G.; Fischer, C.S.; Williams, R. Kaon-box contribution to the anomalous magnetic moment of the muon. *Phys. Rev. D* **2020**, *101*, 054015. [CrossRef]
203. Xu, Y.Z.; Binosi, D.; Cui, Z.F.; Li, B.L.; Roberts, C.D.; Xu, S.S.; Zong, H.S. Elastic electromagnetic form factors of vector mesons. *Phys. Rev. D* **2019**, *100*, 114038. [CrossRef]
204. Xu, Y.Z.; Chen, S.; Yao, Z.Q.; Binosi, D.; Cui, Z.F.; Roberts, C.D. Vector-meson production and vector meson dominance. *Eur. Phys. J. C* **2021**, *81*, 895. [CrossRef]
205. Yao, Z.Q.; Binosi, D.; Cui, Z.F.; Roberts, C.D.; Xu, S.S.; Zong, H.S. Semileptonic decays of $D_{(s)}$ mesons. *Phys. Rev. D* **2020**, *102*, 014007. [CrossRef]
206. Yao, Z.Q.; Binosi, D.; Cui, Z.F.; Roberts, C.D. Semileptonic $B_c \to \eta_c, J/\psi$ transitions. *Phys. Lett. B* **2021**, *818*, 136344. [CrossRef]
207. Yao, Z.Q.; Binosi, D.; Cui, Z.F.; Roberts, C.D. Semileptonic transitions: $B_{(s)} \to \pi(K)$; $D_s \to K$; $D \to \pi, K$; and $K \to \pi$. *Phys. Lett. B* **2022**, *824*, 136793. [CrossRef]
208. Höll, A.; Krassnigg, A.; Roberts, C.D. Pseudoscalar meson radial excitations. *Phys. Rev. C* **2004**, *70*, 042203. [CrossRef]

209. Höll, A.; Krassnigg, A.; Maris, P.; Roberts, C.D.; Wright, S.V. Electromagnetic properties of ground and excited state pseudoscalar mesons. *Phys. Rev. C* **2005**, *71*, 065204. [CrossRef]
210. Fischer, C.S.; Williams, R. Probing the gluon self-interaction in light mesons. *Phys. Rev. Lett.* **2009**, *103*, 122001. [CrossRef]
211. Krassnigg, A. Survey of J=0,1 mesons in a Bethe–Salpeter approach. *Phys. Rev. D* **2009**, *80*, 114010. [CrossRef]
212. Qin, S.X.; Chang, L.; Liu, Y.X.; Roberts, C.D.; Wilson, D.J. Interaction model for the gap equation. *Phys. Rev. C* **2011**, *84*, 042202. [CrossRef]
213. Qin, S.X.; Chang, L.; Liu, Y.X.; Roberts, C.D.; Wilson, D.J. Investigation of rainbow-ladder truncation for excited and exotic mesons. *Phys. Rev. C* **2012**, *85*, 035202. [CrossRef]
214. Blank, M.; Krassnigg, A. Bottomonium in a Bethe–Salpeter-equation study. *Phys. Rev. D* **2011**, *84*, 096014.
215. Hilger, T.; Popovici, C.; Gomez-Rocha, M.; Krassnigg, A. Spectra of heavy quarkonia in a Bethe–Salpeter-equation approach. *Phys. Rev. D* **2015**, *91*, 034013. [CrossRef]
216. Fischer, C.S.; Kubrak, S.; Williams, R. Spectra of heavy mesons in the Bethe–Salpeter approach. *Eur. Phys. J. A* **2015**, *51*, 10. [CrossRef]
217. Eichmann, G.; Fischer, C.S.; Sanchis-Alepuz, H. Light baryons and their excitations. *Phys. Rev. D* **2016**, *94*, 094033. [CrossRef]
218. Qin, S.X.; Roberts, C.D.; Schmidt, S.M. Spectrum of light- and heavy-baryons. *Few Body Syst.* **2019**, *60*, 26. [CrossRef]
219. Chang, L.; Roberts, C.D. Tracing masses of ground state light quark mesons. *Phys. Rev. C* **2012**, *85*, 052201. [CrossRef]
220. Williams, R.; Fischer, C.S.; Heupel, W. Light mesons in QCD and unquenching effects from the 3PI effective action. *Phys. Rev. D* **2016**, *93*, 034026. [CrossRef]
221. Qin, S.X.; Roberts, C.D. Resolving the Bethe–Salpeter kernel. *Chin. Phys. Lett. Express* **2021**, *38*, 071201. [CrossRef]
222. Xu, Z.N.; Yao, Z.Q.; Qin, S.X.; Cui, Z.F.; Roberts, C.D. Bethe–Salpeter kernel and properties of strange-quark mesons. *arXiv* **2022**, arXiv:2208.13903.
223. Qin, S.X.; Roberts, C.D.; Schmidt, S.M. Ward–Green–Takahashi identities and the axial-vector vertex. *Phys. Lett. B* **2014**, *733*, 202–208. [CrossRef]
224. Bhagwat, M.S.; Höll, A.; Krassnigg, A.; Roberts, C.D.; Tandy, P.C. Aspects and consequences of a dressed quark-gluon vertex. *Phys. Rev. C* **2004**, *70*, 035205. [CrossRef]
225. Chang, L.; Liu, Y.X.; Roberts, C.D.; Shi, Y.M.; Sun, W.M.; Zong, H.S. Chiral susceptibility and the scalar Ward identity. *Phys. Rev. C* **2009**, *79*, 035209. [CrossRef]
226. Singh, J. Anomalous magnetic moment of light quarks and dynamical symmetry breaking. *Phys. Rev. D* **1985**, *31*, 1097–1108. [CrossRef]
227. Bicudo, P.J.A.; Ribeiro, J.E.F.T.; Fernandes, R. The anomalous magnetic moment of quarks. *Phys. Rev. C* **1999**, *59*, 1107–1112. [CrossRef]
228. Chang, L.; Liu, Y.X.; Roberts, C.D. Dressed-quark anomalous magnetic moments. *Phys. Rev. Lett.* **2011**, *106*, 072001. [CrossRef]
229. Bashir, A.; Bermúdez, R.; Chang, L.; Roberts, C.D. Dynamical chiral symmetry breaking and the fermion–gauge-boson vertex. *Phys. Rev. C* **2012**, *85*, 045205. [CrossRef]
230. Kızılersü, A.; Oliveira, O.; Silva, P.J.; Skullerud, J.I.; Sternbeck, A. Quark-gluon vertex from Nf=2 lattice QCD. *Phys. Rev. D* **2021**, *103*, 114515. [CrossRef]
231. Williams, R. The quark-gluon vertex in Landau gauge bound state studies. *Eur. Phys. J. A* **2015**, *51*, 57. [CrossRef]
232. Yin, P.L.; Cui, Z.F.; Roberts, C.D.; Segovia, J. Masses of positive- and negative-parity hadron ground states, including those with heavy quarks. *Eur. Phys. J. C* **2021**, *81*, 327. [CrossRef]
233. Xu, Z.N.; Cui, Z.F.; Roberts, C.D.; Xu, C. Heavy + light pseudoscalar meson semileptonic transitions. *Eur. Phys. J. C* **2021**, *81*, 1105. [CrossRef]
234. Gutiérrez-Guerrero, L.X.; Paredes-Torres, G.; Bashir, A. Mesons and baryons: Parity partners. *Phys. Rev. D* **2021**, *104*, 094013. [CrossRef]
235. Höll, A.; Maris, P.; Roberts, C.D.; Wright, S.V. Schwinger functions and light quark bound states, and sigma terms. *Nucl. Phys. Proc. Suppl.* **2006**, *161*, 87–94. [CrossRef]
236. Santowsky, N.; Eichmann, G.; Fischer, C.S.; Wallbott, P.C.; Williams, R. The σ-meson: Four-quark vs. two-quark components and decay width in a Bethe–Salpeter approach. *Phys. Rev. D* **2020**, *102*, 056014. [CrossRef]
237. Nakanishi, N. A General survey of the theory of the Bethe–Salpeter equation. *Prog. Theor. Phys. Suppl.* **1969**, *43*, 1–81. [CrossRef]
238. Cui, Z.F.; Binosi, D.; Roberts, C.D.; Schmidt, S.M. Hadron and light nucleus radii from electron scattering. *Chin. Phys. C* **2022**, in press. [CrossRef]
239. Binosi, D.; Chang, L.; Ding, M.; Gao, F.; Papavassiliou, J.; Roberts, C.D. Distribution Amplitudes of Heavy-Light Mesons. *Phys. Lett. B* **2019**, *790*, 257–262. [CrossRef]
240. McNeile, C.; Michael, C. The decay constant of the first excited pion from lattice QCD. *Phys. Lett. B* **2006**, *642*, 244–247. [CrossRef]
241. Ballon-Bayona, A.; Krein, G.; Miller, C. Decay constants of the pion and its excitations in holographic QCD. *Phys. Rev. D* **2015**, *91*, 065024. [CrossRef]
242. Chen, M.; Chang, L. A pattern for the flavor dependent quark-antiquark interaction. *Chin. Phys. C* **2019**, *43*, 114103. [CrossRef]
243. Qin, P.; Qin, S.X.; Liu, Y.X. Heavy-light mesons beyond the ladder approximation. *Phys. Rev. D* **2020**, *101*, 114014. [CrossRef]
244. Burden, C.J.; Pichowsky, M.A. J^{PC} exotic mesons from the Bethe–Salpeter equation. *Few Body Syst.* **2002**, *32*, 119–126. [CrossRef]

245. Hilger, T.; Gomez-Rocha, M.; Krassnigg, A. Masses of $J^{PC} = 1^{-+}$ exotic quarkonia in a Bethe–Salpeter-equation approach. *Phys. Rev. D* **2015**, *91*, 114004. [CrossRef]
246. Xu, S.S.; Cui, Z.F.; Chang, L.; Papavassiliou, J.; Roberts, C.D.; Zong, H.S. New perspective on hybrid mesons. *Eur. Phys. J. A (Lett.)* **2019**, *55*, 113. [CrossRef]
247. Meyers, J.; Swanson, E.S. Spin Zero Glueballs in the Bethe–Salpeter Formalism. *Phys. Rev. D* **2013**, *87*, 036009. [CrossRef]
248. Souza, E.V.; Narciso Ferreira, M.; Aguilar, A.C.; Papavassiliou, J.; Roberts, C.D.; Xu, S.S. Pseudoscalar glueball mass: a window on three-gluon interactions. *Eur. Phys. J. A (Lett.)* **2020**, *56*, 25. [CrossRef]
249. Kaptari, L.P.; Kämpfer, B. Mass spectrum of pseudo-scalar glueballs from a Bethe–Salpeter approach with the rainbow-ladder truncation. *Few Body Syst.* **2020**, *61*, 28. [CrossRef]
250. Huber, M.Q.; Fischer, C.S.; Sanchis-Alepuz, H. Higher spin glueballs from functional methods. *Eur. Phys. J. C* **2021**, *81*, 1083. [CrossRef]
251. Ablikim, M.; Achasov, M.N.; Adlarson, P.; Ahmed, S.; Albrecht, M.; Alekseev, M.; Amoroso, A.; An, F.F.; An, Q.; Bai, Y.; et al. Future Physics Programme of BESIII. *Chin. Phys. C* **2020**, *44*, 040001. [CrossRef]
252. Khalek, R.A.; Accardi, A.; Adam, J.; Adamiak, D.; Akers, W.; Albaladejo, M.; Blin, A.H. Science Requirements and Detector Concepts for the Electron-Ion Collider: EIC Yellow Report. *Nucl. Phys. A* **2022**, *1026*, 122447. [CrossRef]
253. Pauli, P. Accessing glue through photoproduction measurements at GlueX. *Rev. Mex. Fis. Suppl.* **2022**, *3*, 0308002. [CrossRef]
254. Eichmann, G.; Alkofer, R.; Krassnigg, A.; Nicmorus, D. Nucleon mass from a covariant three-quark Faddeev equation. *Phys. Rev. Lett.* **2010**, *104*, 201601. [CrossRef] [PubMed]
255. Barabanov, M.; Bedolla, M.; Brooks, W.; Cates, G.; Chen, C.; Chen, Y.; Cisbani, E.; Ding, M.; Eichmann, G.; Ent, R.; et al. Diquark Correlations in Hadron Physics: Origin, Impact and Evidence. *Prog. Part. Nucl. Phys.* **2021**, *116*, 103835. [CrossRef]
256. Cahill, R.T.; Roberts, C.D.; Praschifka, J. Baryon structure and QCD. *Austral. J. Phys.* **1989**, *42*, 129–145. [CrossRef]
257. Burden, C.J.; Cahill, R.T.; Praschifka, J. Baryon Structure and QCD: Nucleon Calculations. *Austral. J. Phys.* **1989**, *42*, 147–159. [CrossRef]
258. Reinhardt, H. Hadronization of Quark Flavor Dynamics. *Phys. Lett. B* **1990**, *244*, 316–326. [CrossRef]
259. Efimov, G.V.; Ivanov, M.A.; Lyubovitskij, V.E. Quark—Diquark approximation of the three quark structure of baryons in the quark confinement model. *Z. Phys. C* **1990**, *47*, 583–594. [CrossRef]
260. Gutiérrez-Guerrero, L.X.; Bashir, A.; Cloet, I.C.; Roberts, C.D. Pion form factor from a contact interaction. *Phys. Rev. C* **2010**, *81*, 065202. [CrossRef]
261. Gutiérrez-Guerrero, L.X.; Bashir, A. Contact Interaction Studies of Hadron Observables. *Particles* **2022**, *5*, 69–138.
262. Brown, Z.S.; Detmold, W.; Meinel, S.; Orginos, K. Charmed bottom baryon spectroscopy from lattice QCD. *Phys. Rev. D* **2014**, *90*, 094507. [CrossRef]
263. Mathur, N.; Padmanath, M.; Mondal, S. Precise predictions of charmed-bottom hadrons from lattice QCD. *Phys. Rev. Lett.* **2018**, *121*, 202002. [CrossRef] [PubMed]
264. Bahtiyar, H.; Can, K.U.; Erkol, G.; Gubler, P.; Oka, M.; Takahashi, T.T. Charmed baryon spectrum from lattice QCD near the physical point. *Phys. Rev. D* **2020**, *102*, 054513. [CrossRef]
265. Chen, C.; El-Bennich, B.; Roberts, C.D.; Schmidt, S.M.; Segovia, J.; Wan, S. Structure of the nucleon's low-lying excitations. *Phys. Rev. D* **2018**, *97*, 034016. [CrossRef]
266. Chen, C.; Krein, G.I.; Roberts, C.D.; Schmidt, S.M.; Segovia, J. Spectrum and structure of octet and decuplet baryons and their positive-parity excitations. *Phys. Rev. D* **2019**, *100*, 054009. [CrossRef]
267. Liu, L.; Chen, C.; Lu, Y.; Roberts, C.D.; Segovia, J. Composition of low-lying $J = \frac{3}{2}^{\pm}$ Δ-baryons. *Phys. Rev. D* **2022**, *105*, 114047. [CrossRef]
268. Liu, L.; Chen, C.; Roberts, C.D. Wave functions of $(I, J^P) = (\frac{1}{2}, \frac{3}{2}^{\mp})$ baryons. *arXiv* **2022**, arXiv:2208.12353.
269. Crede, V.; Roberts, W. Progress towards understanding baryon resonances. *Rept. Prog. Phys.* **2013**, *76*, 076301. [CrossRef]
270. Alkofer, R.; Höll, A.; Kloker, M.; Krassnigg, A.; Roberts, C.D. On nucleon electromagnetic form-factors. *Few Body Syst.* **2005**, *37*, 1–31. [CrossRef]
271. Segovia, J.; Cloet, I.C.; Roberts, C.D.; Schmidt, S.M. Nucleon and Δ elastic and transition form factors. *Few Body Syst.* **2014**, *55*, 1185–1222. [CrossRef]
272. Chen, C.; Lu, Y.; Binosi, D.; Roberts, C.D.; Rodríguez-Quintero, J.; Segovia, J. Nucleon-to-Roper electromagnetic transition form factors at large Q^2. *Phys. Rev. D* **2019**, *99*, 034013. [CrossRef]
273. Lu, Y.; Chen, C.; Cui, Z.F.; Roberts, C.D.; Schmidt, S.M.; Segovia, J.; Zong, H.S. Transition form factors: $\gamma^* + p \to \Delta(1232)$, $\Delta(1600)$. *Phys. Rev. D* **2019**, *100*, 034001. [CrossRef]
274. Chen, C.; Fischer, C.S.; Roberts, C.D.; Segovia, J. Form Factors of the Nucleon Axial Current. *Phys. Lett. B* **2021**, *815*, 136150. [CrossRef]
275. Chen, C.; Fischer, C.S.; Roberts, C.D.; Segovia, J. Nucleon axial-vector and pseudoscalar form factors and PCAC relations. *Phys. Rev. D* **2022**, *105*, 094022. [CrossRef]
276. Chen, C.; Roberts, C.D. Nucleon axial form factor at large momentum transfers. *Eur. Phys. J. A* **2022**, *58*, 206. [CrossRef]
277. Klempt, E.; Metsch, B.C. Multiplet classification of light quark baryons. *Eur. Phys. J. A* **2012**, *48*, 127. [CrossRef]
278. Weinberg, S. Precise relations between the spectra of vector and axial vector mesons. *Phys. Rev. Lett.* **1967**, *18*, 507–509. [CrossRef]
279. Lane, K.D. Asymptotic Freedom and Goldstone Realization of Chiral Symmetry. *Phys. Rev. D* **1974**, *10*, 2605. [CrossRef]

280. Politzer, H.D. Effective Quark Masses in the Chiral Limit. *Nucl. Phys. B* **1976**, *117*, 397. [CrossRef]
281. Pagels, H. Dynamical Chiral Symmetry Breaking in Quantum Chromodynamics. *Phys. Rev. D* **1979**, *19*, 3080. [CrossRef]
282. Higashijima, K. Dynamical Chiral Symmetry Breaking. *Phys. Rev. D* **1984**, *29*, 1228. [CrossRef]
283. Pagels, H.; Stokar, S. The Pion Decay Constant, Electromagnetic Form-Factor and Quark Electromagnetic Selfenergy in QCD. *Phys. Rev. D* **1979**, *20*, 2947. [CrossRef]
284. Cahill, R.T.; Roberts, C.D. Soliton Bag Models of Hadrons from QCD. *Phys. Rev. D* **1985**, *32*, 2419. [CrossRef] [PubMed]
285. Munczek, H.J. Infrared Effects on the Axial Gauge Quark Propagator. *Phys. Lett. B* **1986**, *175*, 215–218. [CrossRef]
286. Efimov, G.V.; Ivanov, M.A. Confinement and quark structure of light hadrons. *Int. J. Mod. Phys. A* **1989**, *4*, 2031–2060. [CrossRef]
287. Hecht, M.B.; Oettel, M.; Roberts, C.D.; Schmidt, S.M.; Tandy, P.C.; Thomas, A.W. Nucleon mass and pion loops. *Phys. Rev. C* **2002**, *65*, 055204. [CrossRef]
288. Sanchis-Alepuz, H.; Fischer, C.S.; Kubrak, S. Pion cloud effects on baryon masses. *Phys. Lett. B* **2014**, *733*, 151–157. [CrossRef]
289. Julia-Diaz, B.; Lee, T.S.H.; Matsuyama, A.; Sato, T. Dynamical coupled channel model of pi N scattering in the $W \leq$ 2-GeV nucleon resonance region. *Phys. Rev. C* **2007**, *76*, 065201. [CrossRef]
290. Suzuki, N.; Julia-Diaz, B.; Kamano, H.; Lee, T.S.H.; Matsuyama, A.; Sato, T. Disentangling the Dynamical Origin of P-11 Nucleon Resonances. *Phys. Rev. Lett.* **2010**, *104*, 042302. [CrossRef]
291. Rönchen, D.; Döring, M.; Huang, F.; Haberzettl, H.; Haidenbauer, J.; Hanhart, C.; Krewald, S.; Meissner, U.G.; Nakayama, K. Coupled-channel dynamics in the reactions $\pi N \to \pi N, \eta N, K\Lambda, K\Sigma$. *Eur. Phys. J. A* **2013**, *49*, 44. [CrossRef]
292. Kamano, H.; Nakamura, S.X.; Lee, T.S.H.; Sato, T. Nucleon resonances within a dynamical coupled channels model of πN and γN reactions. *Phys. Rev. C* **2013**, *88*, 035209. [CrossRef]
293. García-Tecocoatzi, H.; Bijker, R.; Ferretti, J.; Santopinto, E. Self-energies of octet and decuplet baryons due to the coupling to the baryon-meson continuum. *Eur. Phys. J. A* **2017**, *53*, 115. [CrossRef]
294. Eichmann, G.; Alkofer, R.; Cloet, I.C.; Krassnigg, A.; Roberts, C.D. Perspective on rainbow-ladder truncation. *Phys. Rev. C* **2008**, *77*, 042202. [CrossRef]
295. Eichmann, G.; Cloet, I.C.; Alkofer, R.; Krassnigg, A.; Roberts, C.D. Toward unifying the description of meson and baryon properties. *Phys. Rev. C* **2009**, *79*, 012202. [CrossRef]
296. Roberts, H.L.L.; Chang, L.; Cloet, I.C.; Roberts, C.D. Masses of ground and excited state hadrons. *Few Body Syst.* **2011**, *51*, 1–25. [CrossRef]
297. Isupov, E.L.; Burkert, V.D.; Carman, D.S.; Gothe, R.W.; Hicks, K.; Ishkhanov, B.S; Mokeev, V.I.; Adhikari, K.P.; Adhikari, S.; Adikaram, D.; et al. Measurements of $ep \to e'\pi^+\pi^-p'$ Cross Sections with CLAS at 1.40 Gev $< W <$ 2.0 GeV and 2.0 GeV^2 $< Q^2 <$ 5.0 GeV^2. *Phys. Rev. C* **2017**, *96*, 025209. [CrossRef]
298. Trivedi, A. Measurement of New Observables from the $\pi^+\pi^-$ p Electroproduction Off the Proton. *Few Body Syst.* **2019**, *60*, 5. [CrossRef]
299. Raya, K.; Gutiérrez-Guerrero, L.X.; Bashir, A.; Chang, L.; Cui, Z.F.; Lu, Y.; Roberts, C.D.; Segovia, J. Dynamical diquarks in the $\gamma^{(*)}p \to N(1535)\frac{1}{2}^-$ transition. *Eur. Phys. J. A* **2021**, *57*, 266. [CrossRef]
300. Aznauryan, I.G.; Burkert, V.D.; Biselli, A.S.; Egiyan, H.; Joo, K.; Kim, W.; Park, K.; Smith, L.C.; Ungaro, M.; Adhikari, K.P.; et al. Electroexcitation of nucleon resonances from CLAS data on single pion electroproduction. *Phys. Rev. C* **2009**, *80*, 055203. [CrossRef]
301. Mokeev, V.I.; Burkert, V.D.; Carman, D.S.; Elouadrhiri, L.; Fedotov, G.V.; Golovatch, E.N.; Gothe, R.W.; Hicks, K.; Ishkhanov, B.S.; Isupov, E.L.; et al. New Results from the Studies of the $N(1440)1/2^+$, $N(1520)3/2^-$, and $\Delta(1620)1/2^-$ Resonances in Exclusive $ep \to e'p'\pi^+\pi^-$ Electroproduction with the CLAS Detector. *Phys. Rev. C* **2016**, *93*, 025206. [CrossRef]
302. Mokeev, V.I. Two Pion Photo- and Electroproduction with CLAS. *EPJ Web Conf.* **2020**, *241*, 03003. [CrossRef]
303. Carman, D.; Joo, K.; Mokeev, V. Strong QCD Insights from Excited Nucleon Structure Studies with CLAS and CLAS12. *Few Body Syst.* **2020**, *61*, 29. [CrossRef]
304. Mokeev, V.I.; Carman, D.S. New baryon states in exclusive meson photo-/electroproduction with CLAS. *Rev. Mex. Fis. Suppl.* **2022**, *3*, 0308024. [CrossRef]
305. Mokeev, V.I.; Carman, D.S. Photo- and Electrocouplings of Nucleon Resonances. *Few Body Syst.* **2022**, *63*, 59. [CrossRef]
306. Chang, L.; Cloet, I.C.; Roberts, C.D.; Schmidt, S.M.; Tandy, P.C. Pion electromagnetic form factor at spacelike momenta. *Phys. Rev. Lett.* **2013**, *111*, 141802. [CrossRef]
307. Raya, K.; Chang, L.; Bashir, A.; Cobos-Martinez, J.J.; Gutiérrez-Guerrero, L.X.; Roberts, C.D.; Tandy, P.C. Structure of the neutral pion and its electromagnetic transition form factor. *Phys. Rev. D* **2016**, *93*, 074017. [CrossRef]
308. Raya, K.; Ding, M.; Bashir, A.; Chang, L.; Roberts, C.D. Partonic structure of neutral pseudoscalars via two photon transition form factors. *Phys. Rev. D* **2017**, *95*, 074014. [CrossRef]
309. Gao, F.; Chang, L.; Liu, Y.X.; Roberts, C.D.; Tandy, P.C. Exposing strangeness: Projections for kaon electromagnetic form factors. *Phys. Rev. D* **2017**, *96*, 034024. [CrossRef]
310. Lepage, G.P.; Brodsky, S.J. Exclusive Processes in Quantum Chromodynamics: Evolution Equations for Hadronic Wave Functions and the Form-Factors of Mesons. *Phys. Lett. B* **1979**, *87*, 359–365. [CrossRef]
311. Efremov, A.V.; Radyushkin, A.V. Factorization and Asymptotical Behavior of Pion Form- Factor in QCD. *Phys. Lett. B* **1980**, *94*, 245–250. [CrossRef]

312. Lepage, G.P.; Brodsky, S.J. Exclusive Processes in Perturbative Quantum Chromodynamics. *Phys. Rev. D* **1980**, *22*, 2157–2198. [CrossRef]
313. Brodsky, S.J.; Schmidt, I.A.; de Teramond, G.F. Nuclear-bound quarkonium. *Phys. Rev. Lett.* **1990**, *64*, 1011. [CrossRef]
314. Tarrús Castellà, J.; Krein, G. Effective field theory for the nucleon-quarkonium interaction. *Phys. Rev. D* **2018**, *98*, 014029. [CrossRef]
315. Luke, M.E.; Manohar, A.V.; Savage, M.J. A QCD Calculation of the interaction of quarkonium with nuclei. *Phys. Lett. B* **1992**, *288*, 355–359. [CrossRef]
316. Kharzeev, D. Quarkonium interactions in QCD. *Proc. Int. Sch. Phys. Fermi* **1996**, *130*, 105–131.
317. Krein, G.; Peixoto, T.C. Femtoscopy of the Origin of the Nucleon Mass. *Few Body Syst.* **2020**, *61*, 49. [CrossRef]
318. Aaij, R.; Adeva, B.; Adinolfi, M.; Affolder, A.; Ajaltouni, Z.; Akar, S.; Albrecht, J.; Alessio, F.; Alexander, M.; Ali, S.; et al. Observation of $J/\psi p$ Resonances Consistent with Pentaquark States in $\Lambda_b^0 \to J/\psi K^- p$ Decays. *Phys. Rev. Lett.* **2015**, *115*, 072001. [CrossRef]
319. Ali, A.; Amaryan, M.; Anassontzis, E.G.; Austregesilo, A.; Baalouch, M.; Barbosa, F.; Barlow, J.; Barnes, A.; Barriga, E.; Beattie, T.D.; et al. First Measurement of Near-Threshold J/ψ Exclusive Photoproduction off the Proton. *Phys. Rev. Lett.* **2019**, *123*, 072001. [CrossRef]
320. Sakurai, J.J. Theory of strong interactions. *Annals Phys.* **1960**, *11*, 1–48. [CrossRef]
321. Sakurai, J.J. *Currents and Mesons*; University of Chicago Press: Chicago, IL, USA, 1969.
322. Fraas, H.; Schildknecht, D. Vector-meson dominance and diffractive electroproduction of vector mesons. *Nucl. Phys. B* **1969**, *14*, 543–565. [CrossRef]
323. Sun, P.; Tong, X.B.; Yuan, F. Near threshold heavy quarkonium photoproduction at large momentum transfer. *Phys. Rev. D* **2022**, *105*, 054032. [CrossRef]
324. Ivanov, M.A.; Kalinovsky, Yu.L.; Roberts, C.D. Survey of heavy-meson observables. *Phys. Rev. D* **1999**, *60*, 034018. [CrossRef]
325. Du, M.L.; Baru, V.; Guo, F.K.; Hanhart, C.; Meißner, U.G.; Nefediev, A.; Strakovsky, I. Deciphering the mechanism of near-threshold J/ψ photoproduction. *Eur. Phys. J. C* **2020**, *80*, 1053. [CrossRef]
326. Lee, T.S.H.; Sakinah, S.; Oh, Y. Models of J/Ψ photo-production reactions on the nucleon. *arXiv* **2022**, arXiv:2210.02154.
327. Eichmann, G.; Fischer, C.S.; Heupel, W.; Santowsky, N.; Wallbott, P.C. Four-quark states from functional methods. *Few Body Syst.* **2020**, *61*, 38. [CrossRef]
328. Pichowsky, M.A.; Lee, T.S.H. Exclusive diffractive processes and the quark substructure of mesons. *Phys. Rev. D* **1997**, *56*, 1644–1662. [CrossRef]
329. Shi, C.; Xie, Y.P.; Li, M.; Chen, X.; Zong, H.S. Light front wave functions and diffractive electroproduction of vector mesons. *Phys. Rev. D* **2021**, *104*, L091902. [CrossRef]
330. Roberts, C.D. Electromagnetic pion form-factor and neutral pion decay width. *Nucl. Phys. A* **1996**, *605*, 475–495. [CrossRef]
331. Miramontes, A.; Bashir, A.; Raya, K.; Roig, P. Pion and Kaon box contribution to a_μ^{HLbL}. *Phys. Rev. D* **2022**, *105*, 074013. [CrossRef]
332. Rodríguez-Quintero, J. Gravitational form factors of Nambu–Goldstone bosons. In Proceedings of the ECT* Workshop: Revealing Emergent Mass through Studies of Hadron Spectra and Structure, Online, 12–16 September 2022.
333. Eichmann, G. Nucleon electromagnetic form factors from the covariant Faddeev equation. *Phys. Rev. D* **2011**, *84*, 014014. [CrossRef]
334. Eichmann, G.; Fischer, C.S. Nucleon axial and pseudoscalar form factors from the covariant Faddeev equation. *Eur. Phys. J. A* **2012**, *48*, 9. [CrossRef]
335. Sanchis-Alepuz, H.; Williams, R.; Alkofer, R. Delta and Omega electromagnetic form factors in a three-body covariant Bethe–Salpeter approach. *Phys. Rev. D* **2013**, *87*, 095015. [CrossRef]
336. Mokeev, V.I. Insight into EHM from results on electroexcitation of $\Delta(1600)3/2^+$ resonance. In Proceedings of the Workshop on Receiving the Emergence of Hadron Mass through AMBER @ CERN-VII, Geneva, Switzerland, 10–13 May 2022.
337. Perdrisat, C.F.; Punjabi, V.; Vanderhaeghen, M. Nucleon electromagnetic form factors. *Prog. Part. Nucl. Phys.* **2007**, *59*, 694–764. [CrossRef]
338. Mosel, U. Neutrino Interactions with Nucleons and Nuclei: Importance for Long-Baseline Experiments. *Ann. Rev. Nucl. Part. Sci.* **2016**, *66*, 171–195. [CrossRef]
339. Alvarez-Ruso, L.; Athar, M.S.; Barbaro, M.B.; Cherdack, D.; Christy, M.; Coloma, P.; Donnelly, T.; Dytman, S.; de Gouvêa, A.; Hill, R.; et al. NuSTEC White Paper: Status and challenges of neutrino–nucleus scattering. *Prog. Part. Nucl. Phys.* **2018**, *100*, 1–68. [CrossRef]
340. Hill, R.J.; Kammel, P.; Marciano, W.J.; Sirlin, A. Nucleon Axial Radius and Muonic Hydrogen — A New Analysis and Review. *Rept. Prog. Phys.* **2018**, *81*, 096301. [CrossRef]
341. Gysbers, P.; Hagen, G.; Holt, J.D.; Jansen, G.R.; Morris, T.D.; Navrátil, P.; Papenbrock, T.; Quaglioni, S.; Schwenk, A.; Stroberg, S.R.; et al. Discrepancy between experimental and theoretical β-decay rates resolved from first principles. *Nat. Phys.* **2019**, *15*, 428–431. [CrossRef]
342. Lovato, A.; Carlson, J.; Gandolfi, S.; Rocco, N.; Schiavilla, R. Ab initio study of (ν_ℓ, ℓ^-) and $(\bar{\nu}_\ell, \ell^+)$ inclusive scattering in ^{12}C: confronting the MiniBooNE and T2K CCQE data. *Phys. Rev. X* **2020**, *10*, 031068.
343. King, G.B.; Andreoli, L.; Pastore, S.; Piarulli, M.; Schiavilla, R.; Wiringa, R.B.; Carlson, J.; Gandolfi, S. Chiral Effective Field Theory Calculations of Weak Transitions in Light Nuclei. *Phys. Rev. C* **2020**, *102*, 025501. [CrossRef]

344. Simons, D.; Steinberg, N.; Lovato, A.; Meurice, Y.; Rocco, N.; Wagman, M. Form factor and model dependence in neutrino-nucleus cross section predictions. *arXiv* **2022**, arXiv:2210.02455.
345. Oettel, M.; Pichowsky, M.; von Smekal, L. Current conservation in the covariant quark–diquark model of the nucleon. *Eur. Phys. J. A* **2000**, *8*, 251–281. [CrossRef]
346. Meyer, A.S.; Betancourt, M.; Gran, R.; Hill, R.J. Deuterium target data for precision neutrino-nucleus cross sections. *Phys. Rev. D* **2016**, *93*, 113015. [CrossRef]
347. Jang, Y.C.; Gupta, R.; Yoon, B.; Bhattacharya, T. Axial Vector Form Factors from Lattice QCD that Satisfy the PCAC Relation. *Phys. Rev. Lett.* **2020**, *124*, 072002. [CrossRef]
348. Andreev, V.A.; Banks, T.I.; Carey, R.M.; Case, T.A.; Clayton, S.M.; Crowe, K.M.; Deutsch, J.; Egger, J.; Freedman, S.J.; Ganzha, V.A.; et al. Measurement of Muon Capture on the Proton to 1% Precision and Determination of the Pseudoscalar Coupling g_P. *Phys. Rev. Lett.* **2013**, *110*, 012504. [CrossRef]
349. Andreev, V.A.; Banks, T.I.; Carey, R.M.; Case, T.A.; Clayton, S.M.; Crowe, K.M.; Deutsch, J.; Egger, J.; Freedman, S.J.; Ganzha, V.A.; et al. Measurement of the Formation Rate of Muonic Hydrogen Molecules. *Phys. Rev. C* **2015**, *91*, 055502. [CrossRef]
350. Bernard, V.; Elouadrhiri, L.; Meissner, U.G. Axial structure of the nucleon: Topical Review. *J. Phys. G* **2002**, *28*, R1–R35. [CrossRef]
351. Baru, V.; Hanhart, C.; Hoferichter, M.; Kubis, B.; Nogga, A.; Phillips, D.R. Precision calculation of threshold $\pi^- d$ scattering, πN scattering lengths, and the GMO sum rule. *Nucl. Phys. A* **2011**, *872*, 69–116. [CrossRef]
352. Navarro Pérez, R.; Amaro, J.E.; Ruiz Arriola, E. Precise Determination of Charge Dependent Pion-Nucleon-Nucleon Coupling Constants. *Phys. Rev. C* **2017**, *95*, 064001. [CrossRef]
353. Reinert, P.; Krebs, H.; Epelbaum, E. Precision determination of pion–nucleon coupling constants using effective field theory. *Phys. Rev. Lett.* **2021**, *126*, 092501. [CrossRef]
354. Bloch, J.C.R.; Roberts, C.D.; Schmidt, S.M. Selected nucleon form-factors and a composite scalar diquark. *Phys. Rev. C* **2000**, *61*, 065207. [CrossRef]
355. Cheng, P.; Serna, F.E.; Yao, Z.Q.; Chen, C.; Cui, Z.F.; Roberts, C.D. Contact interaction analysis of octet baryon axial-vector and pseudoscalar form factors. *Phys. Rev. D* **2022**, *106*, 054031. [CrossRef]
356. Haidenbauer, J.; Meissner, U.G. The Julich hyperon-nucleon model revisited. *Phys. Rev. C* **2005**, *72*, 044005. [CrossRef]
357. Rijken, T.A.; Nagels, M.M.; Yamamoto, Y. Baryon-baryon interactions: Nijmegen extended-soft-core models. *Prog. Theor. Phys. Suppl.* **2010**, *185*, 14–71. [CrossRef]
358. Chang, L.; Roberts, C.D.; Schmidt, S.M. Dressed-quarks and the nucleon's axial charge. *Phys. Rev. C* **2013**, *87*, 015203. [CrossRef]
359. Cabibbo, N.; Swallow, E.C.; Winston, R. Semileptonic hyperon decays. *Ann. Rev. Nucl. Part. Sci.* **2003**, *53*, 39–75. [CrossRef]
360. Ledwig, T.; Martin Camalich, J.; Geng, L.S.; Vicente Vacas, M.J. Octet-baryon axial-vector charges and SU(3)-breaking effects in the semileptonic hyperon decays. *Phys. Rev. D* **2014**, *90*, 054502. [CrossRef]
361. Aidala, C.A.; Bass, S.D.; Hasch, D.; Mallot, G.K. The Spin Structure of the Nucleon. *Rev. Mod. Phys.* **2013**, *85*, 655–691. [CrossRef]
362. Deur, A.; Brodsky, S.J.; De Téramond, G.F. The Spin Structure of the Nucleon. *Rept. Prog. Phys.* **2019**, *82*, 655. [CrossRef]
363. Del Guerra, A.; Giazotto, A.; Giorgi, M.A.; Stefanini, A.; Botterill, D.R.; Montgomery, H.E.; Norton, P.R.; Matone, G. Threshold π^+ electroproduction at high momentum transfer: a determination of the nucleon axial vector form-factor. *Nucl. Phys. B* **1976**, *107*, 65–81. [CrossRef]
364. Park, K.; Gothe, R.W.; Adhikari, K.P.; Adikaram, D.; Anghinolfi, M.; Baghdasaryan, H.; Ball, J.; Battaglieri, M.; Batourine, V.; Bedlinskiy, I.; et al. Measurement of the generalized form factors near threshold via $\gamma^* p \to n\pi^+$ at high Q^2. *Phys. Rev. C* **2012**, *85*, 035208. [CrossRef]
365. Cui, Z.F.; Chen, C.; Binosi, D.; de Soto, F.; Roberts, C.D.; Rodríguez-Quintero, J.; Schmidt, S.M.; Segovia, J. Nucleon elastic form factors at accessible large spacelike momenta. *Phys. Rev. D* **2020**, *102*, 014043. [CrossRef]
366. Xing, H.Y.; Xu, Z.N.; Cui, Z.F.; Roberts, C.D.; Xu, C. Heavy + heavy and heavy + light pseudoscalar to vector semileptonic transitions. *Eur. Phys. J. C* **2022**, *82*, 889. [CrossRef]
367. Isgur, N.; Wise, M.B. Weak Transition Form Factors between Heavy Mesons. *Phys. Lett. B* **1990**, *237*, 527–530. [CrossRef]
368. Amhis, Y.; Banerjee, S.; Ben-Haim, E.; Bernlochner, F.U.; Bona, M.; Bozek, A.; Bozzi, C.; Brodzicka, M.; Chrzaszcz, M.; Dingfelder, J.; et al. Averages of b-hadron, c-hadron, and τ-lepton properties as of 2018. *Eur. Phys. J. C* **2021**, *81*, 226.
369. Glattauer, R. Measurement of the decay $B \to D\ell\nu_\ell$ in fully reconstructed events and determination of the Cabibbo–Kobayashi–Maskawa matrix element $|V_{cb}|$. *Phys. Rev. D* **2016**, *93*, 032006. [CrossRef]
370. Ablikim, M.; Achasov, M.N.; Ahmed, S.; Albrecht, M.; Alekseev, M.; Amoroso, A.; An, F.F.; An, Q.; Bai, Y.; Bakina, O.; et al. First Measurement of the Form Factors in $D_s^+ \to K^0 e^+ \nu_e$ and $D_s^+ \to K^{*0} e^+ \nu_e$ Decays. *Phys. Rev. Lett.* **2019**, *122*, 061801. [CrossRef]
371. LHCb Collaboration; Alfonso Albero, A.; Badalov, A.; Camboni, A.; Coquereau, S.; Fernandez, G.; Garrido Beltrán, L. Measurement of the ratio of branching fractions $\mathcal{B}(B_c^+ \to J/\psi \tau^+ \nu_\tau)/\mathcal{B}(B_c^+ \to J/\psi \mu^+ \nu_\mu)$. *Phys. Rev. Lett.* **2018**, *120*, 121801.
372. Tran, C.T.; Ivanov, M.A.; Körner, J.G.; Santorelli, P. Implications of new physics in the decays $B_c \to (J/\psi, \eta_c)\tau\nu$. *Phys. Rev. D* **2018**, *97*, 054014. [CrossRef]
373. Issadykov, A.; Ivanov, M.A. The decays $B_c \to J/\psi + \bar{\ell}\nu_\ell$ and $B_c \to J/\psi + \pi(K)$ in covariant confined quark model. *Phys. Lett. B* **2018**, *783*, 178–182. [CrossRef]
374. Wang, W.; Zhu, R. Model independent investigation of the $R_{J/\psi,\eta_c}$ and ratios of decay widths of semileptonic B_c decays into a P-wave charmonium. *Int. J. Mod. Phys. A* **2019**, *34*, 1950195. [CrossRef]

375. Leljak, D.; Melic, B.; Patra, M. On lepton flavour universality in semileptonic $B_c \to \eta_c, J/\psi$ decays. *J. High Energy Phys.* **2019**, *5*, 094. [CrossRef]
376. Hu, X.Q.; Jin, S.P.; Xiao, Z.J. Semileptonic decays $B_c \to (\eta_c, J/\psi) l \bar{\nu}_l$ in the "PQCD + Lattice" approach. *Chin. Phys. C* **2020**, *44*, 023104. [CrossRef]
377. Zhou, T.; Wang, T.; Jiang, Y.; Tan, X.Z.; Li, G.; Wang, G.L. Relativistic calculations of $R(D^{(*)})$, $R(D_s^{(*)})$, $R(\eta_c)$ and $R(J/\psi)$. *Int. J. Mod. Phys. A* **2020**, *35*, 2050076. [CrossRef]
378. Harrison, J.; Davies, C.T.H.; Lytle, A. $R(J/\psi)$ and $B_c^- \to J/\psi \ell^- \bar{\nu}_\ell$ Lepton Flavor Universality Violating Observables from Lattice QCD. *Phys. Rev. Lett.* **2020**, *125*, 222003. [CrossRef] [PubMed]
379. Harrison, J.; Davies, C.T.H.; Lytle, A. $B_c \to J/\psi$ form factors for the full q^2 range from lattice QCD. *Phys. Rev. D* **2020**, *102*, 094518. [CrossRef]
380. Widhalm, L.; Abe, K.; Abe, K.; Adachi, I.; Aihara, H.; Arinstein, K.; Asano, Y.; Aushev, T.; Bakich, A.M.; Balagura, V.; et al. Measurement of $D^0 \to \pi l \nu(K l \nu)$ Form Factors and Absolute Branching Fractions. *Phys. Rev. Lett.* **2006**, *97*, 061804. [CrossRef] [PubMed]
381. Besson, D.; Pedlar, T.K.; Xavier, J.; Cronin-Hennessy, D.; Gao, K.Y.; Hietala, J.; Kubota, Y.; Klein, T.; Poling, R.; Scott, A.W.; et al. Improved measurements of D meson semileptonic decays to π and K mesons. *Phys. Rev. D* **2009**, *80*, 032005. [CrossRef]
382. Ablikim, M.; Achasov, M.N.; Ahmed, S.; Albrecht, M.; Alekseev, M.; Amoroso, A.; An, F.F.; An, Q.; Bai, J.Z.; Bai, Y.; et al. Study of Dynamics of $D^0 \to K^- e^+ \nu_e$ and $D^0 \to \pi^- e^+ \nu_e$ Decays. *Phys. Rev. D* **2015**, *92*, 072012. [CrossRef]
383. Ablikim, M.; Achasov, M.N.; Ahmed, S.; Ai, X.C.; Albayrak, O.; Albrecht, M.; Ambrose, D.J.; Amoroso, A.; An, F.F.; An, Q.; et al. Analysis of $D^+ \to \bar{K}^0 e^+ \nu_e$ and $D^+ \to \pi^0 e^+ \nu_e$ semileptonic decays. *Phys. Rev. D* **2017**, *96*, 012002. [CrossRef]
384. Lubicz, V.; Riggio, L.; Salerno, G.; Simula, S.; Tarantino, C. Scalar and vector form factors of $D \to \pi(K) \ell \nu$ decays with $N_f = 2 + 1 + 1$ twisted fermions. *Phys. Rev. D* **2017**, *96*, 054514; Erratum: *Phys. Rev. D* **2019**, *99*, 099902. [CrossRef]
385. Del Amo Sanchez, P.; Lees, J.P.; Poireau, V.; Prencipe, E.; Tisserand, V.; Garra Tico, J.; Grauges, E.; Martinelli, M.; Palano, A.; Pappagallo, M.; et al. Study of $B \to \pi \ell \nu$ and $B \to \rho \ell \nu$ Decays and Determination of $|V_{ub}|$. *Phys. Rev. D* **2011**, *83*, 032007. [CrossRef]
386. Ha, H.; Won, E.; Adachi, I.; Aihara, H.; Aziz, T.; Bakich, A.M.; Balagura, V.; Barberio, E.; Bay, A.; Belous, K.; et al. Measurement of the decay $B^0 \to \pi^- \ell^+ \nu$ and determination of $|V_{ub}|$. *Phys. Rev. D* **2011**, *83*, 071101. [CrossRef]
387. Lees, J.P.; Poireau, V.; Tisserand, V.; Garra Tico, J.; Grauges, E.; Palano, A.; Eigen, G.; Stugu, B.; Brown, D.N.; Kerth, L.T.; et al. Branching fraction and form-factor shape measurements of exclusive charmless semileptonic B decays, and determination of $|V_{ub}|$. *Phys. Rev. D* **2012**, *86*, 092004. [CrossRef]
388. Sibidanov, A.; Varvell, K.E.; Adachi, I.; Aihara, H.; Asner, D.M.; Aulchenko, V.; Aushev, T.; Bakich, A.M.; Bala, A.; Bozek, A.; et al. Study of Exclusive $B \to X_u \ell \nu$ Decays and Extraction of $\|V_{ub}\|$ using Full Reconstruction Tagging at the Belle Experiment. *Phys. Rev. D* **2013**, *88*, 032005. [CrossRef]
389. Bouchard, C.M.; Lepage, G.P.; Monahan, C.; Na, H.; Shigemitsu, J. $B_s \to K \ell \nu$ form factors from lattice QCD. *Phys. Rev. D* **2014**, *90*, 054506. [CrossRef]
390. Flynn, J.M.; Izubuchi, T.; Kawanai, T.; Lehner, C.; Soni, A.; Van de Water, R.S.; Witzel, O. $B \to \pi \ell \nu$ and $B_s \to K \ell \nu$ form factors and $|V_{ub}|$ from 2+1-flavor lattice QCD with domain wall light quarks and relativistic heavy quarks. *Phys. Rev. D* **2015**, *91*, 074510. [CrossRef]
391. Aaij, R.; Abellán Beteta, C.; Ackernley, T.; Adeva, B.; Adinolfi, M.; Afsharnia, H.; Aidala, C.A.; Aiola, S.; Ajaltouni, Z.; Akar, S.; et al. First observation of the decay $B_s^0 \to K^- \mu^+ \nu_\mu$ and Measurement of $|V_{ub}|/|V_{cb}|$. *Phys. Rev. Lett.* **2021**, *126*, 081804. [CrossRef]
392. Bazavov, A.; Bernard, C.; DeTar, C.; Du, J.; El-Khadra, A.X.; Freeland, E.D.; Gámiz, E.; Gelzer, Z.; Gottlieb, S.; Heller, U.M.; et al. $B_s \to K \ell \nu$ decay from lattice QCD. *Phys. Rev. D* **2019**, *100*, 034501. [CrossRef]
393. Wu, Y.L.; Zhong, M.; Zuo, Y.B. $B_{(s)}, D_{(s)} \to \pi, K, \eta, \rho, K^*, \omega, \phi$ Transition Form Factors and Decay Rates with Extraction of the CKM parameters $|V_{ub}|, |V_{cs}|, |V_{cd}|$. *Int. J. Mod. Phys. A* **2006**, *21*, 6125–6172. [CrossRef]
394. Faustov, R.N.; Galkin, V.O. Charmless weak B_s decays in the relativistic quark model. *Phys. Rev. D* **2013**, *87*, 094028. [CrossRef]
395. Xiao, Z.J.; Fan, Y.Y.; Wang, W.F.; Cheng, S. The semileptonic decays of B/B_s meson in the perturbative QCD approach: A short review. *Chin. Sci. Bull.* **2014**, *59*, 3787–3800. [CrossRef]
396. Gonzàlez-Solís, S.; Masjuan, P.; Rojas, C. Padé approximants to $B \to \pi \ell \nu_\ell$ and $B_s \to K \ell \nu_\ell$ and determination of $|V_{ub}|$. arXiv **2021**, arXiv:2110.06153.
397. Melikhov, D. Dispersion approach to quark binding effects in weak decays of heavy mesons. *Eur. Phys. J. Direct C* **2002**, *4*, 1–154. [CrossRef]
398. Faessler, A.; Gutsche, T.; Ivanov, M.A.; Korner, J.; Lyubovitskij, V.E. The Exclusive rare decays $B \to K(K^*) \bar{\ell}\ell$ and $B_c \to D(D^*) \bar{\ell}\ell$ in a relativistic quark model. *Eur. Phys. J. Direct C* **2002**, *4*, 18.
399. Ebert, D.; Faustov, R.N.; Galkin, V.O. Weak decays of the B_c meson to B_s and B mesons in the relativistic quark model. *Eur. Phys. J. C* **2003**, *32*, 29–43. [CrossRef]
400. Ball, P.; Zwicky, R. New results on $B \to \pi, K, \eta$ decay form factors from light-cone sum rules. *Phys. Rev. D* **2005**, *71*, 014015. [CrossRef]
401. Khodjamirian, A.; Mannel, T.; Offen, N. Form-factors from light-cone sum rules with B-meson distribution amplitudes. *Phys. Rev. D* **2007**, *75*, 054013. [CrossRef]
402. Lü, C.D.; Wang, W.; Wei, Z.T. Heavy-to-light form factors on the light cone. *Phys. Rev. D* **2007**, *76*, 014013. [CrossRef]

403. Ivanov, M.A.; Körner, J.G.; Kovalenko, S.G.; Roberts, C.D. B- to light-meson transition form factors. *Phys. Rev. D* **2007**, *76*, 034018. [CrossRef]
404. El-Bennich, B.; Ivanov, M.A.; Roberts, C.D. Strong D* -> D+pi and B* -> B+pi couplings. *Phys. Rev. C* **2011**, *83*, 025205. [CrossRef]
405. Ji, C.R.; Maris, P. $K_{\ell 3}$ transition form-factors. *Phys. Rev.* **2001**, *D64*, 014032.
406. Chen, C.; Chang, L.; Roberts, C.D.; Wan, S.L.; Schmidt, S.M.; Wilson, D.J. Features and flaws of a contact interaction treatment of the kaon. *Phys. Rev. C* **2013**, *87*, 045207. [CrossRef]
407. Holt, R.J.; Roberts, C.D. Distribution Functions of the Nucleon and Pion in the Valence Region. *Rev. Mod. Phys.* **2010**, *82*, 2991–3044. [CrossRef]
408. Taylor, R.E. Deep inelastic scattering: The Early years. *Rev. Mod. Phys.* **1991**, *63*, 573–595. [CrossRef]
409. Kendall, H.W. Deep inelastic scattering: Experiments on the proton and the observation of scaling. *Rev. Mod. Phys.* **1991**, *63*, 597–614. [CrossRef]
410. Friedman, J.I. Deep inelastic scattering: Comparisons with the quark model. *Rev. Mod. Phys.* **1991**, *63*, 615–629. [CrossRef]
411. Friedman, J.I.; Kendall, H.W.; Taylor, R.E. Deep inelastic scattering: Acknowledgements. *Rev. Mod. Phys.* **1991**, *63*, 629. [CrossRef]
412. Brodsky, S.J.; Burkardt, M.; Schmidt, I. Perturbative QCD constraints on the shape of polarized quark and gluon distributions. *Nucl. Phys. B* **1995**, *441*, 197–214. [CrossRef]
413. Yuan, F. Generalized parton distributions at $x \to 1$. *Phys. Rev. D* **2004**, *69*, 051501. [CrossRef]
414. Dokshitzer, Y.L. Calculation of the Structure Functions for Deep Inelastic Scattering and e+ e- Annihilation by Perturbation Theory in Quantum Chromodynamics. *Sov. Phys. JETP* **1977**, *46*, 641–653. (In Russian)
415. Lipatov, L.N. The parton model and perturbation theory. *Sov. J. Nucl. Phys.* **1975**, *20*, 94–102.
416. Altarelli, G.; Parisi, G. Asymptotic Freedom in Parton Language. *Nucl. Phys. B* **1977**, *126*, 298–318. [CrossRef]
417. Aicher, M.; Schäfer, A.; Vogelsang, W. Soft-Gluon Resummation and the Valence Parton Distribution Function of the Pion. *Phys. Rev. Lett.* **2010**, *105*, 252003. [CrossRef] [PubMed]
418. Ball, R.D.; Nocera, E.R.; Rojo, J. The asymptotic behaviour of parton distributions at small and large x. *Eur. Phys. J. C* **2016**, *76*, 383. [CrossRef] [PubMed]
419. Hou, T.-J.; Gao, J.; Hobbs, T.J.; Xie, K.; Dulat, S.; Guzzi, M.; Huston, J.; Nadolsky, P.; Pumplin, J.; Schmidt, C.; et al. New CTEQ global analysis of quantum chromodynamics with high-precision data from the LHC. *Phys. Rev. D* **2021**, *103*, 014013. [CrossRef]
420. Bailey, S.; Cridge, T.; Harland-Lang, L.A.; Martin, A.D.; Thorne, R.S. Parton distributions from LHC, HERA, Tevatron and fixed target data: MSHT20 PDFs. *Eur. Phys. J. C* **2021**, *81*, 341. [CrossRef]
421. Novikov, I.; Abdolmaleki, H.; Britzger, D.; Cooper-Sarkar, A.; Giuli, F.; Glazov, A.; Kusina, A.; Luszczak, A.; Olness, F.; Starovoitov, P.; et al. Parton Distribution Functions of the Charged Pion Within The xFitter Framework. *Phys. Rev. D* **2020**, *102*, 014040. [CrossRef]
422. Barry, P.C.; Ji, C.R.; Sato, N.; Melnitchouk, W. Global QCD Analysis of Pion Parton Distributions with Threshold Resummation. *Phys. Rev. Lett.* **2021**, *127*, 232001. [CrossRef]
423. Chang, L.; Cloet, I.C.; Cobos-Martinez, J.J.; Roberts, C.D.; Schmidt, S.M.; Tandy, P.C. Imaging dynamical chiral symmetry breaking: Pion wave function on the light front. *Phys. Rev. Lett.* **2013**, *110*, 132001. [CrossRef]
424. Lee, T.D.; Yang, C.N. Charge Conjugation, a New Quantum Number G, and Selection Rules Concerning a Nucleon Anti-nucleon System. *Nuovo Cim.* **1956**, *10*, 749–753. [CrossRef]
425. Gao, F.; Chang, L.; Liu, Y.X.; Roberts, C.D.; Schmidt, S.M. Parton distribution amplitudes of light vector mesons. *Phys. Rev. D* **2014**, *90*, 014011. [CrossRef]
426. Lu, Y.; Binosi, D.; Ding, M.; Roberts, C.D.; Xing, H.Y.; Xu, C. Distribution amplitudes of light diquarks. *Eur. Phys. J A (Lett)* **2021**, *57*, 115. [CrossRef]
427. Alexandrou, C.; Constantinou, M.; Drach, V.; Hatziyiannakou, K.; Jansen, K.; Kallidonis, C.; Koutsou, G.; Leontiou, T.; Vaquero, A. Nucleon Structure using lattice QCD. *Nuovo Cim. C* **2013**, *036*, 111–120.
428. Lin, H.W.; Chen, J.W.; Zhang, R. Lattice Nucleon Isovector Unpolarized Parton Distribution in the Physical-Continuum Limit. *arXiv* **2020**, arXiv:2011.14971.
429. Alexandrou, C.; Constantinou, M.; Hadjiyiannakou, K.; Jansen, K.; Manigrasso, F. Flavor decomposition of the nucleon unpolarized, helicity, and transversity parton distribution functions from lattice QCD simulations. *Phys. Rev. D* **2021**, *104*, 054503. [CrossRef]
430. Ji, X.; Liu, Y.S.; Liu, Y.; Zhang, J.H.; Zhao, Y. Large-momentum effective theory. *Rev. Mod. Phys.* **2021**, *93*, 035005. [CrossRef]
431. Xu, S.S.; Chang, L.; Roberts, C.D.; Zong, H.S. Pion and kaon valence quark parton quasidistributions. *Phys. Rev. D* **2018**, *97*, 094014. [CrossRef]
432. Field, R.D.; Feynman, R.P. Quark Elastic Scattering as a Source of High Transverse Momentum Mesons. *Phys. Rev. D* **1977**, *15*, 2590–2616. [CrossRef]
433. Fan, Z.; Lin, H.W. Gluon parton distribution of the pion from lattice QCD. *Phys. Lett. B* **2021**, *823*, 136778. [CrossRef]
434. Accardi, A.; Brady, L.T.; Melnitchouk, W.; Owens, J.F.; Sato, N. Constraints on large-x parton distributions from new weak boson production and deep-inelastic scattering data. *Phys. Rev. D* **2016**, *93*, 114017. [CrossRef]
435. Ball, R.D.; Bertone, V.; Carrazza, S.; Del Debbio, L.; Forte, S.; Groth-Merrild, P.; Guffanti, A.; Hartland, N.P.; Kassabov, Z.; Latorre, J.I.; et al. Parton distributions from high-precision collider data. *Eur. Phys. J. C* **2017**, *77*, 663. [CrossRef] [PubMed]

436. Courtoy, A.; Nadolsky, P.M. Testing momentum dependence of the nonperturbative hadron structure in a global QCD analysis. *Phys. Rev. D* **2021**, *103*, 054029. [CrossRef]
437. Gottfried, K. Sum rule for high-energy electron - proton scattering. *Phys. Rev. Lett.* **1967**, *18*, 1174. [CrossRef]
438. Brock, R.; Collins, J.C.; Huston, J.; Kuhlmann, S.; Mishra, S.; Morfin, J.G.; Olness, F.I.; Owens, J.F.; Pumplin, J.; Qiu, J.-W.; et al. Handbook of perturbative QCD: Version 1.0. *Rev. Mod. Phys.* **1995**, *67*, 157–248.
439. Amaudruz, P.; Arneodo, M.; Arvidson, A.; Badelek, B.; Baum, G.; Beaufays, J.; Bird, I.G.; Botje, M.; Broggini, C.; Bruckner, W.; et al. The Gottfried sum from the ratio F_2^n/F_2^p. *Phys. Rev. Lett.* **1991**, *66*, 2712–2715. [CrossRef]
440. Arneodo, M.; Arvidson, A.; Badelek, B.; Ballintijn@f, M.; Baum@f, G.; Beaufays, J.; Bird, I.G.; Björkholm, P.; Botje, M.; Broggini, C.; et al. A Reevaluation of the Gottfried sum. *Phys. Rev. D* **1994**, *50*, R1–R3. [CrossRef]
441. Baldit, A.; Barrière, C.; Castor, J.; Chambon, T.; Devaux, A.; Espagnon, B.; Fargeix, J.; Force, P.; Landaud, G.; Saturnini, P.; et al. Study of the isospin symmetry breaking in the light quark sea of the nucleon from the Drell-Yan process. *Phys. Lett. B* **1994**, *332*, 244–250. [CrossRef]
442. Towell, R.S.; McGaughey, P.L.; Awes, T.C.; Beddo, M.E.; Brooks, M.L.; Geesaman, D.F.; Kaufman, S.B.; Makins, N.; Mueller, B.A.; Reimer, P.E.; et al. Improved measurement of the $\bar{d}/\bar{u}$ asymmetry in the nucleon sea. *Phys. Rev. D* **2001**, *64*, 052002. [CrossRef]
443. Dove, J.; Kerns, B.; McClellan, R.E.; Miyasaka, S.; Morton, D.H.; Nagai, K.; Prasad, S.; Sanftl, F.; Scott, M.B.C.; Tadepalli, A.S.; et al. The asymmetry of antimatter in the proton. *Nature* **2021**, *590*, 561–565. [CrossRef]
444. Tkachenko, S.; Baillie, N.; Kuhn, S.E.; Zhang, J.; Arrington, J.; Bosted, P.; Bültmann, S.; Christy, M.E.; Dutta, D.; Ent, R.; et al. Measurement of the structure function of the nearly free neutron using spectator tagging in inelastic ^{2}H(e, e'p)X scattering with CLAS. *Phys. Rev. C* **2014**, *89*, 045206; Addendum: *Phys. Rev. C* **2014**, *90*, 059901. [CrossRef]
445. Abrams, D.; Albataineh, H.; Aljawrneh, B.S.; Alsalmi, S.; Aniol, K.; Armstrong, W.; Arrington, J.; Atac, H.; Averett, T.; Ayerbe Gayoso, C.; et al. Measurement of the Nucleon F_2^n/F_2^p Structure Function Ratio by the Jefferson Lab MARATHON Tritium/Helium-3 Deep Inelastic Scattering Experiment. *Phys. Rev. Lett.* **2022**, *128*, 132003. [CrossRef]
446. Farrar, G.R.; Jackson, D.R. Pion and Nucleon Structure Functions Near $x = 1$. *Phys. Rev. Lett.* **1975**, *35*, 1416. [CrossRef]
447. Brodsky, S.J.; Lepage, G.P. Perturbative Quantum Chromodynamics. *Prog. Math. Phys.* **1979**, *4*, 255–422.
448. Roberts, C.D.; Holt, R.J.; Schmidt, S.M. Nucleon spin structure at very high x. *Phys. Lett. B* **2013**, *727*, 249–254. [CrossRef]
449. Close, F.E.; Thomas, A.W. The Spin and Flavor Dependence of Parton Distribution Functions. *Phys. Lett. B* **1988**, *212*, 227. [CrossRef]
450. Xu, S.S.; Chen, C.; Cloet, I.C.; Roberts, C.D.; Segovia, J.; Zong, H.S. Contact-interaction Faddeev equation and, inter alia, proton tensor charges. *Phys. Rev. D* **2015**, *92*, 114034. [CrossRef]
451. Bodek, A.; Breidenbach, M.; Dubin, D.L.; Elias, J.E.; Friedman, J.I.; Kendall, H.W.; Poucher, J.S.; Riordan, E.M.; Sogard, M.R.; Coward, D.H. Comparisons of Deep Inelastic ep and en Cross-Sections. *Phys. Rev. Lett.* **1973**, *30*, 1087. [CrossRef]
452. Poucher, J.S.; Breidenbach, M.; Ditzler, R.; Friedman, J.I.; Kendall, H.W.; Bloom, E.D.; Taylor, R.E. High-Energy Single-Arm Inelastic ep and ed Scattering at 6-Degrees and 10-Degrees. *Phys. Rev. Lett.* **1974**, *32*, 118. [CrossRef]
453. Whitlow, L.W.; Riordan, E.M.; Dasu, S.; Rock, S.; Bodek, A. Precise measurements of the proton and deuteron structure functions from a global analysis of the SLAC deep inelastic electron scattering cross-sections. *Phys. Lett. B* **1992**, *282*, 475–482. [CrossRef]
454. Afnan, I.R.; Bissey, F.R.P.; Gomez, J.; Katramatou, A.T.; Melnitchouk, W.; Petratos, G.G.; Thomas, A.W. Neutron structure function and A = 3 mirror nuclei. *Phys. Lett. B* **2000**, *493*, 36–42. [CrossRef]
455. Pace, E.; Salme, G.; Scopetta, S.; Kievsky, A. Neutron structure function $F_2^n(x)$ from deep inelastic electron scattering off few nucleon systems. *Phys. Rev. C* **2001**, *64*, 055203. [CrossRef]
456. Segarra, E.; Schmidt, A.; Kutz, T.; Higinbotham, D.; Piasetzky, E.; Strikman, M.; Weinstein, L.; Hen, O. Neutron Valence Structure from Nuclear Deep Inelastic Scattering. *Phys. Rev. Lett.* **2020**, *124*, 092002. [CrossRef]
457. Nachtmann, O. Positivity constraints for anomalous dimensions. *Nucl. Phys. B* **1973**, *63*, 237–247. [CrossRef]
458. Cui, Z.F.; Gao, F.; Binosi, D.; Chang, L.; Roberts, C.D.; Schmidt, S.M. Valence quark ratio in the proton. *Chin. Phys. Lett. Express* **2022**, *39*, 041401. [CrossRef]
459. Anselmino, M.; Predazzi, E.; Ekelin, S.; Fredriksson, S.; Lichtenberg, D.B. Diquarks. *Rev. Mod. Phys.* **1993**, *65*, 1199–1234. [CrossRef]
460. Pace, E.; Rinaldi, M.; Salmè, G.; Scopetta, S. The European Muon Collaboration effect in Light-Front Hamiltonian Dynamics. *arXiv* **2022**, arXiv:2206.05485.
461. Joó, B.; Karpie, J.; Orginos, K.; Radyushkin, A.V.; Richards, D.G.; Sufian, R.S.; Zafeiropoulos, S. Pion valence structure from Ioffe-time parton pseudodistribution functions. *Phys. Rev. D* **2019**, *100*, 114512. [CrossRef]
462. Alexandrou, C.; Bacchio, S.; Cloet, I.; Constantinou, M.; Hadjiyiannakou, K.; Koutsou, G.; Lauer, C. Pion and kaon $\langle x^3 \rangle$ from lattice QCD and PDF reconstruction from Mellin moments. *Phys. Rev. D* **2021**, *104*, 054504. [CrossRef]
463. Brodsky, S.J.; Hoyer, P.; Peterson, C.; Sakai, N. The Intrinsic Charm of the Proton. *Phys. Lett. B* **1980**, *93*, 451–455. [CrossRef]
464. Ball, R.D.; Candido, A.; Cruz-Martinez, J.; Forte, S.; Giani, T.; Hekhorn, F.; Kudashkin, K.; Magni, G.; Rojo, J. Evidence for intrinsic charm quarks in the proton. *Nature* **2022**, *608*, 483–487.

Disclaimer/Publisher's Note: The statements, opinions and data contained in all publications are solely those of the individual author(s) and contributor(s) and not of MDPI and/or the editor(s). MDPI and/or the editor(s) disclaim responsibility for any injury to people or property resulting from any ideas, methods, instructions or products referred to in the content.

Review

Gauge Sector Dynamics in QCD

Mauricio Narciso Ferreira *,† and Joannis Papavassiliou *,†

Department of Theoretical Physics and IFIC, University of Valencia and CSIC, E-46100 Valencia, Spain
* Correspondence: ansonar@uv.es (M.N.F.); Joannis.Papavassiliou@uv.es (J.P.)
† These authors contributed equally to this work.

Abstract: The dynamics of the QCD gauge sector give rise to non-perturbative phenomena that are crucial for the internal consistency of the theory; most notably, they account for the generation of a gluon mass through the action of the Schwinger mechanism, the taming of the Landau pole, the ensuing stabilization of the gauge coupling, and the infrared suppression of the three-gluon vertex. In the present work, we review some key advances in the ongoing investigation of this sector within the framework of the continuum Schwinger function methods, supplemented by results obtained from lattice simulations.

Keywords: continuum Schwinger function methods; emergence of hadron mass; gluon mass generation; lattice QCD; non-perturbative quantum field theory; quantum chromodynamics; Schwinger–Dyson equations; Schwinger mechanism

Contents

1. Introduction 313
2. Basic Concepts and General Theoretical Framework 316
3. Schwinger Mechanism in Yang–Mills Theories 321
4. Dynamical formation of Massless Poles 324
5. Generation of the Gluon Mass 327
   5.1. Gluon Mass from the $q_\mu q_\nu$ Component 328
   5.2. Gluon Mass from the $g_{\mu\nu}$ Component: Seagull Identity and Ward Identity Displacement 329
6. Renormalization Group Invariant Interaction Strength 331
7. Three-Gluon Vertex and Its Planar Degeneracy 334
8. Ghost Dynamics from Schwinger–Dyson Equations 336
9. Divergent Ghost Loops and Their Impact on the QCD Green's Functions 341
10. Ward Identity Displacement of the Three-Gluon Vertex 345
11. The Ghost-Gluon Kernel Contribution to the Ward Identity 346
12. Displacement Function from Lattice Inputs 350
13. Conclusions 351
A. Appendix A 352
B. Appendix B 354
References 354

Citation: Ferreira, M.N.; Papavassiliou, J. Gauge sector dynamics in QCD. *Particles* **2023**, *6*, 312–363. https://doi.org/10.3390/particles6010017

Academic Editors: Minghui Ding, Craig Roberts, Sebastian M. Schmidt and Armen Sedrakian

Received: 7 January 2023
Revised: 3 February 2023
Accepted: 9 February 2023
Published: 15 February 2023

Copyright: © 2023 by the authors. Licensee MDPI, Basel, Switzerland. This article is an open access article distributed under the terms and conditions of the Creative Commons Attribution (CC BY) license (https://creativecommons.org/licenses/by/4.0/).

1. Introduction

The systematic exploration of Green's functions (n-point correlation functions) of quantum chromodynamics (QCD) [1] by means of continuous Schwinger function methods [2–9], such as Schwinger–Dyson equations (SDEs) [10–21] and the functional renormalization group [22–31], together with a plethora of gauge-fixed lattice simulations [32–88], has afforded ample access to the dynamical mechanisms responsible for the non-perturbative properties of this remarkable theory. Particularly prominent in this quest is the notion of the emergent hadron mass (EHM) [3,8,9,89–93], together with its three supporting pillars: first, the generation of a gluon mass [18,32,93–126] through the action of the Schwinger mechanism [127,128]; second, the construction of the process-independent effective charge [3,16,20,79,96,129–131], which arises as the QCD analog of the Gell-Mann–Low charge is known from quantum electrodynamics (QED) [132,133], and is associated with a renormalization-group invariant (RGI) scale of about half of the proton mass [20,79]; and third, the dynamical breaking of chiral symmetry and the generation of constituent quark masses [10,17,134–158].

The dynamics of the gauge sector of QCD, which encompasses both gluonic and ghost interactions, is instrumental in the physical picture of the EHM outlined above. In fact, the basic concepts and pivotal mechanisms sustaining the first two pillars of the EHM have their original inception and most genuine realization in the realm of pure Yang–Mills theories [18,93,94,96,109,112,117,159–161]. Therefore, in the present review, we focus precisely on the rich dynamical content of the gauge sector, especially in relation to the generation of a gluon mass scale out of the intricate gluon self-interactions.

The formulation of the non-perturbative QCD physics in terms of Green's functions of the fundamental degrees of freedom, such as gluon and ghost propagators and vertices, provides an intuitive framework for unraveling a wide array of subtle mechanisms; in fact, certain distinctive features of these functions have been inextricably connected with key phenomena such as gluon mass generation, violation of reflection positivity, and confinement, to name a few. Thus, the saturation of the gluon propagator in the deep infrared [37,45–49,52,55–59,63,65–67,77,81] has been interpreted as an unequivocal signal of a gluon mass [32,96–100,103,105,107–109,112,160–166]; and the existence of an inflection point in the same function has been argued to lead to a non-positive gluon spectral density [8], and the ensuing loss of reflection positivity [8,11,13,16,167–171] for the dressed gluons. Similarly, the masslessness of the ghost induces [172] a maximum in the gluon propagator, and a zero crossing in the form factors of the three-gluon vertex [28,50,68,69,71,72,81,84,172–180], followed by an infrared divergence for vanishing momenta. The dynamic origin of these special traits will be the focal point of the analysis presented in the main body of this article.

The integral equations that govern the full momentum evolution of Green's functions, known as SDEs, constitute the indispensable formal and practical instrument for unraveling the special characteristics mentioned above. In their primordial form, the SDEs are rigorously derived from the generating functional of the theory [133,181], and encode all dynamical information on the correlation functions, within the entire range of physical momenta. In practice, due to the enormous complexity of these equations, truncation approximations need to be implemented; but, unlike perturbation theory, no expansion parameter is available in the strongly coupled regime of the theory for carrying out such a task. Despite this intrinsic shortcoming, in recent years, the SDE predictions have become particularly robust, in part due to various theoretical advances, and in part thanks to the intense synergy with gauge-fixed lattice simulations, as will be evidenced in subsequent sections.

Typically, Green's QCD function is defined within the quantization scheme obtained by implementing the linear covariant (R_ξ) gauges [182]. The corresponding SDEs are derived and solved within this same quantization scheme, particularly in the Landau gauge ($\xi = 0$), where lattice simulations are almost exclusively performed; for studies away from the Landau gauge, see e.g., [55,58,66,74,75,110,114,120,183–191]. A great deal may be learned, however, by considering Green's functions and corresponding SDEs formulated within the "PT-BFM" scheme [109,192], namely the framework that arises from the fusion of the pinch technique (PT) [14,96,100,193–195] with the background field method (BFM) [196–206]. The main advantage of the PT-BFM originates from the fact that certain appropriately chosen Green's functions satisfy Abelian Slavnov–Taylor identities (STIs), whose tree-level form does not get modified by quantum corrections. This situation is to be contrasted to the standard STIs [207,208] obtained in the conventional framework of the linear covariant gauges, which are deformed by non-trivial contributions stemming from the gauge sector of the theory. In the present work, we will carry out computations and develop arguments within both frameworks (R_ξ and PT-BFM), and will elaborate on their connection by means of the so-called background-quantum identities (BQIs) [14,209–211].

The article is organized as follows:

- In Section 2, we introduce some basic notations and review certain prominent features of Green's functions within the linear gauges and the PT-BFM formalism [109,192]. We stress, in particular, the properties of the auxiliary function $G(q)$ [16,131,212,213], which relates the gluon propagators with quantum and background gluons, and is intimately connected with the definition of the process-independent and RGI interaction strength [16], to be discussed in detail in Section 6. In addition, we elucidate (with a concrete example) the important property of "block-wise" transversality, displayed by the background gluon self-energy [18,109,112].

- In Section 3, we review the general principles associated with the Schwinger mechanism [127,128] that endows gauge bosons with an effective mass, focusing on the details associated with its realization in the context of Yang–Mills theories. We place particular emphasis on the pivotal requirement that must be satisfied by the fundamental vertices of the theory, namely the appearance of massless poles in their form factors [18,93,109,111–113,117,159,214].

- In Section 4, we examine the dynamical formation of *colored* composite excitations (bound states) of vanishing masses, which provide the required structures in the vertices in order for the Schwinger mechanism to be activated [18,117,159,214]. The formation of these states out of a pair of gluons or a ghost–anti-ghost pair is controlled by a set of coupled Bethe–Salpeter equations (BSEs) [18,117,124,214,215], which are found to have nontrivial solutions for the corresponding Bethe–Salpeter (BS) amplitudes, to be denoted by $\mathbb{C}(r)$ and $\mathcal{C}(r)$, respectively.

- In Section 5, we explain in detail how the presence of the massless poles in the dressed vertices that enter the SDE of the gluon propagator give rise to a gluon mass. The demonstration is carried out separately for the $g_{\mu\nu}$ and $q_\mu q_\nu / q^2$ components of the gluon self-energy. The former case requires the evasion of the so-called "seagull identity" [113,166]; this becomes possible by virtue of the crucial Ward identity (WI) displacement, to be further considered in Section 10.

- In Section 6, we go over the basic notions underpinning the PT [14,96,100,193,194], and show how their application leads naturally to the definition of a dimensionful process-independent RGI interaction strength [3,16,20,79,96,129–131], denoted by $\widehat{d}(q)$. The genuine process independence of this quantity is concretely exemplified by demonstrating its appearance in two processes involving fundamentally different external fields. Next, $\widehat{d}(q)$ is computed by combining lattice data for the gluon propagator and SDE results for the function $G(q)$. Finally, the dimensionless quantity is derived that constitutes the physical definition of the one-gluon exchange interaction appearing in standard bound-state computations [15–17,216–222].

- In Section 7, we focus on the structure of the "transversely projected" three-gluon vertex [126,174,175,223], and discuss briefly the property of planar degeneracy [86], satisfied, at a high level of accuracy [86–88,174,175,223], by the vertex form factors. This special property induces a striking simplification to the structure of this vertex, captured by a particularly compact expression [86], which will be extensively used in some of the following sections.

- In Section 8, we take a close look at the ghost sector of the theory, and solve the coupled system of SDEs governing the ghost propagator and ghost–gluon vertex [85,224–228]; as is well-known, the ghost remains massless, but its dressing function saturates at the origin [21,42,47,49,51,56,62,63,73,79,85,112,178,225,227–233], because the infrared-finite gluon propagator used in the ghost SDE provides an effective infrared cutoff. In the SDE of the ghost–gluon vertex, we employ as central input the compact expression for the three-gluon vertex presented in the previous section. The results are in excellent agreement with the available lattice data for the ghost dressing function [73,85] and the form factor of the ghost–gluon vertex evaluated in the soft-gluon limit [42,43].
- In Section 9, we discuss two important consequences of the masslessness of the ghost propagator, which manifest themselves at the level of both the gluon propagator and the three-gluon vertex. Specifically, the diagrams comprised by a ghost loop induce "unprotected" logarithms, i.e., of the type $\ln q^2$; instead, gluonic loops give rise to "protected" logarithms, of the type $\ln(q^2 + m^2)$, where m is the effective gluon mass [172,234]. As $q^2 \to 0$, the unprotected contributions diverge, driving the appearance of a maximum in the gluon propagator and a divergence in its first derivative, as well as a zero-crossing and a corresponding divergence in the form factors of the three-gluon vertex. As we comment in this section, of particular phenomenological importance [234–240] is the relative suppression that the above features induce to the dominant vertex form factors in the intermediate range of momenta.
- In Section 10, we discuss an outstanding feature of the WI satisfied by the pole-free part of the three-gluon vertex, namely the displacement induced by the presence of the aforementioned massless poles [93,124]. In this context, we introduce the key quantity denominated "displacement function", whose appearance serves as a smoking gun signal of the action of the Schwinger mechanism in QCD; quite interestingly, it coincides [93,124] with the BS amplitude $\mathbb{C}(r)$ for the formation of a massless scalar out of a pair of gluons, introduced in Section 4. In addition, we derive a crucial relation, which ultimately permits the indirect determination of $\mathbb{C}(r)$ from lattice QCD [93,124,126]; an important ingredient in this relation is a partial derivative [124,241], denoted by $\mathcal{W}(r)$, of the ghost–gluon kernel [228], to be determined in the next section.
- In Section 11, we set up and solve the SDE that governs the evolution of $\mathcal{W}(r)$ [124,126,241,242]; the main component of this SDE is a special projection of the three-gluon vertex, which is computed by appealing to formulas established in Section 7, and allows for the accurate determination of $\mathcal{W}(r)$ in the entire range of relevant momenta [126].
- In Section 12, we substitute into the central relation derived in Section 10 the solution for $\mathcal{W}(r)$ found in the previous section, together with the lattice data [84,85] for the gluon propagator, the ghost dressing function, and the form factor of the three-gluon vertex associated with the soft-gluon limit, in order to obtain the form of the displacement function $\mathbb{C}(r)$ [124,126]. As we discuss, the results exclude—with near-absolute certainty—the null hypothesis (absence of Schwinger mechanism, $\mathbb{C}(r) = 0$), and corroborate the action of the Schwinger mechanism in QCD [126]. In addition, we show that the form of $\mathbb{C}(r)$ found is statistically completely compatible with that obtained from the BSE-based analysis presented in Section 4.
- In Section 13, we present our conclusions.
- Finally, in Appendix A, we derive the BQIs related to the displacement functions of the conventional and background vertices, while in Appendix B, we provide details about the renormalization scheme employed in our computations.

2. Basic Concepts and General Theoretical Framework

We start by considering the Lagrangian density of an SU(N) Yang–Mills theory, comprised of the classical part, $\mathcal{L}_{\text{cl}}$, the contribution from the ghosts, $\mathcal{L}_{\text{gh}}$, and the covariant gauge-fixing term, $\mathcal{L}_{\text{gf}}$, namely

$$\mathcal{L}_{\text{YM}} = \mathcal{L}_{\text{cl}} + \mathcal{L}_{\text{gh}} + \mathcal{L}_{\text{gf}}, \tag{1}$$

where

$$\mathcal{L}_{\text{cl}} = -\frac{1}{4} F^a_{\mu\nu} F^{a\mu\nu}, \qquad \mathcal{L}_{\text{gh}} = -\bar{c}^a \partial^\mu D^{ab}_\mu c^b, \qquad \mathcal{L}_{\text{gf}} = \frac{1}{2\xi} (\partial^\mu A^a_\mu)^2. \tag{2}$$

In the above formula, $A^a_\mu(x)$ denotes the gauge field, while $c^a(x)$ and $\bar{c}^a(x)$ represent the ghost and anti-ghost fields, respectively, with $a = 1, \ldots, N^2 - 1$.

In addition,

$$F^a_{\mu\nu} = \partial_\mu A^a_\nu - \partial_\nu A^a_\mu + g f^{abc} A^b_\mu A^c_\nu, \tag{3}$$

is the antisymmetric field tensor, where f^{abc} stands for the totally antisymmetric structure constants of the SU(N) gauge group, and g is the gauge coupling, while

$$D^{ab}_\mu = \partial_\mu \delta^{ac} + g f^{amb} A^m_\mu, \tag{4}$$

denotes the covariant derivative in the adjoint representation. Finally, ξ represents the gauge-fixing parameter; $\xi = 0$ corresponds to the Landau gauge, while $\xi = 1$ specifies the Feynman–'t Hooft gauge.

The transition from the pure Yang–Mills theory of Equation (1) to QCD is implemented by supplementing the corresponding kinetic and interaction terms for the quark fields. However, since throughout this work we do not consider effects due to dynamical quarks, the aforementioned terms will be omitted entirely.

The most fundamental correlation function is the gluon propagator, whose nonperturbative features are inextricably connected with key dynamical properties of the theory. In the *Landau gauge* that we will employ throughout, the gluon propagator, $\Delta^{ab}_{\mu\nu}(q) = -i \delta^{ab} \Delta_{\mu\nu}(q)$, is completely transverse, i.e.,

$$\Delta_{\mu\nu}(q) = \Delta(q) P_{\mu\nu}(q), \qquad P_{\mu\nu}(q) := g_{\mu\nu} - q_\mu q_\nu / q^2. \tag{5}$$

In the continuum, the dynamical properties of the gluon propagator are encoded in the corresponding SDE, given by

$$\Delta^{-1}(q) P_{\mu\nu}(q) = q^2 P_{\mu\nu}(q) + i \Pi_{\mu\nu}(q), \tag{6}$$

where $\Pi_{\mu\nu}(q)$ is the gluon self-energy, shown diagrammatically in the first row of Figure 1. The fully-dressed vertices entering the diagrams are determined by their own SDEs, obtaining finally a tower of coupled integral equations, which, for practical purposes, must be truncated or treated approximately.

Given that, by virtue of the fundamental STI satisfied by the two-point function, the self-energy $\Pi_{\mu\nu}(q)$ is transverse,

$$q^\mu \Pi_{\mu\nu}(q) = 0, \tag{7}$$

we have that

$$\Pi_{\mu\nu}(q) = \Pi(q) P_{\mu\nu}(q), \tag{8}$$

and from Equation (6) follows that

$$\Delta^{-1}(q) = q^2 + i \Pi(q). \tag{9}$$

Of particular importance is the exact way that Equation (7) is enforced at the level of the SDE given in Figure 1 which governs the gluon evolution. In particular, if we were to contract the corresponding diagrams by q^μ, the entire set of diagrams must be considered in order for Equation (7) to emerge from the SDE. This pattern manifests itself already at the one-loop level, where it is known that the ghost loop must be included in order to guarantee the transversality of the self-energy. The main practical drawback stemming from this observation is that truncations, in the form of the omission of certain subsets of graphs, are likely to distort this fundamental property.

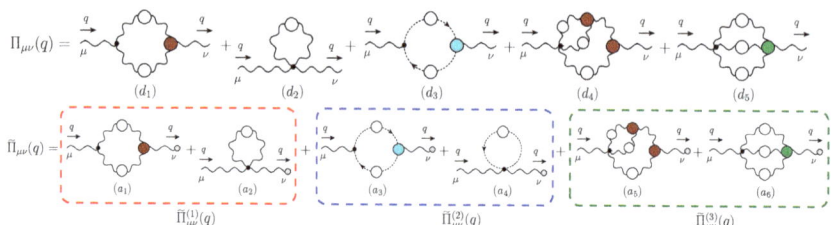

Figure 1. Upper panel: the diagrammatic representation of the conventional gluon self-energy, $\Pi_{\mu\nu}(q)$. Bottom panel: the diagrammatic representation of the $Q_\mu^a(q)B_\nu^b(-q)$, self-energy $\delta^{ab}\widetilde{\Pi}_{\mu\nu}(q)$; the grey circles at the end of the gluon lines indicate a background gluon. The corresponding Feynman rules are given in Appendix B of [14].

Quite interestingly, within the PT-BFM framework the transversality property of Equation (7) is enforced in a very special way, which permits physically meaningful truncations. In what follows we will predominantly employ the language of the BFM; for the basic principles of the PT and its connection with the BFM, the reader is referred to the extended literature on the subject [14,96,100,193,194,211,243], as well as to Section 6 of the present work.

The BFM is a powerful quantization procedure, where the gauge-fixing is implemented without compromising explicit gauge invariance. Within this framework, gauge field A appearing in the classical action is decomposed as $A = B + Q$, where B and Q are the background and quantum (fluctuating) fields, respectively. Note that the variable of integration in the generating functional $Z(J)$ is the quantum field, which couples to the external sources, as $J \cdot Q$. The background field does not appear in loops. Instead, it couples externally to the Feynman diagrams, connecting them with the asymptotic states to form elements of the S-matrix. Then, if the gauge-fixing term

$$\widehat{\mathcal{L}}_{\rm gf} = \frac{1}{2\xi_Q}(\widehat{D}_\mu^{ab}Q^{b\mu})^2, \qquad \widehat{D}_\mu^{ab} = \partial_\mu \delta^{ab} + gf^{amb}B_\mu^m, \qquad (10)$$

is used, the resulting gauge-fixed action retains its invariance under gauge transformations of the background field. As a result of this invariance, when Green's functions are contracted by the momentum carried by a background gluon, they satisfy Abelian (ghost-free) STIs, akin to the Takahashi identities known from QED. In particular, the STIs of the BFM retain their tree-level forms in all orders, in contradistinction to the STIs of the R_ξ gauges, whose forms are modified by contributions stemming from the ghost sector.

Within the BFM, one may consider three kinds of propagators, by choosing the types of incoming and outgoing gluons [244]. In particular, we have:

(i) The propagator $\langle 0|T[Q_\mu^a(q)Q_\nu^b(-q)]|0\rangle$ that connects two quantum gluons. Notice that this propagator coincides with the conventional gluon propagator of the covariant gauges, defined in Equation (5), under the assumption that the corresponding gauge-fixing parameters, ξ and ξ_Q, are identified, i.e., $\xi = \xi_Q$.

(ii) The propagator $\langle 0|T[Q_\mu^a(q)B_\nu^b(-q)]|0\rangle$ that connects a $Q_\mu^a(q)$ with a $B_\nu^b(-q)$, to be denoted by $\widetilde{\Delta}_{\mu\nu}^{ab}(q) = -i\delta^{ab}\widetilde{\Delta}_{\mu\nu}(q)$.

(iii) The propagator $\langle 0|T[B_\mu^a(q)B_\nu^b(-q)]|0\rangle$ that connects a $B_\mu^a(q)$ with a $B_\nu^b(-q)$, to be denoted by $\widehat{\Delta}_{\mu\nu}^{ab}(q) = -i\delta^{ab}\widehat{\Delta}_{\mu\nu}(q)$. Note that its full definition requires an additional gauge-fixing term, with the associated "classical" gauge-fixing parameter, ξ_C [14,202,206].

Given that the relations captured by Equations (5) and (6) apply also in the cases of $\widetilde{\Delta}_{\mu\nu}(q)$ and $\widehat{\Delta}_{\mu\nu}(q)$, one may define the corresponding self-energies $\widetilde{\Pi}_{\mu\nu}(q)$ and $\widehat{\Pi}_{\mu\nu}(q)$, as well as the functions $\widetilde{\Delta}(q)$ and $\widehat{\Delta}(q)$.

Quite interestingly, the three propagators defined in (i)-(iii) are related by a set of exact identities, known as BQIs [14,209–211]. In particular, we have that (see also Table 1)

$$\Delta(q) = [1+G(q)]\widetilde{\Delta}(q) = [1+G(q)]^2\widehat{\Delta}(q), \qquad (11)$$

where the function $G(q)$ is the $g_{\mu\nu}$ component of a particular two-point ghost function, $\Lambda_{\mu\nu}(q)$, given by [209,211,213,245]

$$\Lambda_{\mu\nu}(q) := ig^2 C_A \int_k \Delta_\mu^\rho(k) D(k+q) H_{\nu\rho}(-q,k+q,-k) = g_{\mu\nu} G(q) + \frac{q_\mu q_\nu}{q^2} L(q), \qquad (12)$$

where C_A is the Casimir eigenvalue of the adjoint representation [N for SU(N)], $D^{ab}(q) = i\delta^{ab} D(q)$ is the ghost propagator, and $H_{\nu\mu}(r,p,q)$ denotes the ghost–gluon kernel defined in Figure 2.

Table 1. The different types of gluon propagators of the background field method (BFM), together with their diagrammatic representations, symbols, corresponding self-energies, and the background quantum identities (BQIs) that relate them to the conventional propagator.

External Legs	Diagrammatic Representation	Symbol	Self-Energy	BQI
$Q_\mu^a(q) Q_\nu^b(-q)$		$-i\delta^{ab}\Delta_{\mu\nu}(q)$	$\Pi_{\mu\nu}(q)$	—
$Q_\mu^a(q) B_\nu^b(-q)$		$-i\delta^{ab}\widetilde{\Delta}_{\mu\nu}(q)$	$\widetilde{\Pi}_{\mu\nu}(q)$	$\widetilde{\Delta}(q) = \dfrac{\Delta(q)}{1+G(q)}$
$B_\mu^a(q) B_\nu^b(-q)$		$-i\delta^{ab}\widehat{\Delta}_{\mu\nu}(q)$	$\widehat{\Pi}_{\mu\nu}(q)$	$\widehat{\Delta}(q) = \dfrac{\Delta(q)}{[1+G(q)]^2}$

In the Landau gauge, a special identity relates the form factors of $\Lambda_{\mu\nu}(q)$ to the ghost dressing function, $F(q)$, defined as $F(q) = q^2 D(q)$, namely [16,131,213]

$$F^{-1}(q) = 1 + G(q) + L(q), \qquad (13)$$

which is valid before renormalization. In fact, in this particular gauge, $G(q)$ coincides with the so-called Kugo–Ojima function [212,245–247].

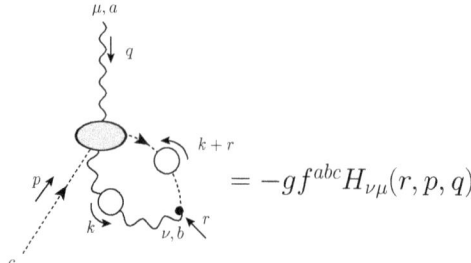

Figure 2. Diagrammatic definition of the ghost–gluon scattering kernel, $H_{\nu\mu}(r,p,q)$. At the tree level, $H^0_{\nu\mu} = g_{\nu\mu}$.

To determine the renormalized form of Equation (13), we introduce the renormalization constants of the conventional Green's functions

$$\Delta_R(q) = Z_A^{-1}\Delta(q), \qquad F_R(q) = Z_c^{-1}F(q),$$
$$\Gamma^R_\mu(r,p,q) = Z_1\Gamma_\mu(r,p,q), \qquad \Gamma^R_{\alpha\mu\nu}(q,r,p) = Z_3\Gamma_{\alpha\mu\nu}(q,r,p),$$
$$g_R = Z_g^{-1}g, \qquad \left[g_{\mu\nu} + \Lambda^R_{\mu\nu}(q)\right] = Z_\Lambda\left[g_{\mu\nu} + \Lambda_{\mu\nu}(q)\right], \quad (14)$$
$$Z_g^{-1} = Z_1^{-1}Z_A^{1/2}Z_c = Z_3^{-1}Z_A^{3/2},$$

where we denote by $\Gamma^{abc}_\mu(r,p,q) = -gf^{abc}\Gamma_\mu(r,p,q)$ and $\Gamma^{abc}_{\alpha\mu\nu}(q,r,p) = gf^{abc}\Gamma_{\alpha\mu\nu}(q,r,p)$ the conventional ghost–gluon $[Q^a_\mu(q)c^c(p)\bar{c}^b(r)]$ and three-gluon $[Q^a_\alpha(q)Q^b_\mu(r)Q^c_\nu(p)]$ vertices, respectively. Note that, by virtue of Taylor's theorem [207], Z_1 is *finite* in the Landau gauge; its precise value depends on the renormalization scheme adopted, see Section 8. Moreover, denoting by $\widehat{Z}_A$ the (wave-function) renormalization constant of $\widehat{\Delta}(q)$, the Abelian STIs of the BFM impose the validity of the pivotal relation [14,202,206]

$$Z_g = \widehat{Z}_A^{-1/2}, \quad (15)$$

which is the non-Abelian analog of the textbook relation $Z_e = Z_A^{-1/2}$ [133], relating the renormalization constants of the electric charge and the photon propagator in QED.

Then, since the BQIs of Equation (11) are direct consequences of the Becchi–Rouet–Stora–Tyutin (BRST) symmetry [248–250] of the theory [209,211,213,245], the form is preserved by renormalization. Hence, by combining Equations (11), (15) and (15), we obtain

$$Z_\Lambda = Z_1^{-1}Z_c, \quad (16)$$

which yields (note that in the original and widely used [3,8,16,20,79,131] version of Equation (17) the renormalization is performed in the so-called Taylor scheme, where $Z_1 = 1$.)

$$Z_1^{-1}F^{-1}(q) = 1 + G(q) + L(q). \quad (17)$$

As has been shown in [131], the dynamical equation governing $L(q)$ yields $L(0) = 0$, provided that the gluon propagator entering it is finite at the origin. Thus, one obtains from Equation (17) the useful identity [212]

$$Z_1^{-1}F^{-1}(0) = 1 + G(0). \quad (18)$$

According to numerous lattice simulations and studies in the continuum (see e.g., [21,42, 47,49,51,56,62,63,73,79,85,112,178,225,227–233]), the ghost dressing function reaches a finite (nonvanishing) value at the origin, which, due to Equation (18), furnishes also the value of $G(0)$.

The final upshot of the above considerations is that one may use the BQIs in Equation (11) to express the SDE given in Equation (6) in terms of the $\widetilde{\Pi}_{\mu\nu}(q)$ or $\widehat{\Pi}_{\mu\nu}(q)$, at the modest cost of introducing in the dynamics the quantities $1 + G(q)$ or $[1 + G(q)]^2$. Focusing on the former possibility, Equation (11) becomes

$$\Delta^{-1}(q) P_{\mu\nu}(q) = \frac{q^2 P_{\mu\nu}(q) + i \widetilde{\Pi}_{\mu\nu}(q)}{1 + G(q)}, \quad (19)$$

where the diagrammatic representation of the self-energy $\widetilde{\Pi}_{\mu\nu}(q)$ is shown in the lower panel of Figure 1.

The principal advantage of this formulation is that the self-energy $\widetilde{\Pi}_{\mu\nu}(q)$ contains fully-dressed vertices with a background gluon of momentum q exiting from them, which satisfy Abelian STIs. In particular, denoting by $\widetilde{\mathbb{\Gamma}}_{\mu\alpha\beta}(q,r,p)$, $\widetilde{\mathbb{\Gamma}}_{\mu}(r,p,q)$, and $\widetilde{\mathbb{\Gamma}}^{mnrs}_{\mu\alpha\beta\gamma}(q,r,p,t)$ the BQQ, Bcc, and BQQQ vertices, respectively, we have that [14,100,109]

$$q^{\mu} \widetilde{\mathbb{\Gamma}}_{\mu\alpha\beta}(q,r,p) = \Delta^{-1}_{\alpha\beta}(r) - \Delta^{-1}_{\alpha\beta}(p), \quad (20)$$

$$q^{\mu} \widetilde{\mathbb{\Gamma}}_{\mu}(r,p,q) = D^{-1}(p) - D^{-1}(r), \quad (21)$$

$$q^{\mu} \widetilde{\mathbb{\Gamma}}^{mnrs}_{\mu\alpha\beta\gamma}(q,r,p,t) = f^{mse} f^{ern} \mathbb{\Gamma}_{\alpha\beta\gamma}(r,p,q+t) + f^{mne} f^{esr} \mathbb{\Gamma}_{\beta\gamma\alpha}(p,t,q+r)$$
$$+ f^{mre} f^{ens} \mathbb{\Gamma}_{\gamma\alpha\beta}(t,r,q+p). \quad (22)$$

In contrast, the conventional three-gluon and ghost–gluon vertices, $\mathbb{\Gamma}_{\alpha\mu\nu}(q,r,p)$ and $\mathbb{\Gamma}_{\alpha}(r,p,q)$, respectively, satisfy the STIs [1,251–255]

$$q^{\alpha} \mathbb{\Gamma}_{\alpha\mu\nu}(q,r,p) = F(q) \left[\Delta^{-1}(p) P^{\sigma}_{\nu}(p) H_{\sigma\mu}(p,q,r) - \Delta^{-1}(r) P^{\sigma}_{\mu}(r) H_{\sigma\nu}(r,q,p) \right], \quad (23)$$

$$q^{\mu} F^{-1}(q) \mathbb{\Gamma}_{\mu}(r,p,q) + p^{\mu} F^{-1}(p) \mathbb{\Gamma}_{\mu}(r,q,p) = -r^2 F^{-1}(r) U(r,q,p), \quad (24)$$

where $U(r,q,p)$ is an interaction kernel containing only ghost fields; its tree-level value is $U^0(r,q,p) = 1$. The STI for the conventional four-gluon vertex is given in Equation (C.24) of [14].

The special STIs listed in Equations (20)–(22) are responsible for the remarkable property of "block-wise" transversality [109,192,244], displayed by $\widetilde{\Pi}_{\mu\nu}(q)$. To appreciate this point, notice that the diagrams comprising $\widetilde{\Pi}_{\mu\nu}(q)$ in Figure 1 were separated into three different subsets (blocks), consisting of (i) one-loop dressed diagrams containing only gluons, (ii) one-loop dressed diagrams containing a ghost loop, and (iii) two-loop dressed diagrams containing only gluons. The corresponding contributions of each block to $\widetilde{\Pi}_{\mu\nu}(q)$ are denoted by $\widetilde{\Pi}^{(i)}_{\mu\nu}(q)$, with $i = 1, 2, 3$.

The block-wise transversality is a stronger version of the standard transversality relation $q^{\mu} \widetilde{\Pi}_{\mu\nu}(q) = 0$; it states that each block of diagrams mentioned above is individually transverse, namely

$$q^{\mu} \widetilde{\Pi}^{(i)}_{\mu\nu}(q) = 0, \quad i = 1, 2, 3. \quad (25)$$

In order to appreciate in detail the reason why the STIs in Equations (20)–(22) are instrumental for the block-wise transversality, we will consider the case of $\widetilde{\Pi}^{(2)}_{\mu\nu}(q)$; the relevant diagrams are enclosed in the blue box of Figure 1.

The diagrams (a_3) and (a_4) are given by

$$(a_3)_{\mu\nu}(q) = g^2 C_A \int_k (k+q)_{\mu} D(k+q) D(k) \widetilde{\mathbb{\Gamma}}_{\nu}(-k, k+q, -q), \quad (26)$$

$$(a_4)_{\mu\nu}(q) = g^2 C_A g_{\mu\nu} \int_k D(k), \quad (27)$$

where a color factor δ^{ab} is suppressed in both expressions. In addition, for the formal manipulations of integrals, we employ dimensional regularization [256]; to that end, we introduce the short-hand notation

$$\int_k := \frac{\mu_0^\epsilon}{(2\pi)^d} \int_{-\infty}^{+\infty} d^d k, \qquad (28)$$

where $d = 4 - \epsilon$ is the dimension of the space-time, and μ_0 denotes the 't Hooft mass.

The contraction of graph $(a_3)_{\mu\nu}(q)$ by q^ν triggers the STI satisfied by $\widetilde{\Gamma}_\nu(-k, k+q, -q)$ [given by Equation (21)], and we obtain

$$\begin{aligned} q^\nu (a_3)_{\mu\nu}(q) &= g^2 C_A \int_k (k+q)_\mu D(k+q) D(k) \left[D^{-1}(k) - D^{-1}(k+q) \right] \\ &= g^2 C_A \int_k (k+q)_\mu [D(k+q) - D(k)] \\ &= -g^2 C_A\, q_\mu \int_k D(k), \end{aligned} \qquad (29)$$

which is precisely the negative of the contraction $q^\nu (a_4)_{\mu\nu}(q)$. Hence,

$$q^\nu \left[(a_3)_{\mu\nu}(q) + (a_4)_{\mu\nu}(q) \right] = 0. \qquad (30)$$

3. Schwinger Mechanism in Yang–Mills Theories

The BRST symmetry of the Yang–Mills Lagrangian given in Equation (1) prohibits the inclusion of a mass term of the form $m^2 A_\mu^2$. Moreover, a symmetry-preserving regularization scheme, such as dimensional regularization, prevents the generation of a mass term at any finite order in perturbation theory. Nonetheless, as affirmed four decades ago [94–99], the non-perturbative Yang–Mills dynamics endow the gluons with an effective mass, which sets the scale for all dimensionful quantities, and tames the instabilities originating from the infrared divergences of the perturbative expansion (e.g., Landau pole). In addition, the presence of this mass causes the effective decoupling (screening) of the gluonic modes beyond a "maximum gluon wavelength" [257], and leads to the dynamical suppression of the Gribov copies, see e.g., [16,258,259] and references therein.

The generation of a gluon mass proceeds through the non-perturbative realization of the Schwinger mechanism [127,128]. Even though the technical details associated with the implementation of this mechanism in a four-dimensional non-Abelian setting are particularly elaborate, the general underlying idea is relatively easy to convey.

To that end, consider the dimensionless vacuum polarization $\mathbf{\Pi}(q)$, defined through $\Pi(q) = q^2 \mathbf{\Pi}(q)$, such that

$$\Delta^{-1}(q) = q^2 [1 + i\mathbf{\Pi}(q)]. \qquad (31)$$

The Schwinger mechanism is based on the fundamental observation that, if $\mathbf{\Pi}(q)$ develops a pole at $q^2 = 0$ (to be referred to as "*massless pole*") then the vector meson (gluon) picks up a mass, regardless of any "prohibition" imposed by the gauge symmetry at the level of the original Lagrangian. Thus, in Euclidean space, the above sequence of ideas leads to

$$\lim_{q \to 0} \mathbf{\Pi}(q) = m^2/q^2 \implies \lim_{q \to 0} \Delta^{-1}(q) = \lim_{q \to 0} (q^2 + m^2) \implies \Delta^{-1}(0) = m^2, \qquad (32)$$

and the gauge boson propagator saturates to a non-zero value at the origin. This effect of infrared saturation of the propagator signifies the generation of a mass, which is identified with the positive residue of the pole.

At this descriptive level, Schwinger's argument is completely general, making no particular reference to the specific dynamics that would lead to the appearance of the required massless pole inside $\mathbf{\Pi}(q)$. In fact, depending on the particular theory, the field-theoretic circumstances that trigger the crucial sequence captured by Equation (32) may be very distinct, see e.g., [260,261]. In the case of Yang–Mills theories, the origin of the massless poles is purely non-perturbative [159]: the strong dynamics produce scalar composite excitations, which carry color and have vanishing masses. These poles are carried by the fully-dressed vertices of the theory; and since these vertices enter the gluon SDE shown in Figure 1 (upper (lower) panel for the QQ (QB) propagator), the massless poles find their way into the gluon self-energy (or, equivalently, the gluon vacuum polarization). The detailed implementation of this idea has been presented in a series of works [18,93,96,112,116–118,159–161,166,189,262], and will be summarized in the rest of this section.

Let us focus for now on the conventional three-gluon and ghost–gluon vertices, $\mathbb{\Gamma}_{\alpha\mu\nu}(q,r,p)$ and $\mathbb{\Gamma}_{\alpha}(r,p,q)$, respectively, introduced above Equation (23). When the formation of massless poles is triggered, these vertices assume the general form (see Figure 3)

$$\begin{aligned}\mathbb{\Gamma}_{\alpha\mu\nu}(q,r,p) &= \Gamma_{\alpha\mu\nu}(q,r,p) + V_{\alpha\mu\nu}(q,r,p),\\ \mathbb{\Gamma}_{\alpha}(r,p,q) &= \Gamma_{\alpha}(r,p,q) + V_{\alpha}(r,p,q),\end{aligned} \quad (33)$$

where $\Gamma_{\alpha\mu\nu}(q,r,p)$ and $\Gamma_{\alpha}(r,p,q)$ are their pole-free components, while $V_{\alpha\mu\nu}(q,r,p)$ and $V_{\alpha}(q,r,p)$ contain *longitudinally coupled* poles, whose special tensorial structure is given by

$$\begin{aligned}V_{\alpha\mu\nu}(q,r,p) &= \frac{q_\alpha}{q^2}C_{\mu\nu}(q,r,p) + \frac{r_\mu}{r^2}A_{\alpha\nu}(q,r,p) + \frac{p_\nu}{p^2}B_{\alpha\mu}(q,r,p),\\ V_{\alpha}(r,p,q) &= \frac{q_\alpha}{q^2}C(r,p,q),\end{aligned} \quad (34)$$

such that

$$P^{\alpha}_{\alpha'}(q)P^{\mu}_{\mu'}(r)P^{\nu}_{\nu'}(p)V_{\alpha\mu\nu}(q,r,p) = 0, \qquad P^{\alpha}_{\alpha'}(q)V_{\alpha}(r,p,q) = 0. \quad (35)$$

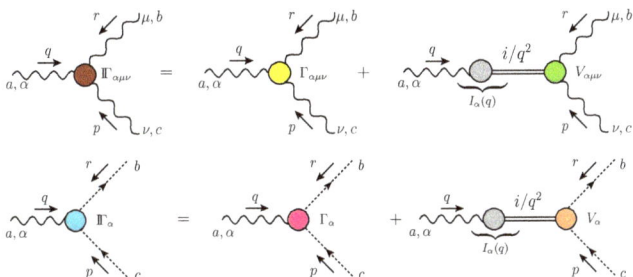

Figure 3. The diagrammatic representation of the three-gluon and ghost–gluon vertices introduced in Equation (33): $\mathbb{\Gamma}_{\alpha\mu\nu}(q,r,p)$ (first row) and $\mathbb{\Gamma}_{\alpha}(r,p,q)$ (second row). The first term on the r.h.s. indicates the pole-free part, $\Gamma_{\alpha\mu\nu}(q,r,p)$ or $\Gamma_{\alpha}(r,p,q)$, while the second denotes the pole term $V_{\alpha\mu\nu}(q,r,p)$ or $V_{\alpha}(r,p,q)$.

We emphasize that the reason why $V_{\alpha\mu\nu}(q,r,p)$ and $V_{\alpha}(q,r,p)$ are longitudinally coupled may be directly inferred from their special decomposition, shown in Figure 3. In particular, let us denote by $I_\alpha(q)$ the transition amplitude that connects a gluon with a massless composite scalar, depicted as a gray circle in Figure 3. Since $I_\alpha(q)$ depends solely on the momentum q, and carries a single Lorentz index, α, its general form is given by $I_\alpha(q) = q_\alpha I(q)$, where $I(q)$ is a scalar form factor [117,214]. This observation accounts directly for the form of $V_\alpha(q,r,p)$ given in Equation (34); to deduce the form of $V_{\alpha\mu\nu}(q,r,p)$, one must, in addition, appeal to Bose symmetry, which imposes the structures r_μ/r^2 and p_ν/p^2 in the remaining two channels.

Returning to the SDE of Figure 1, the component $V_{\alpha\mu\nu}(q,r,p)$ will enter in it through graphs (d_1) and (d_4), while the component $V_\alpha(q,r,p)$ through graph (d_3). Since $V_{\alpha\mu\nu}(q,r,p)$ has poles for each one of its three momenta, let us point out that only the pole associated with the q-channel, i.e., the channel that carries the momentum entering the gluon propagator is relevant for the Schwinger mechanism that will generate mass for $\Delta(q)$. In fact, in the Landau gauge that we employ, the gluon propagators inside the diagrams (d_1) and (d_4) are transverse, leading to a considerable reduction in the number of form factors of $V_{\alpha\mu\nu}(q,r,p)$ that participate actively, since

$$P^\mu_{\mu'}(r)P^\nu_{\nu'}(p)V_{\alpha\mu\nu}(q,r,p) = \frac{q_\alpha}{q^2}P^\mu_{\mu'}(r)P^\nu_{\nu'}(p)C_{\mu\nu}(q,r,p). \tag{36}$$

Consequently, for the ensuing analysis, one requires only the tensorial decomposition of the component $C_{\mu\nu}(q,r,p)$ in Equation (34), which is given by

$$C_{\mu\nu}(q,r,p) = C_1 g_{\mu\nu} + C_2 r_\mu r_\nu + C_3 p_\mu p_\nu + C_4 r_\mu p_\nu + C_5 p_\mu r_\nu, \tag{37}$$

where $C_j := C_j(q,r,p)$. Then, the substitution of Equation (37) into Equation (36), and use of the relation $q + p + r = 0$, reveals that only two form factors survive inside (d_1) and (d_4), namely

$$P^\mu_{\mu'}(r)P^\nu_{\nu'}(p)V_{\alpha\mu\nu}(q,r,p) = \frac{q_\alpha}{q^2}P^\mu_{\mu'}(r)P^\nu_{\nu'}(p)[C_1 g_{\mu\nu} + C_5 q_\mu q_\nu]. \tag{38}$$

Since the main function of the Schwinger mechanism is to make the gluon propagator saturate at the origin, it is important to explore the properties of the structures appearing in Equation (38) near $q = 0$. To that end, we expand the r.h.s. of Equation (38), keeping terms at most linear in q. After noticing that the term proportional to C_5 in Equation (38) is of order $\mathcal{O}(q^2)$, we end up with a single relevant form factor associated with $V_{\alpha\mu\nu}(q,r,p)$, namely $C_1(q,r,p)$, which survives the $q \to 0$ limit of graphs (d_1) and (d_4). As for $V_\alpha(r,p,q)$, its unique component, $C(q,r,p)$, enters directly in (d_3).

The continuation of this analysis entails the Taylor expansion of $C_1(q,r,p)$ and $C(r,p,q)$ around $q = 0$. In carrying out this expansion, one employs the following two key relations,

$$C_1(0,r,-r) = 0, \qquad C(r,-r,0) = 0. \tag{39}$$

The first one follows directly from the Bose symmetry of the three-gluon vertex, which implies that $C_1(q,r,p) = -C_1(q,p,r)$; as we will see in Section 10, it may also be derived in a completely independent way from the fundamental STIs satisfied by the three-gluon vertex. The justification of the second relation in Equation (39) is less straightforward; its derivation, presented in Appendix A, relies on the BQI [14,211] linking the conventional ghost–gluon vertex, $\mathbb{I}_\alpha(r,p,q)$, with its background counterpart, $\widetilde{\mathbb{I}}_\alpha(r,p,q)$.

Thus, after taking Equation (39) into account, the Taylor expansion of $C_1(q,r,p)$ and $C(r,p,q)$ around $q = 0$ yields

$$\lim_{q\to 0} C_1(q,r,p) = 2(q\cdot r)\mathbb{C}(r) + \cdots, \qquad \lim_{q\to 0} C(r,p,q) = 2(q\cdot r)\mathcal{C}(r) + \cdots, \tag{40}$$

with

$$\mathbb{C}(r) := \left[\frac{\partial C_1(q,r,p)}{\partial p^2}\right]_{q=0}, \qquad \mathcal{C}(r) := \left[\frac{\partial C(r,p,q)}{\partial p^2}\right]_{q=0}. \tag{41}$$

The functions $\mathbb{C}(r)$ and $\mathcal{C}(r)$ are of central importance for the rest of this review. In particular, there are three key points related to them that will be elucidated in detail in what follows:

1. $\mathbb{C}(r)$ and $\mathcal{C}(r)$ are the *BS amplitudes* describing the formation of gluon–gluon and ghost–anti-ghost *colored* composite bound states, respectively, see Section 4.
2. The gluon mass is determined by certain integrals that involve $\mathbb{C}(r)$ and $\mathcal{C}(r)$, given explicitly in Section 5.
3. $\mathbb{C}(r)$ and $\mathcal{C}(r)$ lead to smoking-gun displacements of the WIs. In fact, the displacement induced by $\mathbb{C}(r)$, has been confirmed by lattice QCD, by combining judiciously the results of several lattice simulations, see Section 5.2.

We emphasize that the BFM vertices develop poles in exactly the same way as their conventional counterparts. In particular, the main relations Equations (33), (34), (39) and (41) remain valid, with the only modification that all quantities carry hats or tildes; these BFM vertices will be used extensively in Section 5. Note that the conventional and background vertices, including their pole content, are related through appropriate BQIs, see e.g., Equations (A3) and (A6).

We end this section by commenting briefly on the implementation of the Schwinger mechanism away from the Landau gauge, i.e., when the gluon propagator is given by

$$\Delta_{\mu\nu}(q,\xi) = \Delta(q,\xi)P_{\mu\nu}(q) + \xi\, q_\mu q_\nu/q^4, \qquad \xi \neq 0; \tag{42}$$

for further details, the reader is referred to [189].

(*i*) The massless poles remain longitudinally coupled for every value of ξ, i.e., the form of $V_{\alpha\mu\nu}(q,r,p)$ and $V_\alpha(r,p,q)$ given in Equation (34) persists, with the only difference that the form factors comprising $C_{\mu\nu}(q,r,p)$, $A_{\alpha\nu}(q,r,p)$, $B_{\alpha\mu}(q,r,p)$, and $C(r,p,q)$ depend in general on ξ. Indeed, as explained right below Equation (35), the longitudinal nature of the poles is dictated solely by Lorentz invariance, which forces the transition amplitude $I_\alpha(q,\xi)$ to assume the form $I_\alpha(q) = q_\alpha I(q,\xi)$; clearly, this fundamental argument holds for every ξ.

(*ii*) Since the gluon propagators entering the graphs (d_1) and (d_4) of Figure 1 are now given by Equation (42), the l.h.s. of Equation (36) becomes $\Delta^\mu_{\mu'}(r,\xi)\Delta^\nu_{\nu'}(p,\xi) V_{\alpha\mu\nu}(q,r,p)$, and, as a result, the terms in Equation (34) proportional to p_ν/p^2 and r_μ/r^2 are not fully annihilated. Note, however, that the presence of poles in $p^2 \to k^2$ and $r^2 \to (k+q)^2$ poses no problem, given that one integrates over the loop momentum k. Similar observations hold for the BSE discussed in the next section, which acquires a more complicated form, involving not only the $\mathbb{C}(r)$ and $\mathcal{C}(r)$, but also additional form factors [189].

(*iii*) A general property of the massless excitations that trigger the Schwinger mechanism is that they do not induce divergences to physical amplitudes; their contributions are completely vanishing, or, at most, finite [260,261]. As was shown recently in [93], in Landau gauge QCD this property hinges on the validity of Equations (35) and (39). Away from the Landau gauge, Equation (39) persists, because its validity relies on Bose symmetry [189]. However, in Equation (35) the substitution $P_{\mu\nu}(q) \to \Delta_{\mu\nu}(q,\xi)$ must be carried out for all projectors; as a result, the r.h.s. no longer vanishes, but includes ξ-dependent longitudinal contributions. Even though this issue has not been addressed in the literature, the longitudinal nature of the additional terms heralds their cancellation through the same general mechanism that renders physical amplitudes ξ-independent.

4. Dynamical formation of Massless Poles

One crucial aspect of the implementation of the Schwinger mechanism in a Yang–Mills context is that the poles that comprise the components $V_{\alpha\mu\nu}(q,r,p)$ and $V_\alpha(q,r,p)$ in Equation (34) are *not* introduced by hand; rather, they are generated *dynamically*, as massless composite excitations that carry color. In fact, this subtle process is controlled by a system of coupled linear BSEs for the functions $\mathbb{C}(r)$ and $\mathcal{C}(r)$, which play the role of the BS amplitudes for generating composite massless scalars out of two gluons and a ghost–anti-ghost pair, respectively.

The starting points for the derivations of the aforementioned BSEs are the SDEs for $\mathbb{\Gamma}_{\alpha\mu\nu}(q,r,p)$ and $\mathbb{\Gamma}_\alpha(r,p,q)$, shown diagrammatically in Figure 4, and given by [124]

$$\begin{aligned}
\mathbb{\Gamma}^{\alpha\mu\nu} &= \Gamma_0^{\alpha\mu\nu} - \lambda\int_k \mathbb{\Gamma}^{\alpha\beta\gamma}\Delta_{\beta\rho}\Delta_{\gamma\sigma}\mathcal{K}_{11}^{\mu\nu\sigma\rho} + 2\lambda\int_k \mathbb{\Gamma}^\alpha DD\mathcal{K}_{12}^{\mu\nu}, \\
\mathbb{\Gamma}^\alpha &= \Gamma_0^\alpha - \lambda\int_k \mathbb{\Gamma}^{\alpha\beta\gamma}\Delta_{\beta\rho}\Delta_{\gamma\sigma}\mathcal{K}_{21}^{\sigma\rho} - \lambda\int_k \mathbb{\Gamma}^\alpha DD\mathcal{K}_{22},
\end{aligned} \tag{43}$$

where

$$\lambda := ig^2 C_A/2, \tag{44}$$

and the tree-level expressions for the vertices $\Gamma^{\alpha\mu\nu}$ and Γ^α are given by

$$\Gamma_0^{\alpha\mu\nu}(q,r,p) = (q-r)^\nu g^{\alpha\mu} + (r-p)^\alpha g^{\mu\nu} + (p-q)^\mu g^{\nu\alpha}, \qquad \Gamma_0^\alpha(r,p,q) = r^\alpha. \qquad (45)$$

Note that, for compactness, all momentum arguments have been suppressed; they may be easily restored by appealing to Figure 4.

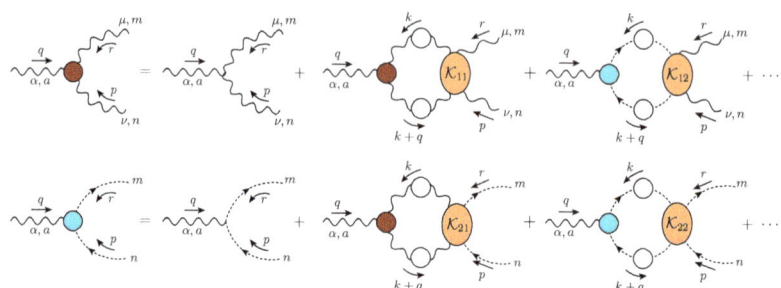

Figure 4. The coupled system of Schwinger–Dyson equations (SDEs) for the three-gluon and ghost-gluon vertices, $\Gamma_{\alpha\mu\nu}(q,r,p)$ and $\Gamma_\alpha(r,p,q)$, respectively. The orange ellipses represent four-point scattering kernels, denoted by $\mathcal{K}_{ij}$. We omit diagrams containing five-point scattering kernels.

The following steps are subsequently implemented:

1. Substitute into both sides of Equation (43) the expressions for the fully-dressed vertices given in Equation (33).
2. In order to exploit Equation (38), multiply the first equation by the factor $P_{\mu'\mu}(r)P_\nu^{\mu'}(p)$.
3. Take the limit of the system as $q \to 0$: this activates Equation (40) and introduces the functions $\mathbb{C}(r)$ and $\mathcal{C}(r)$.
4. Isolate the tensor structures proportional to q^α, and match the terms on both sides.
5. Employ the "one-particle exchange" approximation for the kernels $\mathcal{K}_{ij}$, to be denoted by $\mathcal{K}_{ij}^0$, shown in Figure 5.

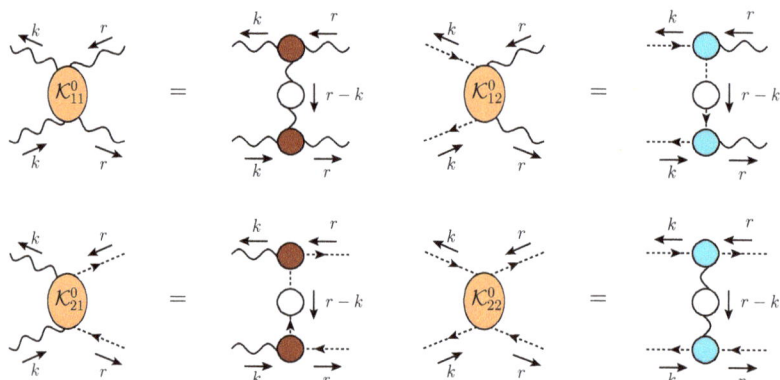

Figure 5. The one-particle exchange approximations, $\mathcal{K}_{ij}^0$, of the kernels $\mathcal{K}_{ij}$ appearing in Figure 4.

Thus, we arrive at a system of homogeneous equations involving $\mathbb{C}(r)$ and $\mathcal{C}(r)$,

$$\begin{aligned}\mathbb{C}(r) &= -\frac{\lambda}{3}\int_k \mathbb{C}(k)\Delta^2(k)P_{\rho\sigma}(k)P_{\mu\nu}(r)\widetilde{\mathcal{K}}_{11}^{\mu\nu\sigma\rho} + \frac{2\lambda}{3}\int_k \mathcal{C}(k)D^2(k)P_{\mu\nu}(r)\widetilde{\mathcal{K}}_{12}^{\mu\nu}, \\ \mathcal{C}(r) &= -\lambda\int_k \mathbb{C}(k)\Delta^2(k)P_{\sigma\rho}(k)\widetilde{\mathcal{K}}_{21}^{\sigma\rho} - \lambda\int_k \mathcal{C}(k)D^2(k)\widetilde{\mathcal{K}}_{22},\end{aligned} \qquad (46)$$

where $\widetilde{\mathcal{K}}_{ij} := (r \cdot k/r^2) \, \mathcal{K}^0_{ij}(r, -r, k, -k)$; the system is diagrammatically depicted in Figure 6.

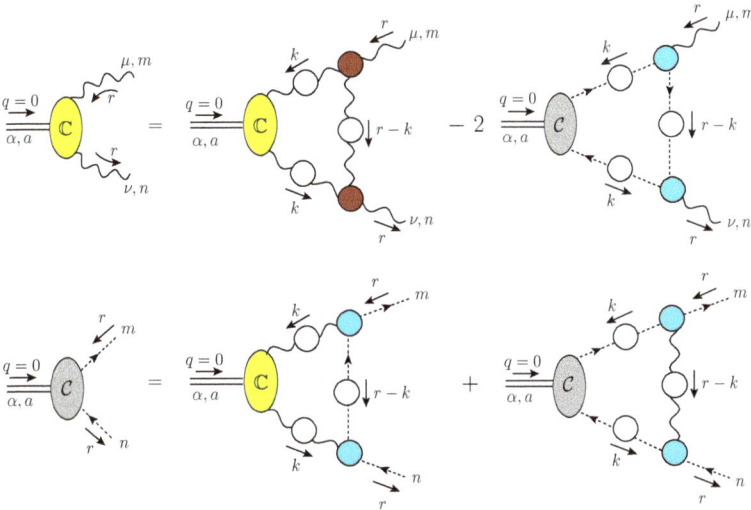

Figure 6. The diagrammatic representation of the coupled system of Bethe–Salpeter equations (BSEs) that governs the evolution of the functions $\mathbb{C}(r^2)$ and $\mathcal{C}(r^2)$.

Before turning to the numerical analysis, the BSE system must be passed to the Euclidean space, following standard conversion rules. In doing so, we note that the integral measure is modified according to $d^4k \to i d^4k_E$; this extra factor of i combines with the λ defined in Equation (44) to give real expressions.

As announced, the system of coupled equations given in Equation (46) represents the BSEs that govern the formation of massless colored bound states out of two gluons and a ghost–anti-ghost pair. The functions $\mathbb{C}(r)$ and $\mathcal{C}(r)$ are the corresponding BS amplitudes; finding nontrivial solutions for them, i.e., something other than $\mathbb{C}(r) = \mathcal{C}(r) = 0$ identically, is crucial for the implementation of the Schwinger mechanism.

The equations in Equation (46) are linear and homogeneous in the unknown functions. There are two main consequences arising from this fact. First, the numerical solution of the system will be reduced to an eigenvalue problem. Second, the overall scale of the solutions is undetermined, since the multiplication of a given solution by an arbitrary real constant produces another solution (The ambiguity originates from considering only leading terms in the expansion around $q = 0$, and may be resolved if further orders in q are kept, see e.g., [219,263,264]).

It turns out that the condition for obtaining nontrivial solutions, when expressed in terms of the strong coupling, $\alpha_s := g^2/4\pi$, states that they exist for $\alpha_s = 0.63$ when the renormalization point $\mu = 4.3$ GeV. The solutions obtained when α_s acquires this special value are shown in Figure 7; they have undergone scale fixing (The scale was fixed by requiring the best possible matching with the result obtained for $\mathbb{C}(r)$ from the WI displacement, see Section 12), and are denoted by $\mathbb{C}_\star(r)$ and $\mathcal{C}_\star(r)$. Observe that $\mathbb{C}_\star(r)$ is significantly larger in magnitude than $\mathcal{C}_\star(r)$, implying that the three-gluon vertex accounts for the bulk of the gluon mass, as originally claimed in [215].

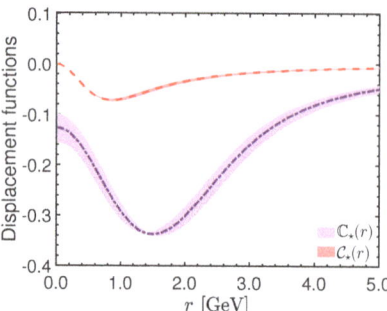

Figure 7. The solutions for $\mathbb{C}_\star(r)$ (purple dot-dashed) and $\mathcal{C}_\star(r)$ (red dashed) obtained from the coupled BSE system of Equation (46).

It is important to compare the value of $\alpha_s = 0.63$, imposed by the BSE eigenvalue, with the expected value for α_s for the renormalization scheme employed: within the *asymmetric* momentum subtraction (MOM) scheme (see Appendix B), we have that $\alpha_s = 0.27$ [71]. This numerical discrepancy in the values of α_s is clearly an artifact of the truncation employed, and concretely of the approximation of the kernels $\mathcal{K}_{ij}$ by their one-particle exchange diagrams, $\mathcal{K}_{ij}^0$. A preliminary analysis reveals that mild modifications of the kernels $\mathcal{K}_{ij}$ lead to considerable variations in the value of α_s, but leave the form of the solutions for $\mathbb{C}_\star(r)$ and $\mathcal{C}_\star(r)$ practically unaltered. This observation suggests that, while a more complete knowledge of the BSE kernels is required in order to bring α_s closer to its MOM value, the solutions obtained with the present approximations should be considered as particularly stable.

5. Generation of the Gluon Mass

We next demonstrate in detail how the presence of the massless poles in the vertices that enter the SDE of the gluon propagator generate a gluon mass.

Since the fundamental STIs of the theory remain intact under the action of the Schwinger mechanism, Equations (7) and (8) remain valid, and the mass term $m^2 = \Delta^{-1}(0)$ will appear in the transverse combination $\Delta^{-1}(0)P_{\mu\nu}(q)$. However, the determination of the mass proportional to $g_{\mu\nu}$ exposes an entirely different array of principles compared to the corresponding computation for the $q_\mu q_\nu / q^2$ component.

The calculation with respect to the $q_\mu q_\nu / q^2$ component is rather direct; since the massless poles in the vertices are themselves longitudinally coupled, their contribution to the $q_\mu q_\nu / q^2$ component of $\Pi_{\mu\nu}(q)$ is easily worked out, as will be illustrated in Section 5.1. In contrast, the emergence of a mass proportional to $g_{\mu\nu}$ is intimately connected with a powerful relation, known as *seagull identity* [113,166], which in the absence of the Schwinger mechanism would enforce the masslessness of the propagator, as will be discussed in Section 5.2. In fact, one main conceptual difference between the two approaches is that in the $g_{\mu\nu}$ case, the use of the PT-BFM-based version of the SDE given in Equation (19) is crucial for the emergence of the correct result.

In order to simplify the technical aspects of the calculation without compromising its conceptual content, we will determine the contribution to the gluon mass due to the pole in the ghost–gluon vertex, namely $V_\alpha(r, p, q)$ in the case of $\mathbb{\Gamma}_\alpha(r, p, q)$, and $\widetilde{V}_\alpha(r, p, q)$ in the case of $\widetilde{\mathbb{\Gamma}}_\alpha(r, p, q)$. To that end, we will focus on the subset of self-energy graphs containing only ghost loops, i.e., graph (d_3) in the case of $\Pi_{\mu\nu}(q)$, and graphs (a_3) and (a_4) in the case of $\widetilde{\Pi}_{\mu\nu}(q)$, shown in the upper and lower row of Figure 1, respectively.

5.1. Gluon Mass from the $q_\mu q_\nu$ Component

Let us calculate the contribution to the gluon mass stemming from the ghost loop, i.e., the diagram (d_3) of Figure 1, which, for general values of q, reads

$$(d_3)_{\mu\nu}(q) = g^2 C_A \int_k (k+q)_\mu D(k+q) D(k) \mathbb{\Gamma}_\nu(-k, k+q, -q). \tag{47}$$

To isolate the $q_\mu q_\nu/q^2$ component of Equation (47) at the origin, we first decompose the full vertex $\mathbb{\Gamma}_\nu(-k, k+q, -q)$ as in Equations (33) and (34), and drop directly the pole-free part since it does not contribute at $q = 0$. Then, denoting by $(d_3^V)_{\mu\nu}(q)$ the contribution of $V_\nu(-k, k+q, -q)$ to $(d_3)_{\mu\nu}(q)$, we obtain

$$(d_3^V)_{\mu\nu}(q) = -g^2 C_A \frac{q_\nu}{q^2} \int_k (k+q)_\mu D(k+q) D(k) \mathcal{C}(-k, k+q, -q). \tag{48}$$

Next, a Taylor expansion around $q = 0$, using Equations (39) and (40), yields

$$(d_3^V)_{\mu\nu}(q) = -2g^2 C_A \frac{q_\nu q^\rho}{q^2} \int_k k_\mu k_\rho D^2(k) \mathcal{C}(k). \tag{49}$$

Evidently, the integral above can only be proportional to $g_{\mu\rho}$, such that

$$(d_3^V)_{\mu\nu}(q) = -\frac{2g^2 C_A}{d} \left(\frac{q_\mu q_\nu}{q^2}\right) \int_k k^2 D^2(k) \mathcal{C}(k), \tag{50}$$

where the tensor structure $q_\mu q_\nu/q^2$ is already isolated.

Then, let us denote by $\Delta_{\text{gh}}^{-1}(0)$ the contribution to the mass originating in the $q_\mu q_\nu/q^2$ of the ghost loop. Noting that the contribution of $(d_3^V)_{\mu\nu}(q)$ to the propagator is i times the negative of its $q_\mu q_\nu/q^2$ form factor, we obtain that

$$\Delta_{\text{gh}}^{-1}(0) = \frac{4\lambda}{d} \int_k k^2 D^2(k) \mathcal{C}(k). \tag{51}$$

At this point, we set $d = 4$ and renormalize Equation (51). This leads to the appearance of the *finite* renormalization constant of the ghost–gluon vertex, Z_1.

Next, we express the result in terms of the ghost dressing function F, pass to Euclidean space, and employ hyperspherical coordinates, to obtain the final expression

$$\Delta_{\text{gh}}^{-1}(0) = \hat{\lambda} Z_1 \int_0^\infty dy \, F^2(y) \, \mathcal{C}(y), \tag{52}$$

where $\hat{\lambda} := C_A \alpha_s / 8\pi$.

The derivation of the contributions from the diagrams (d_1) and (d_4) proceeds in a completely analogous way, but is algebraically more involved, see [166] for details.

It is instructive to consider how the result of Equation (52) emerges in the context of Equation (19). To this end, we consider the ghost block $\widetilde{\Pi}_{\mu\nu}^{(2)}(q)$ of Figure 1, whose diagrams have the expressions given in Equation (27); clearly, only diagram $(a_3)_{\mu\nu}(q)$ can contribute to the $q_\mu q_\nu$ component of $\widetilde{\Pi}_{\mu\nu}^{(2)}(q)$.

Then, we decompose $\widetilde{\mathbb{\Gamma}}_\alpha(r, p, q)$ in complete analogy with Equations (33) and (34), i.e.,

$$\widetilde{\mathbb{\Gamma}}_\alpha(r, p, q) = \widetilde{\Gamma}_\alpha(r, p, q) + \frac{q_\alpha}{q^2} \widetilde{\mathcal{C}}(r, p, q), \tag{53}$$

and expand the $(a_3)_{\mu\nu}(q)$ of Equation (27) around $q = 0$, isolating its $q_\mu q_\nu/q^2$ component. These steps eventually lead to

$$\widetilde{\Delta}_{\text{gh}}^{-1}(0) = \frac{4\lambda}{d} \int_k k^2 D^2(k) \widetilde{\mathcal{C}}(k), \tag{54}$$

where $\widetilde{\mathcal{C}}(q)$ is defined in the exact same way as $\mathcal{C}(q)$, namely through Equation (41) but with tildes over all relevant quantities. It is now easy to establish that Equation (54) is completely equivalent to Equation (51), simply by multiplying both of its sides by $Z_1 F(0)$, and then using Equation (A4) on the r.h.s. and Equations (19) and (18) on the l.h.s.

Hence, when the mass is computed through the $q_\mu q_\nu / q^2$ component of the self-energy, the contributions originating from the ghost diagrams of either the BQ or the QQ propagator furnish the same result. The same is not true for the calculation through the $g_{\mu\nu}$ component, since the ghost diagram $(d_3)_{\mu\nu}$ of the QQ propagator is not by itself transverse, and a meaningful analysis is preferably carried out within the BFM.

5.2. Gluon Mass from the $g_{\mu\nu}$ Component: Seagull Identity and Ward Identity Displacement

The fact that the activation of the Schwinger mechanism is crucial for the self-consistent generation of a gluon mass may be best appreciated in conjunction with the so-called *seagull identity* [113,166]. The content of this identity is that

$$\int_k k^2 \frac{\partial f(k)}{\partial k^2} + \frac{d}{2} \int_k f(k) = 0, \quad (55)$$

for functions $f(k)$ that satisfy Wilson's criterion [265]; the cases of physical interest are $f(k) = \Delta(k), D(k)$. The general demonstration of the validity of Equation (55) has been given in [166]; for a detailed discussion of how Equation (55) prevents the photon from acquiring a mass in scalar electrodynamics, see [18].

What is so special about Equation (55) is that, within the PT-BFM formalism, the l.h.s. of Equation (55) coincides with the contributions of loop diagrams to the $g_{\mu\nu}$ component of the gluon mass. Therefore, Equation (55) enforces the non-perturbative masslessness of the gluon in the absence of the Schwinger mechanism: even if a massive gluon propagator (made "massive" through a procedure other than the Schwinger mechanism) were to be substituted inside Equation (55), one would obtain zero as a contribution to the gluon mass! For example, the simple choice $f = (k^2 - m^2)^{-1}$, reduces the l.h.s of Equation (55) to (dimensionally regularized) textbook integrals, which add up to give precisely zero [18].

In order to appreciate in some detail how the seagull identity prevents the $g_{\mu\nu}$ component of the propagator from acquiring a mass in the absence of the Schwinger mechanism, let us consider once again the ghost block $\widetilde{\Pi}^{(2)}_{\mu\nu}(q)$ of Figure 1; now both graphs, (a_3) and (a_4), contribute to the $g_{\mu\nu}$ component.

Let us assume that the Schwinger mechanism is turned off; at the level of the Bcc vertex this means that $\widehat{V}_\alpha(r, p, q)$ vanishes identically, and $\widetilde{\mathbb{\Gamma}}_\alpha(r, p, q) = \widetilde{\Gamma}_\alpha(r, p, q)$. Consequently, $\widetilde{\Gamma}_\alpha(r, p, q)$ saturates the STI of Equation (21),

$$q^\alpha \widetilde{\Gamma}_\alpha(r, p, q) = D^{-1}(p) - D^{-1}(r). \quad (56)$$

Since the form factors of the vertex $\widetilde{\Gamma}_\alpha(r, p, q)$ do not contain any poles, the derivation from Equation (56) of the corresponding WI proceeds in the standard textbook way: both sides of Equation (56) undergo a Taylor expansion around $q = 0$, and terms at most linear in q are retained. Thus, one arrives at the simple QED-like WI

$$\widetilde{\Gamma}_\alpha(r, -r, 0) = \frac{\partial D^{-1}(r)}{\partial r^\alpha} \implies D^2(r)\widetilde{\Gamma}_\alpha(r, -r, 0) = -2r_\alpha \frac{\partial D(r)}{\partial r^2}. \quad (57)$$

We now compute the $g_{\mu\nu}$ component of $\widetilde{\Pi}^{(2)}_{\mu\nu}(q)$ at $q = 0$, or, equivalently, $\widetilde{\Delta}^{-1}_{\text{gh}}(0)$. From Equation (27), we see that $(a_4)_{\mu\nu}$ is proportional to $g_{\mu\nu}$ in its entirety. On the other hand, $(a_3)_{\mu\nu}(q)$ contains both $g_{\mu\nu}$ and $q_\mu q_\nu$ components; however, the latter vanishes in the limit $q \to 0$ if the vertex is pole-free. Then, it is straightforward to show that, as $q \to 0$,

$$\widetilde{\Delta}^{-1}_{\text{gh}}(0) = \frac{2\lambda}{d} \left[\int_k k_\mu D^2(k) \widetilde{\Gamma}^\mu(-k, k, 0) + d \int_k D(k) \right]. \quad (58)$$

At this point, employing the WI of Equation (57) (with $r \to -k$), we get

$$\widetilde{\Delta}_{\text{gh}}^{-1}(0) = \frac{4\lambda}{d} \underbrace{\left[\int_k k^2 \frac{\partial D^{-1}(k)}{\partial k^2} + \frac{d}{2} \int_k D(k) \right]}_{\text{seagull identity}} = 0. \tag{59}$$

Hence, the WI satisfied by the vertex in the absence of the Schwinger mechanism triggers the seagull identity, which, in turn, enforces the masslessness of the propagator.

When the Schwinger mechanism is activated, the STIs that are satisfied by the vertices of the theory retain their original forms but are resolved through the nontrivial participation of the terms containing the massless poles [96,112,159–161,166,262,266]. In particular, the full vertex $\widetilde{\mathbb{\Gamma}}_\alpha(r,p,q)$ precisely satisfies Equation (21), namely

$$\begin{aligned} q^\alpha \widetilde{\mathbb{\Gamma}}_\alpha(r,p,q) &= q^\alpha \widetilde{\Gamma}_\alpha(r,p,q) + \widetilde{C}(r,p,q) \\ &= D^{-1}(p) - D^{-1}(r). \end{aligned} \tag{60}$$

Notice in particular that the contraction of $\widetilde{\mathbb{\Gamma}}_\alpha(r,p,q)$ by q^α cancels the massless pole in q^2, leading to a completely pole-free result. Therefore, the WI obeyed by $\widetilde{\Gamma}_\alpha(r,p,q)$ may be derived as before, through a standard Taylor expansion, leading to

$$q^\alpha \widetilde{\Gamma}_\alpha(r,-r,0) = -\widetilde{C}(r,-r,0) + q^\alpha \left\{ \frac{\partial D^{-1}(r)}{\partial r^\alpha} - \left[\frac{\partial \widetilde{C}(r,p,q)}{\partial q^\alpha} \right]_{q=0} \right\}. \tag{61}$$

Evidently, the unique zeroth-order contribution appearing in Equation (61), namely $\widetilde{C}(r,-r,0)$, must vanish,

$$\widetilde{C}(r,-r,0) = 0. \tag{62}$$

Note that this particular property may be independently derived from the antisymmetry of $\widetilde{C}(r,p,q)$ under $r \leftrightarrow p$, $\widetilde{C}(r,p,q) = -\widetilde{C}(p,r,q)$, which is a consequence imposed by the ghost–anti-ghost symmetry of the $B(q)\bar{c}(r)c(p)$ vertex. The above result, together with Equation (A3), is used to prove Equation (39) in Appendix A.

Thus, Equation (61) becomes

$$q^\alpha \widetilde{\Gamma}_\alpha(r,-r,0) = q^\alpha \left\{ \frac{\partial D^{-1}(r)}{\partial r^\alpha} - 2r_\alpha \widetilde{\mathcal{C}}(r) \right\}, \quad \widetilde{\mathcal{C}}(r) := \left[\frac{\partial \widetilde{C}(r,p,q)}{\partial p^2} \right]_{q=0}, \tag{63}$$

and the matching of the terms linear in q yields the WI

$$\widetilde{\Gamma}_\alpha(r,-r,0) = \frac{\partial D^{-1}(r)}{\partial r^\alpha} - \underbrace{2r_\alpha \widetilde{\mathcal{C}}(r)}_{\text{WI displacement}}. \tag{64}$$

Comparing Equations (57) and (64), it becomes clear that the Schwinger mechanism induces a characteristic displacement to the WIs that are satisfied by the pole-free parts of the vertices [166].

Returning to Equation (58), but now substituting in it the displaced version of Equation (57), namely

$$D^2(k)\widetilde{\Gamma}^\mu(-k,k,0) = 2k^\mu \left[\frac{\partial D(k)}{\partial k^2} + D^2(k)\widetilde{\mathcal{C}}(k) \right]. \tag{65}$$

When Equation (65) is substituted into Equation (58), the first term of its r.h.s. triggers the seagull identity and vanishes, exactly as before; however, the second term survives, precisely furnishing the result given in Equation (54).

Completely analogous procedures may be applied to the remaining two blocks, $\widetilde{\Pi}^{(1)}_{\mu\nu}(q)$ and $\widetilde{\Pi}^{(3)}_{\mu\nu}(q)$, by exploiting the Abelian STIs of Equations (20) and (22), respectively [161].

6. Renormalization Group Invariant Interaction Strength

The PT-BFM formalism provides the natural framework for the construction of the RGI version of the naive one-gluon exchange interaction.

To fix the ideas, recall that in QED, the one-photon exchange interaction, defined as $\alpha \Delta_A(q)$, where $\alpha := e^2/4\pi$ is the hyper-fine structure constant and $\Delta_A(q)$ the photon propagator, is an RGI combination, by virtue of the relation $Z_e = Z_A^{-1/2}$; see comments following Equation (15). Moreover, this particular combination is universal (process-independent) because it may be identified within any two-to-two scattering process, regardless of the nature of the initial and final states (electrons, muons, taus, etc). Instead, in QCD, the corresponding combination $\alpha_s \Delta(q)$ is (trivially) universal but not RGI. When the vertices that connect the gluon to the external particles are "dressed" ($\Gamma_0 \to \Gamma$), the combination $\Gamma \alpha_s \Delta \Gamma$ becomes RGI; however, it is no longer process-independent, because the vertices Γ contain information on the characteristics of the external particles, e.g., the Γ is not the same if the external particles are quarks or gluons. This apparent conundrum may be resolved by resorting to the PT, which reconciles harmoniously the notions of RGI and process independence.

Within the PT framework, the starting point of the construction involves "on-shell" processes [14,96,100,193,194], such as those depicted in Figure 8. The fundamental observation is that the dressed vertices appearing there contain propagator-like contributions, which may be unambiguously identified by means of a well-defined diagrammatic procedure. After discarding terms that vanish on the shell, the contributions extracted from a vertex have a two-fold effect: (i) the genuine vertex contributions left behind form a new vertex, $\widetilde{\Gamma}$, which satisfies Abelian STIs, and (ii) when the propagator-like pieces from both vertices are allotted to the conventional propagator, $\Delta_{\mu\nu}(q)$, the resulting effective propagator, $\widehat{\Delta}_{\mu\nu}(q)$, captures all RG logarithms associated with the running of the coupling; for example, at one loop and for large q^2, one has

$$\widehat{\Delta}^{-1}(q) \approx q^2 \left[1 + bg^2 \ln(q^2/\mu^2)\right], \tag{66}$$

where $b = 11 C_A/48\pi^2$ is the first coefficient of the Yang–Mills β function. We emphasize that the PT construction goes through all orders in perturbation theory, as well as non-perturbatively, and all key properties of the PT Green's function persist unaltered [194,195].

The correspondence between the PT and the BFM may be summarized by stating that the PT rearrangement outlined above amounts effectively to replacing the Q-type gluon that is being exchanged (carrying momentum q) by a B-type gluon [193,267–269]; external (on-shell) fields are always of the Q-type. Thus, the notation used above for the PT effective Green's functions ("tildes" and "hats") corresponds precisely to the BFM notation introduced in Section 2. Note that the formal expression of all PT rearrangements implemented diagrammatically are the BQIs that relate conventional Green's functions to their BFM counterparts [14]. For example, in the case of the quark–gluon vertex, we have that the vertices $\Gamma_\mu(q, k_1, -k_2)$ [with external fields $Q_\mu^a(q) q^b(k_1) \bar{q}^c(-k_2)$] and $\widetilde{\Gamma}_\mu(q, k_1, -k_2)$ [$B_\mu^a(q) q^b(k_1) \bar{q}^c(-k_2)$] are related by the BQI [270]

$$\widetilde{\Gamma}_\mu(q, k_1, -k_2) = [1 + G(q)] \Gamma_\mu(q, k_1, -k_2) + \cdots, \tag{67}$$

where the ellipsis denotes terms that vanish on the shell. Similarly, the BQI of Equation (A5), when evaluated on-shell, yields a completely analogous result, to wit,

$$\widetilde{\Gamma}_{\mu\alpha\rho}(q, k_1, -k_2) = [1 + G(q)] \Gamma_{\mu\alpha\rho}(q, k_1, -k_2) + \cdots. \tag{68}$$

It is now clear how the PT gives rise to a process-independent propagator-like component: regardless of the process (i.e., the type of vertex connecting the internal gluon to the external states), each vertex contributes to the conventional $\Delta(q)$ a factor of $[1 + G(q)]^{-1}$, finally leading to the BQI of Equation (11) [16].

The culmination of the above sequence of ideas is reached by noting that, by virtue of Equation (15), the combination

$$\widehat{d}(q) := \alpha_s \widehat{\Delta}(q) = \frac{\alpha_s \Delta(q)}{[1 + G(q)]^2}, \quad (69)$$

is RGI: it retains exactly the same form before and after renormalization, and, consequently, does not depend on the renormalization point μ [96]. The quantity $\widehat{d}(q)$ has a mass dimension of -2, and is known in the literature as the "RGI running interaction strength" [16].

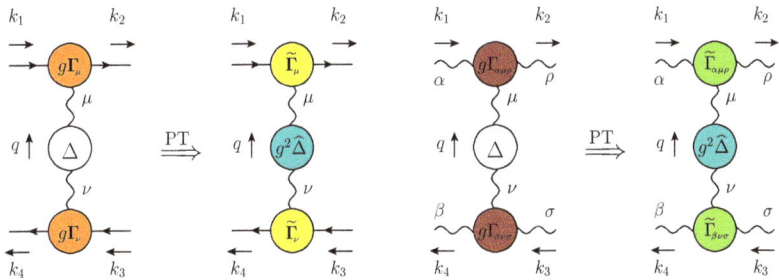

Figure 8. Diagrammatic representation of the basic PT rearrangement in the case of quark–antiquark scattering, corresponding to the S-matrix element $\mathcal{T}_{q\bar{q} \to q\bar{q}}$ of Equation (70) (**left**), and gluon–gluon scattering, corresponding to $\mathcal{T}_{gg \to gg}$ of Equation (71) (**right**).

The steps leading to the natural appearance of $\widehat{d}(q)$ within any given process may be summarized in the case of quark–antiquark, or gluon–gluon scattering.

Consider the S-matrix elements $\mathcal{T}_{q\bar{q} \to q\bar{q}}$, for the scattering of a quark and an antiquark, and $\mathcal{T}_{gg \to gg}$, for the scattering of two gluons. The quark–antiquark scattering is depicted in the left panel of Figure 8. Using the BQI of Equation (11) we obtain

$$\begin{aligned}
\mathcal{T}_{q\bar{q} \to q\bar{q}} &= [g\Gamma_\mu(q, k_1, -k_2)] \Delta(q) P^{\mu\nu}(q) [g\Gamma_\nu(-q, k_3, -k_4)] \\
&\stackrel{\text{PT}}{=} \left\{ g[1 + G(q)]^{-1} \widetilde{\Gamma}_\mu(q, k_1, -k_2) \right\} \Delta(q) P^{\mu\nu}(q) \left\{ g[1 + G(q)]^{-1} \widetilde{\Gamma}_\nu(-q, k_3, -k_4) \right\} \\
&\stackrel{\text{PT}}{=} \widetilde{\Gamma}_\mu(q, k_1, -k_2) \left\{ g^2 [1 + G(q)]^{-2} \Delta(q) \right\} P^{\mu\nu}(q) \widetilde{\Gamma}_\nu(-q, k_3, -k_4) \\
&\stackrel{\text{PT}}{=} \widetilde{\Gamma}_\mu(q, k_1, -k_2) \underbrace{\left[g^2 \widehat{\Delta}(q) \right]}_{4\pi \widehat{d}(q)} P^{\mu\nu}(q) \widetilde{\Gamma}_\nu(-q, k_3, -k_4),
\end{aligned} \quad (70)$$

where we omit color structures.

Similarly, the scattering of two gluons depicted in the right panel of Figure 8, yields

$$\begin{aligned}
\mathcal{T}_{gg \to gg} &= [g\Gamma_{\alpha\mu\rho}(k_1, q, -k_2)] \Delta(q) P^{\mu\nu}(q) [g\Gamma_{\beta\nu\sigma}(k_3, -q, -k_4)] \\
&\stackrel{\text{PT}}{=} \left\{ g[1 + G(q)]^{-1} \widetilde{\Gamma}_{\alpha\mu\rho}(k_1, q, -k_2) \right\} \Delta(q) P^{\mu\nu}(q) \left\{ g[1 + G(q)]^{-1} \widetilde{\Gamma}_{\beta\nu\sigma}(k_3, -q, -k_4) \right\} \\
&\stackrel{\text{PT}}{=} \widetilde{\Gamma}_{\alpha\mu\rho}(k_1, q, -k_2) \left\{ g^2 [1 + G(q)]^{-2} \Delta(q) \right\} P^{\mu\nu}(q) \widetilde{\Gamma}_{\beta\nu\sigma}(k_3, -q, -k_4) \\
&\stackrel{\text{PT}}{=} \widetilde{\Gamma}_{\alpha\mu\rho}(k_1, q, -k_2) \underbrace{\left[g^2 \widehat{\Delta}(q) \right]}_{4\pi \widehat{d}(k)} P^{\mu\nu}(q) \widetilde{\Gamma}_{\beta\nu\sigma}(k_3, -q, -k_4).
\end{aligned} \quad (71)$$

Evidently, the same $\widehat{d}(q)$, defined in Equation (69), appears naturally in both Equations (70) and (71): it is, in that sense, a process-independent RGI interaction capturing faithfully the one-gluon exchange dynamics [3,16,20,79,96,129–131].

The actual determination of $\widehat{d}(q)$ proceeds by means of the second equality in Equation (69), i.e., by combining the standard gluon propagator, $\Delta(q)$, together with the function $1 + G(q)$. In the top left panel of Figure 9 we show lattice data for the conventional gluon propagator from [85] (points) and a physically motivated fit (blue continuous), given by Equation (C11) of [124]. In the top right panel of the same figure, we show the $1 + G(q)$ auxiliary function, which can be computed by contracting Equation (12) with $P^{\mu\nu}(q)/3$ (see e.g., [131]), using the results of [228] for the ghost–gluon kernel, $H_{\nu\mu}(r,p,q)$. Then, in the bottom left panel of Figure 9 we show the $\widehat{d}(q)$ that results from combining the fit for $\Delta(q)$ and the $1 + G(q)$ shown in the top panels of the same figure and using $\alpha_s = 0.27$ [71] and $Z_1 = 0.9333$ [see Section 8].

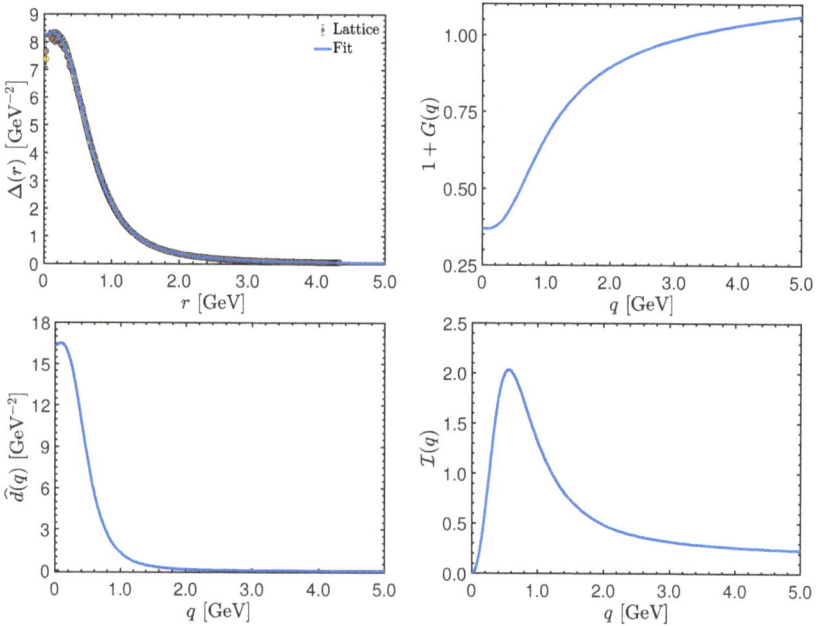

Figure 9. Top left: Gluon propagator, $\Delta(q)$, from lattice simulations of Reference [85] (points) and a fit given by Equation (C11) of [124] (blue continuous). **Top right**: The auxiliary function $1 + G(q)$, defined in Equation (12). **Bottom left**: The renormalization group invariant (RGI) running interaction strength $\widehat{d}(q)$ defined in Equation (69), computed using the $\Delta(q)$ and $1 + G(q)$ shown in the top panels, with $\alpha_s = 0.27$ [71] and $Z_1 = 0.9333$ [see Section 8]. **Bottom right**: The corresponding dimensionless RGI interaction $\mathcal{I}(q)$, defined in Equation (72).

From the $\widehat{d}(q)$ of Equation (69) one may define the dimensionless RGI interaction [16], $\mathcal{I}(q)$,

$$\mathcal{I}(q) := q^2 \widehat{d}(q). \tag{72}$$

As explained in [16], this quantity provides the strength required in order to describe ground-state hadron observables using SDEs in the matter sector of the theory. In that sense, $\mathcal{I}(q)$ bridges a longstanding gap that has existed between non-perturbative continuum QCD and ab initio predictions of basic hadron properties.

7. Three-Gluon Vertex and Its Planar Degeneracy

The three-gluon vertex, $\mathbb{\Gamma}_{\alpha\mu\nu}(q,r,p)$, plays a pivotal role in the dynamics of QCD [234], manifesting its non-Abelian nature through the gluon self-interaction. In fact, the most celebrated perturbative feature of QCD, namely asymptotic freedom, hinges on the properties of this particular interaction vertex. Its importance in the non-perturbative domain has led to an intense effort for unveiling its elaborate features [21,28,33–36,41,50,68,69,71,78,81,86, 87,122,172–180,271]. Indeed, as we have seen in Sections 3 and 4, the pole structure of the three-gluon vertex is crucial for the onset of the Schwinger mechanism and the dynamical generation of a gluon mass. Moreover, its pole-free part provides highly nontrivial contributions to the SDEs of several Green's functions, most notably the gluon propagator (cf. Figure 1), as well as in the Bethe–Salpeter and Faddeev equations that determine the properties of glue balls [235,236,238–240] and hybrid mesons [237], respectively.

For general momenta, $\mathbb{\Gamma}_{\alpha\mu\nu}(q,r,p)$ is a particularly complicated function, comprised by 14 tensor structures and their associated form factors [251]. Fortunately, in the Landau gauge, considerable simplifications take place, making the treatment of the three-gluon vertex less cumbersome. Indeed, in the latter gauge, quantities of interest require only the knowledge of the *transversely projected* three-gluon vertex [126,174,175,223], $\overline{\Gamma}_{\alpha\mu\nu}(q,r,p)$, defined as

$$\overline{\Gamma}_{\alpha\mu\nu}(q,r,p) = \mathbb{\Gamma}^{\alpha'\mu'\nu'}(q,r,p) P_{\alpha'\alpha}(q) P_{\mu'\mu}(r) P_{\nu'\nu}(p)$$
$$= \Gamma^{\alpha'\mu'\nu'}(q,r,p) P_{\alpha'\alpha}(q) P_{\mu'\mu}(r) P_{\nu'\nu}(p). \tag{73}$$

Note that $\overline{\Gamma}_{\alpha\mu\nu}(q,r,p)$ does not contain massless poles, by virtue of Equation (35). Furthermore, $\overline{\Gamma}_{\alpha\mu\nu}(q,r,p)$ can be parameterized in terms of only 4 independent tensor structures, i.e.,

$$\overline{\Gamma}^{\alpha\mu\nu}(q,r,p) = \sum_{i=1}^{4} \widetilde{\Gamma}_i(q^2, r^2, p^2) \widetilde{\lambda}_i^{\alpha\mu\nu}(q,r,p). \tag{74}$$

Due to the Bose symmetry of $\overline{\Gamma}_{\alpha\mu\nu}(q,r,p)$, the $\widetilde{\lambda}_i^{\alpha\mu\nu}(q,r,p)$ can be chosen to be individually Bose symmetric, such that its form factors $\widetilde{\Gamma}_i(q^2, r^2, p^2)$ are symmetric under the exchange of any two arguments [86]. In fact, they can only depend on three totally symmetric combinations of momenta.

Quite remarkably, lattice [86–88] and continuum [174,175,223] studies alike, have demonstrated that, to a very good level of accuracy, the $\widetilde{\Gamma}_i$ depend exclusively on a single judiciously chosen variable. Specifically, the $\widetilde{\Gamma}_i$ computed on the lattice in [86–88] can be parameterized in terms of the special Bose symmetry combination

$$s^2 = \frac{1}{2}\left(q^2 + r^2 + p^2\right). \tag{75}$$

Thus, the $\widetilde{\Gamma}_i$ are the same for any combination of q^2, r^2, and p^2 that fulfils Equation (75) for a given value of s^2. This property has been denominated *planar degeneracy*, because Equation (75) with fixed s defines a plane, normal to the vector $(1,1,1)$, in the first octant of the coordinate system (q^2, r^2, p^2).

In particular, the form factor $\widetilde{\Gamma}_1(q^2, r^2, p^2)$ of the classical tensor structure is rather accurately approximated by

$$\widetilde{\Gamma}_1(q^2, r^2, p^2) \approx \widetilde{\Gamma}_1(s^2, s^2, 0) \approx L_{sg}(s). \tag{76}$$

In the above equation, L_{sg} is the single transverse form factor of the three-gluon vertex in the soft gluon limit [124], and is obtained in lattice simulations as the $q = 0$ limit of the following totally transverse projection [84]

$$L_{sg}(r) = \left. \frac{\Gamma_0^{\alpha\mu\nu}(q,r,p)P_{\alpha\alpha'}(q)P_{\mu\mu'}(r)P_{\nu\nu'}(p)\Gamma^{\alpha'\mu'\nu'}(q,r,p)}{\Gamma_0^{\alpha\mu\nu}(q,r,p)P_{\alpha\alpha'}(q)P_{\mu\mu'}(r)P_{\nu\nu'}(p)\Gamma_0^{\alpha'\mu'\nu'}(q,r,p)} \right|_{q\to 0}. \tag{77}$$

A particular realization of the planar degeneracy property is shown in Figure 10, where we show the classical form factor $\widetilde{\Gamma}_1(q^2, r^2, p^2)$, obtained from the lattice simulation of [86]; we consider three different kinematic configurations, characterized by a single momentum. Specifically, the orange stars correspond to the soft-gluon limit, $q = 0$, which implies $p^2 = r^2$; the green diamonds denote the symmetric limit, where all of the momenta have the same magnitude, $q^2 = p^2 = r^2$; and the purple circles represent points with $p^2 = r^2$ and $q^2 = 2r^2$. When plotted against the momentum r, the three configurations of $\widetilde{\Gamma}_1(q^2, r^2, p^2)$ produce three clearly distinct curves; however, when plotted in terms of the Bose symmetry variable s of Equation (75), they become statistically indistinguishable, manifesting the validity of Equation (76).

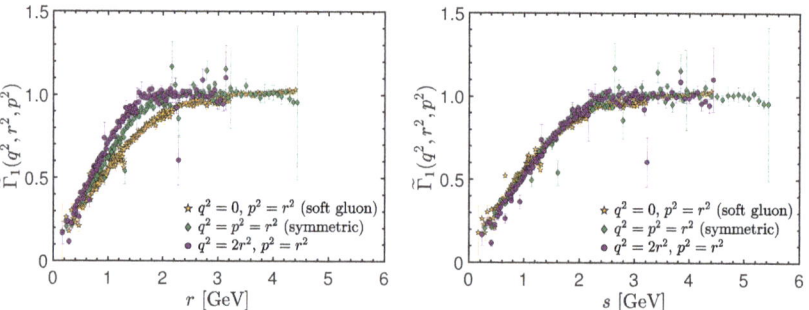

Figure 10. Lattice data from Reference [86] for the classical form factor, $\widetilde{\Gamma}_1(q^2, r^2, p^2)$, of the transversely projected three-gluon vertex in three different kinematic configurations: the soft-gluon ($q = 0$, $p^2 = r^2$, orange stars), the symmetric limit ($q^2 = p^2 = r^2$, green diamonds), and the case $p^2 = r^2$ with $q^2 = 2r^2$ (purple circles). In the left panel $\widetilde{\Gamma}_1(q^2, r^2, p^2)$ is plotted as a function of r, while in the right it is plotted as a function of the Bose symmetry variable s defined in Equation (75).

In addition to the planar degeneracy property, lattice [84,86–88] and continuum [174,175,179,223] results show a clear dominance of the classical form factor $\widetilde{\Gamma}_1$ over the remaining ones. Based on these considerations, the special approximation

$$\overline{\Gamma}^{\alpha\mu\nu}(q,r,p) \approx L_{sg}(s)\overline{\Gamma}_0^{\alpha\mu\nu}(q,r,p), \tag{78}$$

has been put forth, where $\overline{\Gamma}_0^{\alpha\mu\nu}(q,r,p)$ is the tree-level value of $\overline{\Gamma}^{\alpha\mu\nu}(q,r,p)$, i.e., Equation (73) with $\Gamma^{\alpha'\mu'\nu'}(q,r,p) \to \Gamma_0^{\alpha'\mu'\nu'}(q,r,p)$, and the form factor $L_{sg}(s)$ has been defined in Equation (77). We emphasize that the shape of $L_{sg}(r)$ has been very precisely determined through dedicated lattice studies with large-volume simulations [68,71,84,85]. The outcome of this exploration is shown in Figure 11, where we plot the lattice data of [84] for $L_{sg}(r)$, together with a physically motivated fit given by Equation (C12) of [124] (blue continuous curve). The corresponding fitting formula is rather complicated and will not be reported here; note, however, that the simple expression given in Equation (102) captures rather well the qualitative behavior of $L_{sg}(s)$.

Equation (78) provides an accurate and exceptionally compact approximation for $\overline{\Gamma}^{\alpha\mu\nu}(q,r,p)$ in general kinematics. This approximation, with the fit for L_{sg} shown in Figure 11, will be used explicitly in Sections 8 and 11, where the $\overline{\Gamma}^{\alpha\mu\nu}(q,r,p)$ in general kinematics will be needed as input for the determination of other physically important quantities.

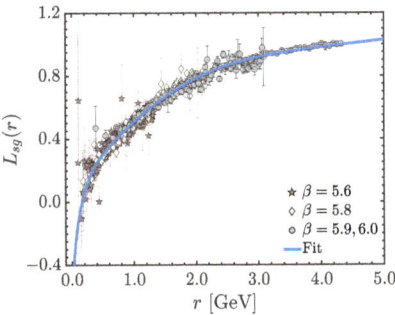

Figure 11. Lattice data from Reference [84] for $L_{sg}(q)$, compared to the fit for it given by Equation (C12) of [124] (blue continuous curve).

8. Ghost Dynamics from Schwinger–Dyson Equations

We next turn our attention to the ghost sector of the theory, whose scrutiny is important for several reasons. First, it has been connected to particular scenarios of color confinement [272,273]. Second, Green's functions associated with the ghost sector appear as ingredients in the SDEs of several key functions, such as the gluon propagator and the three-gluon vertex [41,50,68,69,71,81,122,172–179,274], affecting their non-perturbative behavior in nontrivial ways, as will be discussed in Section 9. Third, the SDEs governing the ghost sector are simpler than their gluonic counterparts because they are comprised by fewer diagrams; in fact, the SDE of the ghost propagator contains a single diagram, see Figure 12. Fourth, in the Landau gauge, the validity of Taylor's theorem [207] facilitates considerably the task of renormalization.

Consequently, the SDEs of the ghost sector are an excellent testing ground for (a) probing the impact of the gluonic Green's functions that contribute to them [85]; (b) assessing the reliability of truncation schemes [275,276]; and (c) testing the agreement between lattice and continuum approaches.

One of the central results of numerous studies in the continuum [21,62,85,112,178, 225,227–233] as well as a variety of lattice simulations [42,47,49,51,56,63,73,79] may be summarized by stating that the ghost propagator, $D(q)$, remains massless, while the corresponding dressing function, $F(q)$, saturates at the origin. As we will discuss in Section 9, the non-perturbative masslessness of the ghost has important implications for the infrared behavior of the gluon propagator and the three-gluon vertex.

In what follows we provide a concrete example of the state-of-the-art SDE analysis of the ghost sector, by solving the coupled system of equations that governs the ghost-dressing function and the ghost–gluon vertex. In order to obtain a closed system of equations, we use lattice results for the gluon propagator, the three-gluon vertex, and the value of the coupling constant in the particular renormalization scheme employed.

The main points of this analysis may be summarized as follows.

(i) We begin by considering the coupled system of SDEs given in Figure 12, which determines the ghost propagator and ghost–gluon vertex. The treatment will be simplified by neglecting the diagram (d_3^ν) of Figure 12, thus eliminating the dependence on the ghost–ghost–gluon–gluon vertex, $\Gamma^{\mu\sigma}$. This is a particularly robust truncation, because the impact of the neglected diagram on the ghost–gluon vertex has been shown to be less than 2% [275].

(ii) Note that due to the fully transverse nature of the gluon propagators in the Landau gauge, in conjunction with the fact that various projections need to be implemented during the treatment of this system, the pole parts V of all fully dressed vertices appearing in Figure 12 will be annihilated; thus, we will have throughout the replacement $I\!\Gamma \to \Gamma$.

(iii) We proceed by decomposing the pole-free part, $\Gamma_\nu(r,q,p)$, of the ghost–gluon vertex into its most general Lorentz structure, namely

$$\Gamma_\nu(r,q,p) = r_\nu B_1(r,q,p) + p_\nu B_2(r,q,p), \qquad (79)$$

whose scalar form factors reduce to $B_1^0 = 1$ and $B_2^0 = 0$ at the tree level. Evidently, due to the transversality of the gluon propagator, only the classical tensor r_ν, accompanied by the form factor B_1, will survive in all SDE diagrams of Figure 12.

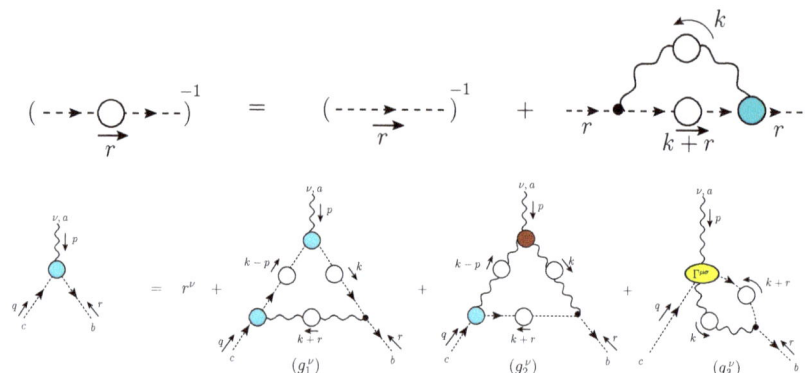

Figure 12. Top: SDE governing the momentum evolution of the ghost propagator. Bottom: SDE for the ghost–gluon vertex, $\Gamma_\nu(r,q,p)$.

(iv) The SDE of Figure 12 is given by

$$F^{-1}(r) = 1 + 2\lambda \int_k f(k,r) B_1(-r, k+r, -k) \Delta(k) D(k+r), \qquad (80)$$

where λ is given by Equation (44), and we define

$$f(k,r) := 1 - \frac{(r \cdot k)^2}{r^2 k^2}. \qquad (81)$$

(v) Next, we note that the form factor $B_1(r,q,p)$ can be extracted from $\Gamma_\nu(r,q,p)$ through the projection

$$B_1(r,q,p) = \varepsilon^\nu \Gamma_\nu(r,q,p), \qquad \varepsilon^\nu := \frac{p^2 r^\nu - (r \cdot p) p^\nu}{r^2 p^2 - (r \cdot p)^2}. \qquad (82)$$

Hence, acting with ε^ν on the diagrams in the second line of Figure 12, we obtain

$$B_1(r,q,p) = 1 - \lambda [a(r,q,p) - b(r,q,p)], \qquad (83)$$

where

$$a(r,q,p) = q^\alpha r^\mu \varepsilon^\nu \int_k D(k) D(k-p) \Delta(k+r) B_1(p-k, q, k+r) B_1(-k, k-p, p) P_{\alpha\mu}(k+r) k_\nu,$$
$$b(r,q,p) = q^\alpha r^\mu \varepsilon^\nu \int_k \Delta(k) \Delta(k-p) D(k+r) B_1(k+r, q, p-k) \overline{\Gamma}_{\nu\mu\alpha}(p, -k, k-p). \qquad (84)$$

(*vi*) At this point, we invoke the property of the planar degeneracy of $\overline{\Gamma}_{\alpha\mu\nu}(q,r,p)$, discussed in Section 7. Employing Equation (78) into the SDE for B_1, the term $b(r,q,p)$ of Equation (84) becomes

$$b(r,q,p) = q^\alpha r^\mu \varepsilon^\nu \int_k \Delta(k)\Delta(k-p)D(k+r)B_1(k+r,q,p-k)\overline{\Gamma}^0_{\nu\mu\alpha}(p,-k,k-p)L_{sg}(\tilde{s}), \quad (85)$$

with $\tilde{s}^2 = p^2 + k^2 - 2(k\cdot p)$.

We emphasize that although Equation (78) constitutes in general an approximation, there is one particular kinematic limit in which the expression for $b(r,q,p)$ given in Equation (85) becomes exact. Specifically, in the soft gluon limit ($p=0$), it can be shown *exactly* that [85]

$$P_\mu^{\mu'}(k)P_\nu^{\nu'}(k)\Gamma_{\alpha\mu'\nu'}(0,k,-k) = 2L_{sg}(k)k_\alpha P_{\mu\nu}(k). \quad (86)$$

Then, starting from either the general expression for $b(r,q,p)$ of Equation (84) and using Equation (86), or the approximate version given by Equation (85), it can easily be shown that the $p=0$ limit is the same. As such, the use of Equation (78) yields not only an excellent approximation in general kinematics, but also the exact soft gluon limit for the contribution of the three-gluon vertex to the form factor B_1.

(*vii*) Now we consider the renormalization of the coupled system of equations. Since the ghost–gluon vertex is finite in the Landau gauge [207], most SDE treatments [85, 224–228] of the ghost sector employ the so-called Taylor renormalization scheme (see Appendix B), defined in such a way that the finite renormalization constant of the ghost–gluon vertex has the exact value $Z_1 = 1$ [54,60,80,85,207].

However, in order to employ Equation (78) most expeditiously, it is more convenient to renormalize in the so-called *asymmetric* MOM scheme, defined in Appendix B, because this is precisely the scheme employed in the lattice calculations of L_{sg} [68,71,84,85]. Past this point, we denote by $\widetilde{Z}_1$ the *finite* value of the ghost–gluon renormalization constant in the asymmetric MOM scheme. Evidently, Equations (15) and (79) imply that $B_1^R = \widetilde{Z}_1 B_1$.

The renormalization of Equations (80) and (83) proceeds by substitution of the unrenormalized quantities by their renormalized counterparts, following Equation (15), and imposing Equation (A8) for $F(\mu^2)$.

Note that, in principle, $\widetilde{Z}_1$, may be determined from the relation $\widetilde{Z}_1 = Z_3 Z_c Z_A^{-1}$, imposed by the corresponding STI [277]; however, these renormalization constants are not available to us, given that Green's functions have been obtained from the lattice. Therefore, $\widetilde{Z}_1$ is treated as an adjustable parameter, whose value is determined by requiring that the solution of the SDE for $F(q)$ reproduces the corresponding lattice data of [73,85] as well as possible.

(*viii*) Finally, we transform Equations (80) and (83) from Minkowski to the Euclidean space, using standard conversion rules. Note that, once in Euclidean space, we will express the functional dependence of $B_1(r,q,p)$ in terms of the squared momenta of the anti-ghost and gluon legs, r^2 and p^2, and the angle, θ, between them, i.e., $B_1(r,q,p) \equiv B_1(r^2,p^2,\theta)$.

The result of these manipulations is that Equations (80) and (83) become

$$F^{-1}(r) = 1 - \frac{\alpha_s C_A \widetilde{Z}_1}{2\pi^2} \int_0^\infty dk^2 k^2 \Delta(k) \int_0^\pi d\phi\, s_\phi^4$$
$$\times \left[B_1(r^2,k^2,\phi)\frac{F(\sqrt{z})}{z} - B_1(\mu^2,k^2,\phi)\frac{F(\sqrt{u})}{u} \right], \quad (87)$$

and

$$B_1(r^2,p^2,\theta) = \widetilde{Z}_1 - \frac{\alpha_s C_A \widetilde{Z}_1}{8\pi^2}\left[\overline{a} + 2\overline{b}\right], \quad (88)$$

respectively, with

$$\bar{a} = \frac{1}{s_\theta} \int_0^\infty dk^2 k^2 F(k) \int_0^\pi d\phi s_\phi^3 \frac{\Delta(\sqrt{z})}{z} \int_0^\pi d\omega s_\omega \frac{F(\sqrt{v})}{v} B_1(k^2, p^2, \alpha) B_1(v, z, \beta) \mathcal{K}_a, \quad (89)$$

$$\bar{b} = \frac{1}{s_\theta} \int_0^\infty dk^2 k^2 \Delta(k) \int_0^\pi d\phi s_\phi^3 \frac{F(\sqrt{z})}{z} \int_0^\pi d\omega s_\omega \frac{\Delta(\sqrt{v})}{v} B_1(z, v, \beta) L_{sg}(\bar{s}) \mathcal{K}_b.$$

In the above equations, we employ the notation $c_x := \cos x$ and $s_x := \sin x$, and define the following variables

$$r \cdot k := rkc_\phi,$$
$$z := r^2 + k^2 + 2rkc_\phi,$$
$$\bar{s}^2 := (p^2 + k^2 + v)/2,$$
$$\alpha := \pi - \cos^{-1}[c_\theta c_\phi + s_\theta s_\phi c_\omega],$$

$$p \cdot k := pk(c_\theta c_\phi + s_\theta s_\phi c_\omega),$$
$$u := \mu^2 + k^2 + 2\mu k c_\phi,$$
$$v := p^2 + k^2 - 2pk(c_\theta c_\phi + s_\theta s_\phi c_\omega),$$
$$\beta := \cos^{-1}\left[\frac{k(pc_\theta c_\phi + ps_\theta s_\phi c_\omega - rc_\phi) + prc_\theta - k^2}{\sqrt{vz}}\right].$$

Finally, the kernels $\mathcal{K}_a$ and $\mathcal{K}_b$ are given by

$$\mathcal{K}_a = (c_\theta c_\omega s_\phi - c_\phi s_\theta)\left[ks_\phi(pc_\theta + r) - pc_\theta c_\omega(kc_\phi + r)\right],$$

$$\mathcal{K}_b = c_\omega \left\{ k^2 pc_\phi \left[c_\theta p\left(s_\theta^2(s_\phi^2 s_\omega^2 - 4s_\phi^2 + 1) + s_\phi^2\right) + r\left(s_\theta^2 - s_\theta^2(2s_\phi^2 + 1)\right)\right] \right.$$
$$- k^3 \left[s_\theta^2\left(rc_\theta - 2ps_\theta^2 + p\right) + ps_\theta^2\right] + kp^2\left[s_\theta^2\left(2s_\theta^2(p - rc_\theta) - rc_\theta - p\right) + s_\theta^2(rc_\theta - p)\right]$$
$$- c_\phi p^3 rs_\theta^2 \right\} + s_\theta s_\phi \left\{ c_\theta p \left[r\left(p^2 - k^2(s_\omega^2 + s_\phi^2 s_\omega^2 - 2s_\phi^2)\right) - c_\phi k(s_\omega^2 - 2)(k^2 + p^2)\right] \right.$$
$$+ k \left[c_\phi k^2 r - c_\phi p^2 r\left(s_\theta^2(s_\omega^2 - 2) + s_\omega^2\right) + kp^2\left(3s_\theta^2 s_\phi^2 s_\omega^2 - 2s_\theta^2 s_\omega^2 - 4s_\theta^2 s_\phi^2 + 3s_\theta^2\right.\right.$$
$$\left.\left.\left. + (3 - 2s_\omega^2)s_\phi^2 - 2\right)\right]\right\}.$$

We are now in a position to solve Equations (87) and (88) numerically. We choose the renormalization point at $\mu = 4.3$ GeV and employ for $\Delta(q)$ and $L_{sg}(q)$ the fits to the lattice data shown in Figures 9 and 11, respectively. Note that for large momenta these fits recover the behaviors dictated by the corresponding anomalous dimensions [124]. For the strong coupling, we use the value $\alpha_s(4.3 \text{ GeV}) = 0.27$, determined from the lattice simulations of [71].

Below, we discuss the main results of this analysis:

The value of $\widetilde{Z}_1$ was obtained by solving the SDE system for various values of this constant until the χ^2 of the comparison between the solution for $F(q)$ and the lattice data of [73,85] was minimized. This procedure yields $\widetilde{Z}_1 = 0.9333 \pm 0.0075$.

In the left panel of Figure 13, we show as a blue continuous line the SDE result for $F(q)$, with the above value of $\widetilde{Z}_1$. The result is compared to the lattice data of [73,85], which have been cured from discretization artifacts. As it turns out, the SDE and lattice results for F agree within 1%.

We next consider the form factor B_1. In the right panel of Figure 13 we show $B_1(r^2, p^2, \theta)$ as a surface, for arbitrary values of the magnitudes of the momenta r and p, and for the angle θ formed between them at $\theta = 2\pi/3$. In the same panel, we highlight as a red dot-dashed curve the soft gluon limit $B_1(r^2, 0, 2\pi/3)$ of the general kinematics $B_1(r^2, p^2, 2\pi/3)$ (The soft gluon limit is approached by taking $p \to 0$ in $B_1(r^2, p^2, \theta)$; in the non-perturbative case, this limit is independent of the value of θ).

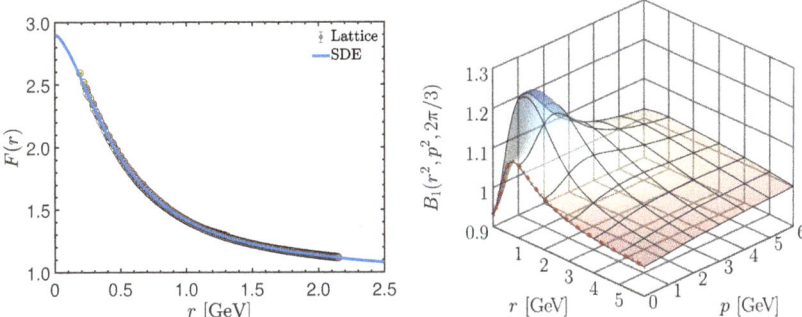

Figure 13. (**Left**): ghost dressing function $F(q)$ obtained from the coupled system of SDEs of Equations (80) and (83) (blue continuous line) compared to the lattice data of Reference [73,85]. (**Right**): The corresponding result for $B_1(r^2, p^2, \theta)$ for arbitrary magnitudes of the anti-ghost and gluon momenta, r and p, respectively, and a representative value of $\theta = 2\pi/3$ for the angle between them. The red dot-dashed curve highlights the soft gluon limit ($p = 0$).

The only available SU(3) lattice data for B_1 were obtained in the soft gluon limit [42,43], and have sizable error bars. Furthermore, they have been computed within the Taylor scheme, while in the present work, we used the asymmetric MOM scheme. Nevertheless, we can meaningfully compare our SDE results with those of the lattice, and perform a statistical analysis to assess their agreement.

Specifically, denoting by B_1^T the Taylor scheme value of the form factor B_1, it can easily be shown that

$$B_1(r^2, p^2, \theta) = \widetilde{Z}_1 B_1^T(r^2, p^2, \theta), \qquad (90)$$

which allows us to carry out the desired comparison.

Then, we use Equation (90) to compute $B_1^T(r^2, 0, \theta)$ from the $B_1(r^2, 0, 2\pi/3)$ slice (red dot-dashed curve) in the right panel of Figure 13, and compare the result to the lattice data of [42,43] (points) in Figure 14. Evidently, the SDE determination agrees with the lattice results.

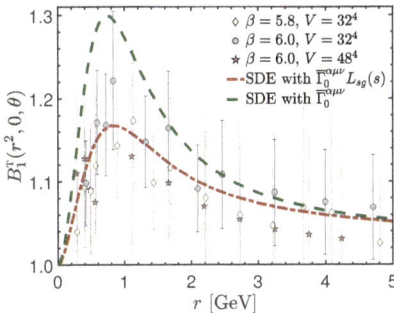

Figure 14. Soft gluon limit, $B_1^T(r^2, 0, \theta)$, of the classical form factor of the ghost–gluon vertex in the Taylor scheme. The points correspond to the lattice data of Reference [42,43]. The red dot-dashed line shows the SDE solution with the three-gluon vertex dressed according to Equation (78), while the green dashed represents the SDE solution with tree-level three-gluon vertex.

In order to quantify this agreement, we next conduct a χ^2 analysis. To this end, we consider only the 22 lattice points r_i in the interval $r_i \in [0.3, 2.5]$ GeV, where the signal is most pronounced. Then, we compute the χ^2 of the data through

$$\chi_j^2 = \sum_i \frac{[B_1^{\text{lat}}(r_i^2, 0, \theta) - g_j(r_i)]^2}{\epsilon_{B_1}(r_i^2, 0, \theta)}, \qquad (91)$$

where $B_1^{\text{lat}}(r_i^2, 0, \theta)$ are the lattice points shown in Figure 14, $\epsilon_{B_1}(r_i^2, 0, \theta)$ are their respective errors, and $g_j(r_i)$ are the three hypotheses that we will compare to the lattice data. Specifically, for the g_j we consider the three cases

$$g_j(r_i) = \begin{cases} 1 & \text{if } j = 1, \\ \text{SDE with } \overline{\Gamma}^{\alpha\mu\nu} = \overline{\Gamma}_0^{\alpha\mu\nu} L_{sg}(s) & \text{if } j = 2, \\ \text{SDE with } \overline{\Gamma}^{\alpha\mu\nu} = \overline{\Gamma}_0^{\alpha\mu\nu} & \text{if } j = 3, \end{cases} \quad (92)$$

i.e., g_1 is the tree-level value of B_1, g_2 is the solution of the SDE using Equation (78) for dressing the three-gluon vertex, corresponding to the red dot-dashed curve of Figure 14, and g_3 is the solution of the SDE obtained by setting the three-gluon vertex to the tree level, which amounts to the substitution $L_{sg} \to 1$ in Equation (88), and is represented by a green dashed curve in Figure 14.

Then, for each χ_j^2 we compute the probability P_j that normally distributed errors would yield a χ^2 at least as large as χ_j^2, through

$$P_j = \int_{\chi_j^2}^{\infty} \chi_{\text{PDF}}^2(22, x) dx = \left. \frac{\Gamma(n_r/2, \chi^2/2)}{\Gamma(n_r/2)} \right|_{n_r=22}^{\chi^2 = \chi_j^2}. \quad (93)$$

In the above equation, $\chi_{\text{PDF}}^2(n, x) = x^{n/2-1} e^{-x/2} / [2^{n/2} \Gamma(n/2)]$ denotes the χ^2 probability distribution function with n degrees of freedom, while $\Gamma(z, x)$ is the incomplete Γ function.

The results of the above analyses are collected in Table 2. We note that the case g_1, i.e., the tree-level value of B_1, is discarded at the 5.1σ confidence level. As for case g_3, it is discarded at the 3.4σ level. On the other hand, the SDE result with dressed three-gluon vertex, g_2, is statistically indistinguishable from the lattice data.

Table 2. Statistical results of the χ^2 analysis for the three hypotheses given in Equation (92) for the form factor B_1. For each case (first column), we give the corresponding χ_j^2 computed from Equation (91) (second column), probability P_j computed from Equation (93) (third row), and the same P_j expressed in terms of confidence levels σ (fourth row).

Case (j)	χ_j^2	P_j	Confidence Level in σ
1	71.37	4.0×10^{-7}	5.1
2	3.399	$1 - 1.8 \times 10^{-6}$	2.2×10^{-6}
3	50.03	5.8×10^{-4}	3.4

Lastly, we point out that for both F and B_1, we find a good qualitative agreement with various related studies [21,29,178,179,224,226–228,278,279], including kinematics other than the soft gluon limit considered in Figure 14.

9. Divergent Ghost Loops and Their Impact on the QCD Green's Functions

The masslessness of the ghost propagator, discussed in Section 8, has important implications for the infrared behavior of Green's functions. Specifically, while the saturation of the gluon propagator renders gluon loops infrared finite, ghost loops furnish infrared divergent contributions [172], akin to those encountered in perturbation theory. In this section, we highlight (with two characteristic examples) how the effects of ghost loops manifest themselves at the level of the two- and three-point functions. Specifically, the ghost loops induce the appearance of a moderate maximum in the gluon propagator and are responsible for the zero-crossing and the logarithmic divergence at the origin displayed by the dominant form factors of the three-gluon vertex.

The basic observation at the level of the gluon SDE shown in Figure 1 is that, the ghost loop of (d_3), due to the masslessness of its ingredients, furnishes "unprotected" logarithms, i.e., terms of the type $\ln q^2$, which diverge as $q^2 \to 0$. Instead, gluonic loops contain infrared finite gluon propagators and, therefore, give rise to contributions that remain finite as $q^2 \to 0$, i.e., they may be described in terms of "protected" logarithms of the type $\ln(q^2 + m^2)$.

The circumstances described above may be modeled by

$$\Delta^{-1}(q) = \underbrace{q^2 + m^2 + c_1 q^2 \ln\left(\frac{q^2 + \rho m^2}{\Lambda^2}\right)}_{f(q)} + c_2 q^2 \ln\left(\frac{q^2}{\Lambda^2}\right), \qquad (94)$$

where m is the gluon mass, Λ the mass scale of QCD, and c_1, c_2, and ρ are constants; note that $\Delta^{-1}(0) = f(0) = m^2$

Differentiating Equation (94) with respect to q^2, we obtain

$$\frac{d\Delta^{-1}(q)}{dq^2} = \frac{df(q)}{dq^2} + c_2 \left[1 + \ln\left(\frac{q^2}{\Lambda^2}\right)\right]. \qquad (95)$$

The second term on the r.h.s. of Equation (95) is infrared divergent, and necessarily dominates the behavior of the derivative of the propagator for sufficiently small q. Moreover, the value of the coefficient c_2 can be computed explicitly by expanding the ghost block $\widetilde{\Pi}_{\mu\nu}^{(2)}(q)$ of Figure 1 around $q = 0$ and using Equation (19), which yields

$$c_2 = \frac{\alpha_s C_A \widetilde{Z}_1^2 F^2(0)}{48\pi}. \qquad (96)$$

Therefore, $d\Delta^{-1}(q)/dq^2$ has the asymptotic behavior

$$\lim_{q \to 0} \frac{d\Delta^{-1}(q)}{dq^2} = \left[\frac{\alpha_s C_A \widetilde{Z}_1^2 F^2(0)}{48\pi}\right] \ln\left(\frac{q^2}{\Lambda^2}\right), \qquad (97)$$

which diverges to $-\infty$ as $q \to 0$. Now, since the gluon propagator is a decreasing function in the ultraviolet, we have that $d\Delta^{-1}(q)/dq^2$ is positive for large momenta. Therefore, there must exist a special momentum, denoted by $q_\star$, such that $[d\Delta(q)/dq^2]_{q=q_\star} = 0$, which corresponds to a maximum of $\Delta(q)$ (Note that $d\Delta^{-1}(q)/dq^2$ is an increasing function since it is negative in the infrared and positive in the ultraviolet, i.e., $d^2\Delta^{-1}(q)/d(q^2)^2 > 0$. Therefore, assuming that $d\Delta^{-1}(q)/dq^2$ only crosses zero once, $q = q_\star$ must be a maximum of $\Delta(q)$).

The maximum of $\Delta(q)$, predicted by means of the simple arguments presented above, is observed in lattice simulations of the gluon propagator [49,56,85]. In particular, it is clearly visible in Figure 15, where the data from the two largest volume lattice setups of [49] are shown. The red dashed lines represent smooth functions, fitted to each of the data sets, in the window $q \in [0, 0.5]$ GeV. For each of the volumes considered, $V = 72^4$ (left panel) and $V = 80^4$ (right panel), the estimate obtained for $q_\star$ is $q_\star = 140$ MeV.

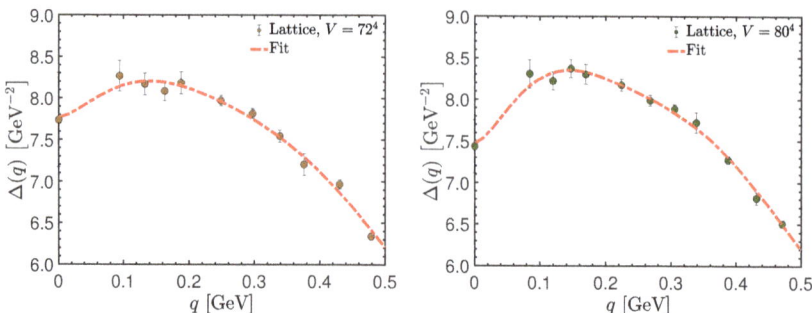

Figure 15. Lattice data for the gluon propagator in the deep infrared. The data displayed correspond to the two lattice setups with the largest volumes of [49], namely, $V = 72^4$ (**left**) and $V = 80^4$ (**right**). The red dashed lines are smooth fits from which the position of the maximum can be estimated.

It is interesting to observe in passing that the existence of a maximum of $\Delta(q)$ has an interesting implication on the form of the spectral function of the gluon propagator [280–285]. In particular, the standard Källén-Lehmann representation [286,287] states that

$$\Delta(q) = \int_0^\infty d\lambda^2 \frac{\rho(\lambda^2)}{q^2 + \lambda^2}, \tag{98}$$

where $\rho(\lambda^2)$ is the gluon spectral function (with a factor $1/\pi$ absorbed in it). Thus, the differentiation of both sides of Equation (98) with respect to q^2 yields

$$\frac{d\Delta(q)}{dq^2} = -\int_0^\infty d\lambda^2 \frac{\rho(\lambda^2)}{(q^2 + \lambda^2)^2}. \tag{99}$$

Then, from Equation (99) follows that the existence of a maximum for $\Delta(q)$ at $q = q_\star$ leads necessarily to the violation of reflection positivity [11,167,168,171], because the condition

$$\int_0^\infty d\lambda^2 \frac{\rho(\lambda^2)}{(q_\star^2 + \lambda^2)^2} = 0, \tag{100}$$

may be fulfilled only if $\rho(\lambda^2)$ reverses its sign. Note that an analogous argument based on the existence of an inflection point has been presented recently in [8].

Turning to the three-gluon vertex, it is well-known that the corresponding ghost loops induce characteristic features to the form factors associated with its classical (tree-level) tensors. There are two complementary continuum descriptions of the dynamics that determine the behavior of these form factors: (*i*) the SDE of the three-gluon vertex [174–176,226], depicted diagrammatically in Figure 16, and (*ii*) the STI of Equation (23) [172], which, in the limit of vanishing gluon momentum, and when the displacement function and the ghost sector are neglected, yields the approximate WI

$$\mathbb{\Gamma}_{\alpha\mu\nu}(0, r, -r) \approx \frac{\partial \Delta_{\mu\nu}^{-1}(r)}{\partial r^\alpha}, \tag{101}$$

which transmits the properties of the propagator derivative to the vertex form factors, as shown schematically in Figure 17.

In the simplified kinematic circumstances where only a single representative momentum is considered, to be denoted by r, the conclusions drawn by either method may be qualitatively described in terms of a simple model, namely

$$L(r) = b_0 + b_{\rm gl} \ln\left(\frac{r^2 + m^2}{\Lambda^2}\right) + b_{\rm gh} \ln\left(\frac{r^2}{\Lambda^2}\right), \tag{102}$$

where $L(r)$ denotes the particular combination of form factors, such that, at tree level, $L_0(r) = 1$, and b_0, b_{gl}, and b_{gh} are positive constants. The model in Equation (102) encompasses two important cases studied on the lattice [68,69,71,81], namely (*i*) *the soft gluon limit*, $L(r) \to L_{sg}(r)$, corresponding to the kinematic choice $q \to 0$, $p = -r$, $\theta := \widehat{pr} = \pi$, defined in Equation (77), and (*ii*) *totally symmetric limit*, $L(r) \to L_{sym}(r)$, corresponding to $q^2 = p^2 = r^2$, $\theta := \widehat{qr} = \widehat{qp} = \widehat{rp} = 2\pi/3$.

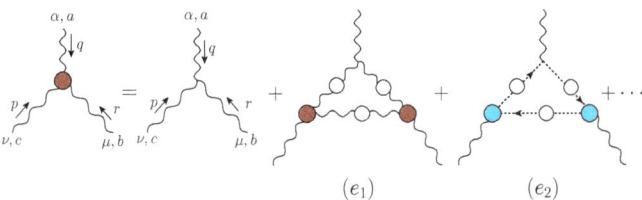

Figure 16. The SDE of the three-gluon vertex at the one-loop dressed level. Diagrams (e_1) and (e_2) are the gluon and the ghost triangle contributions entering the skeleton expansion of the three-gluon vertex.

Upon inspection of Equation (102) we note that, as $r \to 0$, the term with the unprotected logarithm will eventually dominate, forcing $L(r)$ to reverse its sign (zero crossing), and finally display a logarithmic divergence, $L(0) \to -\infty$. Given that, in practice, b_{gl} is considerably larger than b_{gh}, the unprotected logarithm overtakes the protected one rather deep in the infrared: the location of the zero-crossing is at about 160 MeV [71]. Consequently, in the intermediate region of momenta, which is considered relevant for the onset of non-perturbative dynamics, we have $L(r) < 1$; this effect is known in the literature as the infrared suppression of the three-gluon vertex.

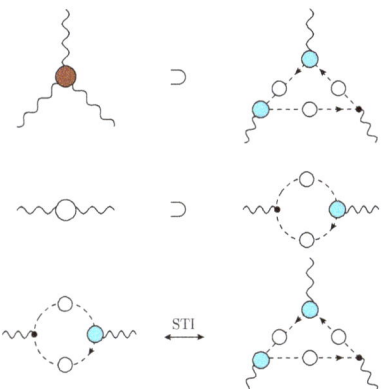

Figure 17. The ghost triangle present in the three-gluon vertex SDE (top) and the ghost loop contributing to the gluon propagator in the corresponding equation (middle). The infrared divergences arising from these diagrams are connected through the Slavnov–Taylor identity (STI) of Equation (23), as shown schematically in the bottom panel.

Most importantly, the special features of infrared suppression, zero-crossing, and logarithmic divergence at the origin have been corroborated through a variety of lattice results [50,68,69,71,72,81,84], as shown, e.g., in Figure 11. The central curve of this figure is presented as the blue line in Figure 18, where the aforementioned characteristics have been explicitly marked for the benefit of the reader. Note the close proximity of the blue curve to the $d\Delta^{-1}(r)/dr^2$ (red dashed line), especially below 1 GeV.

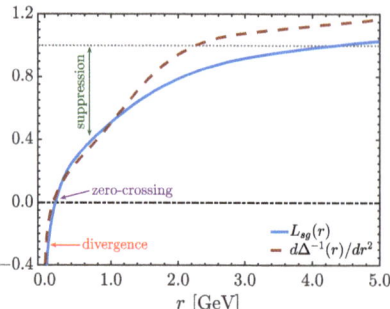

Figure 18. Comparison of $L_{sg}(r)$ (blue continuous) from Figure 11 and $d\Delta^{-1}(r)/dr^2$ (red dashed) resulting from the fit for $\Delta(r)$ of Figure 9. Note that both display the characteristic features of infrared suppression with respect to their tree-level values (which is 1 for both quantities), zero-crossing, and logarithmic divergence at the origin.

We end this section by pointing out that, in the case of Yang–Mills in $d = 3$ [28, 172,223,288–302], the situation is qualitatively similar to the one described above, but the divergences induced due to the masslessness of the ghost are stronger. Specifically, as may be already established at the level of a simple one-loop calculation [302], the first derivative of the gluon propagator diverges at the origin as $1/q$ rather than $\ln q^2$. Consequently, the corresponding effects are significantly enhanced; in particular, the maximum of the gluon propagator is considerably more pronounced, becoming plainly visible on the lattice [53]. Similarly, an abrupt negative divergence is observed in the corresponding vertex form factors [41,82].

10. Ward Identity Displacement of the Three-Gluon Vertex

In complete analogy to the case of the ghost–gluon vertex discussed in Section 5.2, the WI satisfied by the pole-free part of the three-gluon vertex is also displaced in the presence of longitudinally coupled massless poles. Quite importantly, the associated displacement function, $\mathbb{C}(r)$, *coincides* with the BS amplitude that controls the formation of a (colored) scalar bound state with vanishing mass out of a gluon pair. The displacement of the WI circumvents the seagull cancellation involving the gluon propagator [i.e., $f = \Delta$ in Equation (55)], furnishing to the $g_{\mu\nu}$ component the mass originating from graphs (d_1) and (d_4) in Figure 1. In addition, it permits the indirect determination of the displacement function $\mathbb{C}(r)$ from the lattice; this is particularly important, given that, by virtue of Equation (35), the lattice "observables" do not perceive directly the presence of the massless poles.

The starting point of the analysis is the STI satisfied by the three-gluon vertex, $\mathbb{\Gamma}_{\alpha\mu\nu}(q,r,p)$, given by Equation (23). In order to eliminate the poles in r and p, thus isolating the displacement of the WI originating from the pole in the channel q, we contract that equation with $P^\mu_{\mu'}(r) P^\nu_{\nu'}(p)$. Note that this procedure also eliminates any longitudinal pole terms in the ghost kernels $H_{\sigma\mu}(p,q,r)$ and $H_{\sigma\nu}(r,q,p)$; for the diagrammatic definition of the ghost–gluon kernel, see Figure 2.

Then, we decompose $\mathbb{\Gamma}_{\alpha\mu\nu}(q,r,p)$ into pole-free and longitudinally coupled massless pole parts, as in Equation (33), and use Equation (38), to obtain

$$P^\mu_{\mu'}(r) P^\nu_{\nu'}(p) \left[q^\alpha \Gamma_{\alpha\mu\nu}(q,r,p) + g_{\mu\nu} C_1(q,r,p) + q_\mu q_\nu C_5(q,r,p) \right] = P^\mu_{\mu'}(r) P^\nu_{\nu'}(p) R_{\nu\mu}(p,q,r), \quad (103)$$

where

$$R_{\nu\mu}(p,q,r) := F(q) \left[\Delta^{-1}(p) P^\sigma_\nu(p) H_{\sigma\mu}(p,q,r) - \Delta^{-1}(r) P^\sigma_\mu(r) H_{\sigma\nu}(r,q,p) \right]. \quad (104)$$

At this point, we expand Equation (103) around $q = 0$ and match coefficients of equal orders. At the zeroth order in this expansion, we immediately obtain that

$$C_1(0, r, -r) = 0, \tag{105}$$

in exact analogy to Equation (62). Note that we have arrived once again at the result of Equation (39), but through an entirely different path: while Equation (39) is enforced by the Bose symmetry of the three-gluon vertex, Equation (105) is a direct consequence of the STI that this vertex satisfies.

We next gather the terms in the expansion of Equation (103) that are of first order in q. Evidently, the term C_5 does not contribute to this order. Then, the expansion leads to the appearance of derivatives of the gluon propagator, in analogy to Equation (64), but also of the ghost–gluon kernel. Specifically, we obtain for the WI of the three-gluon vertex and its displacement the expression

$$L_{sg}(r) = F(0)\left\{\widetilde{Z}_1 \frac{d\Delta^{-1}(r)}{dr^2} + \frac{\mathcal{W}(r)}{r^2}\Delta^{-1}(r)\right\} - \mathbb{C}(r). \tag{106}$$

In the above equation, $L_{sg}(r)$ is the form factor of the three-gluon vertex defined in Equation (77) and with lattice results shown in Figure 11, while $\mathcal{W}(r)$ is a particular derivative of the ghost–gluon kernel, namely [124,241]

$$\mathcal{W}(r) = -\frac{1}{3r^2}P^{\mu\nu}(r)\left[\frac{\partial H_{\nu\mu}(p,q,r)}{\partial q^\alpha}\right]_{q=0}. \tag{107}$$

For the detailed derivation of Equation (106), we refer to [93,124].

In the following section, we will use Equation (106) to determine the displacement amplitude $\mathbb{C}(r)$ from lattice inputs. To this end, we must first pass to Euclidean space, where we note that

$$\mathbb{C}_E(r_E^2) = -\mathbb{C}(r)|_{r^2=-r_E^2}, \tag{108}$$

with the extra sign originating from the fact that $\mathbb{C}$ is a derivative [see Equation (41)]. Then, suppressing the indices "E" and solving for $\mathbb{C}(r^2)$, we obtain the central relation

$$\mathbb{C}(r) = L_{sg}(r) - F(0)\left\{\frac{\mathcal{W}(r)}{r^2}\Delta^{-1}(r) + \widetilde{Z}_1\frac{d\Delta^{-1}(r)}{dr^2}\right\}. \tag{109}$$

For the determination of $\mathbb{C}(r)$, we use lattice inputs for all the quantities that appear on the r.h.s. of Equation (109), with the exception of the function $\mathcal{W}(r)$, which will be computed from the SDE satisfied by the ghost–gluon kernel derived and analyzed in the next section.

11. The Ghost-Gluon Kernel Contribution to the Ward Identity

In this section, we derive the SDE that determines the key function $\mathcal{W}(r)$; the resulting SDE will be solved using lattice inputs for the various quantities entering it. In addition, the infrared behavior of $\mathcal{W}(r)$ will be analyzed in detail, following an analytic procedure.

Our discussion starts with the SDE of the ghost–gluon kernel, $H_{\mu\nu}(r,q,p)$, shown diagrammatically in Figure 19, from which $\mathcal{W}(r)$ can be obtained using Equation (107).

Note that the similarity between the diagrams shown in Figure 19 and those in the bottom panel of Figure 12, depicting the SDE of the ghost–gluon vertex, is a simple reflection of the fundamental STI relating the ghost–gluon kernel with the ghost–gluon vertex,

$$\Gamma_\nu(r,q,p) = r^\mu H_{\mu\nu}(r,q,p). \tag{110}$$

Specifically, Equation (110) is preserved by the SDEs of $\Gamma_\nu(r,q,p)$ and $H_{\mu\nu}(r,q,p)$; indeed, contraction of each diagram ($h_i^{\mu\nu}$) of Figure 19 by r^μ yields the corresponding diagram

(g_i^ν) of Figure 12 (up to a shift of $k \to -k - r$ for $i = 1$, introduced to simplify certain expressions). Note that, in Figure 19, the diagram corresponding to the (g_3) of Figure 12 has been omitted, for the reason explained in item (*i*) of Section 8.

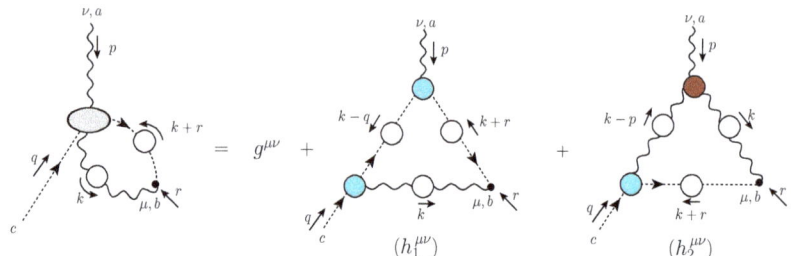

Figure 19. SDE for the ghost–gluon scattering kernel, $H_{\mu\nu}(r,q,p)$. We omit a diagram containing a 1PI four-point function.

It is well-known that, in the Landau gauge, the momentum q of the ghost field in $H_{\mu\nu}(r,q,p)$ factors out of its quantum corrections [1], allowing us to write [124,228,241]

$$H_{\mu\nu}(r,q,p) = g_{\mu\nu} + q^\alpha K_{\mu\nu\alpha}(r,q,p),\qquad(111)$$

where no particular assumptions are made about the structure of the function $K_{\mu\nu\alpha}(r,q,p)$. Following Equation (107), we differentiate the r.h.s. of Equation (111) with respect to q and subsequently set $q = 0$, to obtain

$$\mathcal{W}(r) = -\frac{1}{3} r^\alpha P^{\mu\nu}(r) K_{\mu\nu\alpha}(r,0,-r).\qquad(112)$$

Lastly, the *finite* renormalization of $\mathcal{W}$ in the asymmetric MOM scheme proceeds through the use of Equation (15), which leads to the appearance of an overall factor of $\widetilde{Z}_1$ in the equations.

The outcome of the above steps is that $\mathcal{W}(r)$ can be written as

$$\mathcal{W}(r) = \mathcal{W}_1(r) + \mathcal{W}_2(r),\qquad(113)$$

where the $\mathcal{W}_i(r)$ are the contributions originating from the diagrams ($h_i^{\mu\nu}$) in Figure 19, respectively, and read

$$\mathcal{W}_1(r) = \frac{\lambda \widetilde{Z}_1}{3} \int_k \Delta(k) D(k) D(k+r)(r \cdot k) f(k,r) B_1(k+r,-k,-r) B_1(k,0,-k),$$

$$\mathcal{W}_2(r) = \frac{\lambda \widetilde{Z}_1}{3} \int_k \Delta(k) \Delta(k+r) D(k+r) B_1(k+r,0,-k-r) \mathcal{I}_\mathcal{W}(-r,-k,k+r),\,(114)$$

where $f(k,r)$ is given by Equation (81), and we define the specific contribution of the three-gluon vertex to the kernel of $\mathcal{W}(r^2)$ as

$$\mathcal{I}_\mathcal{W}(q,r,p) := \frac{1}{2}(q-r)^\nu \overline{\Gamma}^\alpha_{\alpha\nu}(q,r,p).\qquad(115)$$

Note that, from Equation (115) and the Bose symmetry of the $\overline{\Gamma}_{\alpha\mu\nu}(q,r,p)$ under the exchange $\{q,\alpha\} \leftrightarrow \{r,\mu\}$, it follows that

$$\mathcal{I}_\mathcal{W}(q,r,p) = \mathcal{I}_\mathcal{W}(r,q,p).\qquad(116)$$

At this point, by capitalizing on the planar degeneracy of $\overline{\Gamma}_{\alpha\mu\nu}(q,r,p)$ discussed in Section 7, we obtain a compact, and yet accurate, approximation for $\mathcal{I}_\mathcal{W}$. Specifically, using Equation (78), we find

$$\mathcal{I}_\mathcal{W}(q,r,p) \approx \mathcal{I}_\mathcal{W}^0(q,r,p) L_{sg}(s), \tag{117}$$

where $\mathcal{I}_\mathcal{W}^0(q,r,p)$ is the tree-level value of $\mathcal{I}_\mathcal{W}$, given by

$$\mathcal{I}_\mathcal{W}^0(q,r,p) := \frac{2f(q,r)}{p^2}\left[2q^2 r^2 - (q^2 + r^2)(q\cdot r) - (q\cdot r)^2\right]. \tag{118}$$

We remark that the approximation given by Equation (117) becomes exact in the limit $p = 0$. Using the above approximation for $\mathcal{I}_\mathcal{W}$, the contribution $\mathcal{W}_2(r)$ reads

$$\mathcal{W}_2(r) = \frac{2\lambda\widetilde{Z}_1}{3}\int_k \Delta(k)\frac{\Delta(k+r)D(k+r)}{(k+r)^2}B_1(k+r,0,-k-r)f(k,r)$$
$$\times \left[2r^2 k^2 - (r^2 + k^2)(r\cdot k) - (r\cdot k)^2\right]L_{sg}(\hat{s}), \tag{119}$$

where we now have $\hat{s}^2 = r^2 + k^2 + (r\cdot k)$.

Lastly, we transform $\mathcal{W}_1$ of Equation (114) and $\mathcal{W}_2$ of Equation (119) to Euclidean space to obtain the final expression to be used for the numerical determination of $\mathcal{W}$,

$$\mathcal{W}_1(r) = -\frac{r\alpha_s C_A \widetilde{Z}_1}{12\pi^2}\int_0^\infty dk^2 k\Delta(k)F(k)B_1(k^2,k^2,\pi)\int_0^\pi d\phi\, s_\phi^4 c_\phi \frac{F(\sqrt{z})}{z}B_1(z,r^2,\chi),$$

$$\mathcal{W}_2(r) = -\frac{r\alpha_s C_A \widetilde{Z}_1}{6\pi^2}\int_0^\infty dk^2\, k^3 \Delta(k)\int_0^\pi d\phi\, s_\phi^4 \Delta(\sqrt{z})B_1(z,z,\pi)\frac{F(\sqrt{z})}{z^2}\left[kr(2+c_\phi^2) - zc_\phi\right]$$
$$\times L_{sg}\left(r^2 + k^2 + rkc_\phi\right), \tag{120}$$

where z has been defined below Equation (89) and

$$\chi := \cos^{-1}\left[-\frac{(r+kc_\phi)}{\sqrt{z}}\right]. \tag{121}$$

We emphasize that for the SDEs of both B_1 and $\mathcal{W}$, given by Equations (88) and (120), respectively, we used the same approximation for the three-gluon vertex, namely Equation (78). Therefore, our analyses of B_1 and $\mathcal{W}$ are self-consistent, in the sense that the STI in Equation (110) is strictly preserved.

Before embarking on the numerical determination of $\mathcal{W}(r)$ for the entire range of Euclidean momenta, we discuss the infrared behavior of this function and demonstrate an important self-consistency proof involving $\mathbb{C}(r)$.

Specifically, as discussed in Section 9, the $L_{sg}(r)$ and $d\Delta^{-1}(r)/dr^2$ that appear in Equation (109) are infrared divergent, due to the massless ghost loops present in their SDEs. Nevertheless, the BSE solutions for the amplitude $\mathbb{C}(r)$ are all found to be finite at $r = 0$, (cf. Figure 7) [117,121,124,215]. Therefore, in order for the WI displacement of Equation (109) to be consistent with the finite $\mathbb{C}(0)$ obtained from BSE solutions, the infrared divergences of the ingredients appearing in Equation (109) must cancel against each other.

Indeed, a careful analysis of diagram (e_2) of Figure 16 yields

$$\lim_{r\to 0} L_{sg}(r) = \left[\frac{\alpha_s C_A \widetilde{Z}_1^3 F^3(0)}{96\pi}\right]\ln\left(\frac{r^2}{\mu^2}\right), \tag{122}$$

up to infrared finite terms (We note that the results identical to Equations (97) and (122) for the infrared divergences of $d\Delta^{-1}(r)/dr^2$ and $L_{sg}(r)$, respectively, have been previously derived within the Curci–Ferrari model [180]). Then, assuming that only $L_{sg}(r)$ and $d\Delta^{-1}/dr^2$

diverge, and using the asymptotic form of $d\Delta(r)/dr^2$ given in Equation (97) to Equation (109), we find that the divergences do not fully cancel. Therefore, the finiteness of $\mathbb{C}(0)$ demands that the term $\mathcal{W}(r)/r^2$ appearing in the WI must be infrared divergent.

Now, it is evident from Equation (120) that $\mathcal{W}(r)$ vanishes as $r \to 0$. Nevertheless, the ratio $\mathcal{W}(r)/r^2$ is found to diverge at the origin. Specifically, expanding Equation (120) around $r = 0$, it can be shown that $\mathcal{W}(r)/r^2$ has the asymptotic behavior

$$\lim_{r \to 0} \frac{\mathcal{W}(r)}{r^2} = -\left[\frac{\alpha_s C_A \widetilde{Z}_1^3 \Delta(0) F^2(0)}{96\pi}\right] \ln\left(\frac{r^2}{\mu^2}\right). \qquad (123)$$

Then, combining Equations (97), (122) and (123) we find that the infrared divergences in Equation (109) cancel out exactly, leaving a finite $\mathbb{C}(0)$, in full agreement with the BSE results.

We finish the discussion of the infrared finiteness of $\mathbb{C}(0)$ with a remark. In the absence of the Schwinger mechanism, i.e., for an identically zero $\mathbb{C}(r)$, the infrared divergences of $L_{sg}(r)$, $\mathcal{W}(r)/r^2$, and $d\Delta^{-1}(r)/dr^2$ must also cancel in Equation (109). For instance, this cancellation can be explicitly verified at the one-loop level (in the perturbative realization of Equation (109) $F(0)$ also diverges, participating in the overall cancellation of infrared divergences), where, evidently, $\mathbb{C}(r) = 0$. In that case, however, the gluon propagator is also massless, causing the gluonic loops that contribute to the functions that enter Equation (109) to also diverge, such that the cancellation occurs among *all* radiative diagrams. In contrast, in the presence of a gluon mass, the cancellation of the remaining infrared divergences takes place at the level of the ghost loops only, as illustrated diagrammatically in Figure 20.

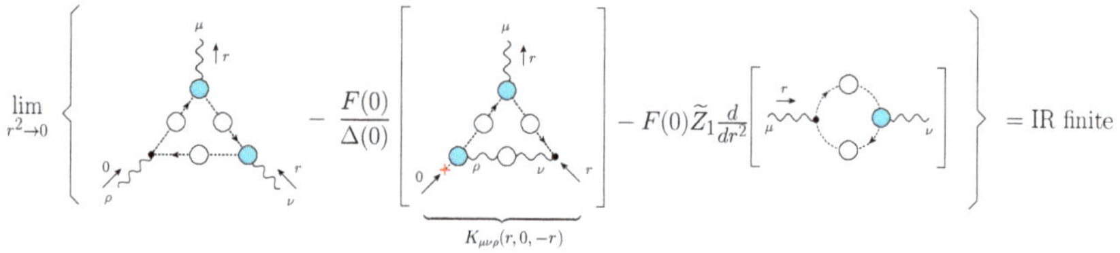

Figure 20. Diagrammatic representation of the cancellation of the infrared divergences originating from massless ghost loops in Equation (109) to yield a finite $\mathbb{C}(0)$. The red cross indicates that the overall ghost momentum is factored out before being set to zero.

We now return to the numerical determination of $\mathcal{W}(r)$ from Equation (120). To this end, we employ the fits to the lattice data of [84] for $\Delta(q)$ and $L_{sg}(q)$, shown in Figures 9 and 11, respectively, and the SDE solution for $F(q)$ is shown in the left panel of Figure 13. All of the fits employed are constructed so as to reproduce the correct ultraviolet behavior of Green's functions. For the value of the coupling in the asymmetric MOM scheme, defined in Appendix B, we employ $g^2 = 4\pi\alpha_s$, with $\alpha_s(4.3 \text{ GeV}) = 0.27$, as determined in the lattice study of [71]. Lastly, for B_1 we use the SDE result of Section 8, shown in the right panel of Figure 13, which reproduces accurately the available lattice data for the ghost–gluon vertex.

Using the above ingredients in Equation (120) we obtain the $\mathcal{W}(r)$ shown as the blue solid curve in Figure 21. The blue band in Figure 21 represents the error estimate on our results; the procedures followed to obtain it are described in detail in [126].

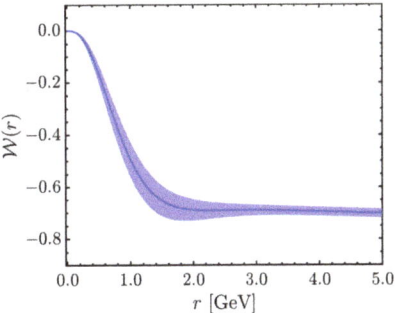

Figure 21. $\mathcal{W}(r)$ obtained using the approximation Equation (117) based on the observed planar degeneracy of the three-gluon vertex in its kernel (blue solid curve) together with uncertainty estimate (blue band).

12. Displacement Function from Lattice Inputs

In this section, we determine $\mathbb{C}(r)$ from the main relation given in Equation (109).

For $\mathcal{W}(r)$ we use the result shown in Figure 21, together with the curves for $L_{sg}(r)$ from Figure 11, $\Delta(r)$ and $d\Delta^{-1}(r)/dr^2$ from Figures 9 and 18, respectively, and the $F(r)$ of Figure 13. The $\mathbb{C}(r)$ obtained is shown as a black solid curve in the left panel of Figure 22. In the same panel, we show as points the estimates of $\mathbb{C}(r)$ obtained by using into Equation (109) the lattice data points of [84] *directly*, rather than a fit. Note that these data points, as well as those used for the propagators [85], have been carefully extrapolated to the continuum, through the methods explained in [73,85]. These methods exploit the $H4$ symmetry of the hypercubic lattice, and are quite effective at minimizing discretization artifacts [54,60,62,63,70,71,73,76,80,85]. As a result of this treatment, the systematic errors are expected to be small. To estimate the uncertainty in the resulting $\mathbb{C}(r)$, we combine the error estimate of $\mathcal{W}(r)$, represented by the blue band in Figure 21, with the statistical error of the lattice data points for $L_{sg}(r)$ of [84], and neglect the error in the gluon propagator, which is much smaller than the errors in L_{sg} and $\mathcal{W}$. Then, a conservative error propagation analysis was carried out in [126], which takes into account an observed correlation between the errors in $\mathcal{W}(r)$ and $L_{sg}(r)$; the results of the analysis are the error bars shown in Figure 22.

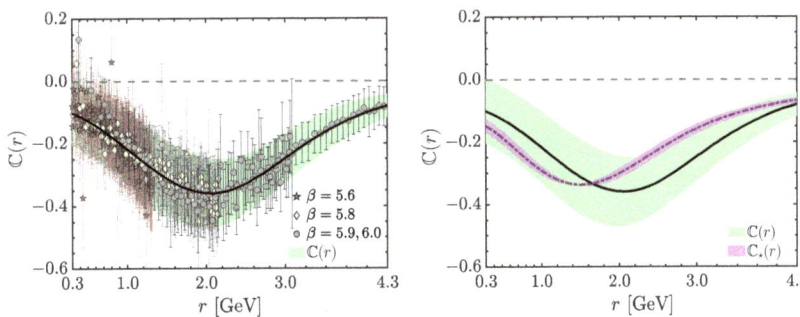

Figure 22. Left: Result for $\mathbb{C}(r)$ (black continuous line) obtained from Equation (109) using the $\mathcal{W}(r)$ shown in Figure 21, the fits to lattice data for $\Delta(r)$ and $L_{sg}(r)$ are shown in Figures 9 and 18, respectively, and the SDE solution for $F(r)$ shown in Figure 13. The points are obtained using for $L_{sg}(r)$ the data in Reference [84], with error bars denoting the error propagated from L_{sg} and $\mathcal{W}$. The green band is obtained by fitting the upper and lower bounds of the points and guiding the eye to the typical error associated with $\mathbb{C}(r)$. **Right**: The $\mathbb{C}(r)$ of the left panel is compared to the BSE prediction $\mathbb{C}_\star(r)$ (purple dot-dashed and error band) of Figure 7.

At this point, we quantify the significance of the $\mathbb{C}(r)$ obtained above, in comparison to the null hypothesis result; evidently, in the absence of the Schwinger mechanism, this latter quantity, to be denoted by $\mathbb{C}_0$ in what follows, vanishes identically, namely $\mathbb{C}_0 = 0$. To this end, we first compute the χ^2 of our points as

$$\chi^2 = \sum_i \frac{[\mathbb{C}(r_i) - \mathbb{C}_0(r_i)]^2}{\epsilon^2_{\mathbb{C}(r_i)}}, \qquad (124)$$

i.e., the null hypothesis is taken as the estimator for our data. The sum runs over the $n_r = 515$ indices i such that $r_i \in [0.3, 4.3]$ GeV, the interval of momenta for which the systematic error in our calculation of $\mathcal{W}(r)$ is best known, and $\epsilon_{\mathbb{C}(r_i)}$ denotes the error estimate of $\mathbb{C}(r_i)$. Then we obtain $\chi^2 = 2630$, corresponding to $\chi^2_{\text{d.o.f.}} = 5.11$. The probability $P_{\mathbb{C}_0}$ that our result for $\mathbb{C}$ is consistent with the null hypothesis s vanishingly small, given by the formula

$$P_{\mathbb{C}_0} = \int_{\chi^2=2630}^{\infty} \chi^2_{\text{PDF}}(515, x)\,dx = \left.\frac{\Gamma(n_r/2, \chi^2/2)}{\Gamma(n_r/2)}\right|_{n_r=515}^{\chi^2=2630} = 7.3 \times 10^{-280}. \qquad (125)$$

Naturally, further correlations in the input data, as well as residual systematic errors, may have escaped the analysis leading to the error estimates shown in Figure 7 for $\mathbb{C}(r)$. Since $P_{\mathbb{C}_0}$ changes rapidly with χ^2, these unknown errors can substantially alter its value. As such, Equation (125) is to be understood as meaning that in the absence of additional uncertainties, the null hypothesis $\mathbb{C}_0$ is excluded. Moreover, it is apparent in Figure 22 that even if the errors had been significantly underestimated, the null hypothesis $\mathbb{C}_0$ would still be unlikely. In fact, even if the errors in *all data points* for $\mathbb{C}(r)$ were 95% larger, i.e., nearly doubled, we could still discard $\mathbb{C}_0$ at the 5σ confidence level.

In the right panel of Figure 22, we compare $\mathbb{C}(r)$ to the BSE prediction, $\mathbb{C}_\star(r)$, of Figure 7, shown as a purple dot-dashed curve and corresponding error band. In that panel, we observe an excellent qualitative agreement between the two results. The most noticeable quantitative difference is in the position of the minimum. Specifically, $\mathbb{C}$ reaches the minimum value of -0.36 ± 0.11 at $r = 1.93^{+0.09}_{-0.06}$ GeV, while the minimum of $\mathbb{C}_\star$ is -0.341 ± 0.003 at $r = 1.5 \pm 0.1$.

Nevertheless, it is clear in the right panel of Figure 22 that the BSE prediction lies within the error estimate of the lattice-derived $\mathbb{C}(r)$. In fact, defining a χ^2 measure for the discrepancy between $\mathbb{C}$ and $\mathbb{C}_\star$ as

$$\chi^2_\star = \sum_i \frac{[\mathbb{C}(r_i) - \mathbb{C}_\star(r_i)]^2}{\epsilon^2_{\mathbb{C}(r_i)}}, \qquad (126)$$

we obtain $\chi^2_\star = 258.5$, which is smaller than the number of degrees of freedom. Then, this value of $\chi^2_\star$ amounts to a probability of

$$P_{\mathbb{C}_\star} = \left.\frac{\Gamma(n_r/2, \chi^2_\star/2)}{\Gamma(n_r/2)}\right|_{n_r=515}^{\chi^2_\star=258.5} = 1 - 2.0 \times 10^{-23}, \qquad (127)$$

showing that $\mathbb{C}_\star$ is statistically compatible with the lattice-derived $\mathbb{C}$, with probability extremely near the unit.

13. Conclusions

The gauge sector of QCD is host to a wide array of subtle mechanisms that are of vital importance for the self-consistency and infrared stability of the theory. In the present work, we offered a comprehensive review of the intricate dynamics that account for some of the most prominent infrared phenomena, such as the generation of a gluon mass through the action of the Schwinger mechanism, the non-perturbative masslessness of the ghost, and the characteristic features induced by this particular mass pattern to the form factors of the three-gluon vertex.

The SDEs, supplemented by the judicious use of certain key results from lattice QCD, provide a robust continuum framework for carrying out such a demanding investigation. In fact, the results obtained from the SDEs are increasingly reliable, passing successfully all sorts of tests imposed on them. A particularly impressive, and certainly not isolated, case of such a success has been outlined in detail in Section 6.

Symmetry and dynamics are tightly interwoven; therefore, the information encoded in the STIs and WIs of the theory is particularly decisive for unraveling basic dynamical patterns. A striking manifestation of the profound connection between symmetry and dynamics is provided by the dual role played by the function $\mathbb{C}(r)$, i.e., the BS amplitudes of the massless states composed by a pair of gluons, and the quantity that embodies the displacement induced to the WIs by the presence of these states.

In our opinion, the determination of $\mathbb{C}(r)$ described in Section 12 represents a major success of the entire set of concepts and techniques surrounding the generation of a gluon mass through the action of the Schwinger mechanism. Thus, fifty years after the genesis of QCD, we seem to be closing in on the mechanism that the theory uses for curing the infrared instabilities endemic to perturbation theory. We hope to be able to report further progress in this direction in the near future.

Funding: The authors are supported by the Spanish MICINN grant PID2020-113334GB-I00. M.N.F. acknowledges financial support from Generalitat Valenciana through contract CIAPOS/2021/74. J.P. also acknowledges funding from the regional Prometeo/2019/087 from Generalitat Valenciana.

Data Availability Statement: Not applicable.

Acknowledgments: The authors thank A.C. Aguilar, D. Binosi, D. Ibánez, J. Pawlowski, C.D. Roberts, and J. Rodríguez-Quintero for the collaborations.

Conflicts of Interest: The authors declare no conflicts of interest.

Abbreviations

The following abbreviations are used in this work:

BFM	background field method
BQI	background-quantum identity
BRST	Becchi–Rouet–Stora–Tyutin
BS	Bethe–Salpeter
BSE	Bethe–Salpeter equation
EHM	emergent hadron mass
MOM	momentum subtraction (renormalization scheme)
PT	pinch technique
QCD	quantum chromodynamics
QED	quantum electrodynamics
RGI	renormalization group invariant
SDE	Schwinger–Dyson equation
STI	Slavnov–Taylor identity
WI	Ward identity

Appendix A. BQIs for the BSE Amplitudes

In this appendix, we use two BQIs in order to relate the displacement functions ($\mathcal{C}$ and $\mathbb{C}$) with their BFM counterparts, i.e., $\widetilde{\mathcal{C}}$ and $\widetilde{\mathbb{C}}$, respectively.

The ghost–gluon vertices $\mathbb{\Gamma}_\mu(r,p,q)$ and $\widetilde{\mathbb{\Gamma}}_\mu(r,p,q)$ are related via a BQI [14], which reads

$$\widetilde{\mathbb{\Gamma}}_\mu(r,p,q) = \left\{[1+G(q)]g_\mu^\nu + L(q)\frac{q_\mu q^\nu}{q^2}\right\}\mathbb{\Gamma}_\nu(r,p,q)$$
$$+ F^{-1}(p)p^\nu K_{\mu\nu}(r,q,p) + r^2 F^{-1}(r) K_\mu(r,q,p), \qquad (A1)$$

where K_μ and $K_{\mu\nu}$ are two auxiliary functions, shown diagrammatically in Figure A1, while $G(q)$ and $L(q)$ are the form factors of $\Lambda_{\mu\nu}(q)$, defined in Equation (12).

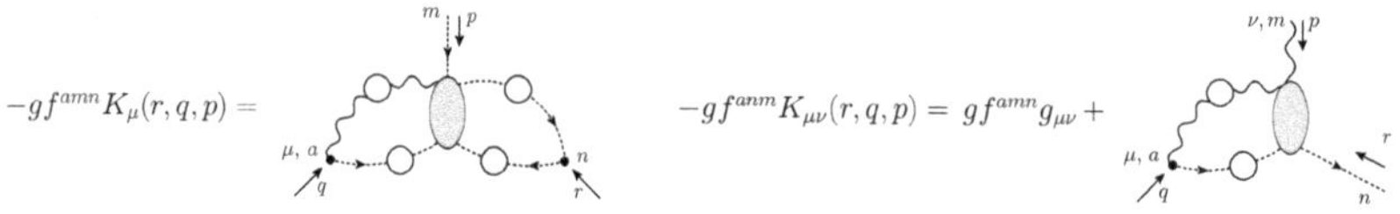

Figure A1. The auxiliary functions $K_\mu(q,r,p)$ and $K_{\mu\nu}(q,r,p)$ in the BQI of Equation (A1).

Next, we decompose the $\widetilde{\mathbb{\Gamma}}_\mu(r,p,q)$ and $\mathbb{\Gamma}_\mu(r,p,q)$ in Equation (A1) into their regular and pole parts, using Equations (33) and (53), respectively. Note that the second and third terms in Equation (A1) do not contain poles in q^2; this is so because $K_{\mu\nu}(r,q,p)$ can contain (longitudinally coupled) poles only in the p_ν channel, whereas $K_\mu(r,q,p)$ has no external gluon legs (and, hence, no poles).

Then, multiplying Equation (A1) by q^2 we obtain

$$q_\mu \widetilde{C}(r,p,q) = q_\mu[1+G(q)+L(q)]C(r,p,q) + \mathcal{O}(q^2). \qquad (A2)$$

Setting $q=0$ in Equation (A2) and using Equation (18), we find

$$C(r,-r,0) = Z_1 F(0)\widetilde{C}(r,-r,0). \qquad (A3)$$

Hence, using Equation (62), we obtain the result in Equation (39).

Then, expanding Equation (A2) to first order in q, using Equation (41) for $C(r,p,q)$ and Equation (63) for $\widetilde{C}(r,p,q)$, entails

$$\mathcal{C}(r) = Z_1 F(0)\widetilde{\mathcal{C}}(r), \qquad (A4)$$

which is one of the main results of this appendix.

A relation identical to Equation (A4) can be obtained for $\mathbb{C}(r)$ and its BFM analog, $\widetilde{\mathbb{C}}(r)$. The starting point of the derivation is the BQI [14]

$$\widetilde{\mathbb{\Gamma}}_{\alpha\mu\nu}(q,r,p) = \left\{[1+G(q)]g_\alpha^\rho + L(q)\frac{q_\alpha q^\rho}{q^2}\right\}\mathbb{\Gamma}_{\rho\mu\nu}(q,r,p) \qquad (A5)$$
$$+ K_{\rho\nu\alpha}(r,q,p)P_\mu^\rho(r)\Delta^{-1}(r) - K_{\rho\mu\alpha}(p,q,r)P_\nu^\rho(p)\Delta^{-1}(p),$$

where $K_{\mu\nu\alpha}(r,q,p)$ is the function defined in Equation (111).

Then, we note that the only longitudinal poles at $q=0$ present in Equation (A5) are those contained in the $\mathbb{\Gamma}_{\alpha\mu\nu}(q,r,p)$ and $\widetilde{\mathbb{\Gamma}}_{\alpha\mu\nu}(q,r,p)$ vertices, with the auxiliary functions $K_{\alpha\nu\rho}(q,p,r)$ having poles only in the r_μ and p_ν channels. As such, isolating the $q_\alpha g_{\mu\nu}/q^2$ pole and expanding around $q=0$, one eventually finds

$$\widetilde{C}_1(0,r,-r) = Z_1^{-1} F^{-1}(0) C_1(0,r,-r) = 0, \qquad (A6)$$

and

$$\mathbb{C}(r) = Z_1 F(0) \widetilde{\mathbb{C}}(r), \qquad (A7)$$

where $\widetilde{C}_1(q,r,p)$ and $\widetilde{\mathbb{C}}(r^2)$ are defined in analogy to Equations (37) and (41), and we used Equation (39).

Appendix B. The Asymmetric MOM Scheme

In this appendix, we provide a brief overview of the asymmetric MOM scheme [68,71,84,85,241] that we employ throughout this work.

The set of boundary conditions imposed on the renormalized quantities defines the renormalization scheme employed. Within the MOM schemes [277], propagators assume their tree-level values at the subtraction point μ, namely

$$\Delta_R^{-1}(\mu) = \mu^2, \qquad F_R(\mu) = 1. \qquad (A8)$$

Past this point, the various MOM schemes are differentiated according to the way the three-point functions are renormalized.

In Landau gauge, a common choice of renormalization prescription is the so-called "Taylor scheme" [54,60,80,85,207]. This scheme capitalizes on the Taylor theorem [207], i.e., the observation that the unrenormalized ghost–gluon vertex in the Landau gauge reduces to its tree-level form in the soft-ghost configuration,

$$\Gamma_\nu(r, 0, -r) = r_\nu. \qquad (A9)$$

The Taylor scheme is defined by requiring Equation (A9) to hold after renormalization [54,60,80,85,207]. Using Equation (15), this requirement yields $Z_1 = 1$.

Alternatively, in lattice simulations of the three-gluon vertex, it is convenient to impose a renormalization prescription for its classical tensor structure. For example, one may choose the classical form factor to reduce to tree-level in the symmetric point, $q^2 = r^2 = p^2 = \mu^2$. This condition defines the "symmetric scheme" [68,71,84].

In the present work, the classical form factor of the three-gluon vertex in the soft-gluon limit, which is denoted by $L_{sg}(r)$ and defined in Equation (77), plays a key role. Indeed, it is the central ingredient in the approximation of the three-gluon vertex given by Equation (78), which is used in the SDE analysis of the ghost–gluon vertex and kernel in Sections 8 and 11, respectively. Moreover, $L_{sg}(r)$ is one of the inputs necessary for the determination of the displacement amplitude $\mathbb{C}(r)$ in Section 12, which signals the activation of the Schwinger mechanism. As such, it is convenient to employ throughout the scheme where $L_{sg}(r)$ is most readily renormalized in lattice simulations, which is the so-called "asymmetric scheme" [68,71,84,85,241].

The asymmetric MOM scheme is defined by imposing that $L_{sg}(r)$ reduces to the tree-level at $q^2 = \mu^2$, i.e.,

$$L_{sg}(\mu) = 1. \qquad (A10)$$

Note that in this scheme the finite renormalization constant of the ghost–gluon vertex is no longer equal to 1 [85,241]. Instead, the special value of Z_1 in the asymmetric scheme is denoted by $\widetilde{Z}_1$, and is determined to be $\widetilde{Z}_1 = 0.9333 \pm 0.0075$ [126], at $\mu = 4.3$ GeV, through the SDE analysis discussed in Section 8.

References

1. Marciano, W.J.; Pagels, H. Quantum Chromodynamics: A Review. *Phys. Rep.* **1978**, *36*, 137. [CrossRef]
2. Qin, S.X.; Roberts, C.D. Impressions of the Continuum Bound State Problem in QCD. *Chin. Phys. Lett.* **2020**, *37*, 121201. [CrossRef]
3. Roberts, C.D. Empirical Consequences of Emergent Mass. *Symmetry* **2020**, *12*, 1468. [CrossRef]
4. Cui, Z.F.; Ding, M.; Gao, F.; Raya, K.; Binosi, D.; Chang, L.; Roberts, C.D.; Rodríguez-Quintero, J.; Schmidt, S.M. Kaon and pion parton distributions. *Eur. Phys. J. C* **2020**, *80*, 1064. [CrossRef]
5. Chang, L.; Roberts, C.D. Regarding the Distribution of Glue in the Pion. *Chin. Phys. Lett.* **2021**, *38*, 081101. [CrossRef]
6. Cui, Z.F.; Ding, M.; Morgado, J.M.; Raya, K.; Binosi, D.; Chang, L.; De Soto, F.; Roberts, C.D.; Rodríguez-Quintero, J.; Schmidt, S.M. Emergence of pion parton distributions. *Phys. Rev. D* **2022**, *105*, L091502. [CrossRef]
7. Lu, Y.; Chang, L.; Raya, K.; Roberts, C.D.; Rodríguez-Quintero, J. Proton and pion distribution functions in counterpoint. *Phys. Lett. B* **2022**, *830*, 137130. [CrossRef]
8. Ding, M.; Roberts, C.D.; Schmidt, S.M. Emergence of Hadron Mass and Structure. *Particles* **2023**, *6*, 57–120. [CrossRef]
9. Roberts, C.D. Origin of the Proton Mass. *arXiv* **2022**, arXiv:2211.09905.

10. Roberts, C.D.; Williams, A.G. Dyson-Schwinger equations and their application to hadronic physics. *Prog. Part. Nucl. Phys.* **1994**, *33*, 477–575. [CrossRef]
11. Alkofer, R.; von Smekal, L. The Infrared behavior of QCD Green's functions: Confinement dynamical symmetry breaking, and hadrons as relativistic bound states. *Phys. Rep.* **2001**, *353*, 281. [CrossRef]
12. Fischer, C.S. Infrared properties of QCD from Dyson-Schwinger equations. *J. Phys. G* **2006**, *32*, R253–R291. [CrossRef]
13. Roberts, C.D. Hadron Properties and Dyson-Schwinger Equations. *Prog. Part. Nucl. Phys.* **2008**, *61*, 50–65. [CrossRef]
14. Binosi, D.; Papavassiliou, J. Pinch Technique: Theory and Applications. *Phys. Rep.* **2009**, *479*, 1–152. [CrossRef]
15. Bashir, A.; Chang, L.; Cloet, I.C.; El-Bennich, B.; Liu, Y.X.; Roberts, C.D.; Tandy, P.C. Collective perspective on advances in Dyson-Schwinger Equation QCD. *Commun. Theor. Phys.* **2012**, *58*, 79–134. [CrossRef]
16. Binosi, D.; Chang, L.; Papavassiliou, J.; Roberts, C.D. Bridging a gap between continuum-QCD and ab initio predictions of hadron observables. *Phys. Lett.* **2015**, *B742*, 183–188. [CrossRef]
17. Cloet, I.C.; Roberts, C.D. Explanation and Prediction of Observables using Continuum Strong QCD. *Prog. Part. Nucl. Phys.* **2014**, *77*, 1–69. [CrossRef]
18. Aguilar, A.C.; Binosi, D.; Papavassiliou, J. The Gluon Mass Generation Mechanism: A Concise Primer. *Front. Phys. (Beijing)* **2016**, *11*, 111203. [CrossRef]
19. Binosi, D.; Chang, L.; Papavassiliou, J.; Qin, S.X.; Roberts, C.D. Symmetry preserving truncations of the gap and Bethe–Salpeter equations. *Phys. Rev.* **2016**, *D93*, 096010. [CrossRef]
20. Binosi, D.; Mezrag, C.; Papavassiliou, J.; Roberts, C.D.; Rodriguez-Quintero, J. Process-independent strong running coupling. *Phys. Rev.* **2017**, *D96*, 054026. [CrossRef]
21. Huber, M.Q. Nonperturbative properties of Yang-Mills theories. *Phys. Rep.* **2020**, *879*, 1–92. [CrossRef]
22. Pawlowski, J.M.; Litim, D.F.; Nedelko, S.; von Smekal, L. Infrared behavior and fixed points in Landau gauge QCD. *Phys. Rev. Lett.* **2004**, *93*, 152002. [CrossRef] [PubMed]
23. Pawlowski, J.M. Aspects of the functional renormalisation group. *Ann. Phys.* **2007**, *322*, 2831–2915. [CrossRef]
24. Fischer, C.S.; Maas, A.; Pawlowski, J.M. On the infrared behavior of Landau gauge Yang-Mills theory. *Ann. Phys.* **2009**, *324*, 2408–2437. [CrossRef]
25. Carrington, M.E. Renormalization group flow equations connected to the n-particle-irreducible effective action. *Phys. Rev.* **2013**, *D87*, 045011. [CrossRef]
26. Carrington, M.E.; Fu, W.J.; Pickering, D.; Pulver, J.W. Renormalization group methods and the 2PI effective action. *Phys. Rev. D* **2015**, *91*, 025003. [CrossRef]
27. Cyrol, A.K.; Mitter, M.; Pawlowski, J.M.; Strodthoff, N. Nonperturbative quark, gluon, and meson correlators of unquenched QCD. *Phys. Rev.* **2018**, *D97*, 054006. [CrossRef]
28. Corell, L.; Cyrol, A.K.; Mitter, M.; Pawlowski, J.M.; Strodthoff, N. Correlation functions of three-dimensional Yang-Mills theory from the FRG. *SciPost Phys.* **2018**, *5*, 066. [CrossRef]
29. Huber, M.Q. Correlation functions of Landau gauge Yang-Mills theory. *Phys. Rev. D* **2020**, *101*, 114009. [CrossRef]
30. Dupuis, N.; Canet, L.; Eichhorn, A.; Metzner, W.; Pawlowski, J.M.; Tissier, M.; Wschebor, N. The nonperturbative functional renormalization group and its applications. *Phys. Rep.* **2021**, *910*, 1–114. [CrossRef]
31. Blaizot, J.P.; Pawlowski, J.M.; Reinosa, U. Functional renormalization group and 2PI effective action formalism. *Ann. Phys.* **2021**, *431*, 168549. [CrossRef]
32. Mandula, J.; Ogilvie, M. The Gluon Is Massive: A Lattice Calculation of the Gluon Propagator in the Landau Gauge. *Phys. Lett. B* **1987**, *185*, 127–132. [CrossRef]
33. Parrinello, C. Exploratory study of the three gluon vertex on the lattice. *Phys. Rev.* **1994**, *D50*, R4247–R4251. [CrossRef] [PubMed]
34. Alles, B.; Henty, D.; Panagopoulos, H.; Parrinello, C.; Pittori, C.; Richards, D.G. α_s from the nonperturbatively renormalised lattice three gluon vertex. *Nucl. Phys.* **1997**, *B502*, 325–342. [CrossRef]
35. Parrinello, C.; Richards, D.; Alles, B.; Panagopoulos, H.; Pittori, C. Status of alpha-s determinations from the nonperturbatively renormalized three gluon vertex. *Nucl. Phys. B Proc. Suppl.* **1998**, *63*, 245–247. [CrossRef]
36. Boucaud, P.; Leroy, J.P.; Micheli, J.; Pene, O.; Roiesnel, C. Lattice calculation of alpha(s) in momentum scheme. *J. High Energy Phys.* **1998**, *10*, 017. [CrossRef]
37. Alexandrou, C.; de Forcrand, P.; Follana, E. The gluon propagator without lattice Gribov copies on a finer lattice. *Phys. Rev.* **2002**, *D65*, 114508. [CrossRef]
38. Bowman, P.O.; Heller, U.M.; Williams, A.G. Lattice quark propagator with staggered quarks in Landau and Laplacian gauges. *Phys. Rev. D* **2002**, *66*, 014505. [CrossRef]
39. Skullerud, J.I.; Bowman, P.O.; Kizilersu, A.; Leinweber, D.B.; Williams, A.G. Nonperturbative structure of the quark gluon vertex. *J. High Energy Phys.* **2003**, *4*, 047. [CrossRef]
40. Bowman, P.O.; Heller, U.M.; Leinweber, D.B.; Parappilly, M.B.; Williams, A.G. Unquenched gluon propagator in Landau gauge. *Phys. Rev. D* **2004**, *70*, 034509. [CrossRef]
41. Cucchieri, A.; Maas, A.; Mendes, T. Exploratory study of three-point Green's functions in Landau-gauge Yang-Mills theory. *Phys. Rev.* **2006**, *D74*, 014503. [CrossRef]
42. Ilgenfritz, E.M.; Muller-Preussker, M.; Sternbeck, A.; Schiller, A.; Bogolubsky, I. Landau gauge gluon and ghost propagators from lattice QCD. *Braz. J. Phys.* **2007**, *37*, 193–200. [CrossRef]

43. Sternbeck, A. The Infrared Behavior of Lattice QCD Green's Functions. Ph.D. Thesis, Humboldt-University Berlin, Berlin, Germany, 2006.
44. Furui, S.; Nakajima, H. Unquenched Kogut-Susskind quark propagator in lattice Landau gauge QCD. *Phys. Rev. D* **2006**, *73*, 074503. [CrossRef]
45. Bowman, P.O.; Heller, U.M.; Leinweber, D.B.; Parappilly, M.B.; Sternbeck, A.; von Smekal, L.; Williams, A.G.; Zhang, J.b. Scaling behavior and positivity violation of the gluon propagator in full QCD. *Phys. Rev. D* **2007**, *76*, 094505. [CrossRef]
46. Kamleh, W.; Bowman, P.O.; Leinweber, D.B.; Williams, A.G.; Zhang, J. Unquenching effects in the quark and gluon propagator. *Phys. Rev.* **2007**, *D76*, 094501. [CrossRef]
47. Cucchieri, A.; Mendes, T. What's up with IR gluon and ghost propagators in Landau gauge? A puzzling answer from huge lattices. *PoS* **2007**, *LATTICE2007*, 297. [CrossRef]
48. Cucchieri, A.; Mendes, T. Constraints on the IR behavior of the gluon propagator in Yang-Mills theories. *Phys. Rev. Lett.* **2008**, *100*, 241601. [CrossRef]
49. Bogolubsky, I.; Ilgenfritz, E.; Muller-Preussker, M.; Sternbeck, A. The Landau gauge gluon and ghost propagators in 4D SU(3) gluodynamics in large lattice volumes. *PoS* **2007**, *LATTICE2007*, 290. [CrossRef]
50. Cucchieri, A.; Maas, A.; Mendes, T. Three-point vertices in Landau-gauge Yang-Mills theory. *Phys. Rev.* **2008**, *D77*, 094510. [CrossRef]
51. Cucchieri, A.; Mendes, T. Constraints on the IR behavior of the ghost propagator in Yang-Mills theories. *Phys. Rev. D* **2008**, *78*, 094503. [CrossRef]
52. Cucchieri, A.; Mendes, T. Landau-gauge propagators in Yang-Mills theories at beta = 0: Massive solution versus conformal scaling. *Phys. Rev.* **2010**, *D81*, 016005. [CrossRef]
53. Cucchieri, A.; Mendes, T. Numerical test of the Gribov-Zwanziger scenario in Landau gauge. *PoS* **2009**, *QCD-TNT09*, 026. [CrossRef]
54. Boucaud, P.; De Soto, F.; Leroy, J.P.; Le Yaouanc, A.; Micheli, J.; Pene, O.; Rodriguez-Quintero, J. Ghost-gluon running coupling, power corrections and the determination of Lambda(MS-bar). *Phys. Rev.* **2009**, *D79*, 014508. . [CrossRef]
55. Cucchieri, A.; Mendes, T.; Santos, E.M.S. Covariant gauge on the lattice: A New implementation. *Phys. Rev. Lett.* **2009**, *103*, 141602. [CrossRef]
56. Bogolubsky, I.; Ilgenfritz, E.; Muller-Preussker, M.; Sternbeck, A. Lattice gluodynamics computation of Landau gauge Green's functions in the deep infrared. *Phys. Lett.* **2009**, *B676*, 69–73. [CrossRef]
57. Oliveira, O.; Silva, P. The Lattice infrared Landau gauge gluon propagator: The Infinite volume limit. *PoS* **2009**, *LAT2009*, 226. [CrossRef]
58. Cucchieri, A.; Mendes, T.; Nakamura, G.M.; Santos, E.M.S. Gluon Propagators in Linear Covariant Gauge. *PoS* **2010**, *FACESQCD*, 026. [CrossRef]
59. Oliveira, O.; Bicudo, P. Running Gluon Mass from Landau Gauge Lattice QCD Propagator. *J. Phys. G* **2011**, *G38*, 045003. [CrossRef]
60. Blossier, B.; Boucaud, P.; De soto, F.; Morenas, V.; Gravina, M.; Pene, O.; Rodriguez-Quintero, J. Ghost-gluon coupling, power corrections and $\Lambda_{\overline{MS}}$ from twisted-mass lattice QCD at Nf = 2. *Phys. Rev. D* **2010**, *82*, 034510. [CrossRef]
61. Maas, A. Describing gauge bosons at zero and finite temperature. *Phys. Rep.* **2013**, *524*, 203–300. [CrossRef]
62. Boucaud, P.; Leroy, J.P.; Yaouanc, A.L.; Micheli, J.; Pene, O.; Rodriguez-Quintero, J. The Infrared Behaviour of the Pure Yang-Mills Green Functions. *Few Body Syst.* **2012**, *53*, 387–436. [CrossRef]
63. Ayala, A.; Bashir, A.; Binosi, D.; Cristoforetti, M.; Rodriguez-Quintero, J. Quark flavour effects on gluon and ghost propagators. *Phys. Rev.* **2012**, *D86*, 074512. [CrossRef]
64. Oliveira, O.; Silva, P.J. The lattice Landau gauge gluon propagator: Lattice spacing and volume dependence. *Phys. Rev.* **2012**, *D86*, 114513. [CrossRef]
65. Sternbeck, A.; Müller-Preussker, M. Lattice evidence for the family of decoupling solutions of Landau gauge Yang-Mills theory. *Phys. Lett. B* **2013**, *726*, 396–403. [CrossRef]
66. Bicudo, P.; Binosi, D.; Cardoso, N.; Oliveira, O.; Silva, P.J. Lattice gluon propagator in renormalizable ζ gauges. *Phys. Rev.* **2015**, *D92*, 114514. [CrossRef]
67. Duarte, A.G.; Oliveira, O.; Silva, P.J. Lattice Gluon and Ghost Propagators, and the Strong Coupling in Pure SU(3) Yang-Mills Theory: Finite Lattice Spacing and Volume Effects. *Phys. Rev. D* **2016**, *94*, 014502. [CrossRef]
68. Athenodorou, A.; Binosi, D.; Boucaud, P.; De Soto, F.; Papavassiliou, J.; Rodriguez-Quintero, J.; Zafeiropoulos, S. On the zero crossing of the three-gluon vertex. *Phys. Lett.* **2016**, *B761*, 444–449. [CrossRef]
69. Duarte, A.G.; Oliveira, O.; Silva, P.J. Further Evidence For Zero Crossing On The Three Gluon Vertex. *Phys. Rev.* **2016**, *D94*, 074502. [CrossRef]
70. Oliveira, O.; Kizilersu, A.; Silva, P.J.; Skullerud, J.I.; Sternbeck, A.; Williams, A.G. Lattice Landau gauge quark propagator and the quark-gluon vertex. *Acta Phys. Pol. Suppl.* **2016**, *9*, 363–368. [CrossRef]
71. Boucaud, P.; De Soto, F.; Rodríguez-Quintero, J.; Zafeiropoulos, S. Refining the detection of the zero crossing for the three-gluon vertex in symmetric and asymmetric momentum subtraction schemes. *Phys. Rev.* **2017**, *D95*, 114503. [CrossRef]
72. Sternbeck, A.; Balduf, P.H.; Kizilersu, A.; Oliveira, O.; Silva, P.J.; Skullerud, J.I.; Williams, A.G. Triple-gluon and quark-gluon vertex from lattice QCD in Landau gauge. *PoS* **2017**, *LATTICE2016*, 349. [CrossRef]

73. Boucaud, P.; De Soto, F.; Raya, K.; Rodríguez-Quintero, J.; Zafeiropoulos, S. Discretization effects on renormalized gauge-field Green's functions, scale setting, and the gluon mass. *Phys. Rev.* **2018**, *D98*, 114515. [CrossRef]
74. Cucchieri, A.; Dudal, D.; Mendes, T.; Oliveira, O.; Roelfs, M.; Silva, P.J. Lattice Computation of the Ghost Propagator in Linear Covariant Gauges. *PoS* **2018**, *LATTICE2018*, 252. [CrossRef]
75. Cucchieri, A.; Dudal, D.; Mendes, T.; Oliveira, O.; Roelfs, M.; Silva, P.J. Faddeev-Popov Matrix in Linear Covariant Gauge: First Results. *Phys. Rev. D* **2018**, *98*, 091504. [CrossRef]
76. Oliveira, O.; Silva, P.J.; Skullerud, J.I.; Sternbeck, A. Quark propagator with two flavors of O(a)-improved Wilson fermions. *Phys. Rev. D* **2019**, *99*, 094506. [CrossRef]
77. Dudal, D.; Oliveira, O.; Silva, P.J. High precision statistical Landau gauge lattice gluon propagator computation vs. the Gribov–Zwanziger approach. *Ann. Phys.* **2018**, *397*, 351–364. [CrossRef]
78. Vujinovic, M.; Mendes, T. Probing the tensor structure of lattice three-gluon vertex in Landau gauge. *Phys. Rev.* **2019**, *D99*, 034501. [CrossRef]
79. Cui, Z.F.; Zhang, J.L.; Binosi, D.; de Soto, F.; Mezrag, C.; Papavassiliou, J.; Roberts, C.D.; Rodríguez-Quintero, J.; Segovia, J.; Zafeiropoulos, S. Effective charge from lattice QCD. *Chin. Phys. C* **2020**, *44*, 083102. [CrossRef]
80. Zafeiropoulos, S.; Boucaud, P.; De Soto, F.; Rodríguez-Quintero, J.; Segovia, J. Strong Running Coupling from the Gauge Sector of Domain Wall Lattice QCD with Physical Quark Masses. *Phys. Rev. Lett.* **2019**, *122*, 162002. [CrossRef]
81. Aguilar, A.C.; De Soto, F.; Ferreira, M.N.; Papavassiliou, J.; Rodríguez-Quintero, J.; Zafeiropoulos, S. Gluon propagator and three-gluon vertex with dynamical quarks. *Eur. Phys. J.* **2020**, *C80*, 154. [CrossRef]
82. Maas, A.; Vujinović, M. More on the three-gluon vertex in SU(2) Yang-Mills theory in three and four dimensions. *SciPost Phys. Core* **2022**, *5*, 019. [CrossRef]
83. Kızılersü, A.; Oliveira, O.; Silva, P.J.; Skullerud, J.I.; Sternbeck, A. Quark-gluon vertex from Nf = 2 lattice QCD. *Phys. Rev. D* **2021**, *103*, 114515. [CrossRef]
84. Aguilar, A.C.; De Soto, F.; Ferreira, M.N.; Papavassiliou, J.; Rodríguez-Quintero, J. Infrared facets of the three-gluon vertex. *Phys. Lett. B* **2021**, *818*, 136352. [CrossRef]
85. Aguilar, A.C.; Ambrósio, C.O.; De Soto, F.; Ferreira, M.N.; Oliveira, B.M.; Papavassiliou, J.; Rodríguez-Quintero, J. Ghost dynamics in the soft gluon limit. *Phys. Rev. D* **2021**, *104*, 054028. [CrossRef]
86. Pinto-Gómez, F.; De Soto, F.; Ferreira, M.N.; Papavassiliou, J.; Rodríguez-Quintero, J. Lattice three-gluon vertex in extended kinematics: Planar degeneracy. *Phys. Lett. B* **2023**, *838*, 137737. [CrossRef]
87. Pinto-Gomez, F.; de Soto, F. Three-gluon vertex in Landau-gauge from quenched-lattice QCD in general kinematics. In Proceedings of the 15th Conference on Quark Confinement and the Hadron Spectrum, Stavanger, Norway, 1–6 August 2022.
88. Pinto-Gómez, F.; de Soto, F.; Ferreira, M.N.; Papavassiliou, J.; Rodríguez-Quintero, J. General kinematics of the three-gluon vertex from quenched lattice QCD. *arXiv* **2022**, arXiv:2212.11894.
89. Roberts, C.D.; Schmidt, S.M. Reflections upon the emergence of hadronic mass. *Eur. Phys. J. ST* **2020**, *229*, 3319–3340. [CrossRef]
90. Roberts, C.D. On Mass and Matter. *AAPPS Bull.* **2021**, *31*, 6. [CrossRef]
91. Roberts, C.D.; Richards, D.G.; Horn, T.; Chang, L. Insights into the emergence of mass from studies of pion and kaon structure. *Prog. Part. Nucl. Phys.* **2021**, *120*, 103883. [CrossRef]
92. Binosi, D. Emergent Hadron Mass in Strong Dynamics. *Few Body Syst.* **2022**, *63*, 42. [CrossRef]
93. Papavassiliou, J. Emergence of mass in the gauge sector of QCD*. *Chin. Phys. C* **2022**, *46*, 112001. [CrossRef]
94. Cornwall, J.M. Quark Confinement and Vortices in Massive Gauge Invariant QCD. *Nucl. Phys.* **1979**, *B157*, 392. [CrossRef]
95. Parisi, G.; Petronzio, R. On Low-Energy Tests of QCD. *Phys. Lett.* **1980**, *B94*, 51. [CrossRef]
96. Cornwall, J.M. Dynamical Mass Generation in Continuum QCD. *Phys. Rev. D* **1982**, *26*, 1453. [CrossRef]
97. Bernard, C.W. Monte Carlo Evaluation of the Effective Gluon Mass. *Phys. Lett. B* **1982**, *108*, 431–434. [CrossRef]
98. Bernard, C.W. Adjoint Wilson Lines and the Effective Gluon Mass. *Nucl. Phys. B* **1983**, *219*, 341–357. [CrossRef]
99. Donoghue, J.F. The Gluon 'Mass' in the Bag Model. *Phys. Rev. D* **1984**, *29*, 2559. [CrossRef]
100. Cornwall, J.M.; Papavassiliou, J. Gauge Invariant Three Gluon Vertex in QCD. *Phys. Rev. D* **1989**, *40*, 3474. [CrossRef]
101. Lavelle, M. Gauge invariant effective gluon mass from the operator product expansion. *Phys. Rev. D* **1991**, *44*, 26–28. [CrossRef]
102. Halzen, F.; Krein, G.I.; Natale, A.A. Relating the QCD pomeron to an effective gluon mass. *Phys. Rev.* **1993**, *D47*, 295–298. [CrossRef]
103. Wilson, K.G.; Walhout, T.S.; Harindranath, A.; Zhang, W.M.; Perry, R.J.; Glazek, S.D. Nonperturbative QCD: A Weak coupling treatment on the light front. *Phys. Rev.* **1994**, *D49*, 6720–6766. [CrossRef]
104. Mihara, A.; Natale, A.A. Dynamical gluon mass corrections in heavy quarkonia decays. *Phys. Lett.* **2000**, *B482*, 378–382. [CrossRef]
105. Philipsen, O. On the nonperturbative gluon mass and heavy quark physics. *Nucl. Phys.* **2002**, *B628*, 167–192. [CrossRef]
106. Kondo, K.I. Vacuum condensate of mass dimension 2 as the origin of mass gap and quark confinement. *Phys. Lett.* **2001**, *B514*, 335–345. [CrossRef]
107. Aguilar, A.C.; Natale, A.A.; Rodrigues da Silva, P.S. Relating a gluon mass scale to an infrared fixed point in pure gauge QCD. *Phys. Rev. Lett.* **2003**, *90*, 152001. [CrossRef]
108. Aguilar, A.C.; Natale, A.A. A Dynamical gluon mass solution in a coupled system of the Schwinger-Dyson equations. *J. High Energy Phys.* **2004**, *8*, 057. [CrossRef]

109. Aguilar, A.C.; Papavassiliou, J. Gluon mass generation in the PT-BFM scheme. *J. High Energy Phys.* **2006**, *12*, 012. [CrossRef]
110. Epple, D.; Reinhardt, H.; Schleifenbaum, W.; Szczepaniak, A.P. Subcritical solution of the Yang-Mills Schroedinger equation in the Coulomb gauge. *Phys. Rev.* **2008**, *D77*, 085007. [CrossRef]
111. Aguilar, A.C.; Papavassiliou, J. On dynamical gluon mass generation. *Eur. Phys. J.* **2007**, *A31*, 742–745. [CrossRef]
112. Aguilar, A.C.; Binosi, D.; Papavassiliou, J. Gluon and ghost propagators in the Landau gauge: Deriving lattice results from Schwinger-Dyson equations. *Phys. Rev.* **2008**, *D78*, 025010. [CrossRef]
113. Aguilar, A.C.; Papavassiliou, J. Gluon mass generation without seagull divergences. *Phys. Rev.* **2010**, *D81*, 034003. [CrossRef]
114. Campagnari, D.R.; Reinhardt, H. Non-Gaussian wave functionals in Coulomb gauge Yang–Mills theory. *Phys. Rev.* **2010**, *D82*, 105021. [CrossRef]
115. Fagundes, D.A.; Luna, E.G.S.; Menon, M.J.; Natale, A.A. Aspects of a Dynamical Gluon Mass Approach to Elastic Hadron Scattering at LHC. *Nucl. Phys. A* **2012**, *886*, 48–70. [CrossRef]
116. Aguilar, A.C.; Binosi, D.; Papavassiliou, J. The dynamical equation of the effective gluon mass. *Phys. Rev.* **2011**, *D84*, 085026. [CrossRef]
117. Aguilar, A.C.; Ibanez, D.; Mathieu, V.; Papavassiliou, J. Massless bound-state excitations and the Schwinger mechanism in QCD. *Phys. Rev.* **2012**, *D85*, 014018. [CrossRef]
118. Aguilar, A.C.; Binosi, D.; Papavassiliou, J. Gluon mass through ghost synergy. *J. High Energy Phys.* **2012**, *01*, 050. [CrossRef]
119. Aguilar, A.C.; Binosi, D.; Papavassiliou, J. Gluon mass generation in the presence of dynamical quarks. *Phys. Rev.* **2013**, *D88*, 074010. [CrossRef]
120. Glazek, S.D.; Gómez-Rocha, M.; More, J.; Serafin, K. Renormalized quark–antiquark Hamiltonian induced by a gluon mass ansatz in heavy-flavor QCD. *Phys. Lett.* **2017**, *B773*, 172–178. [CrossRef]
121. Binosi, D.; Papavassiliou, J. Coupled dynamics in gluon mass generation and the impact of the three-gluon vertex. *Phys. Rev.* **2018**, *D97*, 054029. [CrossRef]
122. Aguilar, A.C.; Ferreira, M.N.; Figueiredo, C.T.; Papavassiliou, J. Gluon mass scale through nonlinearities and vertex interplay. *Phys. Rev. D* **2019**, *100*, 094039. [CrossRef]
123. Eichmann, G.; Pawlowski, J.M.; Silva, J.M. Mass generation in Landau-gauge Yang-Mills theory. *Phys. Rev. D* **2021**, *104*, 114016. [CrossRef]
124. Aguilar, A.C.; Ferreira, M.N.; Papavassiliou, J. Exploring smoking-gun signals of the Schwinger mechanism in QCD. *Phys. Rev. D* **2022**, *105*, 014030. [CrossRef]
125. Horak, J.; Ihssen, F.; Papavassiliou, J.; Pawlowski, J.M.; Weber, A.; Wetterich, C. Gluon condensates and effective gluon mass. *SciPost Phys.* **2022**, *13*, 042. [CrossRef]
126. Aguilar, A.C.; De Soto, F.; Ferreira, M.N.; Papavassiliou, J.; Pinto-Gómez, F.; Roberts, C.D.; Rodríguez-Quintero, J. Schwinger mechanism for gluons from lattice QCD. *arXiv* **2022**, arXiv:2211.12594.
127. Schwinger, J.S. Gauge Invariance and Mass. *Phys. Rev.* **1962**, *125*, 397–398. [CrossRef]
128. Schwinger, J.S. Gauge Invariance and Mass. 2. *Phys. Rev.* **1962**, *128*, 2425–2429. [CrossRef]
129. Watson, N.J. The gauge-independent QCD effective charge. *Nucl. Phys.* **1997**, *B494*, 388–432. [CrossRef]
130. Binosi, D.; Papavassiliou, J. The QCD effective charge to all orders. *Nucl. Phys. Proc. Suppl.* **2003**, *121*, 281–284. [CrossRef]
131. Aguilar, A.C.; Binosi, D.; Papavassiliou, J.; Rodriguez-Quintero, J. Non-perturbative comparison of QCD effective charges. *Phys. Rev.* **2009**, *D80*, 085018. [CrossRef]
132. Gell-Mann, M.; Low, F.E. Quantum electrodynamics at small distances. *Phys. Rev.* **1954**, *95*, 1300–1312. [CrossRef]
133. Itzykson, C.; Zuber, J.B. *Quantum Field Theory*; International Series in Pure and Applied Physics; Mcgraw-Hill: New York, NY, USA, 1980; 705p.
134. Nambu, Y.; Jona-Lasinio, G. Dynamical model of elementary particles based on an analogy with superconductivity. I. *Phys. Rev.* **1961**, *122*, 345–358. [CrossRef]
135. Lane, K.D. Asymptotic Freedom and Goldstone Realization of Chiral Symmetry. *Phys. Rev.* **1974**, *D10*, 2605. [CrossRef]
136. Politzer, H.D. Effective Quark Masses in the Chiral Limit. *Nucl. Phys.* **1976**, *B117*, 397. [CrossRef]
137. Miransky, V.A.; Fomin, P.I. Chiral symmetry breakdown and the spectrum of pseudoscalar mesons in quantum chromodynamics. *Phys. Lett.* **1981**, *B105*, 387–391. [CrossRef]
138. Atkinson, D.; Johnson, P.W. Chiral Symmetry Breaking in QCD. 2. Running Coupling Constant. *Phys. Rev.* **1988**, *D37*, 2296–2299. [CrossRef]
139. Brown, N.; Pennington, M.R. Studies of confinement: How quarks and gluons propagate. *Phys. Rev.* **1988**, *D38*, 2266. [CrossRef]
140. Williams, A.G.; Krein, G.; Roberts, C.D. Quark propagator in an Ansatz approach to QCD. *Ann. Phys.* **1991**, *210*, 464–485. [CrossRef]
141. Papavassiliou, J.; Cornwall, J.M. Coupled fermion gap and vertex equations for chiral symmetry breakdown in QCD. *Phys. Rev.* **1991**, *D44*, 1285–1297. [CrossRef]
142. Hawes, F.T.; Roberts, C.D.; Williams, A.G. Dynamical chiral symmetry breaking and confinement with an infrared vanishing gluon propagator. *Phys. Rev.* **1994**, *D49*, 4683–4693. [CrossRef]
143. Natale, A.A.; Rodrigues da Silva, P.S. Critical coupling for dynamical chiral-symmetry breaking with an infrared finite gluon propagator. *Phys. Lett.* **1997**, *B392*, 444–451. [CrossRef]

144. Fischer, C.S.; Alkofer, R. Nonperturbative propagators, running coupling and dynamical quark mass of Landau gauge QCD. *Phys. Rev.* **2003**, *D67*, 094020. [CrossRef]
145. Maris, P.; Roberts, C.D. Dyson-Schwinger equations: A Tool for hadron physics. *Int. J. Mod. Phys.* **2003**, *E12*, 297–365. [CrossRef]
146. Aguilar, A.C.; Nesterenko, A.; Papavassiliou, J. Infrared enhanced analytic coupling and chiral symmetry breaking in QCD. *J. Phys.* **2005**, *G31*, 997. [CrossRef]
147. Bowman, P.O.; Heller, U.M.; Leinweber, D.B.; Parappilly, M.B.; Williams, A.G.; Zhang, J.b. Unquenched quark propagator in Landau gauge. *Phys. Rev.* **2005**, *D71*, 054507. [CrossRef]
148. Sauli, V.; Adam, J., Jr.; Bicudo, P. Dynamical chiral symmetry breaking with integral Minkowski representations. *Phys. Rev.* **2007**, *D75*, 087701. [CrossRef]
149. Cornwall, J.M. Center vortices, the functional Schrodinger equation, and CSB. In Proceedings of the 419th WE-Heraeus-Seminar: Approaches to Quantum Chromodynamics, Oberwoelz, Austria, 7–13 September 2022.
150. Alkofer, R.; Fischer, C.S.; Llanes-Estrada, F.J.; Schwenzer, K. The Quark-gluon vertex in Landau gauge QCD: Its role in dynamical chiral symmetry breaking and quark confinement. *Ann. Phys.* **2009**, *324*, 106–172. [CrossRef]
151. Aguilar, A.C.; Papavassiliou, J. Chiral symmetry breaking with lattice propagators. *Phys. Rev.* **2011**, *D83*, 014013. [CrossRef]
152. Rojas, E.; de Melo, J.; El-Bennich, B.; Oliveira, O.; Frederico, T. On the Quark-Gluon Vertex and Quark-Ghost Kernel: Combining Lattice Simulations with Dyson-Schwinger equations. *J. High Energy Phys.* **2013**, *10*, 193. [CrossRef]
153. Mitter, M.; Pawlowski, J.M.; Strodthoff, N. Chiral symmetry breaking in continuum QCD. *Phys. Rev.* **2015**, *D91*, 054035. [CrossRef]
154. Braun, J.; Fister, L.; Pawlowski, J.M.; Rennecke, F. From Quarks and Gluons to Hadrons: Chiral Symmetry Breaking in Dynamical QCD. *Phys. Rev.* **2016**, *D94*, 034016. [CrossRef]
155. Heupel, W.; Goecke, T.; Fischer, C.S. Beyond Rainbow-Ladder in bound state equations. *Eur. Phys. J.* **2014**, *A50*, 85. [CrossRef]
156. Binosi, D.; Chang, L.; Papavassiliou, J.; Qin, S.X.; Roberts, C.D. Natural constraints on the gluon-quark vertex. *Phys. Rev.* **2017**, *D95*, 031501. [CrossRef]
157. Aguilar, A.C.; Cardona, J.C.; Ferreira, M.N.; Papavassiliou, J. Quark gap equation with non-abelian Ball-Chiu vertex. *Phys. Rev.* **2018**, *D98*, 014002. [CrossRef]
158. Gao, F.; Papavassiliou, J.; Pawlowski, J.M. Fully coupled functional equations for the quark sector of QCD. *Phys. Rev. D* **2021**, *103*, 094013. [CrossRef]
159. Eichten, E.; Feinberg, F. Dynamical Symmetry Breaking of Nonabelian Gauge Symmetries. *Phys. Rev. D* **1974**, *10*, 3254–3279. [CrossRef]
160. Smit, J. On the Possibility That Massless Yang-Mills Fields Generate Massive Vector Particles. *Phys. Rev. D* **1974**, *10*, 2473. [CrossRef]
161. Binosi, D.; Iba nez, D.; Papavassiliou, J. The all-order equation of the effective gluon mass. *Phys. Rev.* **2012**, *D86*, 085033. [CrossRef]
162. Tissier, M.; Wschebor, N. Infrared propagators of Yang-Mills theory from perturbation theory. *Phys. Rev. D* **2010**, *82*, 101701. [CrossRef]
163. Serreau, J.; Tissier, M. Lifting the Gribov ambiguity in Yang-Mills theories. *Phys. Lett.* **2012**, *B712*, 97–103. [CrossRef]
164. Peláez, M.; Tissier, M.; Wschebor, N. Two-point correlation functions of QCD in the Landau gauge. *Phys. Rev. D* **2014**, *90*, 065031. [CrossRef]
165. Siringo, F. Analytical study of Yang–Mills theory in the infrared from first principles. *Nucl. Phys.* **2016**, *B907*, 572–596. [CrossRef]
166. Aguilar, A.C.; Binosi, D.; Figueiredo, C.T.; Papavassiliou, J. Unified description of seagull cancellations and infrared finiteness of gluon propagators. *Phys. Rev.* **2016**, *D94*, 045002. [CrossRef]
167. Osterwalder, K.; Schrader, R. Axioms for Euclidean Green'S Functions. *Commun. Math. Phys.* **1973**, *31*, 83–112. [CrossRef]
168. Osterwalder, K.; Schrader, R. Axioms for Euclidean Green's Functions. 2. *Commun. Math. Phys.* **1975**, *42*, 281. [CrossRef]
169. Glimm, J.; Jaffe, A.M. *Quantum Physics. A Functional Integral Point of View*; Springer: New York, NY, USA, 1981.
170. Krein, G.; Roberts, C.D.; Williams, A.G. On the implications of confinement. *Int. J. Mod. Phys.* **1992**, *A7*, 5607–5624. [CrossRef]
171. Cornwall, J.M. Positivity violations in QCD. *Mod. Phys. Lett.* **2013**, *A28*, 1330035. [CrossRef]
172. Aguilar, A.C.; Binosi, D.; Iba nez, D.; Papavassiliou, J. Effects of divergent ghost loops on Green's functions of QCD. *Phys. Rev.* **2014**, *D89*, 085008. [CrossRef]
173. Pelaez, M.; Tissier, M.; Wschebor, N. Three-point correlation functions in Yang-Mills theory. *Phys. Rev.* **2013**, *D88*, 125003. [CrossRef]
174. Blum, A.; Huber, M.Q.; Mitter, M.; von Smekal, L. Gluonic three-point correlations in pure Landau gauge QCD. *Phys. Rev.* **2014**, *D89*, 061703. [CrossRef]
175. Eichmann, G.; Williams, R.; Alkofer, R.; Vujinovic, M. The three-gluon vertex in Landau gauge. *Phys. Rev.* **2014**, *D89*, 105014. [CrossRef]
176. Williams, R.; Fischer, C.S.; Heupel, W. Light mesons in QCD and unquenching effects from the 3PI effective action. *Phys. Rev.* **2016**, *D93*, 034026. [CrossRef]
177. Blum, A.L.; Alkofer, R.; Huber, M.Q.; Windisch, A. Unquenching the three-gluon vertex: A status report. *Acta Phys. Pol. Suppl.* **2015**, *8*, 321. [CrossRef]
178. Cyrol, A.K.; Fister, L.; Mitter, M.; Pawlowski, J.M.; Strodthoff, N. Landau gauge Yang-Mills correlation functions. *Phys. Rev.* **2016**, *D94*, 054005. [CrossRef]

179. Aguilar, A.C.; Ferreira, M.N.; Figueiredo, C.T.; Papavassiliou, J. Nonperturbative Ball-Chiu construction of the three-gluon vertex. *Phys. Rev.* **2019**, *D99*, 094010. [CrossRef]
180. Barrios, N.; Peláez, M.; Reinosa, U. Two-loop three-gluon vertex from the Curci-Ferrari model and its leading infrared behavior to all loop orders. *Phys. Rev. D* **2022**, *106*, 114039. [CrossRef]
181. Rivers, R.J. *Path Integral Methods in Quantum Field Theory*; Cambridge Monographs on Mathematical Physics. Cambridge University Press: Cambridge, UK, 1988. [CrossRef]
182. Fujikawa, K.; Lee, B.W.; Sanda, A.I. Generalized Renormalizable Gauge Formulation of Spontaneously Broken Gauge Theories. *Phys. Rev. D* **1972**, *6*, 2923–2943. [CrossRef]
183. Aguilar, A.C.; Papavassiliou, J. Infrared finite ghost propagator in the Feynman gauge. *Phys. Rev.* **2008**, *D77*, 125022. [CrossRef]
184. Huber, M.Q.; Schwenzer, K.; Alkofer, R. On the infrared scaling solution of SU(N) Yang-Mills theories in the maximally Abelian gauge. *Eur. Phys. J. C* **2010**, *68*, 581–600. [CrossRef]
185. Siringo, F. Gluon propagator in Feynman gauge by the method of stationary variance. *Phys. Rev. D* **2014**, *90*, 094021. [CrossRef]
186. Aguilar, A.C.; Binosi, D.; Papavassiliou, J. Yang-Mills two-point functions in linear covariant gauges. *Phys. Rev.* **2015**, *D91*, 085014. [CrossRef]
187. Huber, M.Q. Gluon and ghost propagators in linear covariant gauges. *Phys. Rev.* **2015**, *D91*, 085018. [CrossRef]
188. Capri, M.A.L.; Fiorentini, D.; Guimaraes, M.S.; Mintz, B.W.; Palhares, L.F.; Sorella, S.P.; Dudal, D.; Justo, I.F.; Pereira, A.D.; Sobreiro, R.F. Exact nilpotent nonperturbative BRST symmetry for the Gribov-Zwanziger action in the linear covariant gauge. *Phys. Rev. D* **2015**, *92*, 045039. [CrossRef]
189. Aguilar, A.C.; Binosi, D.; Papavassiliou, J. Schwinger mechanism in linear covariant gauges. *Phys. Rev.* **2017**, *D95*, 034017. [CrossRef]
190. De Meerleer, T.; Dudal, D.; Sorella, S.P.; Dall'Olio, P.; Bashir, A. Landau-Khalatnikov-Fradkin Transformations, Nielsen Identities, Their Equivalence and Implications for QCD. *Phys. Rev. D* **2020**, *101*, 085005. [CrossRef]
191. Napetschnig, M.; Alkofer, R.; Huber, M.Q.; Pawlowski, J.M. Yang-Mills propagators in linear covariant gauges from Nielsen identities. *Phys. Rev. D* **2021**, *104*, 054003. [CrossRef]
192. Binosi, D.; Papavassiliou, J. Gauge-invariant truncation scheme for the Schwinger-Dyson equations of QCD. *Phys. Rev.* **2008**, *D77*, 061702. [CrossRef]
193. Pilaftsis, A. Generalized pinch technique and the background field method in general gauges. *Nucl. Phys. B* **1997**, *487*, 467–491.
194. Binosi, D.; Papavassiliou, J. The Pinch technique to all orders. *Phys. Rev. D* **2002**, *66*, 111901. [CrossRef]
195. Binosi, D.; Papavassiliou, J. Pinch technique selfenergies and vertices to all orders in perturbation theory. *J. Phys. G* **2004**, *G30*, 203. [CrossRef]
196. DeWitt, B.S. Quantum Theory of Gravity. 2. The Manifestly Covariant Theory. *Phys. Rev.* **1967**, *162*, 1195–1239. [CrossRef]
197. 't Hooft, G. Renormalizable Lagrangians for Massive Yang-Mills Fields. *Nucl. Phys. B* **1971**, *35*, 167–188. [CrossRef]
198. Honerkamp, J. The Question of invariant renormalizability of the massless Yang-Mills theory in a manifest covariant approach. *Nucl. Phys. B* **1972**, *48*, 269–287. [CrossRef]
199. Kallosh, R.E. The Renormalization in Nonabelian Gauge Theories. *Nucl. Phys. B* **1974**, *78*, 293–312. [CrossRef]
200. Kluberg-Stern, H.; Zuber, J.B. Renormalization of Nonabelian Gauge Theories in a Background Gauge Field. 1. Green Functions. *Phys. Rev. D* **1975**, *12*, 482–488. [CrossRef]
201. Arefeva, I.Y.; Faddeev, L.D.; Slavnov, A.A. Generating Functional for the s Matrix in Gauge Theories. *Teor. Mat. Fiz.* **1974**, *21*, 311–321. [CrossRef]
202. Abbott, L. The Background Field Method Beyond One Loop. *Nucl. Phys. B* **1981**, *185*, 189–203. [CrossRef]
203. Weinberg, S. Effective Gauge Theories. *Phys. Lett. B* **1980**, *91*, 51–55. [CrossRef]
204. Abbott, L.F. Introduction to the Background Field Method. *Acta Phys. Polon.* **1982**, *B13*, 33.
205. Shore, G.M. Symmetry Restoration and the Background Field Method in Gauge Theories. *Ann. Phys.* **1981**, *137*, 262. [CrossRef]
206. Abbott, L.F.; Grisaru, M.T.; Schaefer, R.K. The Background Field Method and the S Matrix. *Nucl. Phys. B* **1983**, *229*, 372–380. [CrossRef]
207. Taylor, J. Ward Identities and Charge Renormalization of the Yang-Mills Field. *Nucl. Phys. B* **1971**, *33*, 436–444. [CrossRef]
208. Slavnov, A. Ward Identities in Gauge Theories. *Theor. Math. Phys.* **1972**, *10*, 99–107. [CrossRef]
209. Grassi, P.A.; Hurth, T.; Steinhauser, M. Practical algebraic renormalization. *Ann. Phys.* **2001**, *288*, 197–248. [CrossRef]
210. Grassi, P.A.; Hurth, T.; Steinhauser, M. The Algebraic method. *Nucl. Phys. B* **2001**, *610*, 215–250. [CrossRef]
211. Binosi, D.; Papavassiliou, J. Pinch technique and the Batalin-Vilkovisky formalism. *Phys. Rev.* **2002**, *D66*, 025024. [CrossRef]
212. Aguilar, A.C.; Binosi, D.; Papavassiliou, J. Indirect determination of the Kugo-Ojima function from lattice data. *J. High Energy Phys.* **2009**, *11*, 066. [CrossRef]
213. Binosi, D.; Quadri, A. AntiBRST symmetry and Background Field Method. *Phys. Rev.* **2013**, *D88*, 085036. [CrossRef]
214. Iba nez, D.; Papavassiliou, J. Gluon mass generation in the massless bound-state formalism. *Phys. Rev.* **2013**, *D87*, 034008. [CrossRef]
215. Aguilar, A.C.; Binosi, D.; Figueiredo, C.T.; Papavassiliou, J. Evidence of ghost suppression in gluon mass scale dynamics. *Eur. Phys. J.* **2018**, *C78*, 181. [CrossRef]

216. Munczek, H. Dynamical chiral symmetry breaking, Goldstone's theorem and the consistency of the Schwinger-Dyson and Bethe–Salpeter Equations. *Phys. Rev.* **1995**, *D52*, 4736–4740. [CrossRef]
217. Bender, A.; Roberts, C.D.; Von Smekal, L. Goldstone theorem and diquark confinement beyond rainbow ladder approximation. *Phys. Lett.* **1996**, *B380*, 7–12. [CrossRef]
218. Maris, P.; Roberts, C.D.; Tandy, P.C. Pion mass and decay constant. *Phys. Lett.* **1998**, *B420*, 267–273. [CrossRef]
219. Maris, P.; Roberts, C.D. Pi- and K meson Bethe–Salpeter amplitudes. *Phys. Rev. C* **1997**, *56*, 3369–3383. [CrossRef]
220. Chang, L.; Roberts, C.D. Sketching the Bethe–Salpeter kernel. *Phys. Rev. Lett.* **2009**, *103*, 081601. [CrossRef]
221. Chang, L.; Roberts, C.D.; Tandy, P.C. Selected highlights from the study of mesons. *Chin. J. Phys.* **2011**, *49*, 955–1004.
222. Qin, S.X.; Roberts, C.D. Resolving the Bethe–Salpeter Kernel. *Chin. Phys. Lett.* **2021**, *38*, 071201. [CrossRef]
223. Huber, M.Q. Correlation functions of three-dimensional Yang-Mills theory from Dyson-Schwinger equations. *Phys. Rev. D* **2016**, *93*, 085033. [CrossRef]
224. Schleifenbaum, W.; Maas, A.; Wambach, J.; Alkofer, R. Infrared behaviour of the ghost–gluon vertex in Landau gauge Yang-Mills theory. *Phys. Rev. D* **2005**, *72*, 014017. [CrossRef]
225. Boucaud, P.; Leroy, J.; A., L.Y.; Micheli, J.; Pène, O.; Rodríguez-Quintero, J. On the IR behaviour of the Landau-gauge ghost propagator. *J. High Energy Phys.* **2008**, *06*, 099. [CrossRef]
226. Huber, M.Q.; von Smekal, L. On the influence of three-point functions on the propagators of Landau gauge Yang-Mills theory. *J. High Energy Phys.* **2013**, *04*, 149. [CrossRef]
227. Aguilar, A.C.; Iba nez, D.; Papavassiliou, J. Ghost propagator and ghost–gluon vertex from Schwinger-Dyson equations. *Phys. Rev.* **2013**, *D87*, 114020. [CrossRef]
228. Aguilar, A.C.; Ferreira, M.N.; Figueiredo, C.T.; Papavassiliou, J. Nonperturbative structure of the ghost–gluon kernel. *Phys. Rev.* **2019**, *D99*, 034026. [CrossRef]
229. Dudal, D.; Gracey, J.A.; Sorella, S.P.; Vandersickel, N.; Verschelde, H. A refinement of the Gribov-Zwanziger approach in the Landau gauge: Infrared propagators in harmony with the lattice results. *Phys. Rev.* **2008**, *D78*, 065047. [CrossRef]
230. Boucaud, P.; Leroy, J.P.; Le Yaouanc, A.; Micheli, J.; Pene, O.; Rodríguez-Quintero, J. IR finiteness of the ghost dressing function from numerical resolution of the ghost SD equation. *J. High Energy Phys.* **2008**, *06*, 012. [CrossRef]
231. Kondo, K.I. Infrared behavior of the ghost propagator in the Landau gauge Yang-Mills theory. *Prog. Theor. Phys.* **2010**, *122*, 1455–1475. [CrossRef]
232. Pennington, M.R.; Wilson, D.J. Are the Dressed Gluon and Ghost Propagators in the Landau Gauge presently determined in the confinement regime of QCD? *Phys. Rev. D* **2011**, *84*, 094028; Erratum in *Phys. Rev. D* **2011**, *84*, 119901. [CrossRef]
233. Dudal, D.; Oliveira, O.; Rodriguez-Quintero, J. Nontrivial ghost–gluon vertex and the match of RGZ, DSE and lattice Yang-Mills propagators. *Phys. Rev. D* **2012**, *D86*, 105005; Erratum in *Phys. Rev. D* **2012**, *86*, 109902. [CrossRef]
234. Papavassiliou, J.; Aguilar, A.C.; Ferreira, M.N. Theory and phenomenology of the three-gluon vertex. *Rev. Mex. Fis. Suppl.* **2022**, *3*, 0308112. [CrossRef]
235. Meyers, J.; Swanson, E.S. Spin Zero Glueballs in the Bethe–Salpeter Formalism. *Phys. Rev.* **2013**, *D87*, 036009. [CrossRef]
236. Sanchis-Alepuz, H.; Fischer, C.S.; Kellermann, C.; von Smekal, L. Glueballs from the Bethe–Salpeter equation. *Phys. Rev.* **2015**, *D92*, 034001. [CrossRef]
237. Xu, S.S.; Cui, Z.F.; Chang, L.; Papavassiliou, J.; Roberts, C.D.; Zong, H.S. New perspective on hybrid mesons. *Eur. Phys. J.* **2019**, *A55*, 113. [CrossRef]
238. Souza, E.V.; Ferreira, M.N.; Aguilar, A.C.; Papavassiliou, J.; Roberts, C.D.; Xu, S.S. Pseudoscalar glueball mass: A window on three-gluon interactions. *Eur. Phys. J. A* **2020**, *56*, 25. [CrossRef]
239. Huber, M.Q.; Fischer, C.S.; Sanchis-Alepuz, H. Spectrum of scalar and pseudoscalar glueballs from functional methods. *Eur. Phys. J. C* **2020**, *80*, 1077. [CrossRef] [PubMed]
240. Huber, M.Q.; Fischer, C.S.; Sanchis-Alepuz, H. Higher spin glueballs from functional methods. *Eur. Phys. J. C* **2021**, *81*, 1083; Erratum in *Eur. Phys. J. C* **2022**, *82*, 38. [CrossRef]
241. Aguilar, A.C.; Ferreira, M.N.; Papavassiliou, J. Novel sum rules for the three-point sector of QCD. *Eur. Phys. J. C* **2020**, *80*, 887. [CrossRef]
242. Aguilar, A.C.; Ferreira, M.N.; Papavassiliou, J. Gluon dynamics from an ordinary differential equation. *Eur. Phys. J. C* **2021**, *81*, 54. [CrossRef]
243. Cornwall, J.M.; Papavassiliou, J.; Binosi, D. *The Pinch Technique and its Applications to Non-Abelian Gauge Theories*; Cambridge University Press: Cambridge, UK, 2010; Volume 31.
244. Binosi, D.; Papavassiliou, J. New Schwinger-Dyson equations for non-Abelian gauge theories. *J. High Energy Phys.* **2008**, *11*, 063. [CrossRef]
245. Grassi, P.A.; Hurth, T.; Quadri, A. On the Landau background gauge fixing and the IR properties of YM Green functions. *Phys. Rev.* **2004**, *D70*, 105014. [CrossRef]
246. Kugo, T. The Universal renormalization factors Z(1) / Z(3) and color confinement condition in nonAbelian gauge theory. *arXiv* **1995**, arXiv:9511033.
247. Kondo, K.I. Kugo-Ojima color confinement criterion and Gribov-Zwanziger horizon condition. *Phys. Lett. B* **2009**, *678*, 322–330. [CrossRef]

248. Becchi, C.; Rouet, A.; Stora, R. Renormalization of the Abelian Higgs-Kibble Model. *Commun. Math. Phys.* **1975**, *42*, 127–162. [CrossRef]
249. Becchi, C.; Rouet, A.; Stora, R. Renormalization of Gauge Theories. *Ann. Phys.* **1976**, *98*, 287–321. [CrossRef]
250. Tyutin, I.V. Gauge Invariance in Field Theory and Statistical Physics in Operator Formalism. *arXiv* **1975**, arXiv:0812.0580.
251. Ball, J.S.; Chiu, T.W. Analytic Properties of the Vertex Function in Gauge Theories. 2. *Phys. Rev. D* **1980**, *22*, 2550; Erratum in *Phys. Rev. D* **1981**, *23*, 3085. [CrossRef]
252. Davydychev, A.I.; Osland, P.; Tarasov, O. Three gluon vertex in arbitrary gauge and dimension. *Phys. Rev. D* **1996**, *54*, 4087–4113; Erratum in *Phys. Rev. D* **1999**, *59*, 109901. [CrossRef]
253. von Smekal, L.; Hauck, A.; Alkofer, R. A Solution to Coupled Dyson–Schwinger Equations for Gluons and Ghosts in Landau Gauge. *Ann. Phys.* **1998**, *267*, 1–60; Erratum in *Ann. Phys.* **1998**, *269*, 182. [CrossRef]
254. Binosi, D.; Papavassiliou, J. Gauge invariant Ansatz for a special three-gluon vertex. *J. High Energy Phys.* **2011**, *03*, 121. [CrossRef]
255. Gracey, J.; Kißler, H.; Kreimer, D. Self-consistency of off-shell Slavnov-Taylor identities in QCD. *Phys. Rev. D* **2019**, *100*, 085001. [CrossRef]
256. Collins, J.C. *Renormalization. An Introduction To Renormalization, The Renormalization Group, And The Operator Product Expansion*; Cambridge University Press: Cambridge, UK, 1986.
257. Brodsky, S.J.; Shrock, R. Maximum Wavelength of Confined Quarks and Gluons and Properties of Quantum Chromodynamics. *Phys. Lett.* **2008**, *B666*, 95–99. [CrossRef]
258. Braun, J.; Gies, H.; Pawlowski, J.M. Quark Confinement from Color Confinement. *Phys. Lett.* **2010**, *B684*, 262–267. [CrossRef]
259. Gao, F.; Qin, S.X.; Roberts, C.D.; Rodriguez-Quintero, J. Locating the Gribov horizon. *Phys. Rev.* **2018**, *D97*, 034010. [CrossRef]
260. Jackiw, R.; Johnson, K. Dynamical Model of Spontaneously Broken Gauge Symmetries. *Phys. Rev. D* **1973**, *8*, 2386–2398. [CrossRef]
261. Jackiw, R. Dynamical Symmetry Breaking. In Proceedings of the 11th International School of Subnuclear Physics: Laws of Hadronic Matter, Erice, Italy, 8–26 July 1973.
262. Papavassiliou, J. Gauge Invariant Proper Selfenergies and Vertices in Gauge Theories with Broken Symmetry. *Phys. Rev. D* **1990**, *41*, 3179. [CrossRef] [PubMed]
263. Nakanishi, N. A General survey of the theory of the Bethe–Salpeter equation. *Prog. Theor. Phys. Suppl.* **1969**, *43*, 1–81. [CrossRef]
264. Blank, M.; Krassnigg, A. Matrix algorithms for solving (in)homogeneous bound state equations. *Comput. Phys. Commun.* **2011**, *182*, 1391–1401. [CrossRef] [PubMed]
265. Wilson, K.G. Quantum field theory models in less than four-dimensions. *Phys. Rev.* **1973**, *D7*, 2911–2926. [CrossRef]
266. Poggio, E.C.; Tomboulis, E.; Tye, S.H.H. Dynamical Symmetry Breaking in Nonabelian Field Theories. *Phys. Rev.* **1975**, *D11*, 2839. [CrossRef]
267. Denner, A.; Weiglein, G.; Dittmaier, S. Gauge invariance of green functions: Background field method versus pinch technique. *Phys. Lett.* **1994**, *B333*, 420–426. [CrossRef]
268. Hashimoto, S.; Kodaira, J.; Yasui, Y.; Sasaki, K. The Background field method: Alternative way of deriving the pinch technique's results. *Phys. Rev. D* **1994**, *50*, 7066–7076. [CrossRef]
269. Papavassiliou, J. On the connection between the pinch technique and the background field method. *Phys. Rev.* **1995**, *D51*, 856–861. [CrossRef]
270. Aguilar, A.C.; Binosi, D.; Ibanez, D.; Papavassiliou, J. New method for determining the quark-gluon vertex. *Phys. Rev.* **2014**, *D90*, 065027. [CrossRef]
271. Alkofer, R.; Fischer, C.S.; Llanes-Estrada, F.J. Vertex functions and infrared fixed point in Landau gauge SU(N) Yang-Mills theory. *Phys. Lett. B* **2005**, *611*, 279–288; Erratum in *Phys. Lett. B* **2009**, *670*, 460–461. [CrossRef]
272. Kugo, T.; Ojima, I. Local Covariant Operator Formalism of Nonabelian Gauge Theories and Quark Confinement Problem. *Prog. Theor. Phys. Suppl.* **1979**, *66*, 1–130. [CrossRef]
273. Nakanishi, N.; Ojima, I. *Covariant Operator Formalism of Gauge Theories and Quantum Gravity*; World Scientific Lectures Notes in Physics: Singapore, 1990; Volume 27. [CrossRef]
274. Alkofer, R.; Huber, M.Q.; Schwenzer, K. Infrared singularities in Landau gauge Yang-Mills theory. *Phys. Rev. D* **2010**, *81*, 105010. [CrossRef]
275. Huber, M.Q. On non-primitively divergent vertices of Yang–Mills theory. *Eur. Phys. J.* **2017**, *C77*, 733. [CrossRef]
276. Aguilar, A.C.; Ferreira, M.N.; Oliveira, B.M.; Papavassiliou, J. Schwinger-Dyson truncations in the all-soft limit: A case study. *Eur. Phys. J. C* **2022**, *82*, 1068. [CrossRef]
277. Celmaster, W.; Gonsalves, R.J. The Renormalization Prescription Dependence of the QCD Coupling Constant. *Phys. Rev.* **1979**, *D20*, 1420.
278. Mintz, B.W.; Palhares, L.F.; Sorella, S.P.; Pereira, A.D. Ghost-gluon vertex in the presence of the Gribov horizon. *Phys. Rev.* **2018**, *D97*, 034020. [CrossRef]
279. Barrios, N.; Peláez, M.; Reinosa, U.; Wschebor, N. The ghost-antighost–gluon vertex from the Curci-Ferrari model: Two-loop corrections. *Phys. Rev. D* **2020**, *102*, 114016. [CrossRef]
280. Cyrol, A.K.; Pawlowski, J.M.; Rothkopf, A.; Wink, N. Reconstructing the gluon. *SciPost Phys.* **2018**, *5*, 065. [CrossRef]
281. Binosi, D.; Tripolt, R.A. Spectral functions of confined particles. *Phys. Lett. B* **2020**, *801*, 135171. [CrossRef]

282. Kern, W.; Huber, M.Q.; Alkofer, R. The spectral dimension as a tool for analyzing non-perturbative propagators. *Phys. Rev.* **2019**, *D100*, 094037. [CrossRef]
283. Horak, J.; Papavassiliou, J.; Pawlowski, J.M.; Wink, N. Ghost spectral function from the spectral Dyson-Schwinger equation. *Phys. Rev. D* **2021**, *104*, 074017. [CrossRef]
284. Horak, J.; Pawlowski, J.M.; Rodríguez-Quintero, J.; Turnwald, J.; Urban, J.M.; Wink, N.; Zafeiropoulos, S. Reconstructing QCD spectral functions with Gaussian processes. *Phys. Rev. D* **2022**, *105*, 036014. [CrossRef]
285. Horak, J.; Pawlowski, J.M.; Wink, N. On the complex structure of Yang-Mills theory. *arXiv* **2022**, arXiv:2202.09333.
286. Kallen, G. On the definition of the Renormalization Constants in Quantum Electrodynamics. *Helv. Phys. Acta* **1952**, *25*, 417. [CrossRef]
287. Lehmann, H. On the Properties of propagation functions and renormalization contants of quantized fields. *Nuovo Cim.* **1954**, *11*, 342–357. [CrossRef]
288. Gross, D.J.; Pisarski, R.D.; Yaffe, L.G. QCD and Instantons at Finite Temperature. *Rev. Mod. Phys.* **1981**, *53*, 43. [CrossRef]
289. Jackiw, R.; Templeton, S. How Superrenormalizable Interactions Cure their Infrared Divergences. *Phys. Rev. D* **1981**, *23*, 2291. [CrossRef]
290. Appelquist, T.; Pisarski, R.D. High-Temperature Yang-Mills Theories and Three-Dimensional Quantum Chromodynamics. *Phys. Rev. D* **1981**, *23*, 2305. [CrossRef]
291. Deser, S.; Jackiw, R.; Templeton, S. Three-Dimensional Massive Gauge Theories. *Phys. Rev. Lett.* **1982**, *48*, 975–978. [CrossRef]
292. Cornwall, J.M. HOW d = 3 QCD RESEMBLES d = 4 QCD. *Physica* **1989**, *A158*, 97–110. [CrossRef]
293. Cornwall, J.M. Exact zero momentum sum rules in d = 3 gauge theory. *Nucl. Phys. B* **1994**, *416*, 335–350. [CrossRef]
294. Alexanian, G.; Nair, V.P. A Selfconsistent inclusion of magnetic screening for the quark - gluon plasma. *Phys. Lett.* **1995**, *B352*, 435–439. [CrossRef]
295. Cornwall, J.M.; Yan, B. String tension and Chern-Simons fluctuations in the vortex vacuum of d = 3 gauge theory. *Phys. Rev. D* **1996**, *53*, 4638–4649. [CrossRef]
296. Cornwall, J.M. On the phase transition in D = 3 Yang-Mills Chern-Simons gauge theory. *Phys. Rev. D* **1996**, *54*, 1814–1825. [CrossRef]
297. Buchmuller, W.; Philipsen, O. Magnetic screening in the high temperature phase of the standard model. *Phys. Lett.* **1997**, *B397*, 112–118. [CrossRef]
298. Jackiw, R.; Pi, S.Y. Seeking an even-parity mass term for 3-D gauge theory. *Phys. Lett.* **1997**, *B403*, 297–303. [CrossRef]
299. Cornwall, J.M. On one-loop gap equations for the magnetic mass in d = 3 gauge theory. *Phys. Rev.* **1998**, *D57*, 3694–3700.
300. Karabali, D.; Kim, C.j.; Nair, V. On the vacuum wave function and string tension of Yang-Mills theories in (2+1)-dimensions. *Phys. Lett.* **1998**, *B434*, 103–109. [CrossRef]
301. Eberlein, F. Two loop gap equations for the magnetic mass. *Phys. Lett.* **1998**, *B439*, 130–136. [CrossRef]
302. Aguilar, A.C.; Binosi, D.; Papavassiliou, J. Nonperturbative gluon and ghost propagators for d=3 Yang-Mills. *Phys. Rev.* **2010**, *D81*, 125025. [CrossRef]

Disclaimer/Publisher's Note: The statements, opinions and data contained in all publications are solely those of the individual author(s) and contributor(s) and not of MDPI and/or the editor(s). MDPI and/or the editor(s) disclaim responsibility for any injury to people or property resulting from any ideas, methods, instructions or products referred to in the content.

 *particles*

Article

Chaos in QCD? Gap Equations and Their Fractal Properties

Thomas Klähn [1,*,†], Lee C. Loveridge [2,*,†] and Mateusz Cierniak [3,†]

1. Department of Physics & Astronomy, California State University Long Beach, Long Beach, CA 90840, USA
2. Los Angeles Pierce College, Woodland Hills, CA 91371, USA
3. Institute of Theoretical Physics, University of Wrocław, 50-204 Wrocław, Poland
* Correspondence: thomas.klaehn@csulb.edu (T.K.); loverilc@piercecollege.edu (L.C.L.)
† These authors contributed equally to this work.

Abstract: In this study, we discuss how iterative solutions of QCD-inspired gap-equations at the finite chemical potential demonstrate domains of chaotic behavior as well as non-chaotic domains, which represent one or the other of the only two—usually distinct—positive mass gap solutions with broken or restored chiral symmetry, respectively. In the iterative approach, gap solutions exist which exhibit restored chiral symmetry beyond a certain dynamical cut-off energy. A chirally broken, non-chaotic domain with no emergent mass poles and hence with no quasi-particle excitations exists below this energy cut-off. The transition domain between these two energy-separated domains is chaotic. As a result, the dispersion relation is that of quarks with restored chiral symmetry, cut at a dynamical energy scale, and determined by fractal structures. We argue that the chaotic origin of the infrared cut-off could hint at a chaotic nature of confinement and the deconfinement phase transition.

Keywords: confinement; dynamical chiral symmetry breaking; quantum chaos; quantum chromodynamics; QCD phase transitions

1. Introduction

In the early 1980s, Benoit Mandelbrot pioneered the methodical study and computational visualization of the iteration of quadratic functions and began to cartograph the emerging fractal landscape [1], which, subsequently, has been named in his honor as the Mandelbrot set. With the advance of personal computers during the mid 1980s, fractals gained broad attention scientifically, as well as in popular science.

In 1986, Leo Kadanoff, in an article with the title "Fractals: Where's the physics?" [2], expressed concerned curiosity about an understanding of fractal properties in physics which goes beyond the identification of fractal dimensions for certain problems. Kadanoff stated that without a better understanding of how physical mechanisms result in a geometrical form, it is difficult to trace *types of questions with interesting answers*. We wish to add that even with a lack of such a deep understanding, it is, of course, possible to find these kinds of questions; as mentioned by Mandelbrot: *"I was asking questions which nobody else had asked before, because nobody else had actually looked at certain structures."* [3].

An example for this explorative approach is Hofstadter's butterfly, which is less publicly known. In 1976, ten years before Kadanoff asked his curious question and four years before Mandelbrot's famous work on the quadratic map, Douglas Hofstadter observed what he called a *recursive structure* in the computed spectrum of electrons in electromagnetic fields [4], which was named after the visual appearance as *Hofstadter's butterfly*. A first experimental confirmation of this theoretical prediction was reported nearly twenty years later in 1997 [5].

There is no strict definition of what a fractal is; however, most people would know one when they see it. Common descriptors of fractals refer to their non-analycity, self-similarity, non-linearity, iterative origin, chaotic behavior, and non-integer (Hausdorff and

---

Citation: Klähn, T.; Loveridge, L.C.; Cierniak, M. Chaos in QCD? Gap Equations and Their Fractal Properties. *Particles* **2023**, *6*, 470–484. https://doi.org/10.3390/particles6020026

Academic Editors: Minghui Ding, Craig Roberts and Sebastian M. Schmidt

Received: 4 February 2023
Revised: 15 March 2023
Accepted: 27 March 2023
Published: 11 April 2023

Copyright: © 2023 by the authors. Licensee MDPI, Basel, Switzerland. This article is an open access article distributed under the terms and conditions of the Creative Commons Attribution (CC BY) license (https://creativecommons.org/licenses/by/4.0/).

other) dimension, to name a few. This paper was motivated by the fact that QCD's gap equations are, by definition, highly non-linear and self-consistent. Self-consistency equates the quantity of interest, or gaps, to a functional which depends on these gaps themselves. QCD's gap equations are organized in a hierarchy of inter-dependencies of an infinite number of n-point Green-functions and it is at the heart of contemporary approaches in this field to identify methods which reduce this infinite number in a manageable way while preserving key features of QCD like dynamical mass generation and confinement. While one can argue how to obtain physically meaningful gap equations, viz. which set of approximations, truncations, etc., is the most reasonable, the self-consistent nature of these equations is not debated. Already at the seemingly simple level of two-point Green functions for a single quark flavor, appropriate truncation schemes allow to one compute the mass spectrum of confined and deconfined quarks. The same methods allow for the computation of meson and baryon spectra. Nothing of this is new and, although neither trivial nor brought to a final solution, it is in a structural sense reasonably well understood and dealt with in Dyson and Schwinger's functional approach, which proved to be a powerful tool to investigate the theories of QCD and QED. We refer to recent reviews for examples and more detailed information [6–13].

Practitioners in the field of Dyson–Schwinger equations frequently deal with problems that can arise from their self-consistent nature. As an example, one technique to solve gap equations is by means of iteration starting from an initial guess. There is no guarantee for the convergence of such an iteration in general nor that the obtained solution is physical. In order to cover 'all possible' solutions in this approach, one would scan over different initial guesses. Typically, one can 'tame' diverging iterations by damping the impact of the iteration itself. Instead of

$$g = F[g] \qquad (1)$$

one can write

$$g = \alpha g + (1-\alpha) F[g] \qquad (2)$$

where g is the gap, F a is functional of the gap, and α a is damping parameter close to but less than one, thus avoiding strong responses of g to the iteration. One can wonder—we claim one should—whether it is justified to apply such an algorithm. It looks innocent in the sense that technically any solution of the original gap equation is a solution of the damped iteration equation. Nevertheless, at identical initial values, both may provide different answers and thus one can claim that the damping parameter might bear unwanted physical significance, as it has been introduced ad hoc. We shall discuss this further in Section 4. What happens if the gap equation is allowed to iterate itself freely? We found only one, recently published, paper which asks exactly this question and comes to a clear conclusion: if the system is strongly coupled, chaos emerges and one can observe an infinite spectrum of 'unexpected' gap solutions with increasing coupling strength [14]. In the paper we present, we provide a brief explanation why these unexpected solutions actually should be expected. Further, we employ a model with momentum-dependent gap solutions. In an iterative and inherently fractal context, this led us on a surprising journey, which answered not all but plenty of the questions we asked and at the end of which we are left to wonder whether looking at QCD as a fractal theory might be a key to understand confinement as an emergent fractal phenomenon. The precise physical significance of our results, if any, are uncertain at this time, but we hope they are hints at future avenues of study. Rather than an attempt to offer new quantitative insights, we consider this work as a first qualitative study to explore features of a QCD-inspired model in a fractal context.

Section 2 briefly motivates how iterative mapping generates new solutions of an equation while preserving the solutions of the non-iterated 'seed' equation; Section 3 reviews the quark matter model by Munczek and Nemirovski (MN) in an extension for dense quark matter. We chose it for our exploration as it exhibits confinement and dynamical chiral symmetry-breaking, while being sufficiently simple to make it well suited for iterative mapping and analytic treatment. The following Section 4 illustrates and cartographs chaotic features which emerge upon iteration of the gap equation. Seeking physical meaning in such iterative chaos must be performed with caution, as chaos is generally a result of the iterative solution method rather than directly a result of the equations, but different solution methods, such as perturbation and lattice approaches, are known to highlight different aspects of the as yet unknown full solution. Thus, Section 5 is a cautious attempt to interpret physical meaning into the interplay of the chaotic and non-chaotic structures we observe. Our study focuses on the structure of the mass pole. The appearance and disappearance of the mass pole are highly driven by chaotic behavior. Further, the mass gap itself is allowed to switch between different, usually distinct solutions. To our surprise, the physical properties of the iterative solutions provide a reasonable picture of how deconfinement could present itself in a model which possesses a gap equation with a single solution only. Finally, we estimate how a finite width gluon interaction could affect the observed behavior of the quark dispersion relation under iteration in Section 6 before we conclude in Section 7.

2. Self-Consistency and the Emergence of New Roots amongst the Old

We investigate the possible consequences of chaos that appears in iterative solutions of non-linear and self-consistent equations in the complex domain. For clarity of what we consider physics and math, we start with the latter and briefly review Mandelbrot's fractal, which is obtained by the iteration $z \xleftarrow{z_0=0; n \to \infty} f(z)$ with the explicit choice $f(z) = z^2 + c$ to obtain the Mandelbrot set. We chose to use the symbol $\xleftarrow{z_0;n}$ to have a distinguished notation for the iterative mapping process—specifying the number of iterations n and the initial value z_0—over the equal sign $=$, which appears in the analytic equation $z = z^2 + c$. It is worthwhile to look at the differences between these two. First, the polynomial equation has exactly two solutions $z_{1,2}$ for any given c, which are defined by the roots of the polynomial $P(z) = f(z) - z = z^2 + c - z$. It is further easily observed that one can determine c for a desired root z_0. For example, $P(z_0 = 0) = 0$ if $c = 0$.

In the iterative approach, each iteration generates a new polynomial,

$$f_1(z,c) = \left(z \xleftarrow{z,1} z^2 + c\right) = \qquad f(z) = z^2 + c,$$

$$f_2(z,c) = \left(z \xleftarrow{z,2} z^2 + c\right) = \qquad f(f(z)) = \left(z^2 + c\right)^2 + c,$$

$$\ldots$$

$$f_m(z,c) = \left(z \xleftarrow{z,m} z^2 + c\right) = \qquad f(f(..f(z)))$$

$$= \qquad f_{m-1}^2(z,c) + c, \qquad (3)$$

etc., ad infinitum. There is one trivial but fascinating property of this infinite set of equations which essentially inspired the presented work. The left-hand side of each of the previous equations was set to $f_i(z,c) = z$ in order to obtain the next iteration $f_{i+1}(z,c) = z$. It is thus safe to state that the roots of $P_1(z,c) = f_1(z,c) - z$ are guaranteed to be roots of $P_2(z,c) = f_2(z,c) - z = f(f_1(z,c)) + c$. As $P_2(z,c)$ is a 4th order polynomial, there are two more roots which, of course, did not appear for $P_1(z,c)$, a second order polynomial. The important lesson to be learned is that for a self-consistent non-linear equation $z = f(z)$, the iteration $z \xleftarrow{z;m} f(z)$ generates a new self-consistent equation. *While the solutions of the non-iterated equation remain a subset of solutions of the iterated equation, the iterated equations can develop additional solutions.*

This is a peculiar, almost awkward situation, if one wishes to assign physical meaning to the original solutions of the equation $f(z) = z$. What makes these roots superior with respect to any of the iterative clones if all, the original and clones, *share* these very same original solutions? Evidently, there is an infinite number of (iterated) functions which share the original roots. Is the original function with *only* these roots a superior or inferior function? Is it worth pondering the meaning of the additional roots of iterated clones? Can we *safely* omit them? Do we miss important information when we ignore the duality of the gap equation as the root-defining equation and mapping rule? We decided to explore and ponder the possible meaning.

3. The Munczek–Nemirovsky Model

One approach to move towards an understanding of QCD is based on evaluating QCD's partition function by testing its response to external sources. This is the Dyson–Schwinger formalism which results in sets of coupled n-point Green functions. Out of these, we are interested in the quark propagator, which is obtained from the gap equation

$$S(p;\mu)^{-1} = i\vec{\gamma}\cdot\vec{p} + i\gamma_4(p_4+i\mu) + m + \Sigma(p;\mu), \qquad (4)$$

with the self-energy

$$\begin{aligned}\Sigma(p;\mu) &= \int \frac{d^4q}{(2\pi)^4} g^2(\mu) D_{\rho\sigma}(p-q;\mu) \\ &\times \frac{\lambda^a}{2}\gamma_\rho S(q;\mu)\Gamma^a_\sigma(q,p;\mu).\end{aligned} \qquad (5)$$

Here, m is the quark bare mass, μ is the quark chemical potential, $D_{\rho\sigma}(p-q;\mu)$ is the dressed gluon propagator and $\Gamma^a_\sigma(q,p;\mu)$ is the dressed quark–gluon vertex. This is the first of an infinite tower of gap equations which, without further approximations, couple back to this one. Further, there are similar equations for the dressed gluon-propagator and the quark–gluon vertex. Note that the gap equation is a self-consistent non-linear (in most cases integral) equation: $S^{-1} = F[S]$.

Within the Munczek–Nemirovsky model [15], the dressed quark–gluon vertex is approximated by the free quark–gluon vertex, $\Gamma^a_\sigma(q,p;\mu) = \frac{\lambda^a}{2}\gamma_\sigma$. Gap equations applying this approximation are referred to as rainbow gap equations. For the dressed gluon propagator, the model is specified by the choice

$$g^2(\mu) D_{\rho\sigma}(k;\mu) = \left(\delta_{\rho\sigma} - \frac{k_\rho k_\sigma}{k^2}\right) 4\pi^4 \eta^2 \delta^4(k). \qquad (6)$$

Due to the δ-function, which in a configuration space corresponds to a constant, this is a very simplified approximation of the gluon–propagator, specified by the coupling strength we set to $\eta = 1.09$ GeV in accordance with [15]. For non-zero relative momentum k, the interaction strength in this model vanishes, thus making it super-asymptotically free. Furthermore, the infrared enhanced δ-function is sufficient to provide for the dynamical chiral symmetry breaking and confinement, both features of QCD which we wish to address. Finally, the δ-function effectively turns the integral gap equation into an algebraic equation which can be solved analytically.

In order to obtain these solutions for the in-medium dressed-quark propagator, one employs the general solution

$$\begin{aligned}S(p;\mu)^{-1} = i\vec{\gamma}\cdot\vec{p}A(\vec{p}^2,p_4) \\ +i\gamma_4(p_4+i\mu)C(\vec{p}^2,p_4) + B(\vec{p}^2,p_4).\end{aligned} \qquad (7)$$

Here, spatial momentum $\vec{p}$ and energy component p_4 of the 4-vector p appear as explicitly distinct degrees of freedom due to the presence of the chemical potential μ. Substitution into the dressed-quark gap-equation and appropriate tracing over the Dirac γ-matrices results in three-coupled gap equations, of which two (for A and C) are identical:

$$A(p,\mu) = 1 + \frac{\eta^2}{2}\frac{A(p,\mu)}{\vec{p}^2 A^2(p,\mu) + B^2(p,\mu)} \qquad (8)$$

$$B(p,\mu) = m + \eta^2 \frac{B(p,\mu)}{\vec{p}^2 A^2(p,\mu) + B^2(p,\mu)}. \qquad (9)$$

We introduced $\tilde{p}^2 = \vec{p}^2 + (p_4 + i\mu)^2$. In the chiral limit ($m = 0$), one finds two distinct sets of solutions; one of them is chirally symmetric and referred to as the Nambu phase,

$$A(p,\mu) = \frac{1}{2}\left(1 \pm \sqrt{1 + \frac{2\eta^2}{\tilde{p}^2}}\right) \qquad (10)$$

$$B(p,\mu) = 0 \qquad (11)$$

whereas for the other solution, the Wigner phase, the chiral symmetry is broken for $\mathcal{R}(\tilde{p}^2) < \eta^2/4$,

$$A(p,\mu) = 2 \qquad (12)$$

$$B(p,\mu) = \sqrt{\eta^2 - 4\tilde{p}^2}. \qquad (13)$$

If the real part $\mathcal{R}(\tilde{p}^2) > \eta^2/4$, the gap solution of the Wigner phase coincides with the Nambu solution. Note that these solutions are obtained in the Euclidean metric, but hold in the Minkowski metric after a simple transformation, $\tilde{p}_E^2 \to \tilde{p}_M^2$, with $\tilde{p}_E^2 = \vec{p}^2 + (p_4 + i\mu)^2$ and $\tilde{p}_M^2 = \vec{p}^2 - (p_4 + i\mu)^2$. Due to our interest in particle mass poles, our investigation of the model is performed in the Minkowski metric. For the next section, however, the specific metric is not relevant; we will only work with the fact that $\tilde{p}^2$ is complex-valued and thus can be decomposed into a real and imaginary part, viz. $\tilde{p}^2 = z_R^2 + i z_I^2$. We chose to label the real and imaginary part with squared quantities as a reminder that they come in units of the energy square.

4. Iterative Chaos

Gap Equations (8) and (9) lead to fourth-order polynomial equations with up to four distinct and complex valued solutions at a given $\tilde{p}^2$ for each gap.

Generally, this is the whole solution space one would consider; the only task left is to identify the one physical solution. However, the self-consistent nature of (8) and (9) is evident and, according to our reasoning in the previous section, there is a possibility for iterated functions with the same four and additional solutions.

Before we discuss our analysis, a few comments should be made. Defined by the contact interaction in a momentum space, we chose a very simple model for the effective gluon propagator. For dressed-gluon propagators with finite width in momentum space, the corresponding gap equations turn into integral equations. Thus, the momenta couple and the simplicity of the MN model, which we take advantage of for this exploration, is lost. We address this issue in more detail in the last section of this paper.

As sketched in the introduction, for models with a sophisticated non-trivial interaction-kernel, the iteration is a practical path to find gap solutions. We outlined before that this leaves us with the possibility that the iteration generates new functions which possess roots that correspond to solutions of the original gap equation and potentially an infinite number of additional roots.

We start our iteration from the non-interacting solution ($B_0 = m$, $A_0 = 1$) and treat $\tilde{p}^2 = z_R^2 + i z_I^2$, as one would consider the constant c for the Mandelbrot set $z \leftarrow z^2 + c$. For the moment, this reduces the number of independent variables from three ($\tilde{p}^2, p_4, \mu$) to two. The result of such an iteration is shown in Figure 1 for the real part of the scalar gap B at two different bare-quark masses of 10 MeV and 100 MeV, respectively.

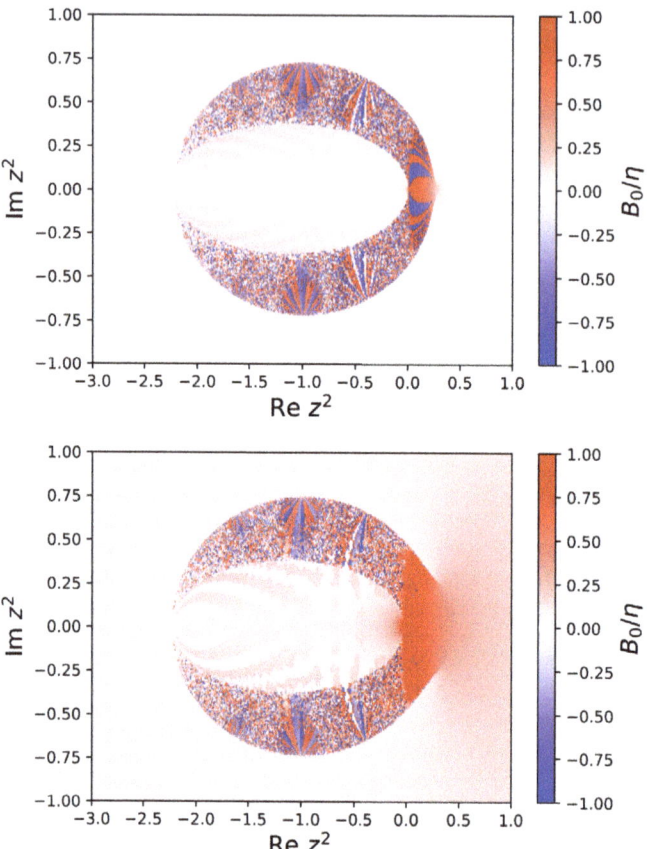

Figure 1. Real part of the scalar gap B after 300 iterations starting from $A_0 = 1$ (**top, bottom**), and $B_0 = m = (10 \text{ MeV}$ (**top**), 100 MeV (**bottom**)).

Unlike the Mandelbrot fractal, this fractal does not diverge; chaos exhibits in domains in which the gaps for infinitesimal changes of energy and momentum take vastly different but finite values in a seemingly random pattern. This fractal region is contained within an almost perfectly shaped ellipsoid, which we fit accordingly with

$$\left(\frac{z_R^2 + z_{R,0}^2}{R_R^2}\right)^2 + \left(\frac{z_I^2}{R_I^2}\right)^2 = 1. \tag{14}$$

$(z_{R,0}^2, R_R^2, R_I^2)$ differs slightly for $m = 10$ MeV $(0.98, 1.26, 0.77)$ MeV^2 and for $m = 100$ MeV $(0.99, 1.28, 0.80)$ MeV^2. The inner almond shape with the less obvious chaotic behavior is well approximated by the same function with $(1.115, 1.085, 0.310)$ MeV^2 for $m = 10$ MeV, and $(1.150, 1.100, 0.340)$ MeV^2 for $m = 100$ MeV. As for the Mandelbrot set, one would be ill advised to understand these figures as a valid representation of the fractal; the appearance of the fractal changes with each new iteration. We identify regions with identical periodicity, ranging from a stable, period one solution in the region outside of the covering ellipse over a period two region within the almond shape, up to higher and higher periodicity in between these two regions. This is illustrated in Figure 2 for $m = 100$ MeV for a periodicity of up to ten.

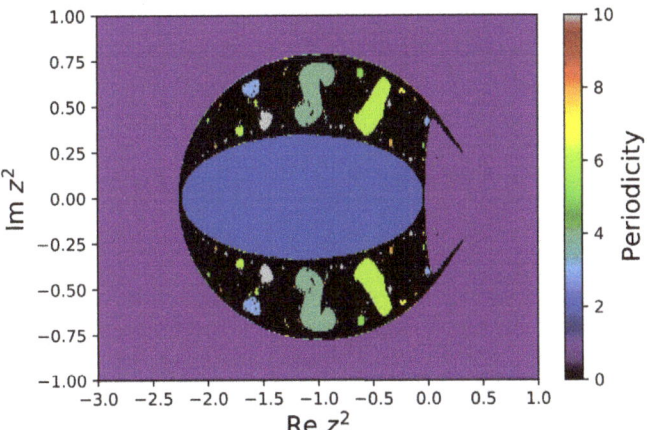

Figure 2. Periodicity of the iterative mass gap solution at $m = 100$ MeV. The outer, indigo region of the plot are absolutely stable under iteration, the inner almond shape has periodicity two, and the area in between exhibits chaos with increasing periodicity. For this plot, areas with periodicity larger than ten are plotted in black.

Keeping in mind that the analytic gap equations possess four distinct solutions, it seems interesting that there is an extended stable domain (periodicity one) which favors one, and only one solution. We follow the gap solution along a vertical path at fixed z_R^2 and vary z_I^2. For reasons which become more clear at a later stage, we chose $z_R^2 = 0.1$ MeV^2. Along this line, one notices that one passes from an outer stable region into an inner stable region by traversing a small chaotic domain. This is illustrated in Figure 3. As within this chaotic domain the value of the gap function can change with each iteration, we plot all obtained values of $\mathcal{R}(B)$ over 300 iterations in gray scale according to how frequently a particular solution has been obtained. Evidently, there is a transition between two distinct analytic solutions of the non-iterated fourth-order polynomial gap equations. This result seems remarkable if one recalls how one would usually deal with different gap solutions for a given model: each solution is understood as a distinct phase, then one examines the stability of each individual solution and picks the energetically favored solution as the physical one. Upon iteration, we are lead to a different conclusion. Although each of the analytic solutions indeed is a solution of the gap equations, only one of them can be stable upon iteration at a given energy and momentum. However, the stable iterative solution over a finite range of energies can switch between distinct analytic solutions. It is further remarkable that the iteratively stable solution is massive (similar to the Wigner solution) when low and massless (similar to the Nambu solution) at high energy. Amongst all the possibilities chaos seems to offer, this seems a very reasonable one. While the exact meaning is unclear, it seems unlikely to be coincidence that iteration favors massive and massless solutions in precisely the energy regimes where confinement and asymptotic freedom are required.

The notion of analytic solutions describing different phases, however, is not supported from an iterative perspective; there is one, and only one, iterative solution to the gap equation.

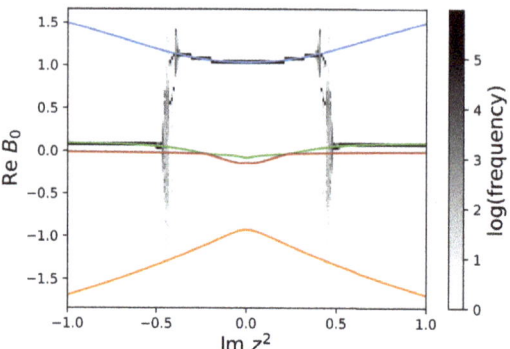

Figure 3. Real part of the mass gap B at $z_R^2 = 0.1$ MeV^2. The color coding indicates how frequently a solution has been found over 300 iterations after the first 100 iterations which are sufficient to shape the fractal as seen. For reference, all analytic solutions to the polynomial gap equations are plotted in color. Iteration switches from massive solutions (blue) at small $\mathcal{I}(z^2)$ to bare-mass solutions (green) at larger values. Except for the chaotic transition domain, the iterative approach picks positive mass-gap solutions, only. Note, that the chaotic domain has solutions of periodicity of two and higher; it is truly unstable. Hence, we add a gray scale to measure the frequency of a particular solution over the final 300 iterations.

Before we go into a further interpretation of what this result implies, we wish to address a question related to the previous paragraph. Initially, we remarked that our iteration starts from the non-interacting solution $A = 1, B = m$. As we try to proceed as carefully as possible, let us investigate the iterative stability of the four analytic gap solutions as plotted in the upper panels of Figure 4, where we demonstrate again the real part of the mass gap. The lower panel of Figure 4 shows the result after 300 iterations of these algebraic solutions as an initial value. It is safe to say that none of them is stable under iteration. Further, there is a visibly favored solution at large values of z_R^2, which does not depend on the initial gap that seeded the iteration. From a global perspective, the fractal keeps the general shape but shows differences for each different seed solution. This is to be expected and would happen in a similar fashion to the Mandelbrot set if the initial value was arbitrarily changed.

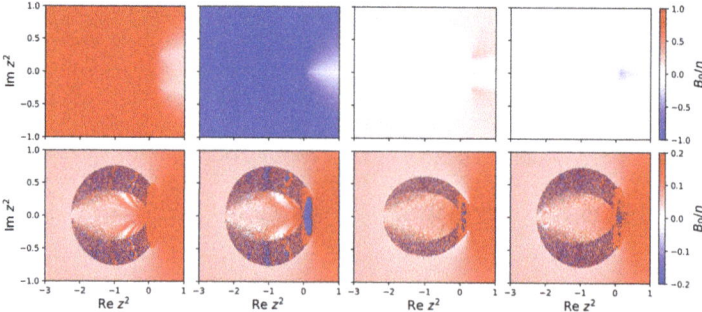

Figure 4. Upper panel: Solutions of the polynomial gap equations for $m = 100$ MeV. Each is plotted on a scale that most accentuates its structure. Solution 1 and 3 (from the left) are stable in some, mutually exclusive domains under iteration, as illustrated in Figure 3. Lower panel: After 300 iterations, using the corresponding solution of the polynomial gap equations from the upper panel as initial seed for the iteration. In the outer, non-chaotic domain, all four cases produce nearly identical results with positive mass gap only.

Comparing the iterations to the algebraic solutions of the gap equations in the upper panel of Figure 4, one can graphically identify which of them is stable under the iteration and in which domain. As observed, this is the case only for the positive mass-gap solutions 1 and 3 from Figure 4, as illustrated in Figure 5. In other words, although the chaotic domain will vary, the described features of Figure 3, with respect to the analytic gap solutions, do not critically depend on the chosen initial gap.

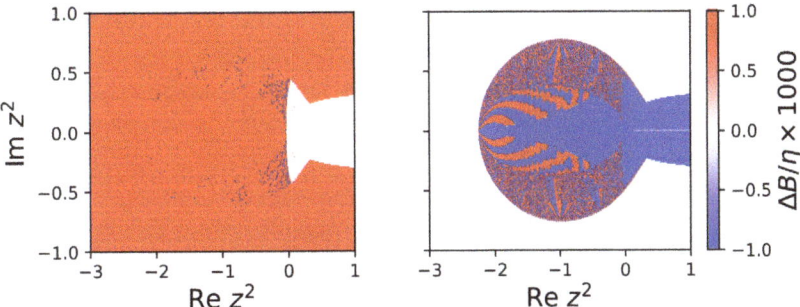

Figure 5. Difference between gap solution 1 and 3 (from the left) in the top panel of Figure 4 and iterative solutions seeded with the non-interacting solution ($A = 1$, $B = m$) after 500 iterations. White domains show no difference between iterative solutions seeded with an analytical model solution or seeded with the non-interacting solution. Solution 2 and 4 show no agreement anywhere in the stable domain of periodicity one (not shown).

This iterative preference for one solution over the others seems to illustrate a case where Equation (1) favors a particular solution, while Equation (2) can be tuned to converge to any of the four analytic solutions. We take a moment, therefore, to discuss this further.

Precisely at an analytic solution, Equation (1) should be an identity so that with infinite precision, all of the solutions should stay precisely at their analytical value. However, any real solution will have at least some error, so that our numerical approximation is only in the neighborhood of the analytical solution. That is

$$g_n = g_a + \epsilon \tag{15}$$

where g_a and g_n represent the analytical solution and its numerical approximation, respectively. If we input g_n into Equation (1), we obtain

$$\begin{aligned} g_n &= F[g_n] \\ &= F[g_a] + \frac{\delta F[g]}{\delta g}\Big|_{g=g_a} \epsilon \\ &= g_a + F'[g_a]\epsilon \end{aligned} \tag{16}$$

Hence, the solution will be stable if and only if the functional's derivative has a magnitude of less than 1.

$$|F'[g_a]| < 1 \tag{17}$$

Equation (2) resolves this so that if $\alpha = 1 - \Delta$ then after iteration

$$g_n = g_a + (1 - \Delta + \Delta F'[g_a])\epsilon. \tag{18}$$

In this situation, we can always choose the sign (or phase) of Δ such that $1 - \Delta + \Delta F'[g_a] < 1$ near a specific analytical solution. Hence, we can make any of the analytical solutions stable by using Equation (2), but at most one such solution is stable under Equation (1), and that solution changes abruptly between very massive and nearly massless behavior in just the locations where we expect massive and massless behavior of the quarks. Whether this is a lucky coincidence, an actual effect of chaos, or a hint at something else in the true and final solution, is yet to be determined.

5. Mass Poles

Up to this point, we refrained from searching for meaning in our study. In spite of the fact that iterative mass gap solutions result in a large domain of chaotic behavior, which may or may not hide future surprises, we cannot help but wonder whether the switching between massive and mass-less gap solutions in the stable domains offers meaning. Before we go further, we want to recall that MN is considered to be a confining model. This is observed by the fact that the inverse propagator has no roots in the chirally broken phase and, therefore, the integration over the four-momentum does not pick up weight to generate a finite particle number. Hence, although confined quarks generate mass via chiral symmetry breaking, the absence of a mass pole results in the absence of a dispersion, viz., there is no explicit relation between specific momenta and energy. For the vacuum MN model, this is easily understood by the realization that in the Minkowski metric, $p^2 + M^2(p^2)$ has no real root if at any p^2, $M^2 > -p^2 = p_4^2 - \vec{p}^2$. This running away of the mass in the chirally broken phase is exactly what happens in the MN model. However, as we have demonstrated in the previous section, the iteration erases the distinction between chirally broken and restored phases and suggests that instead there might be a discontinuous gap solution, which is confinement-like affected by dynamical chiral symmetry at small momenta, and at large momenta chirally unconfined-like and chirally restored. The transition between these domains is characterized by chaotic and unstable solutions (see Figure 3).

At finite chemical potential, the real poles of the propagator $A^2(\vec{p}^2 - (p_4 + i\mu)^2) + B^2$ are represented by $\vec{p}^2 - p_4^2 + \mu^2 + \Re(M^2) = 0$, with $M = B/A$. We note that the shift of the pole due to the chemical potential should not be confused with the physical mass pole of the particle. This becomes evident if one considers an ideal non-interacting gas, with M being constant and real-valued. For the purpose of this study, we refer to the physical mass pole, defined by $\vec{p}^2 - p_4^2 + \Re(M^2) = 0$. From the definition $\tilde{p}^2 = z_R^2 + iz_I^2$, we identify the pole position in our contour plots as $z_R^2 = \mu^2$ and $z_I^2 = -2p_4\mu$ for an ideal particle with constant and real M. This represents a vertical line in our plots, which does not depend on momentum and measures energy with increasing distance from the real axis. It shifts to higher z_R^2 with an increasing chemical potential.

In Figure 6, we trace the physical mass pole in the Minkowski metric by plotting the logarithm of the quantity $(p_3^2 + M^2 - p_4^2)^2$, which gives zero and hence a large negative logarithm at the physical mass pole. As the vertical axis does not depend on the mass ($z_I^2 = 2p_4\mu$), a vertical pole line indicates constant dressed quark masses. We observe the absence of such a well ordered pole structure within the chaotic domain. Since the vertical axis is a measure of the particle energy at a fixed chemical potential, one can conclude that the transition to the massive solution (Figure 3) suppresses quasi-particle behavior in the infrared domain of the model. Again, we can trace the physical pole indicated by the vertical line and find $z_R^2 = \mu^2 - M^2$, since the pole is found at $p_4^2 = p_3^2 + M^2$. Following our elliptic fit of the outer boundary of the fractal domain, this allows one to determine the critical chemical potential where the infrared energy gap entirely disappears

$$\mu_{C,IR} = \sqrt{m^2 + R_R^2 - z_{R,0}^2} \quad . \tag{19}$$

We find $\mu_{C,IR} \approx 625$ MeV for $m = 100$ MeV and $\mu_{C,IR} \approx 540$ MeV for $m = 10$ MeV. At these chemical potentials and beyond, quarks can be considered as completely chirally restored.

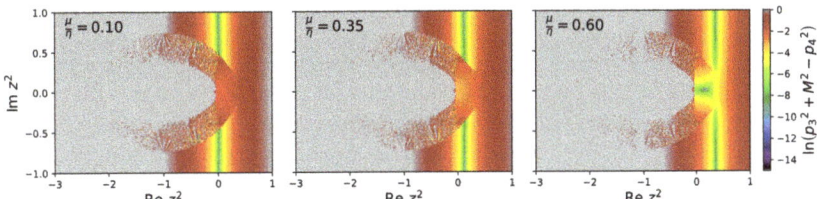

Figure 6. Natural logarithm of $(p_3^2 + M^2 - p_4^2)^2$ for the iterative solution for $\mu = (100, 350, 600)$ MeV (top down) at quark-bare mass $m = 100$ MeV. The vertical line shaped by minimal negative values indicate a physical mass pole, viz. a quasi-particle. In the chaotic domain, this pole structure is absent, viz. the vertical line (or any distinct pole) pattern is absent. This implies an infrared energy gap, below which quarks show no quasi-particle properties. As the chemical potential increases, the quasi-particle pole line moves to the right and simultaneously decreases the gap, viz., the gap region without a pole traces the outer shape of the fractal. Once the chemical potential is sufficiently large, the gap closes entirely. Note that the absence of a mass pole does not imply that there is no mass gap solution, as illustrated in Figure 3.

In order to estimate when mass-pole states can be occupied, we determine at which chemical potential the energy p_4 and the Fermi energy or chemical potential μ turn equal, that is, when $z_I^2 = \mu^2$ on the elliptic boundary of the fractal at the position of the physical mass pole with $M = m$. We choose this scenario, as this is the critical potential starting from where the particle energy is larger than the chemical potential and thus large enough to populate quasi-particle states. This is the case when

$$\left(\frac{\mu^2 - m^2 + z_{R,0}^2}{R_R^2}\right)^2 + \left(\frac{\mu^2}{R_I^2}\right)^2 = 1 \, , \qquad (20)$$

and holds for the light quark with $m = 10$ MeV at $\mu_m \approx 359$ MeV, for the heavy quark with $m = 100$ MeV at $\mu_m \approx 432$ MeV.

Although this is not a rigorous statement, one can roughly relate the critical chemical potential for the transition from a chirally broken mass into the restored phase to the in-vacuum dressed-quark mass. In our case, the situation is a bit different. We estimate a hypothetical chirally broken quark vacuum mass based on the previous estimate of the critical potential for the complete disappearance of the infrared gap by setting them approximately equal. Relating μ_m as the onset of a deconfined chirally restored quark phase with an estimate of the constituent quark mass seems to provide rather reasonable results in comparison to other model calculations. This is interesting, considering that in the MN model, the vacuum mass at zero 4-momentum is defined by the coupling strength η, which is of the order of 1 GeV.

It is noteworthy that our simple approach reproduces quantities related to the effective constituent masses at reasonable values. We state explicitly that in this model, constituent masses are nowhere realized for a physical particle, viz. an entity with a mass pole of that magnitude. We can compare the light quark critical chemical potential $\mu_m \approx 359$ MeV with the deconfinement critical potential obtained within the MN model in a Euclidean metric with a value of 300 MeV [16] or with subsequent work based on a widened version of the effective gluon propagator [17], which predicts deconfinement at a chemical potential of 380 MeV. There is a satisfying agreement of these values with ours. We point out though, that both of these models are defined within a different metric, as slightly different bare quark masses and, most importantly, are based on entirely different assumptions. While the two previous papers employed distinct gap solutions and compare the pressure of the corresponding mass-less Wigner and massive Nambu phase, our approach results in only one gap solution which exhibits a transition from the Nambu to the Wigner phase through a chaotic domain, as depicted in Figure 3. Our quarks are either bare-mass quarks with poles or entities with a chaotic mass function, or a dressed quark mass different from the bare mass with no associated pole. In the latter case, there is a chaotic transition from dressed quark masses to bare quark masses with increasing energy.

6. Finite Interaction Width

We begin the final section of this paper with a plot of the particle pole in an energy momentum space which we obtain by transforming (z_R^2, z_I^2) to $(p_3 = |\vec{p}|, p_4)$ coordinates under explicit choices of the chemical potential, as noted in Figure 7. Although this switch in representation does not provide additional information, we find it instructive to provide an actual dispersion relation obtained from the iterative approach. In this example, at a chemical potential of 700 MeV, no chaotic behavior is visible and the dispersion is exactly that of a free quark at bare-mass 100 MeV. With the decreasing chemical potential chaos, there emerges, at energies higher than that of the expected (now absent), the free particle dispersion. The actual dispersion branch is cut clean at some critical value (as we discussed in the previous section), thus illustrating our interpretation of the fractal boundary as the cause for a dynamical infrared cutoff, below which quarks are mass-pole-free.

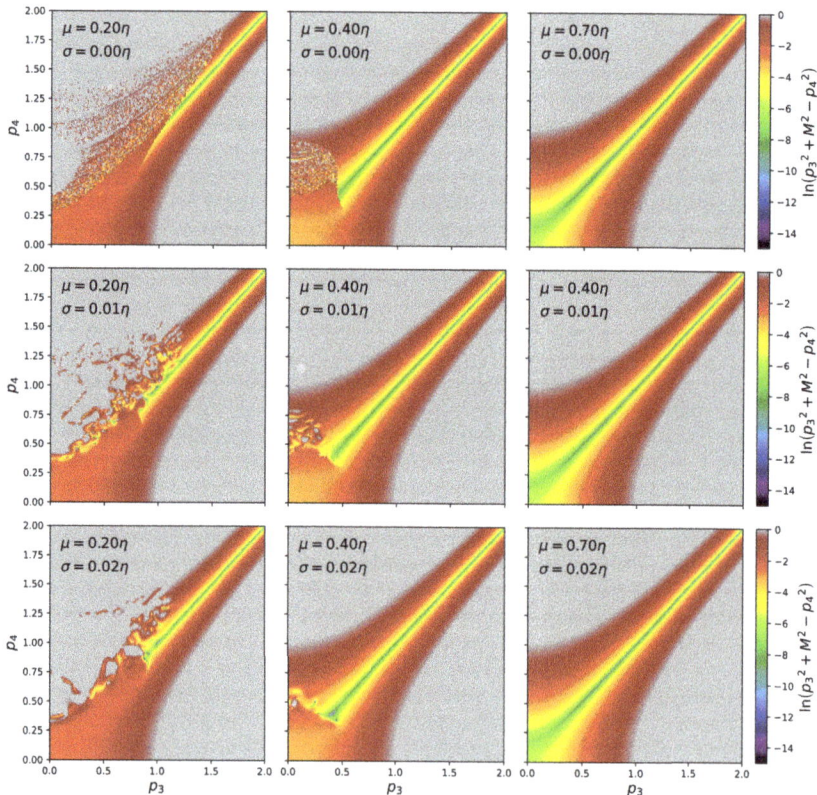

Figure 7. Plotted is the logarithm of the mass-pole condition $\log(|\vec{p}^2 - p_4^2 + \mu^2 + \Re(M^2)|)$, which shows a dispersion relation with distinct, chaos-induced infrared cut-off. With increasing chemical potential ($m = 0.1\eta$; $\mu = (0.2, 0.4, 0.7)\eta$ from **left** to **right**), the infrared cut-off decreases and eventually disappears. With increasing widening ($\sigma = (0.00, 0.01, 0.02)\eta$ from **top** to **bottom**), chaotic domains blur but the observed IR cut-off remains.

Presently, we address a last question which relates to the fact that the MN model is based on the very particular choice of the effective gluon propagator as a δ-function in the four-momentum space. This is the reason that we could easily perform the presented study based on this Ansatz and the subsequent decoupling of momenta, which allows one to iterate point-wise for any given four-momentum without coupling to other momenta. This might raise the suspicion that momentum coupling could destroy the fractal structure we observed. In order to keep the simplicity of the gap equations but still obtain an idea about the stability of the emergent fractal, we averaged each point in our plane after each iteration step and thus mimicked some kind of momentum coupling. The averaging is based on Gaussian weights around a given point according to

$$g^2 D^{\mu\nu}(k) = 3\pi^4 \eta^2 \delta^{\mu\nu} \frac{\exp(k^2/w^2)}{\int \exp(k^2/w^2) d^4 k'}, \qquad (21)$$

where w is the width of the Gaussian. To further simplify, we assumed that the widening only happens in the direction of momentum and energy, i.e., there is no widening perpendicular to the momentum.

As observed in Figure 7, the separation into the chaotic pole-free and non-chaotic mass-pole domains remains, even when we change the momentum dependence of the gluon from a delta function to a Gaussian with a half width as much as 0.02 times the gluon mass. We find numerical evidence that this feature remains even with a width as much as 2% of the gluon mass, ≈ 20 MeV. This corresponds to a spatial width of about 10 fm, which is roughly one order of magnitude larger than the size of a proton. Based on this—certainly simplified—treatment of momentum coupling, we conclude that the statements we make in this paper may indeed survive a more complete treatment involving self-interactions with globally coupled momenta—which has been our main concern, prompting this final analysis.

7. Conclusions

As we have demonstrated, a strictly iterative solution of the MN gap equations results in fractal gap structures which can be characterized by the existence of three qualitatively very different, yet co-existing domains of a single and unique gap solution: a bare-mass quark quasi-particle domain with physical mass poles extending infinitely into the ultraviolet, a dressed-mass quark domain without mass poles and hence no quasi-particle interpretation in the infrared, and a chaotic domain of transition between the first two phases. Remarkably, the two non-chaotic domains correspond to distinct analytic solutions which would usually represent individual phases with either dynamically broken or restored chiral symmetry. The fractal approach offers an alternative to this separation which is rooted in the iterative nature of the gap equation.

Further, it is noteworthy that the iterative mass gap solution is always positive in the smooth, viz. non-chaotic domain of the fractal. The appearance of a chaotic boundary between two qualitatively different domains results in interesting properties:

(I)   The iterative approach provides an ultraviolet cut-off for the massive and mass-pole-free Nambu solution, as this solution appears only within the elliptic region of the $(z_R^2, z_I^2))$ plane. Thus, the approach avoids the appearance of an infinitely increasing dressed-quark mass with increasing momentum and energy. In the MN model, this running mass results in the absence of mass poles for the massive gap solution and thus relates to confinement.

(II) It provides an infrared cut-off for the bare-quark mass Nambu solution and thus ensures that quasi-particle states are not populated at a small chemical potential, although the quark can *virtually* exist as a quasi-particle with well defined dispersion.

(III) Both cutoffs more or less coincide (as observed in Figure 5), although there is a transition region which is chaotic in nature. The resulting effective Nambu-UV/Wigner-IR cutoff depends dynamically on energy, momentum, bare mass, and chemical potential. As a side note, we add that plotting gap solutions in the (z_R^2, z_I^2) plane removes much of the dynamical arbitrariness and leaves the ratio of the bare mass m and coupling constant η as the only 'true' degree of freedom; viz., a change of the chemical potential μ would rescale the plot but cause no qualitative change, whereas plots such as Figure 1 indeed demonstrate 'the' gap solution at an arbitrary chemical potential.

(IV) Sufficiently large chemical potential bare-quark mass-pole states will form at energies which can be populated; thus, physical quarks can exist as quasi-particle excitations.

A mechanism with these properties can be interpreted as a deconfinement mechanism. The appearance of one, and only one, iterative solution of the MN gap equations bears a certain elegance. First, it is by the very fact that there is only one gap solution with expected properties, being the existence of only a positive mass gap, asymptotically restored chiral symmetry, and the absence or appearance of physical mass poles. Next, it builds on distinct solutions which one would obtain in the non-iterative approach but provides a new meaning by slicing them into a single new solution with the aforementioned properties.

A simple treatment of a widened, δ-like gluon interaction indicates that momentum coupling blurs chaotic domains but does not necessarily change the qualitative results we describe if the widening is moderate. As this study is an exploration and qualitative in nature, we look forward to further analyses of this perspective on understanding the confinement and the deconfinement transition as highly non-linear and, to a certain extent, with possibly chaotic phenomena.

Author Contributions: Conceptualization, T.K. and L.C.L.; methodology, T.K., L.C.L. and M.C.; numerical analysis, L.C.L. and M.C.; writing—original draft preparation, T.K. and L.C.L.; writing—review and editing, L.C.L. and M.C.; visualization: L.C.L. All authors have read and agreed to the published version of the manuscript.

Funding: M.C. acknowledges support from the Polish National Science Centre (NCN) under grant no. 2019/33/B/ST9/03059.

Data Availability Statement: No new data were created or analyzed in this study. Data sharing is not applicable to this article.

Acknowledgments: We are grateful to Prashanth Jaikumar, Pok Man Lo, and Craig D. Roberts for helpful comments and discussions, which helped us a great deal to find order in the chaos.

Conflicts of Interest: The authors declare no conflict of interest.

References

1. Mandelbrot, B.B. Fractal Aspects of the Iteration of $z \to \Lambda z(1-z)$ for Complex Λ and z. *Ann. N. Y.Acad. Sci.* **1980**, *357*, 249–259. [CrossRef]
2. Kadanoff, L. Fractals: Where's the Physics? *Phys. Today* **1986**, *39*, 6. [CrossRef]
3. Stories, W.B. Mandelbrot about: Drawing; the Ability to Think in Pictures and Its Continued Influence. Available online: https://www.webofstories.com/play/benoit.mandelbrot/8 (accessed on 25 July 2020).
4. Hofstadter, D.R. Energy levels and wave functions of Bloch electrons in rational and irrational magnetic fields. *Phys. Rev. B* **1976**, *14*, 2239–2249. [CrossRef]
5. Kuhl, U.; Stöckmann, H.J. Microwave Realization of the Hofstadter Butterfly. *Phys. Rev. Lett.* **1998**, *80*, 3232–3235. [CrossRef]
6. Roberts, C.D. Three Lectures on Hadron Physics. *J. Phys. Conf. Ser.* **2016**, *706*, 022003. [CrossRef]
7. Horn, T.; Roberts, C.D. The pion: An enigma within the Standard Model. *J. Phys. G* **2016**, *43*, 073001. [CrossRef]
8. Eichmann, G.; Sanchis-Alepuz, H.; Williams, R.; Alkofer, R.; Fischer, C.S. Baryons as relativistic three-quark bound states. *Prog. Part. Nucl. Phys.* **2016**, *91*, 1–100. [CrossRef]
9. Burkert, V.D.; Roberts, C.D. Colloquium: Roper resonance: Toward a solution to the fifty year puzzle. *Rev. Mod. Phys.* **2019**, *91*, 011003. [CrossRef]
10. Fischer, C.S. QCD at finite temperature and chemical potential from Dyson–Schwinger equations. *Prog. Part. Nucl. Phys.* **2019**, *105*, 1–60. [CrossRef]
11. Roberts, C.D.; Schmidt, S.M. Reflections upon the Emergence of Hadronic Mass. *Eur. Phys. J. Spec. Top.* **2020**, *229*, 3319–3340. [CrossRef]
12. Qin, S.X.; Roberts, C.D. Impressions of the Continuum Bound State Problem in QCD. *Chin. Phys. Lett.* **2020**, *37*, 121201. [CrossRef]
13. Barabanov, M.Y.; Bedolla, M.A.; Brooks, W.K.; Cates, G.D.; Chen, C.; Chen, Y.; Cisbani, E.; Ding, M.; Eichmann, G.; Ent, R.; et al. Diquark Correlations in Hadron Physics: Origin, Impact and Evidence. *Prog. Part. Nucl. Phys.* **2021**, *116*, 103835. [CrossRef]
14. Martínez, A.; Raya, A. Solving the Gap Equation of the NJL Model through Iteration: Unexpected Chaos. *Symmetry* **2019**, *11*, 492. [CrossRef]

15. Munczek, H.J.; Nemirovsky, A.M. Ground-state $q\bar{q}$ mass spectrum in quantum chromodynamics. *Phys. Rev. D* **1983**, *28*, 181–186. [CrossRef]
16. Klahn, T.; Roberts, C.D.; Chang, L.; Chen, H.; Liu, Y.X. Cold quarks in medium: An equation of state. *Phys. Rev. C* **2010**, *82*, 035801. [CrossRef]
17. Chen, H.; Yuan, W.; Chang, L.; Liu, Y.X.; Klahn, T.; Roberts, C.D. Chemical potential and the gap equation. *Phys. Rev. D* **2008**, *78*, 116015. [CrossRef]

Disclaimer/Publisher's Note: The statements, opinions and data contained in all publications are solely those of the individual author(s) and contributor(s) and not of MDPI and/or the editor(s). MDPI and/or the editor(s) disclaim responsibility for any injury to people or property resulting from any ideas, methods, instructions or products referred to in the content.

Review

Generalised Parton Distributions in Continuum Schwinger Methods: Progresses, Opportunities and Challenges

Cédric Mezrag

Institut de Recherche sur les lois Fondamentales de L'univers, Commissariat à L'énergie Atomique et aux Énergies Alternatives, Université Paris-Saclay, F-91191 Gif-sur-Yvette, France; cedric.mezrag@cea.fr

Abstract: This paper review the modelling efforts regarding Generalised Parton Distributions (GPDs) using continuum techniques relying on Dyson–Schwinger and Bethe–Salpeter equations. The definition and main properties of the GPDs are first recalled. Then, we detail the strategies developed in the last decade in the meson sector, highlighting that observables connected to the pion GPDs may be measured at future colliders. We also highlight the challenges one will face when targeting baryons in the future.

Keywords: generalised partons distributions; continuum Schwinger methods; lightfront wave functions

Citation: Mezrag, C. Generalised Parton Distributions in Continuum Schwinger Methods: Progresses, Opportunities and Challenges. Particles 2023, 6, 262–296. https://doi.org/10.3390/particles6010015

Academic Editors: Minghui Ding, Craig Roberts, Sebastian M. Schmidt and Armen Sedrakian

Received: 9 January 2023
Revised: 30 January 2023
Accepted: 2 February 2023
Published: 8 February 2023

Copyright: © 2023 by the author. Licensee MDPI, Basel, Switzerland. This article is an open access article distributed under the terms and conditions of the Creative Commons Attribution (CC BY) license (https://creativecommons.org/licenses/by/4.0/).

1. Introduction

Since the early days of QCD and the major result of Bjorken scaling [1] followed by scaling violations [2–4], hadron structure has been one of the main topics of study regarding the strong interaction. In the 1990s, the family of matrix elements connected to the structure of hadron went from uni-dimensional, Parton Distribution Functions (PDFs) and Electromagnetic Form Factors (EFF), to multi-dimensional with the introduction of Generalised Parton Distributions (GPDs) [5–7] and Transverse Momentum Dependent PDFs (TMDs) [8,9]. The former are connected to the 2D+1D picture of the nucleon [10,11] while the TMDs provide a 3D picture of momentum space. Both are defined from a matrix element of the type

$$\langle p_{\text{out}} | O\left(\frac{-z}{2}, \frac{z}{2}\right) | p_{\text{in}} \rangle, \qquad (1)$$

where, in the case of GPDs, the distance z is lightlike ($z^2 = 0$) and the momentum transfer $\Delta = p_{out} - p_{in}$ is finite. For TMDs, the distance z is off the lightcone ($z^2 \neq 0$) but no momentum transfer is allowed ($\Delta = 0$). They both generalise PDFs for which $z^2 = 0$ and $\Delta = 0$. Similarly, TMDs and GPDs were latter unified as two distinct limits of Generalised Transverse Momentum Dependent PDFs (GTMDs), for which both $z^2 \neq 0$ and $\Delta \neq 0$ [12–17]. GTMDs are thought to provide a 5D picture of hadrons as illustrated in Figure 1.

A wealth of theoretical studies has been dedicated in the last decade to GPDs, TMDs and GTMDs. Since it remains unclear whether some processes could be sensitive to GTMDs (despite some pioneering studies [18]), TMDs and GPDs stand at the core of experimental studies of current and future facilities. The 12 GeV upgrade of the electron beam of the Jefferson Laboratory facility (JLab 12) has been completed. JLab 12 is thus expected to deliver a wealth of very precise data that can be connected to GPDs and TMDs in the forthcoming years (and in fact data release has already started, see, e.g., [19,20]), mostly in the so-called valence region. In the next decade, the experimental community is expected to move from fixed target to collider experiments, with the planned electron-ion colliders both in the US (EIC) and China (EicC).

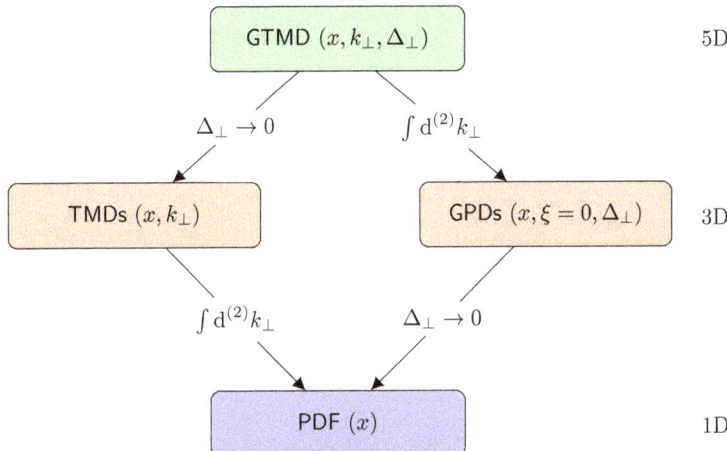

Figure 1. Family of distributions encoding hadron structure, where x is the momentum fraction along the lightcone carried by the active parton, $k_\perp$ its transverse momentum (the Fourier conjugate of $z_\perp$). $\Delta_\perp$ is the transverse momentum transferred between the initial and final hadron state. At this stage, the integrals over $k_\perp$ should be understood as formal only, and need to be regularised, both from TMDs to PDFs [8,21,22] and from GTMDs to GPDs [23]. This picture is valid for vanishing skewness ξ, i.e., no momentum transfer along the lightcone is allowed.

In parallel with experimental facilities, ab initio computational methods have made major progress in the last decade. Lattice-QCD is now able to compute data related directly to the x-dependence of distributions [24–26] instead of solely the first few Mellin moments. This breakthrough has triggered many computing efforts, attempting to perform computations of Distribution Amplitudes (DA), PDFs and GPDs (see, e.g., [27–29]). These steps are very promising but many more remain to be taken before precision studies can be performed on the lattice.

In parallel, non-perturbative studies using continuum techniques such as Bethe–Salpeter equations have also been developed in the past decade, targeting the x-dependence of hadron structure, after the computations of local operators (mostly the EFFs). The main developments came from the use of spectral representations (or Nakanishi representations [30,31]) and analysis of the singularities in the complex plane (see the recent example [32,33]). The breakthrough in the computation of QCD three-point functions [34–36] is also very promising in terms of the ability to reach realistic computations of hadron structure and there interpretation (see, for instance, the case of the gluon mass generation through the Schwinger mechanism [37]). Indeed, continuum techniques based on Dyson–Schwinger and Bethe–Salpeter equations offer a unique window to understand the emergent phenomena in QCD through their ability to select subsets of QCD effects in systematic ways which carefully guarantee that underlying QCD symmetries remain fulfilled.

In this review article, we will focus on GPDs. The latter have already been the main topic of several theoretically oriented [38,39] and phenomenologically oriented [40] review papers and lectures (see, for instance, [41]). Here, we will highlight the efforts in the last decade regarding their computations using continuum formalisms, and thus partly updating them [42]. In Section 2, we will remind the readers the formal definition of GPDs, the properties they should obey from QCD, and explain why phenomenological extractions are challenging. In Section 3, we will discuss the efforts which have been undertaken to compute these GPDs in the meson sector, highlighting the successes and challenges. Then, in Section 4, we will review the pioneering work in the baryon sector, and highlight the future possibilities offered by the continuum formalism.

2. Generalised Parton Distributions

2.1. Formal Definitions and First Properties

Introduced in the 1990s in three distinct series of papers [5–7,43,44], GPDs can be formally defined as off-diagonal matrix elements of a non-local operator in momentum space. The operator is evaluated at light-like distances such that, in the case of the pion, GPDs are defined as:

$$H_\pi^q(x,\xi,t) = \frac{1}{2}\int \frac{dz^-}{2\pi} e^{ixP^+z^-} \langle p_2|\bar{\psi}^q\left(-\frac{z^-}{2}\right)\mathcal{W}\left(-\frac{z^-}{2},\frac{z^-}{2}\right)\gamma^+\psi^q\left(\frac{z^-}{2}\right)|p_1\rangle, \quad (2)$$

$$H_\pi^g(x,\xi,t) = \frac{1}{P^+}\int \frac{dz^-}{2\pi} e^{ixP^+z^-} \langle p_2|G^{+\mu}\left(-\frac{z^-}{2}\right)\mathcal{W}\left(-\frac{z^-}{2},\frac{z^-}{2}\right)G_\mu^+\left(\frac{z^-}{2}\right)|p_1\rangle, \quad (3)$$

where ψ^q is a quark field of flavour q and $G^{\mu\nu}$ is the gluon field strength. The $+$ component indicates the lightcone direction following the standard conventions of:

$$z^\pm = \frac{1}{\sqrt{2}}(z^0 \pm z^3), \quad z^\perp = (z^1, z^2), \quad (4)$$

$$z^2 = 2z^+z^- - z_\perp^2. \quad (5)$$

The average momentum of the hadron P and the momentum transfer Δ are conveniently defined as:

$$P = \frac{p_1+p_2}{2}, \quad \Delta = p_2 - p_1. \quad (6)$$

This allows us to simply express the variables (ξ, t) whose definition is not manifest in Equations (2) and (3) through:

$$\xi = -\frac{\Delta^+}{2P^+}, \quad t = \Delta^2. \quad (7)$$

The definition domain of GPDs in x and ξ is illustrated in Figure 2. Finally, let us mention that $\mathcal{W}$ is the Wilson line defined as:

$$\mathcal{W}\left(-\frac{z^-}{2},\frac{z^-}{2}\right) = P\exp\left[ig\int_{-\frac{z^-}{2}}^{\frac{z^-}{2}} d\zeta^- A^+(\zeta^-)\right] \quad (8)$$

where P is the path ordering between $-z^-/2$ and $z^-/2$ and g is the QCD coupling.

Because of its richer spin structure, more GPDs are necessary in order to parametrise the off-forward matrix element of the nucleon:

$$\frac{1}{2}\int \frac{dz^-}{2\pi} e^{ixP^+z^-} \langle p_2|\bar{\psi}^q\left(-\frac{z^-}{2}\right)\gamma^+\mathcal{W}\left(-\frac{z^-}{2},\frac{z^-}{2}\right)\psi^q\left(\frac{z^-}{2}\right)|p_1\rangle$$

$$= \frac{1}{2P^+}\left(H^q(x,\xi,t)\bar{u}(p_2)\gamma^+u(p_1) + E^q(x,\xi,t)\bar{u}(p_2)\frac{i\sigma^{+\nu}\Delta_\nu}{2M}u(p_1)\right), \quad (9)$$

for quarks and

$$\frac{1}{P^+}\int \frac{dz^-}{2\pi} e^{ixP^+z^-} \langle p_2|G^{+\mu}\left(-\frac{z^-}{2}\right)\mathcal{W}\left(-\frac{z^-}{2},\frac{z^-}{2}\right)G_\mu^+\left(\frac{z^-}{2}\right)|p_1\rangle$$

$$= \frac{1}{2P^+}\left(H^g(x,\xi,t)\bar{u}(p_2)\gamma^+u(p_1) + E^g(x,\xi,t)\bar{u}(p_2)\frac{i\sigma^{+\nu}\Delta_\nu}{2M}u(p_1)\right), \quad (10)$$

for gluons. Quark and gluon polarised distributions can also be defined for the nucleon (see, e.g., [38]). However, since we do not use them in the following, we do not introduce them here. Similarly, transversity GPDs are not introduced in this review paper.

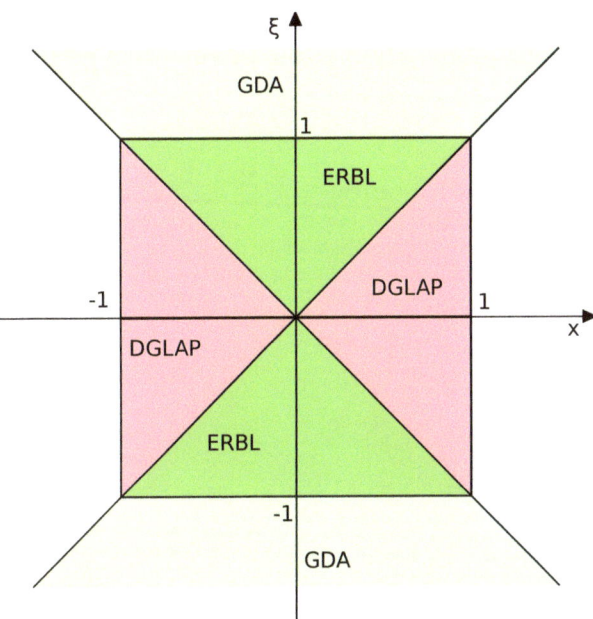

Figure 2. The GPD definition domain in x and ξ. The so-called DGLAP (or outer) region for which $|x| \geq |\xi|$ is shaded in pink, while the ERBL (or inner) region for which $|x| \leq |\xi|$ is shaded in green. The ERBL region can be extended for $|\xi| \geq 1$ (lighter green) where GPDs can be related to Generalised Distribution Amplitude [45–48] through analytic continuations thanks to the crossing symmetry.

The off-forward nature of GPDs triggers interesting consequences, as the incoming and outgoing partons carry a momentum fraction expressed as $x \pm \xi$. Several cases can be drawn, as illustrated in Figure 2. The complete GPD support draws a square such that $(x, \xi) \in [-1, 1]^2$ (see ref. [49] for the derivation of the GPD definition domain). Within this domain, two types of regions can be highlighted: the outer one for which $|x| \geq |\xi|$, and the inner one where $|\xi| \geq |x|$. The former is called the DGLAP region (Dokshitzer, Gribov, Lipatov, Altarelli, Parisi) while the latter is labelled the ERBL region (Efremov–Radyushkin–Brodsky–Lepage). This distinction comes from the fact that the two regions present different physical interpretations when considering probed partons. Indeed, the incoming and outgoing partons momentum fraction can be expressed as $x \pm \xi$. Thus, in the quark sector, depending of the sign of this combination, one probes a quark or an antiquark. Figure 3 illustrates the different possibilities highlighting the specificity of the inner region with respect to the outer ones. There, the interpretation yields an exchange of a pair of quark and anti-quark in the t channel. On the other hand, the outer region can be seen as probing an active quark by taking it out and putting it back within the nucleon. These different interpretations have a major impact as the evolution equations will significantly vary between the two regions, hence the two names from famous evolution equations, DGLAP [2–4] and ERBL [50–53].

The discussion of evolution properties leads us to mention the symmetry of GPDs regarding the lightcone momentum fractions. Pion and nucleon GPDs present interesting symmetry properties: they are even in ξ because of time reversal invariance (see, e.g., [39] for a proof and list of exceptions for higher spin hadrons). In addition, gluon GPDs are even in x within our definition, while quark GPDs have no specific x-parity. Nevertheless, x-even and x-odd combinations are often introduced as:

$$H_q^- = H^q(x,\xi,t) + H^q(-x,\xi,t), \quad (11)$$
$$H_q^+ = H^q(x,\xi,t) - H^q(-x,\xi,t), \quad (12)$$

where $+$ and $-$, respectively, mean singlet and non-singlet combinations. Note that in case of multiple quark flavour, the real singlet component has to be summed over the flavours. These combinations are often introduced as the singlet combination mixes with gluons under evolution, while the non-singlet one does not.

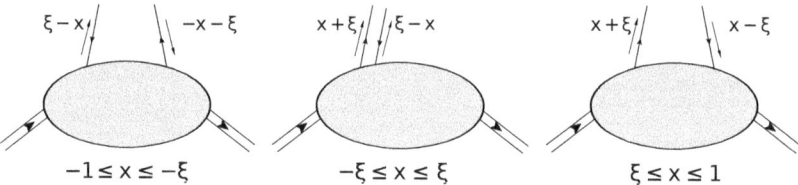

Figure 3. The DGLAP or ERBL regions and their different interpretations in terms of partons. From left to right, one goes through the antiquark interpretation of the probed parton. Then, in the inner region $-\xi < x < \xi$, one recovers the ERBL region and its interpretation as the "extraction" of a quark–antiquark pair from the hadron. Finally, on the right-hand side, $\xi < x < 1$ and we recover the quark interpretation of the DGLAP region.

2.2. Reduction to Unidimensional Distributions

When taking the forward limit of the matrix elements defined in (2) and (3), i.e., when no momentum transfer is allowed, one recovers the standard PDFs as:

$$H^q(x,0,0) = q(x)\theta(x) - \bar{q}(-x)\theta(-x), \quad (13)$$
$$H^g(x,0,0) = xg(x)\theta(x) - xg(-x)\theta(-x), \quad (14)$$

where θ is the Heaviside function. Through the forward limit, PDFs have been a key ingredient in GPD modelling strategies since they were introduced [54–59].

GPDs are also connected with EFFs through integration over the x variable. The quark flavour q contribution to the Dirac form factor F_1 and the Pauli form factor F_2 can be obtained from GPDs as:

$$\int dx\, H^q(x,\xi,t) = F_1^q(t) \quad (15)$$
$$\int dx\, E^q(x,\xi,t) = F_2^q(t) \quad (16)$$

Until now, the EFFs sum rules presented here have been one of the main constraints applied to model the t-dependence of GPDs.

In the pion case, one can highlight an additional property. In the chiral limit, the pion quark GPD can be related to the pion distribution amplitude [46,60]:

$$H_q^-(x,1,0) = \varphi\left(\frac{1+x}{2}\right), \quad (17)$$
$$H_q^+(x,1,0) = 0. \quad (18)$$

This property is usually labelled "soft pion theorem" in the literature. As we mention above, since the singlet and gluon GPDs mix under evolution, Equation (18) implies that the gluon GPD should also vanish in the chiral limit for $(\xi,t) \to (1,0)$.

2.3. Interpretation in Coordinate Space

GPDs are naturally defined in momentum space, as the momentum transfer Δ is the relevant experimental variable to be measured. Nevertheless, one can also define GPDs in the so-called impact parameter space (see, e.g., refs. [10,11]), by introducing the impact parameter $b_\perp$. The latter measures the distance from the centre of +-momentum within the nucleon and is connected to the momentum variable as being the Fourier conjugate of the vector D[11]:

$$D = \frac{p_2}{1-\xi} - \frac{p_1}{1+\xi}. \tag{19}$$

One can then show that a 1+2D probability density $\rho(x, b_\perp)$ can be recovered through the so-called Hankel transform of the GPDs for $\xi = 0$. In this limit, $D_\perp = \Delta_\perp$ and

$$\rho(x, 0, |b_\perp|) = \frac{1}{4\pi} \int_0^\infty d\left(\Delta_\perp^2\right) J_0(|\Delta_\perp||b_\perp|) H(x, 0, t), \tag{20}$$

where J_0 is the standard Bessel function of the first kind of order 0. Note that the impact parameter $b_\perp$ should not be confused with the Fourier conjugate of $k_\perp$ also labelled b by the TMD community.

An illustration of this probabilistic interpretation is given in Figure 4 and comes from model computations (see ref. [60] for details). In principle, it is also possible to extract these 3D distributions from experimental data (see, for instance, [61] for a recent example on the nucleon). However, three main difficulties arise:

- Collinear factorisation allows one to interpret exclusive processes in terms of GPDs for values of t much smaller than the typical hard scale of the system;
- Yet, performing the Fourier transform requires to integrate over t up to infinity, introducing model-dependent extrapolations;
- Furthermore, no experimental data is available for vanishing values of ξ, meaning that additional extrapolations generating more model biases are required.

These difficulties hardly alter the appealing possibilities offered by GPDs to map in 3D the average position of quarks and gluons within hadrons. This explains the enthusiasm in the field towards placing GPDs at the core of the physics cases of running or planned experimental facilities, such as the US [62] and Chinese [63] electron ion colliders.

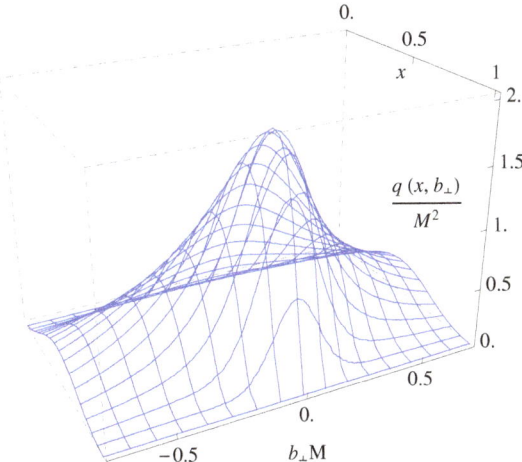

Figure 4. 3D picture of a model computation of the pion quark GPD in the impact parameter space. Figure from [60].

2.4. Connection with the Energy-Momentum Tensor

The ability to perform the 1+2D tomography of the nucleon is not the only exciting feature of GPDs. Indeed, they also allow a unique access to the hadrons' Energy-Momentum Tensor (EMT). Labelling the EMT operator T, one can parametrise the associated matrix element for the pion as [64]:

$$\langle p_2|T_a^{\mu\nu}|p_1\rangle = 2P^\mu P^\nu A^a(t) + \frac{1}{2}\left(\Delta^\mu \Delta^\nu - \eta^{\mu\nu}\Delta^2\right)C^a(t) + 2M^2\eta^{\mu\nu}\bar{C}^a(t). \quad (21)$$

where a is a generic label for quark flavour and gluon contributions, and M is the hadron mass. In the literature, the form factors A, C and $\bar{C}$ may be designated as gravitational form factors. Additional form factors are required for describing the nucleon EMT matrix element [65,66]:

$$\langle p_2|T_a^{\mu\nu}(0)|p_1\rangle = \bar{u}(p_2)\left\{\frac{P^\mu P^\nu}{M}A^a(t) + \frac{\Delta^\mu \Delta^\nu - \eta^{\mu\nu}\Delta^2}{M}C^a(t) + M\eta^{\mu\nu}\bar{C}^a(t)\right.$$
$$\left. + \frac{i(P^\mu\sigma^{\nu\rho} + P^\nu\sigma^{\mu\rho})\Delta_\rho}{4M}[A^a(t) + B^a(t)] + \frac{P^{[\mu}i\sigma^{\nu]\rho}\Delta_\rho}{4M}D^a(t)\right\}u(p_1), \quad (22)$$

where we define the anti-symmetric combination $a^{[\mu}b^{\nu]} = a^\mu b^\nu - a^\nu b^\mu$. These form factors of the EMT have to obey specific constraints, originating from conservation laws [5,67–69]:

$$\sum_f A^{q_f}(0) + A^g(0) = 1 \quad (23)$$

$$\sum_f B^{q_f}(0) + B^g(0) = 0 \quad (24)$$

$$\sum_f \bar{C}^{q_f}(t) + \bar{C}^g(t) = 0 \quad (25)$$

where $\sum_f$ is the sum over the considered quark flavours. Some of the EFF can be related to GPDs through the computation of the first-order Mellin moments:

$$\int_{-1}^{1} dx\, x H^q(x,\xi,t) = A^q(t) + 4\xi^2 C^q(t), \quad (26)$$

$$\int_{-1}^{1} dx\, x E^q(x,\xi,t) = B^q(t) - 4\xi^2 C^q(t). \quad (27)$$

Consequently, two out of three EMT form factors are connected to GPDs in the pion case, while three out of five are in the nucleon one. The form factor $\bar{C}$ remains out of reach for leading twist distributions. In the nucleon case, the form factor D can be related to the nucleon axial form factor and thus to the polarised GPDs [70,71].

Within our conventions, the gluon EMT form factors are related to the gluon GPDs in a very similar way, with a modification of the power of the Mellin moment (consistently with the forward limit of our gluon GPDs):

$$\int_{-1}^{1} dx\, H^g(x,\xi,t) = A^g(t) + 4\xi^2 C^g(t), \quad (28)$$

$$\int_{-1}^{1} dx\, E^g(x,\xi,t) = B^g(t) - 4\xi^2 C^g(t). \quad (29)$$

From these connections between GPDs and the EMT, one can derive the famous Ji sum rule [5], allowing one to decompose the total angular momentum of the nucleon into contributions carried by each quark flavour J^q and gluons J^g:

$$2J^q = A^q(0) + B^q(0) = \int_{-1}^{1} dx\, x(H^q(x,\xi,0) + E^q(x,\xi,0)), \tag{30}$$

$$2J^g = A^g(0) + B^g(0) = \int_{-1}^{1} dx\, (H^g(x,\xi,0) + E^g(x,\xi,0)). \tag{31}$$

The A and B form factors are not the only ones providing an interesting interpretation. The C and $\bar{C}$ ones can be interpreted in terms of pressure and shear forces distributions within the hadrons [64,71–73]. For simplicity, we will only discuss here the quark and gluon contributions to the isotropic pressure p_q and p_g and the pressure anisotropy s_q and s_g. They depend on the form factors C and $\bar{C}$ in the Breit frame through [71]:

$$p_a(\vec{r}) = M \int \frac{d^3\vec{\Delta}}{(2\pi)^3} e^{-i\vec{r}\vec{\Delta}} \left[\frac{2t}{3M} C^i(t) - \bar{C}^i(t)\right], \tag{32}$$

$$s_a(\vec{r}) = \frac{4M}{r^2} \int \frac{d^3\vec{\Delta}}{(2\pi)^3} e^{-i\vec{r}\vec{\Delta}} \frac{t^{-1/2}}{M^2} \frac{d^2}{dt^2}\left[t^{5/2} C^i(t)\right]. \tag{33}$$

We highlight that the index a stands again for quark flavour and gluon contributions and $\vec{r}$ is a 3D spatial vector. One can readily note from Equation (33) that the pressure anisotropy is independent of $\bar{C}$ and thus can be fully extracted from GPDs. In addition, from Equation (32), one can realise that the total isotropic pressure is also independent of $\bar{C}$ thanks to Equation (25), and thus, one can in principle also extract it from GPDs. Finally, let us mention that this discussion is valid both for the pion and the nucleon.

Before concluding this section, we would like to highlight that in the case of the pion, there is an additional constraint on the EMT form factors, thanks to the soft pion theorem of Equation (18). Indeed, at vanishing t in the chiral limit, one has:

$$0 = \int_{-1}^{1} dx\, x H_\xi^q(x,1,0) = A^q(0) + 4C^q(0), \tag{34}$$

connecting the normalisation of C^q and A^q.

Finally, let us stress that H_ξ^q vanishes only if the gluon GPD also vanishes due to mixing of the two under evolution. Since the soft pion theorem is expected to be scale-independent, then Equation (34) can also be written for gluons.

2.5. Double Distribution Representation

The sum rules already encountered previously and connecting GPDs to the EFF and EMT Form Factors can in fact be generalised to higher-order Mellin moments by looking at the tensorial structure of local off-forward operators.

2.5.1. Local Operators Analysis

We define the m-th Mellin moment $\mathcal{M}_m$ of the GPD as:

$$\mathcal{M}_m(\xi,t) = \int_{-1}^{1} dx\, x^m H(x,\xi,t). \tag{35}$$

Through straightforward computations (see, e.g., [41]), one can connect these Mellin moments to local twist-two operators (we remain with the lightcone gauge and the pion case here):

$$\mathcal{M}_m(\xi,t) = \frac{1}{2(P^+)^{m+1}} \langle p_2|\bar{\psi}^q(0)\gamma^+ \left(i\overleftrightarrow{\partial}^+\right)^m \psi^q(0)|p_1\rangle,$$

where we use
$$\overleftrightarrow{\partial}^\mu = \frac{1}{2}\left(\overrightarrow{\partial}^\mu - \overleftarrow{\partial}^\mu\right). \tag{36}$$

The Lorentz structure of the twist-two local operators $O^{\mu\mu_1\cdots\mu_m}$ is given as:
$$O^{\mu\mu_1\cdots\mu_m} = \bar{\psi}^q(0)\gamma^{\{\mu}i\overleftrightarrow{\partial}^{\mu_1}\ldots i\overleftrightarrow{\partial}^{\mu_m\}}\psi^q(0), \tag{37}$$

where $\{\ldots\}$ indicates that the Lorentz indices are symmetrised and traceless. The parametrisation of the associated matrix elements is thus given as:
$$\langle p_2|\bar{\psi}^q(0)\gamma^{\{\mu}i\overleftrightarrow{\partial}^{\mu_1}\ldots i\overleftrightarrow{\partial}^{\mu_m\}}\psi^q(0)|p_1\rangle$$
$$= P^{\{\mu}\sum_{i=1}^{m+1} P^{\mu_1}\ldots P^{\mu_{i-1}}\Delta^{\mu_i}\ldots \Delta^{\mu_m\}}A^q_{i,m}(t) + \Delta^{\{\mu}\Delta^{\mu_1}\ldots\Delta^{\mu_m\}}C^q_{m+1}(t), \tag{38}$$

in the pion case. $A_{i,m}$ and C_{m+1} are called generalised form factors, and only depend on t and μ^2. Moreover, due to discrete symmetries, only even powers of ξ contribute, and the Mellin moments of the pion GPD can be written as:
$$\mathcal{M}_m(\xi,t) = \sum_{i=0}^{\left[\frac{m}{2}\right]} (2\xi)^{2i} A^q_{i,m}(t) + \mathrm{mod}(m,2)(2\xi)^{m+1}C_{m+1}(t), \tag{39}$$

where $[\ldots]$ designates the floor function and $\mathrm{mod}(2,m)$ vanishes if m is even, and is 1 otherwise. Equation (39) is commonly labelled the polynomiality property of GPDs [74,75]. It generalises the sum rules between GPDs, EFFs and EMT form factors, already highlighted in Equations (15) and (26) in the case of the pion. For the nucleon, additional tensorial structures allow one to generalise Equations (15) and (27).

2.5.2. The Radon Transform and the Specific Role of the D-Term

The C_{m+1} generalised form factors play a specific role in the decomposition of the GPD matrix element. They are the moment of an x-odd generating function called the D-term (we highlight that the D-term $D(y,t)$ is connected to the EMT FF $C(t)$ and not $D(t)$; this notation discrepancy is unfortunate but standard in the contemporary literature) and defined through:
$$\int_{-1}^{1} dy\, y^m D(y,t) = (2)^{m+1} C_{m+1}(t). \tag{40}$$

The proof of existence of the function D is not simple and related to the Hausdorff moment problem [76,77]. In the following, we assume that D exists and is unique. The polynomiality property becomes:
$$\sum_{i=0}^{\left[\frac{m}{2}\right]} (2\xi)^{2i} A^q_{i,m}(t) = \int_{-1}^{1} dx\, x^m H(x,\xi,t) - \xi^{m+1}\int_{-1}^{1} dy\, y^m D(y,t). \tag{41}$$

By rescaling the variable $y \to x/\xi$ for $\xi > 0$, one obtains:
$$\sum_{i=0}^{\left[\frac{m}{2}\right]} (2\xi)^{2i} A^q_{i,m}(t) = \int_{-1}^{1} dx\, x^m \left[H(x,\xi,t) - D\left(\frac{x}{\xi},t\right)\mathbb{I}_{-\xi\leq x\leq\xi}\right], \tag{42}$$

where $\mathbb{I}_{-\xi\leq x\leq\xi}$ is 1 for $x \in [-\xi;\xi]$ and 0 otherwise. On top of telling us that the D-term is a function of the ratio x/ξ, Equation (42) highlights that the D-term has support only in the ERBL region. Moreover, the typical polynomial structure of degree n of the nth Mellin moments of $H - D$ is known in mathematics as the the Lugwig–Hegalson consistency condition. As shown by Hertle [78], this indicates that $H - D$ is in the range of the Radon transform $\mathcal{R}$ [79,80]. As a consequence, one can define F such that:

$$H(x,\xi,t) - \mathbb{I}_{-\xi \leq x \leq \xi} D\left(\frac{x}{\xi}, t\right) = \int_\Omega d\beta d\alpha \, \delta(x - \beta - \alpha\xi) F(\beta, \alpha, t) \qquad (43)$$

where $\Omega = \{(\beta, \alpha) \mid |\alpha| + |\beta| \leq 1\}$. Originally, F was introduced as a Double Distribution, independently by D. Mueller [6] and A. Radyushkin [7], while the D-term was originally introduced as a possible complementary tensorial structure to the DD F [81]. F was later identified as the Radon amplitude of the GPDs in ref. [82], and the formalism was further developed in ref. [83]. Finally, Equation (43) can be straightforwardly manipulated to obtain the well-known relation between GPDs and DDs:

$$H(x,\xi,t) = \int_\Omega d\beta d\alpha \, \delta(x - \beta - \alpha\xi)[F(\beta, \alpha, t) + \xi\delta(\beta)D(\alpha, t)]. \qquad (44)$$

This result can be generalised to gluons, up to small modifications. In the nucleon case, the additional spin structure requires the introduction of an additional DD to describe the GPD E, but the D-term is the same between H and E up to a minus sign.

Finally, let us mention that the way to decompose the GPD between a Double Distribution and a D-term is not unique. This is a technical point and it has been discussed in the literature in detail (see [39,41,82,83]).

2.6. Positivity and Lightfront Wave Function Picture

2.6.1. The Lightfront Wave Function Picture

On top of polynomiality, another major property is associated with GPDs, called positivity. In order to properly understand such constraint discussed in several papers, we will introduce it though the Lightfront Wave Function formalism developed in [84].

To do so, we start decomposing the considered hadron states (here—the pion) in terms of LFWFs, labelled here $\Phi_\beta^{i...j}$ such that:

$$|\pi\rangle \propto \sum_\beta \Phi_\beta^{q\bar{q}} |q\bar{q}\rangle + \sum_\beta \Phi_\beta^{q\bar{q},q\bar{q}} |q\bar{q}, q\bar{q}\rangle + \ldots \qquad (45)$$

where the relevant quantum numbers are labelled with β and $i...j$ stands for the partonic content of the considered state. When a N partons state is considered, then the associated LFWFs depend on N lightfront momentum fractions x_i and N 2D transverse momenta $k^i_\perp$. The momentum conservation is guaranteed by three Dirac distributions.

Following the derivation given in great detail in ref. [84], one can express the matrix elements defining GPDs in Equations (2), (3), (9) and (10) in terms of an overlap of LFWFs. The interested reader can also look at examples of explicit computations in at leading Fock states (see, for instance, refs. [42,85]).

We will not reproduce these derivations here, and instead focus on the lightcone interpretation associated with considered kinematic area (see Figure 3), and its consequences on the overlap description. First, looking at the outer or DGLAP region, one obtains an overlap of LFWFs diagonal in terms of parton number N:

$$H^q(x,\xi)|_{x \geq \xi} \propto \sum_N \sqrt{1-\xi^2}^{1-N} \sum_j \delta_{s_j, q} \int \left[dx_i d^2 k^i_\perp\right]_N \left(\Phi^N(\hat{p}_N)\right)^* \Phi^N(\tilde{r}_N) \delta(x - x_j), \qquad (46)$$

where the measure is given as

$$\left[dx_i d^2 k^i_\perp\right]_N = \frac{1}{(16\pi^3)^{N-1}} \left[\prod_{i=1}^N dx_i d^2 k^i_\perp\right] \delta\left(1 - \sum_j x_j\right) \delta^{(2)}\left(\sum_j k^j_\perp\right) \qquad (47)$$

in the $N \to N$ parton case [84]. The label j stands for the active partons, i.e., the one probed by the operator, while s_j is its flavour. The incoming and outgoing momenta degrees of freedom, boosted in the GPDs symmetric frame, are generically labelled with $\check{r}_N$ and $\hat{r}_N$, respectively. More details are provided in ref. [84].

From Equation (46), one realises that the overlap in the DGLAP region only includes convolution of LFWFs with the same number of partons. This is illustrated in Figure 5.

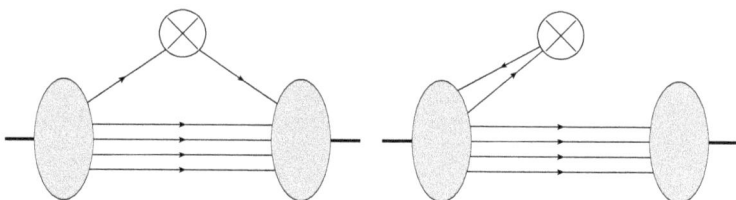

Figure 5. LFWFs decomposition of GPDs. **Left** panel, DGLAP region interpretation conserving parton number between incoming and outgoing states; **right** panel, ERBL region interpretation not conserving parton number as the incoming state emits a pair of quark–antiquark.

This contrasts with the ERBL region. There, a pair of quark–antiquark is extracted from the initial state and and contracted with the operator (see Figure 5). The overlap is off-diagonal in the Fock space and one obtains:

$$H^q(x,\xi)|_{x \leq |\xi|} \propto \sum_N \sqrt{1-\xi}^{2-N} \sqrt{1+\xi}^{-N} \sum_{j,j'} \frac{\delta_{s'_j s_j} \delta_{s_j q}}{\sqrt{n_j n_{j'}}} \int \left[dx_i d^2 k^i_\perp \right]_{N-1}^{N+1} \delta(x - x_j)$$
$$\times \left(\Phi^{N-1}(\hat{r}_{N-1}) \right)^* \Phi^{N+1}(\check{r}_{N+1}). \quad (48)$$

Similarly, the measure $\left[dx_i d^2 k^i_\perp \right]_{N-1}^{N+1}$ corresponds to the $N+1 \to N-1$ transition:

$$\left[dx_i d^2 k^i_\perp \right]_{N-1}^{N+1} = dx_j \prod_{i \neq j,j'}^{N+1} dx_i \delta \left(1 - \xi - \sum_{i \neq j,j'}^{N+1} x_i \right) \frac{d^2 k^j_\perp}{(16\pi^3)^{N-1}} \prod_{i \neq j,j'}^{N+1} dk^i_\perp \delta \left(\frac{\Delta_\perp}{2} - \sum_{i \neq j,j'} k^i_\perp \right). \quad (49)$$

The discrepancy in terms of parton number between the initial and final state triggers ambiguity in the attempt to compute GPDs in the entire kinematic range at a given truncation of the Fock space. Thus, most of the time, only the DGLAP region is computed within models [42,86,87]. Nevertheless, modern solutions allow us to bypass this difficulty as we will discuss later on.

2.6.2. The Positivity Property

The LFWF picture in the DGLAP region allow us to derive an important property of GPDs called positivity. Focusing first on the forward case, when $\Delta \to 0$, one obtains:

$$q(x) \propto \sum_N \sum_j \delta_{s_j,q} \int \left[dx_i d^2 k^i_\perp \right]_N \left| \Phi^N(x_i, k^i_\perp) \right|^2 \delta(x - x_j). \quad (50)$$

The PDF q is expressed as the sum over momenta, partons content and quantum numbers of modulus square of the LFWFs. This is consistent with the probabilistic aspect of the PDF (number density) and a formal interpretation yielding PDF as the norm of a formal vector.

This structure can be generalised beyond the forward limit, to the entire DGLAP region. One formally recovers a scalar product in this parton number conserving kinematic area:

$$H(x,\xi)|_{x \geq \xi} = \langle \Phi_{\text{out}} | \Phi_{\text{in}} \rangle. \tag{51}$$

A simple Cauchy–Schwartz inequality yields:

$$|H(x,\xi)|_{x \geq \xi}| \leq |||\Phi_{\text{in}}\rangle|| \times |||\Phi_{\text{out}}\rangle||, \tag{52}$$

Plugging in the forward limit to describe the incoming and outgoing states, one obtains the positivity condition [84,88–91]:

$$|H^q(x,\xi)|_{x \geq \xi}| \leq \sqrt{q\left(\frac{x-\xi}{1-\xi}\right) q\left(\frac{x+\xi}{1+\xi}\right)}, \tag{53}$$

again, in the case of the pion. Small modifications appear in the nucleon case. The positivity property yields an upper and lower bounds on the GPD in the DGLAP region.

2.7. Scale Dependence and Evolution

2.7.1. Discussion in Momentum Space

As we have mentioned several times already, GPDs obey evolution equations, as the operators they are defined from need to be renormalised. Following ref. [92], the renormalisation of GPDs can be written as follows:

$$H^a(x,\xi,t,\mu^2) = \lim_{\varepsilon \to 0} \int_{-1}^{1} \frac{dy}{|y|} Z^{ab}\left(\frac{x}{y}, \frac{\xi}{x}, \alpha_s(\mu^2), \varepsilon\right) \hat{H}^b(y,\xi,t,\varepsilon), \tag{54}$$

where a,b stand for quark flavours q or gluon contributions. μ is the renormalisation scale, while ε is the regulator, ensuring that the bare operator remains finite. Since we use perturbation theory to describe the renormalisation properties of GPDs, our regularisation will be dimensional, such that $d = 4 - 2\varepsilon$. From Equation (54), one can apply a renormalisation group strategy to obtain:

$$\frac{dH^a}{d \ln \mu^2}(x,\xi,t,\mu^2) = \lim_{\varepsilon \to 0} \int_{-1}^{1} \frac{dy}{|y|} \frac{dZ^{ab}}{d \ln \mu^2}\left(\frac{x}{y}, \frac{\xi}{x}, \alpha_s(\mu^2), \varepsilon\right)$$
$$\times \int_{-1}^{1} \frac{dz}{|z|} (Z^{bc})^{-1}\left(\frac{y}{z}, \frac{\xi}{y}, \alpha_s(\mu^2), \varepsilon\right) H^c(z,\xi,t,\mu^2), \tag{55}$$

i.e., the evolution equation of GPDs with respect to the scale μ. For the sake of completeness, let us highlight that Z^{-1} is defined such that:

$$\delta_{ij}\delta\left(1 - \frac{z}{x}\right) = \lim_{\varepsilon \to 0} \int_{-1}^{1} \frac{dy}{|y|} (Z^{il})^{-1}\left(\frac{z}{y}, \frac{\xi}{z}, \alpha_s(\mu^2), \varepsilon\right) Z^{lj}\left(\frac{y}{x}, \frac{\xi}{y}, \alpha_s(\mu^2), \varepsilon\right). \tag{56}$$

The combination $\frac{dZ}{d \ln \mu^2} Z^{-1}$ can be seen as the momentum-dependent generalisation of the anomalous dimensions. Thus, we introduce the functions $\mathcal{P}$ such that:

$$\mathcal{P}^{ac}\left(\frac{x}{z}, \frac{\xi}{x}, \alpha_s(\mu^2)\right) = \lim_{\varepsilon \to 0} \int_{-1}^{1} \frac{dy}{|y|} \frac{dZ^{ab}}{d \ln \mu^2}\left(\frac{x}{y}, \frac{\xi}{x}, \alpha_s(\mu^2), \varepsilon\right) (Z^{bc})^{-1}\left(\frac{y}{z}, \frac{\xi}{y}, \alpha_s(\mu^2), \varepsilon\right). \tag{57}$$

Note that the function $\mathcal{P}$ is independent of ε, as the singularities are expected to cancel in the product $\frac{dZ}{d \ln \mu^2} Z^{-1}$, as we will explicitly see in the following in a one-loop expansion. Provided that Z can be computed non-perturbatively, one could obtain a description of the scale evolution. However, this has not been achieved yet, and the scale dependence of GPDs is computed perturbatively through the renormalisation group Equation (55). In the

$\overline{MS}$ scheme, letting implicit the factor $S_\varepsilon = \frac{(4\pi)^\varepsilon}{\Gamma(1-\varepsilon)}$ accompanying the poles in order to keep the notation simple, one can write:

$$Z^{ij}\left(\frac{x}{y},\frac{\xi}{x'},\alpha_s(\mu^2),\varepsilon\right) = \delta_{ij}\delta\left(1-\frac{x}{y}\right) + \sum_{n=1}^{\infty} a_s^n(\mu^2) \sum_{p=1}^{n} \frac{1}{\varepsilon^p} Z^{[n,p]}\left(\frac{x}{y},\frac{\xi}{x'}\right), \qquad (58)$$

where $a_s = \alpha_s/4\pi$. The derivative with respect to $\ln \mu^2$ is then given by:

$$\frac{dZ}{d\ln\mu^2}\left(\frac{x}{y},\frac{\xi}{x'},\alpha_s(\mu^2),\varepsilon\right) = \frac{da_s}{d\ln\mu^2}(\mu^2)\frac{dZ}{da_s}\left(\frac{x}{y},\frac{\xi}{x'},\alpha_s(\mu^2),\varepsilon\right)$$
$$= \left(-\varepsilon a_s(\mu^2) + \beta(a_s(\mu^2))\right)\left(\sum_{n=1}^{\infty} n a_s^{n-1}(\mu^2) \sum_{p=1}^{n} \frac{1}{\varepsilon^p} Z^{[n,p]}\left(\frac{x}{y},\frac{\xi}{x'}\right)\right), \qquad (59)$$

where we used the Renormalisation Group Equation (RGE) in $d = 4 - 2\varepsilon$ dimension for the strong running coupling. Consequently, at leading order in a_s, $\mathcal{P}$ is given as:

$$\mathcal{P}^{ac}\left(\frac{x}{z},\frac{\xi}{x'},a_s(\mu^2)\right) = -a_s(\mu^2)Z_{ac}^{[1,1]}\left(\frac{x}{z},\frac{\xi}{x'}\right). \qquad (60)$$

The one-loop result for $\mathcal{P}$ is thus finite, and can be computed in the $\overline{MS}$ scheme only from the pole contribution of the associated Feynman diagrams. Moreover, since $\mathcal{P}$ is only sensitive the UV-diverging part of the GPDs in $\overline{MS}$, it does not depend on the nature of the particle chosen at the level of the matrix element. Thus, one can perform the computations in a "partons-in-partons" approach, as it is performed in ref. [92]. In that case, some of the relevant Feynman diagrams at one loop in the lightcone gauge are shown on Figure 6.

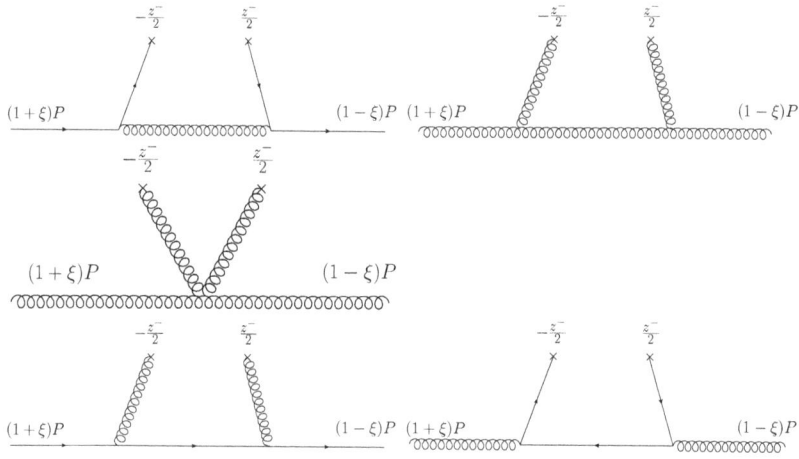

Figure 6. Feynman diagrams used to compute the one-loop anomalous dimension in the lightcone gauge. On the first and second lines, we display non-mixing terms, while the third line display diagrams mixing quark and gluon GPDs. On top of these connected diagrams, disconnected self-energy diagrams need to be added. In covariant gauges, additional contributions come from gluon exchanges with the Wilson line.

2.7.2. Properties of the Momentum-Dependent Anomalous Dimensions

The computation of the one-loop diagram leading to the one loop splitting function can be found in several papers [6,7,43,92]. We will adopt here the presentation of ref. [92]

to feed our discussion. Thus, for convenience, we assume in this section that $x > 0$ and we rewrite the evolution equation as:

$$\frac{dH}{d\ln\mu^2}(x,\xi,\mu^2) = \frac{\alpha_s(\mu)}{4\pi}\int_x^\infty \frac{dy}{y}\mathcal{P}^\pm H^\pm\left(\frac{x}{y},\xi,\mu\right). \tag{61}$$

Here, the $\pm$ index correspond to the singlet and non-singlet combinations, already introduced in Equations (11) and (12), with the difference that the singlet is now a vector encompassing gluons:

$$H^+(x,\xi,\mu^2) = \begin{pmatrix} \sum_q H_q^+(x,\xi,\mu^2) \\ H^g(x,\xi,\mu^2) \end{pmatrix}, \tag{62}$$

and consequently, $\mathcal{P}^+$ is a matrix mixing quarks and gluons contributions:

$$\mathcal{P}^+(y,\kappa) = \begin{pmatrix} \mathcal{P}^+_{q\leftarrow q}(y,\kappa) & \mathcal{P}^+_{q\leftarrow g}(y,\kappa) \\ \mathcal{P}^+_{g\leftarrow q}(y,\kappa) & \mathcal{P}^+_{g\leftarrow g}(y,\kappa) \end{pmatrix}, \tag{63}$$

where we have introduced $\kappa = \xi/x$. One can then decompose $\mathcal{P}^\pm$ depending on the kinematics support of the contributions:

$$\mathcal{P}^\pm(y,\kappa) = \mathcal{P}^\pm_1(y,\kappa)\theta(1-y) + \theta(\kappa-1)\mathcal{P}^\pm_2(y,\kappa). \tag{64}$$

The explicit expressions of $\mathcal{P}^\pm_1$ and $\mathcal{P}^\pm_2$ are given in ref. [92]. Here we see that the $\mathcal{P}^\pm_2$ contributes only for $\xi > x$. Thus, when $\xi = 0$, only $\mathcal{P}^\pm_1$ contributes in the entire y range, and reduces to the DGLAP splitting functions. Hence, the name of the $\xi < x$ kinematic region. In the kinematics region where $x < \xi$ both $\mathcal{P}^\pm_1$ and $\mathcal{P}^\pm_2$ contribute. The transition between the two regions is continuous as $\lim_{\kappa \to 1} \mathcal{P}^\pm_2(y,\kappa) = 0$, but due to the presence of the θ function, it is not smooth. In fact, the evolution kernel will generate a cusp at $x = \pm\xi$ when evolving smooth GPDs functions. This contrasts with the claim sometimes made in the literature that evolution smoothens the behaviour of GPD models on the $|x| = |\xi|$ lines (by rendering a discontinuous GPDs model at low scale continuous at higher scales for instance).

Finally, let us mention that in momentum space, only two evolution codes at leading order are publicly available, the Vinnikov code [93], which is not maintained anymore, and Apfel++ [94–96]. Both of them are integrated through the PARTONS framework [97].

2.7.3. Evolution in Conformal Space

An alternative picture to momentum space evolution is provided by moments, or conformal space evolution. The idea originated from the expansion of a non-local operator into a set of local ones through operator product expansion (OPE). The local operators involved in the expansion need to be renormalised and thus, all of them obtain an "a priori" different renormalisation scale dependence. The original non-local operator and its scale dependence can then be recover by resumming the renormalised local operators.

In the PDF field, this is typically what is done when one performs the evolution through the Mellin moments before applying the inverse Mellin transform. In the GPD case, as the splitting functions are more complicated, the situation is not as simple. The Mellin moments mix under GPD renormalisation, making their evolution and resummation more involved. However, it is possible to diagonalise the basis of local operator for a given order of perturbative evolution. At order α_s, the evolution operator commutes with the conformal symmetry operators, which tells us that both operators can be diagonalised on the same basis [98]. From that, one deduces that the conformal moments, defined in the non-singlet sector as

$$\mathcal{C}_n^-(\xi,t,\mu^2) = \xi^n \int_{-1}^1 dx\, C_n^{3/2}\left(\frac{x}{\xi}\right) H^-(x,\xi,t,\mu^2), \tag{65}$$

where $C_n^{3/2}$ are the 3/2-Gegenbauer polynomials of order n, do not mix among each other. In the limit when $\xi \to 0$, the conformal moments reduce to the Mellin one, consistently with the evolution kernel going to the DGLAP one. In the case when $\xi \to 1$, the Gegenbauer polynomial formed a base in x of the GPD support, and one recovers the diagonalisation of the evolution kernel à la "Efremov-Radyushkin-Brodsky-Lepage". The conformal moments obey an evolution equation of the type of:

$$\frac{dC_n^-}{d\ln \mu^2}(\xi, t, \mu^2) = \frac{\alpha_s(\mu)}{4\pi} \mathcal{V}_n C_n^-(\xi, t, \mu^2), \tag{66}$$

where

$$\mathcal{V}_n = 2C_F \left(\frac{3}{2} + \frac{1}{(n+1)(n+2)} - 2\sum_{k=1}^{n+1} \frac{1}{k} \right) \tag{67}$$

with $C_F = 4/3$. The solution of this equation is given as:

$$C_n^-(\xi, t, \mu^2) = \left(\frac{\alpha_s(\mu^2)}{\alpha_s(\mu_0^2)} \right)^{-\frac{\mathcal{V}_n}{\beta_0}} C_n^-(\xi, t, \mu_0^2) \tag{68}$$

where β_0 is the leading order term of the QCD β function given as $\beta_0 = 11 - 2/3 * n_f$, where n_f is the number of active quark flavours. From Equation (68), one can note the advantage of moments evolution, which is purely multiplicative, and not convoluted as in momentum space. However, the difficulties lie in the reconstruction of the x-dependent function. Indeed, the Gegenbauer polynomials introduced in (65) do not form an orthogonal basis for $x \in [-1, 1]$ but $x \in [-\xi, \xi]$, complicating the evolution of a polynomial expansion of GPDs. Moreover, such an expansion is expected to converge very slowly and thus, resumming techniques have been introduced relying on Mellin–Barnes integrals [99]. The Mellin–Barnes integral transform requires knowledge of the conformal moments for $n \in \mathbb{C}$ and not just integer values. Thus, an analytic continuation of the moments within the complex plane is required, and if the moments are not given by an algebraic formula, this continuation can be challenging. Moreover, its uniqueness is not guaranteed. Nevertheless, models in conformal space have been developed [100,101] and are today among the most successful ones.

In the singlet sector, mixing between quarks and gluons is unavoidable for a given order n of the conformal moments. Gluon conformal moment distribution follows the same definition as in (65), but is computed from 5/2-Gegenbauer polynomials $C_n^{5/2}$. An additional diagonalisation of the 2×2 matrix is necessary. An example is given in the specific case of the D-term in ref. [71].

Finally, let us briefly discuss what happens beyond leading order. Evolution kernels have been derived at two loops (and at three loops in the non-singlet case) [102–107]. In the $\overline{MS}$ scheme, the conformal moments C_n start mixing among each other in such a way that:

$$C_n(\mu^2) = \sum_{j=0}^{n} b_{nj}(\mu^2, \mu_0^2) C_j(\mu_0^2), \tag{69}$$

yielding a triangular matrix at a given order n (see [108]). This complicates the evolution and reconstruction in x-space of NLO evolved GPDs in conformal space. Consequently, in $\overline{MS}$, the strategy is rather to evolve Wilson coefficients in the computation of a process rather then the GPD itself (see, e.g., [100]). An alternative solution is to work in a specific scheme in which the off-diagonal coefficients vanish. This scheme is labelled conformal scheme (or $\overline{CS}$), and is defined so that in the forward limit, one recovers the $\overline{MS}$ scheme results (this is possible as PDFs Mellin moments do not mix under evolution). This specific scheme has allowed early descriptions of DVCS at NNLO accuracy [100], before modern computations in $\overline{MS}$ scheme [109]. From the first studies on scheme dependence [100], the difference between $\overline{\text{CS}}$ and $\overline{\text{MS}}$ is mild for the quarks, and somehow more sizeable for gluons, while NLO and NNLO quark GPDs are indistinguishable.

3. Continuum Results for Mesons

3.1. Impulse Approximation and Its Limitations

In the meson sector, one of the first attempts to compute GPDs (see, e.g., [110–112]) relied on the so-called impulse approximation, graphically represented in Figure 7.

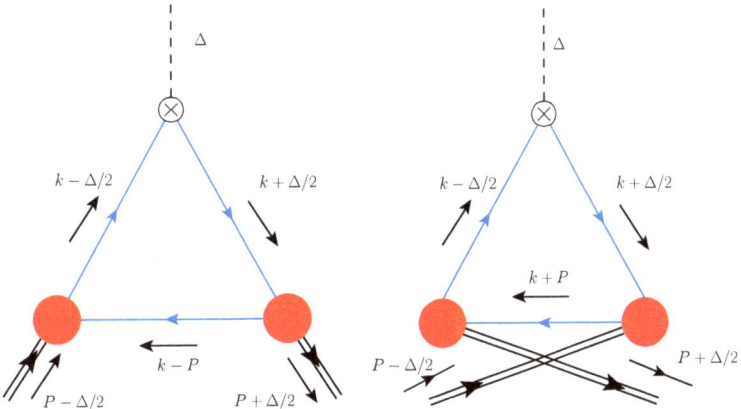

Figure 7. Triangle diagrams used in early computations of meson EFF, PDFs and GPDs.

In this framework, a pion is split into a quark–antiquark pair and is later reformed. In between, one computes the impact of local twist-two operators defined in Equation (37) inserted within a quark line. Using this technique, one gains access to the Mellin moments of the GPDs (including the D-term contribution), and thus, a reconstruction technique is necessary to gain access to the x-dependence of the distribution. If polynomials reconstructions have been tried (see, e.g., refs. [42,113]), direct identifications of Double Distributions has been favoured in the literature [60,112,114].

Such computations rely on two main points:

- The computation (or modelling) of non-perturbative QCD correlation functions such as the Bethe–Salpeter wave function, the quark propagator and the local operator;
- The validity of the impulse approximation.

We will not enter here the discussion about the consistent computation of the non-perturbative QCD correlation function, and instead, we redirect the reader toward a previous review paper [42] and refined studies of the pion Bethe–Salpeter wave function performed since then [115,116]. Rather, we will discuss the limitations of the impulse approximation.

When one chooses an approximation set such that the Bethe–Salpeter amplitude is independent of the relative momentum (k in Figure 7) the forward limit of the GPD, i.e., the PDF, ends up being symmetric with respect to $x \to 1 - x$, as expected from momentum conservation in the two-body case. However, as soon as the Bethe–Salpeter amplitude becomes k dependent, the momentum conservation and the two-body symmetry are lost (see for instance [117]). This is a consequence of missing contributions in the computation of matrix element of local twist-2 operators. Loosely speaking, as the relative momentum dependence kicks in, the Bethe–Salpeter amplitude gains a spatial extension. Thus, the operator can be inserted "within" the amplitude. This idea yields additional contributions as illustrated in Figure 8.

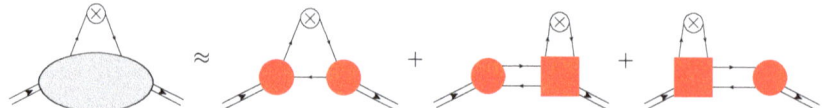

Figure 8. Decomposition of contributions to the local operators when the Bethe–Salpeter amplitude is momentum-dependent. The red circles correspond the standard Bethe–Salpeter amplitudes, while the squares include modifications allowing the insertion of the local operators.

As shown in [42,113,114], in the Rainbow ladder approximation, the new contributions can be written in terms of the Bethe–Salpeter amplitude Γ_π as:

$$= -\frac{1}{2}(k \cdot n)^m n^\nu \frac{\partial \Gamma_\pi}{\partial k^\mu}(k,P), \qquad (70)$$

where n is a lightlike vector so that $k \cdot n = k^+$ and m is the order of the Mellin moment considered.

The impact of the additional terms is shown in Figure 9 in the case of simple algebraic models for the Bethe–Salpeter amplitudes and the quark propagators (see [42,113,114]). Being anti-symmetric with respect to $x \to 1-x$, it does not contribute to the PDF normalisation. It exactly compensates the antisymmetric component of the triangle contribution alone, restoring the symmetry of the complete result. The most advanced computations using this technique can be found in [118].

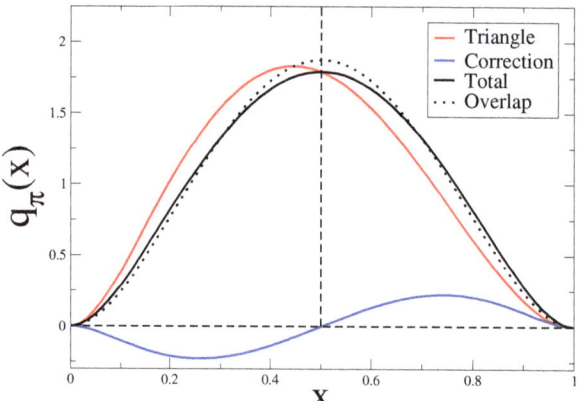

Figure 9. Results for the forward limit after reconstruction of the x-dependence in the case of a simple, algebraic model (see [42,113,114]). The total PDF (solid black) is symmetric under $x \to 1-x$ transformation, but the sole triangle diagram contribution (solid red) is not. The additional terms coming from the square vertices in Equation (70) provide the adequate correction (solid blue).

If this contribution allows one to recover the expected symmetry properties in the forward limit, it remains insufficient for tackling GPD computations in the entire kinematic domain. More precisely, at non-vanishing skewness, the additional vertices in Equation (70) are not able to restore the positivity property. As the forward limit yields a symmetric and thus vanishing contribution at small-x, the positivity property imposes that the GPD is also vanishing at $|x| = |\xi|$. Models employing momentum-dependent Bethe–Salpeter have thus been limited to the vanishing ξ kinematic region [60]. Note that models based on the Nambu–Jona–Lasinio (NJL) approximation are not impacted, since the amplitude is momentum-independent [119–121]. However, in the NJL approximation, the triangle diagram is not sufficient in off-forward kinematics as the meson resonances appear in the *t*-channel, contributing to the *D*-term [119,122,123]. Thus, omitting the resonances leads to a breaking of the soft pion theorem in the chiral limit. To go beyond vanishing skewness, techniques based on LFWFs have been developed.

3.2. From Bethe–Salpeter Wave Funtions to Lightfront Wave Functions

As we highlighted previously, the most natural way to fulfil the positivity property is to start modelling GPDs as overlaps of LFWFs. Different techniques exist to compute the LFWFs, from Lightfront Hamiltonian (see, for instance, [124]) to ADS/QCD (e.g., refs. [86,125]). We will focus here on the connection between the Bethe–Salpeter and Lightfront wave functions, in the case of the pion.

The *N*-body Bethe–Salpeter wave function can be projected out to obtain all the independent *N*-body LFWFs carrying the quantum numbers of interest (mostly the helicity projection of the quark, and thus the orbital angular momentum projection of the pion). If nothing forbids the computation of *N*-body Bethe–Salpeter wave function, in practice, today, only the two-body one is computed as it involves only four-point functions. Thus, we will further restrain ourselves to the lowest Fock state of the pion.

The two-body LFWFs Φ for the pion are given as projection of the Bethe–Salpeter wave function χ_π. The latter is defined by "attaching" quark propagators to the Bethe–Salpeter amplitude: $\chi_\pi = S\Gamma_\pi S$). The connection the the LFWFs is then given through (see, for instance, [126]):

$$P^+ \Phi_{\uparrow\downarrow}(x, k_\perp) = \int \frac{dk^-}{2\pi} \text{Tr}\left[\gamma^+ \gamma_5 \chi_\pi(k, P)\right], \tag{71}$$

$$2ik^i P^+ \Phi_{\uparrow\uparrow}(x, k_\perp) = \int \frac{dk^-}{2\pi} \text{Tr}\left[\sigma^{+i} \gamma_5 \chi_\pi(k, P)\right], \tag{72}$$

where $\uparrow\downarrow$ and $\uparrow\uparrow$ indicate the helicity projection of the quarks. If the projection looks simple, it is in fact made complicated by the lightcone integration over k^-, as the Bethe–Salpeter wave function is typically computed in Euclidean space, using standard euclidean variable. This difficulty is usually bypassed by using spectral, or Nakanishi representation [30,31] for the Bethe–Salpeter amplitude Γ_π in Euclidean space:

$$\Gamma_\pi(k, P) = \int_0^\infty d\omega \int_{-1}^1 dz \frac{\rho(z, \omega)}{\omega + \left(k + \frac{z}{2}P\right)^2}. \tag{73}$$

The advantage of this representation is that the momenta degrees of freedom can be algebraically manipulated, while the non-perturbative information is shifted toward the Nakanishi weight ρ. Different modelling strategies have been performed in the literature, from simple algebraic Ansätze [60,127] to parametric fit [128,129] or even direct computations in the specific case of QED N-point functions [130]. The results obtained in the forward limit are similar to those obtained through the diagrammatic computation, as shown by the dotted line on Figure 9 in the case of the simple algebraic model of [42]. Concerning GPDs computations, only algebraic Ansätze and parametric fits have been used.

As the reader may have noted, the fact that only the two-body LFWFs are available from the two-body Bethe-Salpeter wave function make the computation of the GPDs possible only in the DGLAP region, as the first term in the ERBL one would require the four-body LFWFs. Thus, a priori, if positivity is well fulfilled, it is not possible to complete polynomiality, as DGLAP and ERBL contributions are intertwined to yield polynomials Mellin moments. Such a hole in the kinematic domain has to be filled and we will see how in the next section. Nevertheless, accessing the DGLAP region alone remains highly valuable. Indeed, it gives access to the PDF when $\Delta \to 0$, while EFF can be recovered after integrating over x at $\xi = 0$. Furthermore, the 3D density interpretation is also accessible from the

DGLAP region only. The major missing part is the EMT for factor C of Equation (21), and more generally, the D-term of Equation (40).

As highlighted in [59], under specific assumption on the Bethe–Salpeter wave function, the associated LFWFs ends up being separable between x and $k_\perp$ variable so that:

$$\Phi(x, k_\perp) = \varphi(x)\psi(k_\perp). \tag{74}$$

This separability property triggers interesting simplifications. First, φ yields the pion distribution amplitudes, up to a normalisation constant. Then, the PDF computed as an overlap of LFWFs in Equation (50) becomes proportional to the square of the pion distribution amplitude. Regarding GPDs, this factorisation yields the following functional form [58,85]:

$$H_\pi(x, \xi, t) = \sqrt{q_\pi\left(\frac{x-\xi}{1-\xi}\right) q_\pi\left(\frac{x+\xi}{1+\xi}\right)} \Psi_\pi(x, \xi, t), \tag{75}$$

where $\Psi < 1$. The positivity property is thus manifestly fulfilled, as expected starting from a LFWF description of the nucleon. This formula is used in ref. [58] to introduce new kinds of models, exploiting PDFs computed in the covariant approach, but beyond the impulse approximation [118]. The Ψ can be then modelled or computed following the simple Anzätze or parametric fits available. Under two assumptions regarding the Nakanishi parametrisation: (i) the hadron mass can be neglected and (ii) the ω dependence of the weight ρ can be approximated by a Dirac delta, Ψ can be written as:

$$\Psi(x, \xi, t) = \Psi(z), \quad \text{with} \quad z = -t\frac{(1-x)^2}{1-\xi^2}. \tag{76}$$

Introduced in ref. [85], this simplification has been exploited in refs. [59,129].

The computations in the DGLAP region allow one already to extract valuable information, and in particular, the impact parameter space density $\rho(x, b_\perp)$. Regarding that point, realistic computations start emerging in the literature based one the Nakanishi representation (see refs. [59,87]). Consequently, computations of the 1+2D densities, both for the pion and kaon, have been obtained using different modelling techniques for the Nakanishi weight, or for the LFWFs themselves. The results presented in Figure 11 of ref. [59] highlight the differences that can be expected from different modelling strategies of the LFWFs. They are sizeable, especially in the kaon case.

3.3. The Covariant Extension

Even if the modelling of GPD in the DGLAP region only provides interesting outcomes, it precludes any comparison with potential future experimental data, as part of the kinematic range, the ERBL region, is missing. Since the two-body Bethe–Salpeter wave function allows us only to recover the two-body LFWFs, the standard overlap representation in the ERBL region is out of reach, and would anyway lead to a GPD model violating the polynomiality property. Consequently, another strategy should be sought.

The answer was provided few years ago, exploiting the Radon transform property existing between GPDs and DDs (see Equation (43) and ref. [83]). This technique, labelled covariant extension, allows one to exploit both the LFWFs formalism, and the DDs one, guaranteeing by construction that all GPDs theoretical properties are fulfilled. This work followed a pioneering example given in [131], and can be connected with the Laplace transform [132]. The key point of the covariant extension is to exploit the properties of the inverse Radon transform. More precisely, Boman and Todd-Quinto showed that compactly supported distributions can be reconstructed from a partial knowledge of their Radon transform [133]. In the case of GPDs, this can be translated as such: the knowledge of the GPD in the DGLAP region is sufficient to reconstruct uniquely the Double Distribution

F introduced in Equation (43) [83] (in fact, even an incomplete knowledge of the DGLAP region is enough [134]).

Since the proof presented in refs. [83,133] is rather technical, we will not enter the details here. Rather, we will provide the reader with an intuitive picture relying on tessellation and finite element methods. As illustrated in Equation (43), the Radon transform consists in integrating a function along lines. As illustrated on the left-hand side of Figure 10, the DGLAP and ERBL lines are different to the former cross the $\beta = 0$ axis outside of the definition domain, while the ERBL one hit this line within the the DD support. Looking at the right-hand side of Figure 10, one can discretise the support of the DD (only the upper right corner is shown because of symmetry properties in α and β). A DGLAP-type line is shown, probing a given number of cells. It is easy to realise that each cell can be probed by infinitely many DGLAP lines, and that reducing the size of the cell does not modify this statement. Thus, every point on the DD support, expect the one such that $\beta = 0$, is probed by infinitely many DGLAP lines. This provides the level-arm to recover the DD F from the DGLAP region only. However, it does not allow recovering the D-term, which remains inaccessible (some specific Double Distribution schemes allow one to extract a D-term, but the latter remains ambiguous [83,135]).

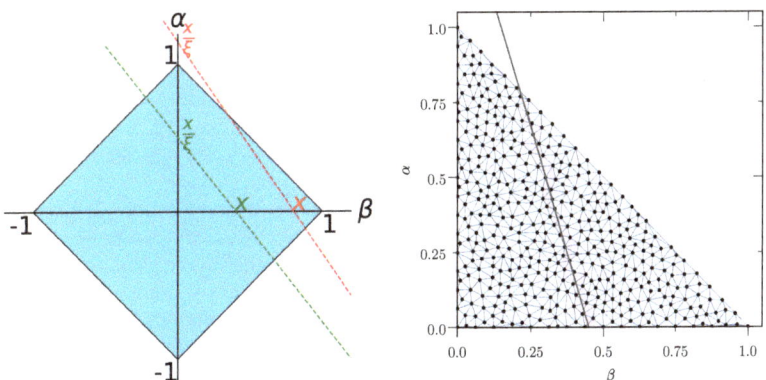

Figure 10. Left: Support of the DD and example of a DGLAP line (red) and ERBL line (green). **Right**: Support of the DD F after tessellation using a Delaunay mesh. An example of a DGLAP line is given. If numerous enough, those lines can probe every cell of the DD support. Figure from [58].

Figure 10 also highlights another difficulty in any attempt to recover the DD from the GPD in the DGLAP region only. Indeed, if one wants to probe the region close to the point $(0,0)$ in the (β, α) space, the associated DGLAP lines become almost parallel, and yields a badly conditioned inverse problem. In fact, the Radon inverse problem is known to be ill-defined in the sense of Hadammard [136–138], since, even before discretisation, the Radon transform is not continuous. The discretisation makes the problem worse, as the existence and uniqueness of a solution is not guaranteed anymore, requiring regularisation techniques to ensure the convergence of reconstruction algorithms. In the following, we will discuss the strategy applied in ref. [58].

The discretised problem can be expressed as looking for a DD vector F such that:

$$\begin{pmatrix} H_i \end{pmatrix} = \begin{pmatrix} \mathcal{R}_{ij} \end{pmatrix} \begin{pmatrix} F_j \end{pmatrix}, \qquad (77)$$

where $\mathcal{R}$ is the matrix of the discretised Radon transform and H_i—a given sample of the GPD in the DGLAP region. The size of the vector F is a function of the number of cells introduced for the discretisation, but also the degree of the polynomials used to approximate locally the Double Distribution. The elements F_j are taken at a position (β_j, α_j), which depends on the polynomial order and is called a node. For instance, a zero-order polynomial interpolation requires a node at the barycentre of each cell, while a degree-one interpolation requires a node at every vertex of the mesh. On the left-hand side, the size of H depends of the available sampling of the GPD in the DGLAP region. At minimum, the size of H_i should be as large as F_j, allowing the rank of the matrix $\mathcal{R}$ to be maximal, and thus, the problem invertible.

However, the solution of Equation (77) might be rather sensitive to the initial conditions such as the shape and size of the grid, or the sampling of the GPD. To overcome this difficulty, one can overconstrain the system by having $\mathcal{R}$ have more lines that columns, and look for a solution through a least-squares strategy. To do so, an adequate strategy is to use the so-called normal equations, although direct least-squares algorithms such as LSMR [139] can be used [83]. Rather than using iterative algorithms to minimise the associated χ^2, the normal equations provide directly the minimal solution F_j. The normal equations transform our linear system (77) into

$$F = \left(\mathcal{R}^T \mathcal{R}\right)^{-1} \mathcal{R}^T H. \tag{78}$$

In addition, it allows to simply propagate uncertainties on the values of H in that DGLAP region onto the ERBL one after the reconstruction through the DD. This is based on computations of the covariant matrix (see [58,140] for details).

In practice, for a simple functional form of DD F from ref. [85]:

$$F(\beta, \alpha) = \frac{15}{2}\theta(\beta)\left(1 - 3(\alpha^2 - \beta^2) - 2\beta\right), \tag{79}$$

an excellent agreement can be obtained through the reconstruction procedure for 780 cells and furst-order Lagrange polynomials. The results are illustrated in Figure 11, highlighting the good control on the uncertainties in the ERBL region through the reconstruction. Note however that even if the functional form is simple, the model is made more complicated by the specific DD scheme labelled Pobylitsa (or P) scheme, in which it is defined.

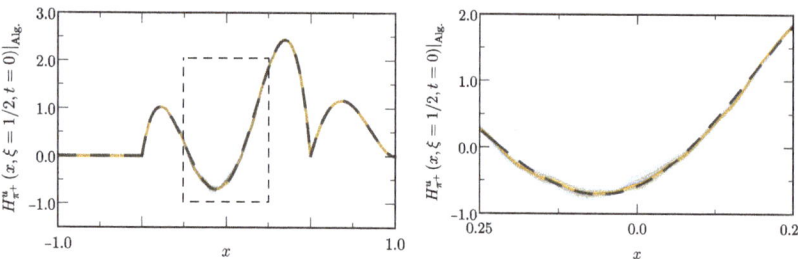

Figure 11. Left: results of the reconstuction of the DD of Equation (79). Black, exact GPD computed from the algebraic DD (79). Blue, reconstructed GPDs with a sample of H_i roughly four times the number of cells. Orange, same but with a sample of 12 times the number of cells. **Right**: zoom on the central region to highlight the uncertainty bands.

This technique, together with the simplified assumption of Equation (75), has been used to build "theoretically complete" models of the pion GPDs, ensuring that both polynomiality and positivity were fulfilled by construction, and based on state-of-the-art computations of the pion PDF using continuum techniques. The next step is to assess whether these sophisticated models of GPDs could one day be compared to experimental data.

3.4. The Sullivan Process

As we mentioned earlier, GPDs can be probed in deep exclusive processes. These processes require a high luminosity of the experimental set up. However, there is no facility today with a high enough luminosity allowing to make exclusive experiments through real pion targets. One thus has to find another solution.

3.4.1. Introduction to the Sullivan Process

Since the 1970s, one of the main ideas for studying the internal structure of the pion is to rely on the so-called Sullivan process [141], i.e., to hit a virtual pion within the meson cloud of the nucleon. It has been used to probe the pion EFF [142–144] and also the pion PDFs [145,146]. In order to probe GPDs, one focuses on the best-understood process, namely, Deep Virtual Compton Scattering (DVCS). The latter describes the exclusive electroproduction of a photon out of a pion, the latter remaining intact. This is illustrated on the left-hand side of Figure 12.

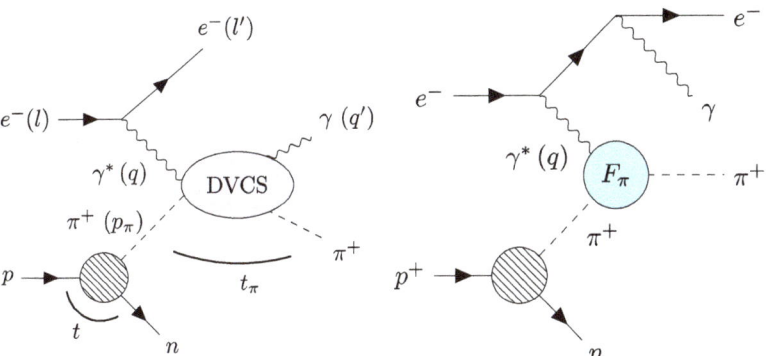

Figure 12. *Left*: Sullivan DVCS $ep \to en\gamma\pi^+$. At small t, one expects the pion pole contribution to be the leading one in such a process. *Right*: Sullivan Bethe–Heitler, interfering with the DVCS.

The differential cross section of the complete process $d\sigma_{Sul}$ depends on multiple kinematic variables [147]. Among the lists are the photon virtuality $Q^2 = -q^2$, the fraction of energy carried by the virtual pion $x_\pi = p_\pi \cdot l/(p \cdot l)$ and the inelasticity of the process $y = p \cdot q/p \cdot l$. Three angles are also necessary to characterise the kinematics in the laboratory frame. The first is ϕ, the angle between the leptonic plane (formed by the incoming and outgoing electrons), and the hadronic plane (formed by the virtual photon and the outgoing pion). The two others are the azimutal angle of the outgoing electron ϕ_e and of the outgoing neutron ϕ_n. The subprocess $e\pi \to e\pi\gamma$ is characterised by related subkinematic variables such as the inelasticity of the subprocess $y_\pi = p_\pi \cdot q/p_\pi \cdot l$ or the Bjorken variable of the subprocess $x_B^\pi = Q^2/(2p_\pi \cdot q)$. Then, one can write the differential cross section of the Sullivan process in terms of the subprocess differential cross section [147,148] as:

$$\frac{d^8\sigma(\lambda, \pm e)}{dydQ^2dt_\pi d\phi d\phi_e dx_\pi dtd\phi_n} = x_\pi \frac{g_{NN\pi}^2}{16\pi^3} F(t, \Lambda^2)^2 \frac{-t}{(m_\pi^2 - t)^2} |J_{x_B^\pi}^{Q^2}| \frac{d^5\sigma^{e\pi \to e\pi\gamma}(\lambda, \pm e)}{dy_\pi dx_B^\pi dt_\pi d\phi d\phi_e}, \quad (80)$$

where λ is the longitudinal polarisation of the electron beam, e—the electron charge, $g_{NN\pi}$—the pion-nucleon coupling, and $F(t, \Lambda)$—the associated form factor. Following ref. [147], a single-parameter functional form is chosen:

$$F(t, \Lambda) = \frac{\Lambda^2 - m_\pi^2}{\Lambda^2 - t} \quad (81)$$

with $\Lambda = 800$ MeV. $|J_{x_B^\pi}^{Q^2}|$ is the Jacobian between Q^2 and x_B^π. To be valid, the Sullivan process requires that the pion pole contribution dominates, and thus, that t remains small enough. Additionally, the Bjorken regime is achieved at the level of the subprocess for both Q^2 and $s_\pi = (p_\pi + q)^2$ large in front of other scales involved.

However, DVCS on a virtual pion is not the only process contributing to the $ep \to en\gamma\pi^+$ cross section. Additional N^* resonances decaying in the $n\pi$ channel can interfere with the Sullivan process. Their impact is reduced by ensuring that the invariant mass between the neutron and pion is large enough, typically larger than 2 GeV. However, one process always interfere with the DVCS one, the Bethe-Heitler process, shown on the right-hand side of Figure 12. Being a QED process and depending only of the pion EFF, the Bethe–Heitler contribution is computable, usually larger than the DVCS one and through the interference term, it magnifies the impact of GPDs. The subprocess cross section is thus decomposed in terms of amplitudes $\mathcal{T}$ as:

$$\frac{d^5\sigma^{e\pi \to e\pi\gamma}(\lambda, \pm e)}{dy_\pi dx_B^\pi dt_\pi d\phi d\phi_e} = \frac{\alpha_{QED}^3 x_B^\pi y_\pi}{16\pi^2 Q^2 \sqrt{1+\epsilon^2}} \frac{|\mathcal{T}^{BH}|^2 + |\mathcal{T}^{DVCS}|^2 \mp \mathcal{I}}{e^6}, \qquad (82)$$

where the interference term is further decomposed in $\mathcal{I} = \mathcal{I}_{unpol} + \lambda \mathcal{I}_{pol}$, α_{QED} is the electromagnetic coupling and $\epsilon = (2m_\pi x_B^\pi)^2/Q^2$.

3.4.2. From the Sullivan Process to GPDs

The amplitude $\mathcal{T}^{DVCS}$ and interference term $\mathcal{I}$ are not directly parametrised in terms of GPDs, but rather in terms of Compton Form Factors (CFFs) $\mathcal{H}$ which are connected to GPDs through:

$$\mathcal{H}(\xi, t, Q^2) = \int_{-1}^{1} \frac{dx}{\xi} C\left(\frac{x}{\xi}, \frac{Q^2}{\mu^2}, \alpha_s(\mu^2)\right) H(x, \xi, t, \mu^2), \qquad (83)$$

where C is a coefficient function, computed in perturbation theory. The decomposition of the DVCS amplitude and the interference term as a function of the CFFs can be found in ref. [149]. Rather than looking at the exact expression, we will assume for a moment that experimental data allow us to extract with a great accuracy $\mathcal{H}$ and ask whether it is possible to inverse the convolution of Equation (83) to regain the GPD H.

The answer to this question has been known for a long time if C is computed and the leading order of α_S only, and it is "no". In fact, at leading order, any GPD such that $H(\xi, \xi, t, \mu^2) = 0$ yields a vanishing CFF, and is thus invisible in experimental data. It was thought that this problem would vanish a once higher order of perturbation theory or evolution equations would be turned on [150]. This has been proven wrong in [151], where the concept of shadow GPDs is introduced. Briefly shadow GPDs are constructed so that below an given order of perturbation theory n, the associated CFFs vanish, and that their forward limits also vanish. Consequently, they are invisible both in DVCS and DIS. Of course, the exact cancellation is valid only at a fixed scale μ_{shadow}, but it was also shown that evolution has little effect on improving the situation [151].

In the case of the nucleon, the way out is to go through multichannel analysis of GPDs, as shadow GPDs are process dependent, i.e., a shadow GPD for DVCS can be visible in another process. In fact, some processes are expected to be free of shadow GPDs, such as Double DVCS [152]. However, these processes are much harder to measure already in the nucleon case, and are thus out of reach for the pion through the Sullivan process. In the case of the pion, rather than extracting GPDs from experimental data, the aim is to challenge existing models through comparison of predictions of the Sullivan process observables, with potential future measurements.

The first step in that direction is to figure out whether or not present and future facilities will be able to measure enough events connected to the Sullivan process. To assess that, we used a theoretically complete model of GPDs, fulfilling all required theoretical properties, and computed the CFFs associated with that model in ref. [58], using the PARTONS software [97] using a next-to-leading order description of the coefficient function, and computed separately quarks and gluon contributions. The result is illustrated in Figure 13, where the theoretically complete model corresponds to the brown curve. One can note an interesting characteristic, the sign of the real and imaginary part of the CFF changes with the inclusion of gluons. This is understood as the quark and gluon contributions to the CFF comes with a relative minus sign. This behaviour is confirmed, and even strengthened at next-to-next-to-leading order (N2LO) [153]. Thus, generally speaking, the sign of the CFF indicates the relative strength between quark and gluon contribution. Accessing such a sign would already favour or disfavour different pion GPD models.

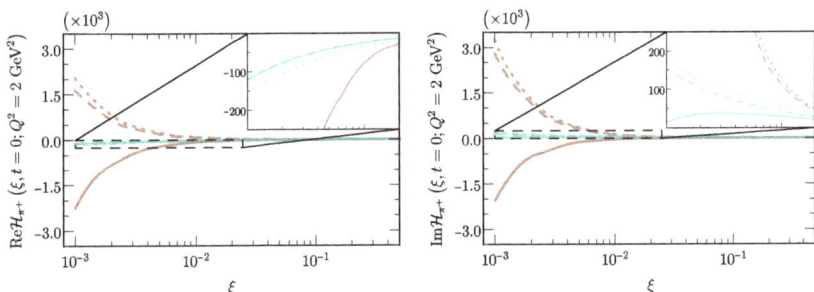

Figure 13. Real (**left**) and imaginary (**right**) parts of DVCS Compton Form Factors computed with theoretically complete model developed in ref. [58] (brown curves). Dotted lines correspond to LO computation, dashed lines—to NLO without gluon contributions and the solid lines to the full NLO computation. The light blue curves are built from standard phenomenological Anzätze in the GPD framework [55], and a phenomenological forward limit [58,154] for details).

3.4.3. A Smoking Gun for Gluons at the EIC and EicC

To access this sign, one needs to look at the interference term introduced in Equation (82) which depends linearly on the CFF [149]. Even better, one can look at the beam spin asymmetry, selecting on the polarised part of the interference and defined as:

$$A_{LU} = \frac{d\sigma^\uparrow - d\sigma^\downarrow}{d\sigma^\uparrow + d\sigma^\downarrow}. \tag{84}$$

At leading-twist, the latter is proportional to the imaginary part of the CFF $\mathcal{H}$. If gluons are strong enough to flip the sign of the polarised interference, it would trigger a flip in the sign of the asymmetry.

Two models were used to assess whether DVCS on a virtual pion would be possible, and if one would see a sign change in the beam spin asymmetry. The results are shown on Figure 14. The first model is a theoretically complete model built in ref. [58] and exploiting the state-of-th-art results on the pion PDF from Continuum Schwinger method techniques [118]. The second one is a phenomenological model, based on the Radyushkin Ansatz [55] and the GRS PDF [155]. The necessary evolution was performed by the Apfel++ software [92,94–96] interfaced with PARTONS [97]. The conclusions are clearly different from the contributions of the sole Bethe–Heitler process at typical EIC kinematics, allowing us to hope that DVCS on virtual pion will be measurable at this future facility. Moreover, the sign of the asymmetry is flipped, highlighting the strength of gluons, and this in both models. Thus, even if pion GPDs will not be extracted from experimental data, the latter will be able to assess the strength of gluons, and scan it with respect to Q^2, making an excellent physics case for the measurement of the Sullivan process at EIC.

In the case of EicC, since the kinematics probe the valence region, one is less sensitive to gluons. Nevertheless, the destructive interferences of quark and gluon trigger a sizeable reduction of the amplitude of the asymmetry, and make it sensitive to the Q^2 value probed [148]. This is also a valuable piece of experimental information that could be delivered in the future regarding pion GPDs.

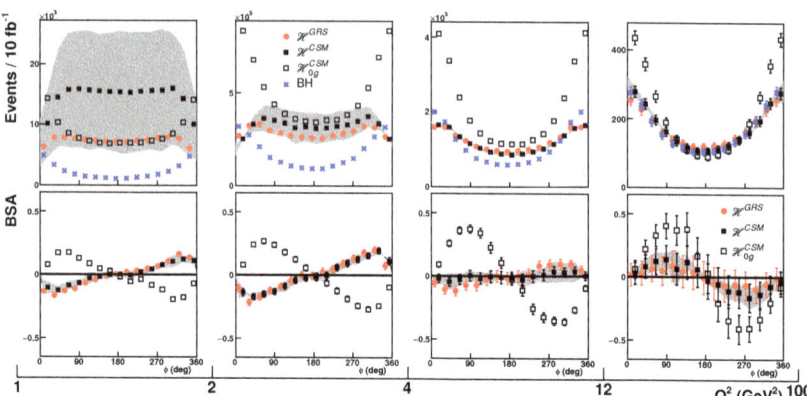

Figure 14. Upper band: number of events assessed as a function of ϕ for different bin in Q^2. The blue crosses correspond to the Bethe–Heitler signal only, the black square—to the theoretically complete model of ref. [58] with a CSM-based forward limit. The open square correspond to the same model but with gluons set to zero in the CFF. The red circles correspond to a phenomenological model with the GRS PDF [155] as an input for the forward limit. The grey band corresponds to the uncertainty of the CSM model associated with the choice of the initial evolution scale. Lower band: associated beam spin asymmetries. The sign flip is visible with the model from ref. [58]. Figure from ref. [148].

3.5. Challenges

The procedure described above and the very encouraging results which have been obtained might make one thing that the problem of computing GPDs for mesons is solved. However, some challenges remain that we will highlight in this section.

3.5.1. The Wilson Line

The first difficulty comes with the Wilson line W introduced in Equation (8). When we discussed the properties of the GPDs, we assumed that we were working in the lightcone gauge, where W is reduced to 1. However, in practice, the Bethe–Salpeter equation is mostly solved in the Landau gauge. There are two main reasons for this: the Landau gauge is the one used for fixed-gauge lattice QCD computation, allowing to compare N-point functions computed from lattice and continuum methods, and because the Landau gauge is a fixed point of the renormalisation group equation, and thus the gauge parameter does not run. Therefore, a fully consistent computation of the GPDs would require computing also the contribution of the Wilson line.

One of the possibility to bypass this problem would be to solve the Bethe–Salpeter and Dyson–Schwinger equations in the lightcone gauge. It is one of the clearest ways to provide a fully consistent description of GPDs through the overlap of LFWFs, avoiding gauge-related difficulties. As we mentioned before, this would trigger additional complications when solving the Dyson–Schwinger and Bethe–Salpeter equations, but remain feasible [156]. However, it would require a significant amount of work as all modern computations would need to be re-adapted for this specific gauge, without any lattice-QCD guidance.

In the case of the impulse approximation, attempts to describe the Wilson line perturbatively have been pursued (see [157]). However, because of the limitations of the impulse approximation already mentioned before, it is unclear that such a procedure could be generalised beyond the NJL model employed there.

3.5.2. Non-Perturbative Renormalisation

Another difficulty lies in the renormalisation properties of the operator already discussed before. One would expect that a two-body description would be valid only at low scale of QCD, a region which cannot be described using standard perturbative QCD. Yet, the renormalisation properties of the twist-two operators are known only perturbatively, generating questions on whether one could use perturbative evolution equations at low scales. Models and prescriptions have been introduced, relying on a saturating coupling in the infrared (see, e.g., [158,159]). However, it is in principle possible to really derive non-perturbative renormalisation group equations for the twist two operators, consistently with the way the Dyson–Schwinger and Bethe–Salpeter equation are renormalised. A way to proceed here would be to look at the local twist-two operators and compute their associated renormalisation constants. However, one would expect that such a procedure would be feasible only for the lowest ones, as in the Landau gauge, the number of gluon fields entering the definition of the operator increases linearly with the order of the operator itself. Nevertheless, renormalising non-perturbatively the lowest-order twist-two operators would already provide insights on the size of the uncertainties associated with perturbative evolution at low scale.

Finally, let us add that for phenomenology purposes, one would need to manipulate objects renormalised in a consistent way, i.e., using the same renormalisation scheme. As most coefficient functions are computed in the $\overline{MS}$ scheme, it is, in principle, necessary to match the non-perturbative scheme to $\overline{MS}$, at a given order of perturbation theory. Since, for the moment, only perturbative evolution is used (though in an improved for with a saturating coupling), this question has not been treated in the continuum literature but will be brought to light once a non-perturbative renormalisation strategy is put in place.

3.5.3. The D-Term

Through the modelling procedure, we have also highlighted a difficulty regarding our ability to extract the so-called D-term when extending GPDs from the DGLAP to ERBL regions. Indeed, the D-term appears as a singularity on the $\beta = 0$ line (see Equation (44)), which remains ambiguous through the covariant extension technique presented here. The soft pion theorem of Equations (17) and (18) provides an anchor for vanishing t in the chiral limit. Moreover, the large t behaviour is known in perturbative QCD [160]. Yet in between, one is left with a hole to fill through modelling, in the absence of better treatment of this specific contribution. It was shown in ref. [135] that the question of the determination of the D-term was equivalent to the assumption of the J = 0 fixed pole in the Regge theory of Compton Scattering and that, for the moment, there is no other way than measuring specific sum rules connected to the D-term in order to tame the ambiguity on the latter.

4. From Mesons to Baryons

Having discussed in detail the results obtained regarding meson GPDs, we provide here some hints at how this can be generalised to baryons, as pioneering studies are still ongoing [161].

4.1. Nucleon LFWFs

The first complication between the pion and the nucleon, is that the lower Fock state of the latter is made of three quarks instead of two. This obviously increases the number of degrees of freedom and thus complicates the computation of three-body Bethe–Salpeter equation (also called the Faddeev equation). Techniques to compute the latter relying either on a quark–diquark picture (see, e.g., [162–164]) or using a three-quark picture [165–167] have been developed. Consequently, it is in principle possible to follow the mesons' path and project the solution of these covariant equations onto the lightfront. The expected result is composed of six independent lightfront wave functions (see [168]) characterising the possible three quarks fluctuations of the nucleon and carrying a given amount of orbital angular momentum.

A first attempt of computing the LFWFs in such a formalism has been pursued in the quark-diquark picture and led to computation of the leading-twist distribution amplitude [169,170]. The computations have highlighted several interesting points. First, it validated the extension of the meson framework toward the baryon ones. Then, it showed that the quark–diquark computations yield very similar results with respect to the lattice QCD simulations of the first Mellin moments, highlighting that this kind of approximation is indeed valuable, as explained in great details in ref. [171]. Finally, a third point of interest is the noticeable difference between the Distribution Amplitude of the nucleon and the Roper resonance highlighted in ref. [169]. Indeed, while the distribution amplitude of the nucleon is positive definite, the one of the Roper resonance changes sign, meaning there is a kinematic area which is forbidden. This behaviour is similar to what one expects in non-relativistic quantum mechanics regarding higher excited states. One may thus expect significant differences between the nucleon GPDs and resonances GPD from such a comparison of Distribution amplitude. We expect that these differences also impact the 1+2D probability densities to find quarks and gluons at a given position of the transverse plane, reinforcing the interest in the computation of excited states GPDs.

4.2. Nucleon GPDs

Once the nucleon LFWFs are at hand, the formalism of overlap of LFWFs can be applied in the same way as for the pion, with some small differences. The first one is that because nucleons are spin 1/2 particles, one needs several GPDs to parametrise the considered leading twist matrix element (see Equation (9)). As a result, the overlap representation will naively yield combinations of GPDs. The latter can be disentangled by playing with the incoming and outgoing nucleon helicity projections (see, for instance, [84]). Multiple GPDs, means multiple DDs to extract as, in the case of the nucleon, Equation (44) is modified to:

$$H(x,\xi,t) = \int_\Omega d\beta d\alpha \, \delta(x - \beta - \alpha\xi)[F(\beta,\alpha,t) + \xi\delta(\beta)D(\alpha,t)], \tag{85}$$

$$E(x,\xi,t) = \int_\Omega d\beta d\alpha \, \delta(x - \beta - \alpha\xi)[K(\beta,\alpha,t) - \xi\delta(\beta)D(\alpha,t)]. \tag{86}$$

Consequently, two DDs, F and K, need to be extracted following the covariant extension procedure. Rather than extracting them independently, it may be better to focus on $H + E$, which is directly a "true" Radon transform, without an additional D-term, and $-E$, as it was suggested in ref. [39,172] and put in practice in phenomenological extractions in ref. [57]. The stability of the incomplete inverse Radon transform remains to be assessed in the case of the nucleon.

This decomposition allows us to highlight that the issue of the D-term is more difficult to handle in the case of the nucleon, with respect to the pion. Indeed, to the best of our knowledge, there is no equivalent in the nucleon case of the soft pion theorem, and thus, a theoretical constrain providing us with information on the D-term is lost. If the large-t expansion in perturbative QCD has been computed (see ref. [160]), at experimentally achievable values of t (and small enough so that collinear factorisation works), there is for the moment no theoretical guidance on the value of this function. It is noticeable though that, at the time of writing, phenomenological extractions of the nucleon D-term based on the DVCS dispersion relations (see [173,174]) provide results which remain compatible with zero [71,175]. This situation might change with the more precise DVCS data provided by the upgraded facilities of the Jefferson Laboratory [19,20].

With the expected wealth of new experimental data coming from Jefferson Laboratory in the following years, the question of a consistent description of experimental data using GPDs computed with continuum technique has never been more relevant. The impact of the Wilson line together with that of non-perturbative renormalisation are expected to be at the core of future theoretical developments. Finally, let us highlight that the overlap of LFWFs will also allow us to compute using continuum techniques the standard nucleon PDFs and the standard nucleon EFFs.

4.3. Transition GPDs

Beyond the nucleon GPDs, we would like to highlight that our formalism is able to handle all spin 1/2 baryons GPDs (providing that computations of the associated Faddeev wave function are available), but also transition GPDs, between two different baryons. Among all possible transitions, the better experimentally accessible one is certainly the nucleon-to-delta transition [176,177]. One can indeed define a transition DVCS process, such that $ep \rightarrow e\Delta^+\gamma$, which factorises between the standard DVCS hard coefficient function and nucleon-to-delta transition GPDs. Such a complementary channel provides an indirect way to probe the internal structure of the delta particle, as in the continuum framework, one would need to compute the transition GPDs as an overlap between the nucleon and delta LFWFs. This will allow to shed light on the delta LFWFs, and thus provide feedback on the solution of the Faddeev equations for baryons other than the nucleon. We expect that it provides guidance in our understanding of the three-body interaction and its consequences in QCD.

5. Conclusions

Concluding this review paper, we would like to highlight again that GPDs provide a unique way to map hadrons in 3D and gain experimental insights into their energy-momentum tensor. However, the path toward reliable models and computations of GPDs remains difficult. We have highlighted how using continuum techniques, one could manage to compute GPDs who fulfil all required theoretical constrains, and more specifically, both positivity and polynomiality. It relies on the projection of the solutions of the Bethe–Salpeter equations (or Faddeev equation for baryons) onto the lightfront, so that one obtains the so-called lightfront wave functions. GPDs can then be expressed as an overlap of lightfront wave functions in the so-called DGLAP region, ensuring the positivity is fulfilled. Then, they are extended to the ERBL region using the covariant extension, which guarantees the polynomiality property.

This technique has been successfully applied to mesons, and it has provided estimates of counting rates for DVCS on virtual pion through the Sullivan process, showing that one can expect such a process to be measurable. Regarding the nucleon, part of the path toward the building of a GPD model has been performed, but some of the work remain to be completed. Today, the baryon sector appears as the target to reach in the forthcoming years.

Funding: This work is supported in part in the framework of the GLUODYNAMICS project funded by the "P2IO LabEx ANR-10-LABX-0038" in the framework "Investissements d'Avenir" (ANR-11-IDEX-0003-01) managed by the Agence Nationale de la Recherche (ANR), France.

Institutional Review Board Statement: Not applicable

Data Availability Statement: Not applicable.

Acknowledgments: I am grateful to José Manuel Morgado Chavez and Michael Riberdy for their comments on the manuscript.

Conflicts of Interest: The authors declare no conflict of interest. The funders had no role in the design of the study; in the collection, analyses, or interpretation of data; in the writing of the manuscript; or in the decision to publish the results.

Abbreviations

The following abbreviations are used in this manuscript:

DDs	Double Distributions
DGLAP	Dokshitzer–Gribov–Lipatov–Altarelli–Parisi
DVCS	Deep Virtual Compton Scattering

EFFs	Electromagnetic Form Factors
EIC	Electron Ion Collider
EicC	Electron Ion collider in China
EMT	Energy Momentum Tensor
ERBL	Efremov–Radyushkin–Brodsky–Lepage
GPDs	Generalised Parton Distribution
GTMDs	Generalised Transverse Momentum dependent Distributions
JLab	Jefferson Laboratory
LFWFs	Lightfront Wave Functions
NJL	Nambu–Jona-Lasinio
PDFs	Parton Distribution Functions
QCD	Quantum Chromodynamics
RGE	Renormalisation Group Equation
TMDs	Transverse Momentum dependent Distributions

References

1. Bjorken, J. Asymptotic Sum Rules at Infinite Momentum. *Phys. Rev.* **1969**, *179*, 1547–1553. [CrossRef]
2. Altarelli, G.; Parisi, G. Asymptotic Freedom in Parton Language. *Nucl. Phys.* **1977**, *B126*, 298. [CrossRef]
3. Gribov, V.; Lipatov, L. Deep inelastic e p scattering in perturbation theory. *Sov. J. Nucl. Phys.* **1972**, *15*, 438–450.
4. Dokshitzer, Y.L. Calculation of the Structure Functions for Deep Inelastic Scattering and e+ e− Annihilation by Perturbation Theory in Quantum Chromodynamics. *Sov. Phys. JETP* **1977**, *46*, 641–653.
5. Ji, X.D. Gauge-Invariant Decomposition of Nucleon Spin. *Phys. Rev. Lett.* **1997**, *78*, 610–613. [CrossRef]
6. Mueller, D.; Robaschik, D.; Geyer, B.; Dittes, F.; Hořejsi, J. Wave functions, evolution equations and evolution kernels from light ray operators of QCD. *Fortsch. Phys.* **1994**, *42*, 101–141. [CrossRef]
7. Radyushkin, A. Nonforward parton distributions. *Phys. Rev.* **1997**, *D56*, 5524–5557. [CrossRef]
8. Collins, J.C.; Soper, D.E.; Sterman, G.F. Transverse Momentum Distribution in Drell-Yan Pair and W and Z Boson Production. *Nucl. Phys. B* **1985**, *250*, 199–224. [CrossRef]
9. Collins, J. *Foundations of Perturbative QCD*; Cambridge University Press: Cambridge, UK, 2013; Volume 32.
10. Burkardt, M. Impact parameter dependent parton distributions and off forward parton distributions for zeta —> 0. *Phys. Rev.* **2000**, *D62*, 071503; Erratum in *Phys. Rev.* **2002**, *D66*, 119903. [CrossRef]
11. Diehl, M. Generalized parton distributions in impact parameter space. *Eur. Phys. J.* **2002**, *C25*, 223–232. [CrossRef]
12. Ji, X.d. Viewing the proton through 'color' filters. *Phys. Rev. Lett.* **2003**, *91*, 062001. [CrossRef] [PubMed]
13. Belitsky, A.V.; Ji, X.d.; Yuan, F. Quark imaging in the proton via quantum phase space distributions. *Phys. Rev.* **2004**, *D69*, 074014. [CrossRef]
14. Meissner, S.; Metz, A.; Schlegel, M. Generalized parton correlation functions for a spin-1/2 hadron. *J. High Energy Phys.* **2009**, *0908*, 056. [CrossRef]
15. Meissner, S.; Metz, A.; Schlegel, M.; Goeke, K. Generalized parton correlation functions for a spin-0 hadron. *J. High Energy Phys.* **2008**, *0808*, 038. [CrossRef]
16. Lorce, C.; Pasquini, B. Quark Wigner Distributions and Orbital Angular Momentum. *Phys. Rev.* **2011**, *D84*, 014015. [CrossRef]
17. Echevarria, M.G.; Idilbi, A.; Kanazawa, K.; Lorcé, C.; Metz, A.; Pasquini, B.; Schlegel, M. Proper definition and evolution of generalized transverse momentum dependent distributions. *Phys. Lett. B* **2016**, *759*, 336–341. [CrossRef]
18. Hatta, Y.; Xiao, B.W.; Yuan, F. Probing the Small- x Gluon Tomography in Correlated Hard Diffractive Dijet Production in Deep Inelastic Scattering. *Phys. Rev. Lett.* **2016**, *116*, 202301. [CrossRef] [PubMed]
19. Georges, F.; Rashad, M.N.H.; Stefanko, A.; Dlamini, M.; Karki, B.; Ali, S.F.; Lin, P.; Ko, H.; Israel, N.; Adikaram, D.; et al. Deeply Virtual Compton Scattering Cross Section at High Bjorken xB. *Phys. Rev. Lett.* **2022**, *128*, 252002. [CrossRef]
20. Christiaens, G.; Defurne, M.; Sokhan, D.; Achenbach, P.; Akbar, Z.; Amaryan, M.J.; Atac, H.; Avakian, H.; Gayoso, C.A.; Baashen, L.; et al. First CLAS12 measurement of DVCS beam-spin asymmetries in the extended valence region. *arXiv* **2022**, arXiv:2211.11274.
21. Catani, S.; de Florian, D.; Grazzini, M. Universality of nonleading logarithmic contributions in transverse momentum distributions. *Nucl. Phys. B* **2001**, *596*, 299–312. [CrossRef]
22. Echevarria, M.G.; Scimemi, I.; Vladimirov, A. Unpolarized Transverse Momentum Dependent Parton Distribution and Fragmentation Functions at next-to-next-to-leading order. *J. High Energy Phys.* **2016**, *2016*, 4. [CrossRef]

23. Bertone, V. Matching generalised transverse-momentum-dependent distributions onto generalised parton distributions at one loop. *Eur. Phys. J. C* **2022**, *82*, 941. [CrossRef]
24. Braun, V.; Müller, D. Exclusive processes in position space and the pion distribution amplitude. *Eur. Phys. J. C* **2008**, *55*, 349–361. [CrossRef]
25. Ji, X. Parton Physics on a Euclidean Lattice. *Phys. Rev. Lett.* **2013**, *110*, 262002. [CrossRef]
26. Radyushkin, A.V. Quasi-parton distribution functions, momentum distributions, and pseudo-parton distribution functions. *Phys. Rev. D* **2017**, *96*, 034025. [CrossRef]
27. Alexandrou, C.; Cichy, K.; Constantinou, M.; Jansen, K.; Scapellato, A.; Steffens, F. Light-Cone Parton Distribution Functions from Lattice QCD. *Phys. Rev. Lett.* **2018**, *121*, 112001. [CrossRef]
28. Egerer, C.; Edwards, R.G.; Kallidonis, C.; Orginos, K.; Radyushkin, A.V.; Richards, D.G.; Romero, E.; Zafeiropoulos, S. Towards high-precision parton distributions from lattice QCD via distillation. *J. High Energy Phys.* **2021**, *11*, 148. [CrossRef]
29. Constantinou, M.; Courtoy, A.; Ebert, M.A.; Engelhardt, M.; Giani, T.; Hobbs, T.; Hou, T.-J.; Kusina, A.; Kutak, K.; Liang, J.; et al. Parton distributions and lattice-QCD calculations: Toward 3D structure. *Prog. Part. Nucl. Phys.* **2021**, *121*, 103908. [CrossRef]
30. Nakanishi, N. A General survey of the theory of the Bethe-Salpeter equation. *Prog. Theor. Phys. Suppl.* **1969**, *43*, 1–81. [CrossRef]
31. Nakanishi, N. *Graph Theory and Feynman Integrals*; Gordon and Breach: New York, NY, USA, 1971.
32. Binosi, D.; Tripolt, R.A. Spectral functions of confined particles. *Phys. Lett. B* **2020**, *801*, 135171, [CrossRef]
33. Eichmann, G.; Ferreira, E.; Stadler, A. Going to the light front with contour deformations. *Phys. Rev. D* **2022**, *105*, 034009. [CrossRef]
34. Aguilar, A.C.; Cardona, J.C.; Ferreira, M.N.; Papavassiliou, J. Non-Abelian Ball-Chiu vertex for arbitrary Euclidean momenta. *Phys. Rev.* **2017**, *D96*, 014029. [CrossRef]
35. Aguilar, A.C.; Ferreira, M.N.; Figueiredo, C.T.; Papavassiliou, J. Nonperturbative Ball-Chiu construction of the three-gluon vertex. *Phys. Rev. D* **2019**, *99*, 094010. [CrossRef]
36. Aguilar, A.C.; De Soto, F.; Ferreira, M.N.; Papavassiliou, J.; Rodríguez-Quintero, J.; Zafeiropoulos, S. Gluon propagator and three-gluon vertex with dynamical quarks. *Eur. Phys. J. C* **2020**, *80*, 154. [CrossRef]
37. Aguilar, A.C.; Ferreira, M.N.; Papavassiliou, J. Exploring smoking-gun signals of the Schwinger mechanism in QCD. *Phys. Rev. D* **2022**, *105*, 014030. [CrossRef]
38. Diehl, M. Generalized parton distributions. *Phys. Rep.* **2003**, *388*, 41–277. [CrossRef]
39. Belitsky, A.; Radyushkin, A. Unraveling hadron structure with generalized parton distributions. *Phys. Rep.* **2005**, *418*, 1–387. [CrossRef]
40. Kumericki, K.; Liuti, S.; Moutarde, H. GPD phenomenology and DVCS fitting. *Eur. Phys. J.* **2016**, *A52*, 157. [CrossRef]
41. Mezrag, C. An Introductory Lecture on Generalised Parton Distributions. *Few Body Syst.* **2022**, *63*, 62. [CrossRef]
42. Mezrag, C.; Moutarde, H.; Rodriguez-Quintero, J. From Bethe–Salpeter Wave functions to Generalised Parton Distributions. *Few Body Syst.* **2016**, *57*, 729–772. [CrossRef]
43. Ji, X.D. Deeply virtual Compton scattering. *Phys. Rev.* **1997**, *D55*, 7114–7125. [CrossRef]
44. Radyushkin, A. Asymmetric gluon distributions and hard diffractive electroproduction. *Phys. Lett.* **1996**, *B385*, 333–342. [CrossRef]
45. Diehl, M.; Gousset, T.; Pire, B.; Teryaev, O. Probing partonic structure in gamma* gamma —> pi pi near threshold. *Phys. Rev. Lett.* **1998**, *81*, 1782–1785. [CrossRef]
46. Polyakov, M.V. Hard exclusive electroproduction of two pions and their resonances. *Nucl. Phys.* **1999**, *B555*, 231. [CrossRef]
47. Kivel, N.; Mankiewicz, L.; Polyakov, M.V. NLO corrections and contribution of a tensor gluon operator to the process gamma* gamma —> pi pi. *Phys. Lett. B* **1999**, *467*, 263–270. [CrossRef]
48. Diehl, M.; Gousset, T.; Pire, B. Exclusive production of pion pairs in gamma* gamma collisions at large Q**2. *Phys. Rev.* **2000**, *D62*, 073014. [CrossRef]
49. Diehl, M.; Gousset, T. Time ordering in off diagonal parton distributions. *Phys. Lett.* **1998**, *B428*, 359–370. [CrossRef]
50. Efremov, A.; Radyushkin, A. Asymptotical Behavior of Pion Electromagnetic Form-Factor in QCD. *Theor. Math. Phys.* **1980**, *42*, 97–110.[CrossRef]
51. Efremov, A.; Radyushkin, A. Factorization and Asymptotical Behavior of Pion Form-Factor in QCD. *Phys. Lett.* **1980**, *B94*, 245–250. [CrossRef]
52. Lepage, G.P.; Brodsky, S.J. Exclusive Processes in Quantum Chromodynamics: Evolution Equations for Hadronic Wave Functions and the Form-Factors of Mesons. *Phys. Lett.* **1979**, *B87*, 359–365. [CrossRef]
53. Lepage, G.P.; Brodsky, S.J. Exclusive Processes in Perturbative Quantum Chromodynamics. *Phys. Rev.* **1980**, *D22*, 2157. [CrossRef]

54. Vanderhaeghen, M.; Guichon, P.A.; Guidal, M. Deeply virtual electroproduction of photons and mesons on the nucleon: Leading order amplitudes and power corrections. *Phys. Rev.* **1999**, *D60*, 094017. [CrossRef]
55. Musatov, I.; Radyushkin, A. Evolution and models for skewed parton distributions. *Phys. Rev.* **2000**, *D61*, 074027. [CrossRef]
56. Goloskokov, S.; Kroll, P. Vector meson electroproduction at small Bjorken-x and generalized parton distributions. *Eur. Phys. J.* **2005**, *C42*, 281–301. [CrossRef]
57. Mezrag, C.; Moutarde, H.; Sabatié, F. Test of two new parameterizations of the Generalized Parton Distribution H. *Phys. Rev.* **2013**, *D88*, 014001. [CrossRef]
58. Chavez, J.M.M.; Bertone, V.; De Soto Borrero, F.; Defurne, M.; Mezrag, C.; Moutarde, H.; Rodríguez-Quintero, J.; Segovia, J. Pion generalized parton distributions: A path toward phenomenology. *Phys. Rev. D* **2022**, *105*, 094012. [CrossRef]
59. Raya, K.; Cui, Z.F.; Chang, L.; Morgado, J.M.; Roberts, C.D.; Rodriguez-Quintero, J. Revealing pion and kaon structure via generalised parton distributions. *Chin. Phys. C* **2022**, *46*, 013105. [CrossRef]
60. Mezrag, C.; Chang, L.; Moutarde, H.; Roberts, C.D.; Rodríguez-Quintero, J.; Sabatié, F.; Schmidt, S.M. Sketching the pion's valence-quark generalised parton distribution. *Phys. Lett.* **2014**, *B741*, 190–196. [CrossRef]
61. Moutarde, H.; Sznajder, P.; Wagner, J. Border and skewness functions from a leading order fit to DVCS data. *Eur. Phys. J.* **2018**, *C78*, 890. [CrossRef]
62. Khalek, R.A.; Accardi, A.; Adam, J.; Adamiak, D.; Akers, W.; Albaladejo, M.; Al-bataineh, A.; Alexeev, M.G.; Ameli, F.; Antonioli, P.; et al. Science Requirements and Detector Concepts for the Electron-Ion Collider: EIC Yellow Report. *arXiv* **2021**, arXiv:2103.05419.
63. Anderle, D.P.; Bertone, V.; Cao, X.; Chang, L.; Chang, N.; Chen, G.; Chen, X.; Chen, Z.; Cui, Z.; Dai, L.; et al. Electron-ion collider in China. *Front. Phys.* **2021**, *16*, 64701. [CrossRef]
64. Polyakov, M.V.; Schweitzer, P. Forces inside hadrons: pressure, surface tension, mechanical radius, and all that. *Int. J. Mod. Phys. A* **2018**, *33*, 1830025. [CrossRef]
65. Bakker, B.L.G.; Leader, E.; Trueman, T.L. A Critique of the angular momentum sum rules and a new angular momentum sum rule. *Phys. Rev. D* **2004**, *70*, 114001. [CrossRef]
66. Leader, E.; Lorcé, C. The angular momentum controversy: What's it all about and does it matter? *Phys. Rep.* **2014**, *541*, 163–248. [CrossRef]
67. Brodsky, S.J.; Hwang, D.S.; Ma, B.Q.; Schmidt, I. Light cone representation of the spin and orbital angular momentum of relativistic composite systems. *Nucl. Phys. B* **2001**, *593*, 311–335. [CrossRef]
68. Lowdon, P.; Chiu, K.Y.J.; Brodsky, S.J. Rigorous constraints on the matrix elements of the energy-momentum tensor. *Phys. Lett. B* **2017**, *774*, 1–6. [CrossRef]
69. Lorcé, C.; Lowdon, P. Universality of the Poincaré gravitational form factor constraints. *Eur. Phys. J. C* **2020**, *80*, 207. [CrossRef]
70. Lorcé, C.; Mantovani, L.; Pasquini, B. Spatial distribution of angular momentum inside the nucleon. *Phys. Lett. B* **2018**, *776*, 38–47. [CrossRef]
71. Dutrieux, H.; Lorcé, C.; Moutarde, H.; Sznajder, P.; Trawiński, A.; Wagner, J. Phenomenological assessment of proton mechanical properties from deeply virtual Compton scattering. *Eur. Phys. J. C* **2021**, *81*, 300, . [CrossRef]
72. Polyakov, M.V. Generalized parton distributions and strong forces inside nucleons and nuclei. *Phys. Lett.* **2003**, *B555*, 57–62. [CrossRef]
73. Lorcé, C.; Moutarde, H.; Trawiński, A.P. Revisiting the mechanical properties of the nucleon. *Eur. Phys. J. C* **2019**, *79*, 89. [CrossRef]
74. Ji, X.D. Off forward parton distributions. *J. Phys.* **1998**, *G24*, 1181–1205. [CrossRef]
75. Radyushkin, A. Symmetries and structure of skewed and double distributions. *Phys. Lett.* **1999**, *B449*, 81–88. [CrossRef]
76. Hausdorff, F. Summationsmethoden und Momentfolgen. I. *Math. Z.* **1921**, *9*, 74–109. [CrossRef]
77. Hausdorff, F. Summationsmethoden und Momentfolgen. II. *Math. Z.* **1921**, *9*, 280–299. [CrossRef]
78. Hertle, A. Continuity of the Radon transform and its inverse on euclidean space. *Math. Z.* **1983**, *184*, 165–192. [CrossRef]
79. Radon, J. Über die Bestimmung von Funktionen durch ihre Integralwerte längs gewisser Mannigfaltigkeiten. *Akad. Wiss.* **1917**, *69*, 262.
80. Radon, J. On the determination of functions from their integral values along certain manifolds. *IEEE Trans. Med. Imaging* **1986**, *5*, 170–176. [CrossRef]
81. Polyakov, M.V.; Weiss, C. Skewed and double distributions in pion and nucleon. *Phys. Rev.* **1999**, *D60*, 114017. [CrossRef]
82. Teryaev, O. Crossing and radon tomography for generalized parton distributions. *Phys. Lett.* **2001**, *B510*, 125–132. [CrossRef]
83. Chouika, N.; Mezrag, C.; Moutarde, H.; Rodríguez-Quintero, J. Covariant Extension of the GPD overlap representation at low Fock states. *Eur. Phys. J.* **2017**, *C77*, 906. [CrossRef]
84. Diehl, M.; Feldmann, T.; Jakob, R.; Kroll, P. The Overlap representation of skewed quark and gluon distributions. *Nucl. Phys.* **2001**, *B596*, 33–65. [CrossRef]

85. Chouika, N.; Mezrag, C.; Moutarde, H.; Rodríguez-Quintero, J. A Nakanishi-based model illustrating the covariant extension of the pion GPD overlap representation and its ambiguities. *Phys. Lett.* **2018**, *B780*, 287–293. [CrossRef]
86. Rinaldi, M. GPDs at non-zero skewness in ADS/QCD model. *Phys. Lett. B* **2017**, *771*, 563–567. [CrossRef]
87. Shi, C.; Bednar, K.; Cloët, I.C.; Freese, A. Spatial and Momentum Imaging of the Pion and Kaon. *Phys. Rev. D* **2020**, *101*, 074014. [CrossRef]
88. Radyushkin, A. Double distributions and evolution equations. *Phys. Rev.* **1999**, *D59*, 014030, [CrossRef]
89. Pire, B.; Soffer, J.; Teryaev, O. Positivity constraints for off-forward parton distributions. *Eur. Phys. J.* **1999**, *C8*, 103–106. [CrossRef]
90. Pobylitsa, P. Disentangling positivity constraints for generalized parton distributions. *Phys. Rev.* **2002**, *D65*, 114015. [CrossRef]
91. Pobylitsa, P. Inequalities for generalized parton distributions H and E. *Phys. Rev.* **2002**, *D65*, 077504, [CrossRef]
92. Bertone, V.; Dutrieux, H.; Mezrag, C.; Morgado, J.M.; Moutarde, H. Revisiting evolution equations for generalised parton distributions. *Eur. Phys. J. C* **2022**, *82*, 888. [CrossRef]
93. Vinnikov, A. Code for prompt numerical computation of the leading order GPD evolution. *arXiv* **2006**, arXiv:hep-ph/hep-ph/0604248.
94. Bertone, V.; Carrazza, S.; Rojo, J. APFEL: A PDF Evolution Library with QED corrections. *Comput. Phys. Commun.* **2014**, *185*, 1647–1668. [CrossRef]
95. Bertone, V.; Carrazza, S.; Hartland, N.P. APFELgrid: a high performance tool for parton density determinations. *Comput. Phys. Commun.* **2017**, *212*, 205–209. [CrossRef]
96. Bertone, V. APFEL++: A new PDF evolution library in C++. *PoS* **2018**, *DIS2017*, 201. [CrossRef]
97. Berthou, B.; Binosi, D.; Chouika, N.; Colaneri, L.; Guidal, M.; Mezrag, C.; Moutarde, H.; Rodríguez-Quintero, J.; Sabatié, F.; Sznajder, P.; et al. PARTONS: PARtonic Tomography Of Nucleon Software. A computing framework for the phenomenology of Generalized Parton Distributions. *Eur. Phys. J.* **2018**, *C78*, 478. [CrossRef]
98. Ohrndorf, T. Constraints From Conformal Covariance on the Mixing of Operators of Lowest Twist. *Nucl. Phys. B* **1982**, *198*, 26–44. [CrossRef]
99. Mueller, D.; Schafer, A. Complex conformal spin partial wave expansion of generalized parton distributions and distribution amplitudes. *Nucl. Phys.* **2006**, *B739*, 1–59. [CrossRef]
100. Kumericki, K.; Mueller, D.; Passek-Kumericki, K. Towards a fitting procedure for deeply virtual Compton scattering at next-to-leading order and beyond. *Nucl. Phys.* **2008**, *794*, 244–323. [CrossRef]
101. Müller, D.; Lautenschlager, T.; Passek-Kumericki, K.; Schaefer, A. Towards a fitting procedure to deeply virtual meson production—The next-to-leading order case. *Nucl. Phys.* **2014**, *B884*, 438–546. [CrossRef]
102. Belitsky, A.V.; Mueller, D. Next-to-leading order evolution of twist-2 conformal operators: The Abelian case. *Nucl. Phys.* **1998**, *B527*, 207–234. [CrossRef]
103. Belitsky, A.V.; Mueller, D. Exclusive evolution kernels in two loop order: Parity even sector. *Phys. Lett.* **1999**, *B464*, 249–256. [CrossRef]
104. Belitsky, A.V.; Mueller, D.; Freund, A. Reconstruction of nonforward evolution kernels. *Phys. Lett.* **1999**, *B461*, 270–279. [CrossRef]
105. Belitsky, A.V.; Freund, A.; Mueller, D. Evolution kernels of skewed parton distributions: Method and two loop results. *Nucl. Phys.* **2000**, *B574*, 347–406. [CrossRef]
106. Braun, V.M.; Manashov, A.N.; Moch, S.; Strohmaier, M. Two-loop evolution equations for flavor-singlet light-ray operators. *J. High Energy Phys.* **2019**, *02*, 191. [CrossRef]
107. Braun, V.M.; Manashov, A.N.; Moch, S.; Strohmaier, M. Three-loop evolution equation for flavor-nonsinglet operators in off-forward kinematics. *J. High Energy Phys.* **2017**, *06*, 037. [CrossRef]
108. Mueller, D. The Evolution of the pion distribution amplitude in next-to-leading-order. *Phys. Rev.* **1995**, *D51*, 3855–3864. [CrossRef]
109. Braun, V.M.; Manashov, A.N.; Moch, S.; Schoenleber, J. Two-loop coefficient function for DVCS: vector contributions. *J. High Energy Phys.* **2020**, *09*, 117. [CrossRef]
110. Pobylitsa, P.V. Integral representations for nonperturbative GPDs in terms of perturbative diagrams. *Phys. Rev. D* **2003**, *67*, 094012. [CrossRef]
111. Theussl, L.; Noguera, S.; Vento, V. Generalized parton distributions of the pion in a Bethe-Salpeter approach. *Eur. Phys. J.* **2004**, *A20*, 483–498. [CrossRef]
112. Tiburzi, B.; Miller, G. Generalized parton distributions and double distributions for q anti-q pions. *Phys. Rev.* **2003**, *D67*, 113004. [CrossRef]
113. Mezrag, C.; Moutarde, H.; Rodríguez-Quintero, J.; Sabatié, F. Towards a Pion Generalized Parton Distribution Model from Dyson-Schwinger Equations. *arXiv* **2014**, arXiv:1406.7425.
114. Chang, L.; Mezrag, C.; Moutarde, H.; Roberts, C.D.; Rodriguez-Quintero, J.; Tandy, P.C. Basic features of the pion valence-quark distribution function. *Phys. Lett.* **2014**, *B737*, 23–29. [CrossRef]

115. Binosi, D.; Chang, L.; Papavassiliou, J.; Qin, S.X.; Roberts, C.D. Symmetry preserving truncations of the gap and Bethe-Salpeter equations. *Phys. Rev. D* **2016**, *93*, 096010. [CrossRef]
116. Qin, S.X.; Roberts, C.D. Resolving the Bethe–Salpeter Kernel. *Chin. Phys. Lett.* **2021**, *38*, 071201. [CrossRef]
117. Nguyen, T.; Bashir, A.; Roberts, C.D.; Tandy, P.C. Pion and kaon valence-quark parton distribution functions. *Phys. Rev.* **2011**, *C83*, 062201. [CrossRef]
118. Ding, M.; Raya, K.; Binosi, D.; Chang, L.; Roberts, C.D.; Schmidt, S.M. Symmetry, symmetry breaking, and pion parton distributions. *Phys. Rev. D* **2020**, *101*, 054014. [CrossRef]
119. Broniowski, W.; Ruiz Arriola, E.; Golec-Biernat, K. Generalized parton distributions of the pion in chiral quark models and their QCD evolution. *Phys. Rev.* **2008**, *D77*, 034023. [CrossRef]
120. Freese, A.; Cloët, I.C. Impact of dynamical chiral symmetry breaking and dynamical diquark correlations on proton generalized parton distributions. *Phys. Rev. C* **2020**, *101*, 035203. [CrossRef]
121. Freese, A.; Cloët, I.C. Quark spin and orbital angular momentum from proton generalized parton distributions. *Phys. Rev. C* **2021**, *103*, 045204. [CrossRef]
122. Zhang, J.L.; Cui, Z.F.; Ping, J.; Roberts, C.D. Contact interaction analysis of pion GTMDs. *Eur. Phys. J. C* **2021**, *81*, 6 . [CrossRef]
123. Xing, Z.; Ding, M.; Raya, K.; Chang, L. A fresh look at the generalized parton distributions of light pseudoscalar mesons. *arXiv* **2023**, arXiv:2301.02958.
124. Lan, J.; Mondal, C.; Jia, S.; Zhao, X.; Vary, J.P. Pion and kaon parton distribution functions from basis light front quantization and QCD evolution. *Phys. Rev. D* **2020**, *101*, 034024. [CrossRef]
125. de Teramond, G.F.; Liu, T.; Sufian, R.S.; Dosch, H.G.; Brodsky, S.J.; Deur, A. Universality of Generalized Parton Distributions in Light-Front Holographic QCD. *Phys. Rev. Lett.* **2018**, *120*, 182001. [CrossRef]
126. Burkardt, M.; Ji, X.d.; Yuan, F. Scale dependence of hadronic wave functions and parton densities. *Phys. Lett.* **2002**, *B545*, 345–351. [CrossRef]
127. Chang, L.; Cloet, I.; Cobos-Martinez, J.; Roberts, C.; Schmidt, S.; Tandy, P.C. Imaging dynamical chiral symmetry breaking: pion wave function on the light front. *Phys. Rev. Lett.* **2013**, *110*, 132001. [CrossRef] [PubMed]
128. Xu, S.S.; Chang, L.; Roberts, C.D.; Zong, H.S. Pion and kaon valence-quark parton quasidistributions. *Phys. Rev. D* **2018**, *97*, 094014. [CrossRef]
129. Albino, L.; Higuera-Angulo, I.M.; Raya, K.; Bashir, A. Pseudoscalar mesons: Light front wave functions, GPDs, and PDFs. *Phys. Rev. D* **2022**, *106*, 034003. [CrossRef]
130. Mezrag, C.; Salmè, G. Fermion and Photon gap-equations in Minkowski space within the Nakanishi Integral Representation method. *Eur. Phys. J. C* **2021**, *81*, 34. [CrossRef]
131. Hwang, D.; Mueller, D. Implication of the overlap representation for modelling generalized parton distributions. *Phys. Lett.* **2008**, *B660*, 350–359. [CrossRef]
132. Müller, D. Double distributions and generalized parton distributions from the parton number conserved light front wave function overlap representation. *arXiv* **2017**, arXiv:1711.09932.
133. Boman, J.; Quinto, E.T. Support theorems for real-analytic Radon transforms. *Duke Math. J.* **1987**, *55*, 943–948. [CrossRef]
134. Chavez, J.M.M.; Dall'Olio P. ; De Soto Borrero, F. ; Mezrag, C. ; Moutarde, H. ; Rodriguez Quintero, J. ; Sznajder, P. ; Segovia, J. Reconstruction of Double distributions from limited GPD knowledge. 2023, *in preparation*.
135. Müller, D.; Semenov-Tian-Shansky, K.M. $J = 0$ fixed pole and D-term form factor in deeply virtual Compton scattering. *Phys. Rev.* **2015**, *D92*, 074025. [CrossRef]
136. Hadamard, J. Sur les problèmes aux dérivées partielles et leur signification physique. *Princet. Univ. Bull.* **1902**, *13*, 49–52.
137. Maz'ya, V.; Shaposhnikova, T. *Jacques Hadamard, a Universal Mathematician. History of Mathematics 14*; American Mathematical Society: Providence, RI, USA, 1998.
138. Natterer, F. *The Mathematics of Computerized Tomography*; Classics in Applied Mathematics; Society for Industrial and Applied Mathematics: Philadelphia, PA, USA, 2001.
139. Fong, D.; Saunders, M. LSMR: An iterative algorithm for sparse least-squares problems. *arXiv* **2010**, arXiv:1006.0758.
140. Press, W.H.; Flannery, B.P.; Teukolsky, S.A.; Vetterling, W.T. *Numerical Recipes*; Cambridge University Press: Cambridge, UK, 1989.
141. Sullivan, J.D. One pion exchange and deep inelastic electron—Nucleon scattering. *Phys. Rev. D* **1972**, *5*, 1732–1737. [CrossRef]
142. Bebek, C.J.; Brown, C.N. ; Holmes, S.D. ; Kline, R.V. ; Pipkin, F.M. ; Raither, S. ; Sisterson, L.K. ; Browman, A. ; Hanson, K.M. ; Larson, D.; et al. Electroproduction of single pions at low epsilon and a measurement of the pion form-factor up to $q^2 = 10$ GeV^2. *Phys. Rev. D* **1978**, *17*, 1693. [CrossRef]
143. Volmer, J.; Abbott, D. ; Anklin, H. ; Armstrong, C. ; Arrington, J. ; Assamagan, K. ; Avery, S. ; Baker, O.K. ; Blok, H.P. ; Bochna, C. ; Brash, E.J. ; et al. Measurement of the Charged Pion Electromagnetic Form-Factor. *Phys. Rev. Lett.* **2001**, *86*, 1713–1716. [CrossRef] [PubMed]
144. Huber, G.; Blok, H.P. ; Horn, T. ; Beise, E.J. ; Gaskell, D. ; Mack, D.J. ; Tadevosyan, V. ; Volmer, J. ; Abbott, D. ; Aniol, K. ; Anklin, H. ; Armstrong, C. ; Arrington, J. ; et al. Charged pion form-factor between Q**2 = 0.60-GeV**2 and 2.45-GeV**2. II. Determination of, and results for, the pion form-factor. *Phys. Rev.* **2008**, *C78*, 045203. [CrossRef]
145. Barry, P.C.; Sato, N.; Melnitchouk, W.; Ji, C.R. First Monte Carlo Global QCD Analysis of Pion Parton Distributions. *Phys. Rev. Lett.* **2018**, *121*, 152001. [CrossRef]

146. Barry, P.C.; Ji, C.R.; Sato, N.; Melnitchouk, W. Global QCD Analysis of Pion Parton Distributions with Threshold Resummation. *Phys. Rev. Lett.* **2021**, *127*, 232001. [CrossRef]
147. Amrath, D.; Diehl, M.; Lansberg, J.P. Deeply virtual Compton scattering on a virtual pion target. *Eur. Phys. J.* **2008**, *C58*, 179–192. [CrossRef]
148. Chávez, J.M.M.; Bertone, V.; De Soto Borrero, F.; Defurne, M.; Mezrag, C.; Moutarde, H.; Rodríguez-Quintero, J.; Segovia, J. Accessing the Pion 3D Structure at US and China Electron-Ion Colliders. *Phys. Rev. Lett.* **2022**, *128*, 202501. [CrossRef] [PubMed]
149. Belitsky, A.V.; Mueller, D. Refined analysis of photon leptoproduction off spinless target. *Phys. Rev.* **2009**, *D79*, 014017. [CrossRef]
150. Freund, A. On the extraction of skewed parton distributions from experiment. *Phys. Lett. B* **2000**, *472*, 412–419. [CrossRef]
151. Bertone, V.; Dutrieux, H.; Mezrag, C.; Moutarde, H.; Sznajder, P. The deconvolution problem of deeply virtual Compton scattering. *Phys. Rev. D* **2021**, *103*, 114019. [CrossRef]
152. Guidal, M.; Vanderhaeghen, M. Double deeply virtual Compton scattering off the nucleon. *Phys. Rev. Lett.* **2003**, *90*, 012001. [CrossRef] [PubMed]
153. Braun, V.M.; Ji, Y.; Schoenleber, J. Deeply Virtual Compton Scattering at Next-to-Next-to-Leading Order. *Phys. Rev. Lett.* **2022**, *129*, 172001. [CrossRef] [PubMed]
154. Novikov, I.; Abdolmaleki, H.; Britzger, D.; Cooper-Sarkar, A.; Giuli, F.; Glazov, A.; Kusina, A.; Luszczak, A.; Olness, F.; Starovoitov, P.; et al. Parton Distribution Functions of the Charged Pion Within The xFitter Framework. *Phys. Rev. D* **2020**, *102*, 014040. [CrossRef]
155. Gluck, M.; Reya, E.; Schienbein, I. Pionic parton distributions revisited. *Eur. Phys. J. C* **1999**, *10*, 313–317. [CrossRef]
156. Cornwall, J.M. Dynamical Mass Generation in Continuum QCD. *Phys. Rev. D* **1982**, *26*, 1453. [CrossRef]
157. Costa, C.S.R.; Freese, A.; Cloët, I.C.; El-Bennich, B.; Krein, G.a.; Tandy, P.C. Intrinsic glue and Wilson lines within dressed quarks. *Phys. Rev. C* **2021**, *104*, 045201. [CrossRef]
158. Ji, C.R.; Sill, A.F.; Lombard, R.M. Leading Order Perturbative QCD Calculation of Nucleon Dirac Form-factors. *Phys. Rev.* **1987**, *D36*, 165. [CrossRef] [PubMed]
159. Rodríguez-Quintero, J.; Binosi, D.; Mezrag, C.; Papavassiliou, J.; Roberts, C.D. Process-independent effective coupling. From QCD Green's functions to phenomenology. *Few Body Syst.* **2018**, *59*, 121. [CrossRef]
160. Hoodbhoy, P.; Ji, X.d.; Yuan, F. Probing quark distribution amplitudes through generalized parton distributions at large momentum transfer. *Phys. Rev. Lett.* **2004**, *92*, 012003. [CrossRef]
161. Riberdy, M.; Mezrag, C. ; Segovia, J. Computing 3D nucleonic orbitals:an exploratory path with continuum QCD methods. 2023, *Under preparation*.
162. Segovia, J.; Cloet, I.C.; Roberts, C.D.; Schmidt, S.M. Nucleon and Δ elastic and transition form factors. *Few Body Syst.* **2014**, *55*, 1185–1222. [CrossRef]
163. Gutiérrez-Guerrero, L.X.; Paredes-Torres, G.; Bashir, A. Mesons and baryons: Parity partners. *Phys. Rev. D* **2021**, *104*, 094013 . [CrossRef]
164. Yin, P.L.; Cui, Z.F.; Roberts, C.D.; Segovia, J. Masses of positive- and negative-parity hadron ground-states, including those with heavy quarks. *Eur. Phys. J. C* **2021**, *81*, 327 . [CrossRef]
165. Eichmann, G.; Alkofer, R.; Krassnigg, A.; Nicmorus, D. Nucleon mass from a covariant three-quark Faddeev equation. *Phys. Rev. Lett.* **2010**, *104*, 201601. [CrossRef]
166. Eichmann, G.; Sanchis-Alepuz, H.; Williams, R.; Alkofer, R.; Fischer, C.S. Baryons as relativistic three-quark bound states. *Prog. Part. Nucl. Phys.* **2016**, *91*, 1–100. [CrossRef]
167. Wang, Q.W.; Qin, S.X.; Roberts, C.D.; Schmidt, S.M. Proton tensor charges from a Poincaré-covariant Faddeev equation. *Phys. Rev.* **2018**, *D98*, 054019. [CrossRef]
168. Ji, X.d.; Ma, J.P.; Yuan, F. Three quark light cone amplitudes of the proton and quark orbital motion dependent observables. *Nucl. Phys.* **2003**, *B652*, 383–404. [CrossRef]
169. Mezrag, C.; Segovia, J.; Chang, L.; Roberts, C.D. Parton distribution amplitudes: Revealing correlations within the proton and Roper. *Phys. Lett.* **2018**, *B783*, 263–267. [CrossRef]
170. Mezrag, C; Segovia, J.; Ding, M.; Chang, L.; Roberts, C.D. Nucleon Parton Distribution Amplitude: A scalar diquark picture. In Proceedings of the 22nd International Conference on Few-Body Problems in Physics (FB22), Caen, France, 9–13 July 2018.
171. Barabanov, M.Y.; Bedolla, M.A.; Brooks, W.K.; Cates, G.D.; Chen, C.; Chen, Y.; Cisbani, E.; Ding, M.; Eichmann, G.; Ent, R.; et al. Diquark correlations in hadron physics: Origin, impact and evidence. *Prog. Part. Nucl. Phys.* **2021**, *116*, 103835. [CrossRef]
172. Radyushkin, A. Modeling Nucleon Generalized Parton Distributions. *Phys. Rev.* **2013**, *D87*, 096017. [CrossRef]
173. Anikin, I.V.; Teryaev, O.V. Dispersion relations and subtractions in hard exclusive processes. *Phys. Rev. D* **2007**, *76*, 056007. [CrossRef]
174. Diehl, M.; Ivanov, D.Y. Dispersion representations for hard exclusive processes: beyond the Born approximation. *Eur. Phys. J. C* **2007**, *52*, 919–932. [CrossRef]
175. Kumerički, K. Measurability of pressure inside the proton. *Nature* **2019**, *570*, E1–E2. [CrossRef]

176. Guichon, P.A.M.; Mossé, L.; Vanderhaeghen, M. Pion production in deeply virtual Compton scattering. *Phys. Rev. D* **2003**, *68*, 034018. [CrossRef]
177. Guidal, M.; Bouchigny, S.; Didelez, J.P.; Hadjidakis, C.; Hourany, E.; Vanderhaeghen, M. Generalized parton distributions and nucleon resonances. *Nucl. Phys. A* **2003**, *721*, 327–332. [CrossRef]

Disclaimer/Publisher's Note: The statements, opinions and data contained in all publications are solely those of the individual author(s) and contributor(s) and not of MDPI and/or the editor(s). MDPI and/or the editor(s) disclaim responsibility for any injury to people or property resulting from any ideas, methods, instructions or products referred to in the content.

Review

Impact of Multiple Phase Transitions in Dense QCD on Compact Stars

Armen Sedrakian [1,2]

[1] Frankfurt Institute for Advanced Studies, D-60438 Frankfurt am Main, Germany; sedrakian@fias.uni-frankfurt.de or armen.sedrakian@uwr.edu.pl
[2] Institute of Theoretical Physics, University of Wrocław, 50-204 Wrocław, Poland

Abstract: This review covers several recent developments in the physics of dense QCD with an emphasis on the impact of multiple phase transitions on astrophysical manifestations of compact stars. To motivate the multi-phase modeling of dense QCD and delineate the perspectives, we start with a discussion of the structure of its phase diagram and the arrangement of possible color-superconducting and other phases. It is conjectured that pair-correlated quark matter in β-equilibrium is within the same universality class as spin-imbalanced cold atoms and the isospin asymmetrical nucleonic matter. This then implies the emergence of phases with broken space symmetries and tri-critical (Lifshitz) points. The beyond-mean-field structure of the quark propagator and its non-trivial implications are discussed in the cases of two- and three-flavor quark matter within the Eliashberg theory, which takes into account the frequency dependence (retardation) of the gap function. We then construct an equation of state (EoS) that extends the two-phase EoS of dense quark matter within the constant speed of sound parameterization by adding a conformal fluid with a speed of sound $c_{\rm conf.} = 1/\sqrt{3}$ at densities $\geq 10\, n_{\rm sat}$, where $n_{\rm sat}$ is the saturation density. With this input, we construct static, spherically symmetrical compact hybrid stars in the mass–radius diagram, recover such features as the twins and triplets, and show that the transition to conformal fluid leads to the spiraling-in of the tracks in this diagram. Stars on the spirals are classically unstable with respect to the radial oscillations but can be stabilized if the conversion timescale between quark and nucleonic phases at their interface is larger than the oscillation period. Finally, we review the impact of a transition from high-temperature gapped to low-temperature gapless two-flavor phase on the thermal evolution of hybrid stars.

Keywords: QCD matter; phase diagram; compact stars

Citation: Sedrakian, A. Impact of Multiple Phase Transitions in Dense QCD on Compact Stars. *Particles* 2023, 6, 713–730. https://doi.org/10.3390/particles6030044

Academic Editors: Sebastian M. Schmidt, Minghui Ding and Craig Roberts

Received: 25 March 2023
Revised: 9 June 2023
Accepted: 11 July 2023
Published: 14 July 2023

Copyright: © 2023 by the author. Licensee MDPI, Basel, Switzerland. This article is an open access article distributed under the terms and conditions of the Creative Commons Attribution (CC BY) license (https:// creativecommons.org/licenses/by/ 4.0/).

1. Introduction

The astrophysics of compact stars entered the era of multimessenger astronomy in 2017 with the discovery of the neutron star binary merger event GW170817 [1]. Combined with radio observations of massive pulsars in binaries with white dwarfs [2] and X-ray observations of nearby solitary neutron stars [3,4], compact star astrophysics nowadays offers important insights into their global properties and potentially into the phase structure of dense matter [5–13]. Studies of matter at high densities are fundamental to our understanding of the strong force of the Standard Model and underlying concepts such as confinement, spontaneous chiral symmetry breaking, and dynamical mass generation [14].

This work studies the impact of multiple phase transitions in dense QCD matter on the physics of compact stars. We partially review the relevant physics but also provide a novel discussion of the mass–radius diagram of compact stars in the case where a conformal fluid is added to the two-phase, constant speed of sound parametrization of the EoS of quark matter [15]. This is motivated by the recent work that showed that even though the high-central-density stars are typically unstable toward radial oscillation (see Ref. [16], hereafter BTM), i.e., when $dM/d\rho_c < 0$, where M is the star's gravitational mass and ρ_c is the central density, they can be stabilized if the conversion between nucleonic and quark phases is slow compared to the characteristic period of radial oscillations [17–20].

To motivate the modeling, Section 2 provides a brief overview of the phase diagram of dense QCD matter as we understand it from the studies of the thermodynamics of various high-density phases, such as the color-superconducting phases [6–10] or quarkionic phases [21–25]. Utilizing the knowledge gained from the studies of imbalanced cold atoms [26] and isospin asymmetrical nuclear matter [27,28], the possible phase structure of pair-correlated quark matter is conjectured based on the universal features of imbalanced pair-correlated fermionic systems. Section 3 discusses the computations of Green's functions in the two- and three-flavor phases [29,30] and potential new effects arising beyond the adiabatic (frequency-independent) approximation of the gap. In Section 4, the constant speed-of-sound parameterization of the EoS of quark matter phases [31,32] is used to explore the mass–radius (M-R) diagram of compact stars with deconfinement and multiple phase transitions. For two-phase transitions, one from nucleonic to two-flavor quark matter and another from two-flavor to three-flavor phase of quark matter, we recover the *fourth family of compact stars*, which is separated from the third family by the instability region [15]. Here we show that a high-density phase of conformal fluid at densities $\geq 10\, n_{\text{sat}}$, where $n_{\text{sat}} = 0.16$ fm^{-3} is the saturation density, modifies the classically unstable tracks in the M-R diagram compared to the case when such transition does not occur [33]. Such modification is phenomenologically important because of the possible stabilization mechanism of radial oscillation modes of hybrid stars [17–20], which is discussed in Section 6. In Section 5, we discuss the cooling of compact stars with quark cores. We then simulate the thermal evolution of these stars on a time scale on the order of million years with a focus on the impact of the phase transition from the gapped to the gapless phase of 2SC matter in the core of the star. Our conclusions are given in Section 7.

2. A Brief Review of the Phase Diagram of Dense QCD

Matter in compact stars covers the large number density ($n \geq n_{\text{sat}}$), large isospin, and relatively low-temperature ($0 \leq T \leq 100$ MeV) portion of the phase diagram of strongly interacting matter. The extremely low temperature ($T \leq 0.1$) The MeV regime is relevant for mature compact stars, whereas the higher temperature domain is relevant for supernovas and binary neutron star mergers. The complexity of the phase diagram arises due to the multiple order parameters describing (interrelated) phenomena, which include deconfinement phase transition (with the Polyakov loop as the order parameter of the center symmetry), chiral phase transition (and its condensate as the order parameter), the color-superconducting phases (with the anomalous correlator as the order parameter). Depending on the non-zero value of one or several order parameters, distinct phases may arise: an extensively studied case is color-superconducting phases with various pairing patterns [6–10]. A more recent suggestion is a confined but chirally symmetric quarkyonic phase at compact star densities [21,23,24]. A crude sketch of the phase diagram of strongly interacting matter is shown Figure 1, along with the regions that are covered by current and future facilities (RHIC, NICA, and FAIR). The low-density and low-temperature region of the phase diagram contains nuclei embedded into a sea of charge-neutralizing electrons and neutrons at higher densities. As the density increases, a first-order phase transition to bulk nuclear matter occurs at around $0.5\, n_{\text{sat}}$. A further increase in density can lead to the deconfinement of nucleons to form quark matter for $n \geq (2-3) \times n_{\text{sat}}$.

The transition from nuclear matter to deconfined quark matter could be of the first or second order, or a crossover [6–10]. The first-order phase transition leaves a marked imprint on the macroscopic properties of compact stars because the EoS contains a density jump, which may give rise to new stable branches of compact stars (i.e., their third family) separated from nucleonic stars by a region of instability [34–37]. Smooth crossover without changes in the values of the order parameter or the wave function of the three-quark states would be a less dramatic change in the slope of the EoS, best visualized in terms of the speed of sound [10,22,38]. As mentioned above, two sequential first-order phase transitions can lead to the appearance of a new branch of compact stars—fourth family—separated from the third family by an instability region [15,33,39,40], assuming the classical stability criterion $dM/d\rho_c > 0$ is valid. In the case of slow phase transition between the nuclear and quark matter phases, the two families are not separated by an instability region; i.e., they form a continuous branch where the regions with $dM/d\rho_c < 0$ are stabilized [20] (see Section 6).

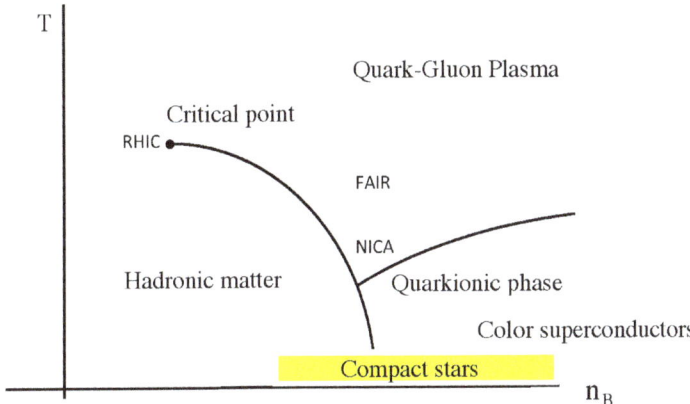

Figure 1. Sketch of the phase diagram of strongly interacting matter in the temperature and baryonic density plain. Compact stars cover the low-temperature and high-density regimes of this phase diagram. The parameter ranges covered by the FAIR, NICA, and RHIC facilities are also indicated.

Deconfined quark matter at low temperature and high density is expected to have characteristic features of degenerate Fermi systems, which are familiar from condensed matter physics. Therefore, emergent phenomena such as (color) superconductivity are expected in channels where gluon exchange is attractive [6–8]. Various color superconducting phases may arise, depending on the number of flavors and colors involved in the pairing, the ratio of $\Delta/\delta\mu$, where Δ is the gap in the quasiparticle spectrum, $\delta\mu = (\mu_d - \mu_u)/2$ is the difference in the chemical potentials of down (d) and up (u) quarks, and the mass of strange quark m_s in the three-flavor quark matter. The two-flavor candidate phases classified according to these parameters are as follows:

(a) The "2SC" phase (where the abbreviation refers to two-superconducting-colors) [6]

$$\Delta_{2SC} \propto \langle \psi^T(x) C \gamma_5 \tau_2 \lambda_2 \psi(x) \rangle \neq 0, \quad 0 \leq \delta\mu < \Delta/\sqrt{2}, \tag{1}$$

where $C = i\gamma^2\gamma^0$ is the charge conjugation operator, τ_2 is the second component of the Pauli matrix acting in the SU(2)$_f$ flavor space, and λ_A is the antisymmetric Gell–Mann matrix acting in the SU(3)$_c$ color space. The properties of the 2SC phase resemble that of the ordinary BCS theory, including vanishing resistivity and vanishing heat capacity because the quarks near the Fermi surface remain gapped.

(b) Phases with broken space symmetries, which are associated with a finite momentum of the condensate [6,8] (hereafter FF phase) or deformation of the Fermi surface [41] (hereafter DFS phase):

$$\Delta_{2SC} \neq 0, \quad \delta\mu > \Delta/\sqrt{2}, \quad \vec{P} \neq 0 \quad \text{(FF)} \tag{2}$$
$$\Delta_{2SC} \neq 0, \quad \delta\mu > \Delta/\sqrt{2}, \quad \delta\epsilon \neq 0 \quad \text{(DFS)} \tag{3}$$

where $\vec{P}$ is the center of mass momentum of a Cooper pair and $\delta\epsilon$ quantifies the quadrupole deformation of the Fermi surfaces of u and d quarks.

(c) Mixed phase(s) [42]

$$\Delta_{2SC} \propto \langle \psi^T(x) C \gamma_5 \tau_2 \lambda_2 \psi(x) \rangle \neq 0, \quad \delta\mu = 0, \quad 0 \leq x_s \leq 1, \quad (4)$$

which corresponds to a mixture between a perfectly symmetrical "2SC" superconductor and a normal system accommodating the excess number of d quarks. Here, x_s is the filling factor defined as the ratio of the superconducting and total volumes. (We assume that there is an excess of d over u quarks, as is expected in quark matter in compact stars under β-equilibrium.)

The color-flavor-locked (CFL) phase [43] is expected to be the ground state of three-flavor quark matter at asymptotically large densities where the strange quark is massless, Fermi surfaces of quarks coincide, and, therefore, the pairing among quarks occurs in a particularly symmetrical manner. At densities relevant for neutron stars, the perfect CFL phase is unlikely to be realized; rather, some of its variants have $m_s \neq 0$ and/or $\delta\mu \neq 0$—chemical potential shifts between various flavors of quarks [44]. Therefore, the phases listed above can be replicated with an allowance of additional non-zero us and ds pairings

$$\Delta_{ud} \neq 0, \quad \Delta_{sd} \neq 0, \quad \Delta_{su} \neq 0, \quad (m_s \neq 0; \delta\mu \neq 0). \quad (5)$$

A complete phase diagram of quark matter that includes most, if not all, of the phases mentioned above, is not available to date. However, various imbalanced superfluids, such as cold atoms, isospin asymmetrical nuclear matter, and flavor-imbalanced quark matter show a high degree of universality. Thus, possible structures of the phase diagram of quark matter can be conjectured by extrapolating from the detailed studies of the phase diagrams of cold atomic gases [26] and isospin asymmetrical nuclear matter [27]. These are, clearly, speculative and need to be confirmed using explicit computations of relevant quark phases.

Figure 2 shows two schematic phase diagrams of color-superconducting matter in the density–temperature plane. For sufficiently large temperatures, the unpaired normal phase is the preferred state of matter, ignoring any other correlation beyond the pairing. The phases with broken symmetries, the FF and the DFS phases, are preferable in temperature–density strips at low temperatures and high densities. At lower temperatures, the PS phase is the preferred one. At higher temperatures, the spatially symmetric 2SC phase dominates. It is seen that the phase diagram contains two tri-critical points, i.e., the points where three different phases coexist. The critical point, which has the FF state at the intersection, is a Lifshitz point as, per construction, it is a meeting point of the modulated (FF), ordered (PS/2SC), and disordered (unpaired) states. Of course, this is the case if the transition temperature to the CFL phase is below the tri-critical temperature; otherwise, the unpaired state should be replaced by a variant of the CFL phase. Note that depending on the parameters of the model, two or one tricritical points may be located on the unpairing line or the line of transition to the CFL phase, as illustrated in Figure 2, left and right panels, respectively. The model can be tuned to produce a four-critical point if both points coincide. We also note that the low-density limit corresponds to the strong coupling regime where the pairs are tightly bound, whereas the high-density limit corresponds to the weak-coupling regime. Therefore, one can anticipate signatures of BCS–BEC crossover. These can be seen by examining several characteristic quantities, for example, the ratio of the coherence length to the interparticle distance ξ/d, where $\xi/d \gg 1$ corresponds to the BCS and $\xi/d \ll 1$ corresponds to the BEC limit, or the ratio of the gap to the (average) chemical potential Δ/μ, where $\Delta/\mu \ll 1$ corresponds to the BCS and $\Delta/\mu \gg 1$ corresponds to the BEC limit. For discussions of BCS=-BEC crossover in dense quark matter, see Refs. [45–48]. This phenomenon shows a high degree of universality as well; see for example, the studies of nuclear matter [49–51] and cold atoms [26,52].

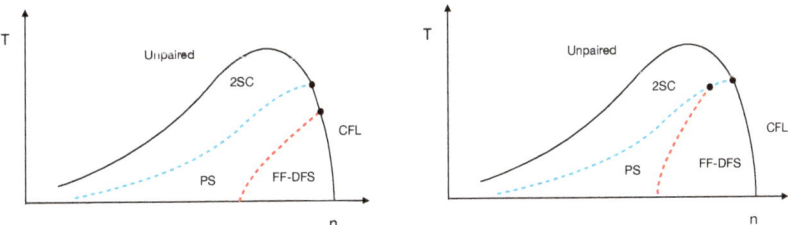

Figure 2. Sketch of the phase diagram of strongly interacting matter in the temperature and baryonic density plain, including (collectively indicated) modulated FF-phase and deformed Fermi surface DFS phase. The tri-critical points are shown with dots; the Lifshitz point is adjacent to the FF, unpaired/CFL phases, and homogenous (PS) phases. In the left panel, it is located on the unpairing or CFL-transition (solid line). The dashed lines correspond to the phase-separation lines among various phases. Signatures of BCS–BEC crossover/transition may emerge when moving from high to low densities.

3. Structure of Green Functions in Two-Flavor Quark Matter

Order-by-order computations of the magnitude of the gaps in the superconducting phases can be carried out in the weak-coupling (extreme high-density) regime, where the one-gluon exchange is the dominant interaction. Approximate Eliashberg-type equations for the flavor-symmetric 2SC phase were solved within one-gluon exchange approximation in Refs. [53,54], showing that the pairing gap scales with the coupling g as $\Delta \sim \mu g^{-5} \exp(-1/g)$. Such a scaling also applies to the high-density CFL phase, where the perturbative approach is more reliable than at densities relevant to the 2SC phase. More recently, Eliashberg-type equations were solved for two-flavor [29] and three-flavor superconductors [30]. The first study used the quark–meson coupling model, keeping only the frequency dependence of the gap, whereas the second study kept frequency and momentum dependences but ignored the imaginary part of the pairing gap. These theories not only improve the description of quark matter but also lead to phenomenologically important implications, such as the presence of electrons in the CFL phase [30], which are not allowed when the gap is constant [55].

We now briefly outline these approaches following Ref. [29]. The inverse Nambu–Gorkov quark propagator is given by

$$S^{-1}(q) = \begin{pmatrix} \slashed{q} + \mu\gamma_0 - m & \bar{\Delta} \\ \Delta & (\slashed{q} - \mu\gamma_0 + m)^T \end{pmatrix}, \qquad (6)$$

where q is the four-momentum and Δ is the gap with $\bar{\Delta} \equiv \gamma_0 \Delta^\dagger \gamma_0$. Equation (6) is written for the case of equal number densities of up and down quarks with a common chemical potential μ and mass m. The bare quark–meson vertices $\Gamma^i_\pi(q)$ and $\Gamma_\sigma(q)$ are given by

$$\Gamma^i_\pi(q) = \begin{pmatrix} \frac{\tau^i}{2}\gamma_5 & 0 \\ 0 & -(\frac{\tau^i}{2}\gamma_5)^T \end{pmatrix}, \quad \Gamma_\sigma(q) = \begin{pmatrix} \mathbb{I} & 0 \\ 0 & -\mathbb{I} \end{pmatrix}, \qquad (7)$$

where pions couple to quarks using a pseudo-scalar coupling, whereas σs couple via a scalar coupling, with $\mathbb{I}$ being a unit matrix in the Dirac and isospin spaces. Their propagators are given by

$$D_\pi(q) = \frac{1}{q_0^2 - q^2 - m_\pi^2}, \quad D_\sigma(q) = \frac{1}{q_0^2 - q^2 - m_\sigma^2}, \qquad (8)$$

where $m_{\pi/\sigma}$ values are the meson masses. The equation for the gap in the Fock approximation is given via

$$\Delta(k) = ig_\pi^2 \int \frac{d^4q}{(2\pi)^4} \left(-\frac{\tau^i}{2}\gamma_5\right)^T S_{21}(q)\frac{\tau^j}{2}\gamma_5\delta_{ij}D_\pi(q-k)$$
$$+ ig_\sigma^2 \int \frac{d^4q}{(2\pi)^4}(-\mathbb{I})^T S_{21}(q)\mathbb{I}D_\sigma(q-k), \qquad (9)$$

where g_π and g_σ are the coupling constants. Adopting the color–flavor structure of the gap function corresponding to a 2SC superconductor, one then finds

$$\Delta_{ij}^{ab}(k) = (\lambda_2)^{ab}(\tau_2)_{ij}C\gamma_5[\Delta_+(k)\Lambda^+(k) + \Delta_-(k)\Lambda^-(k)], \qquad (10)$$

where $a,b\ldots$ refer to the color space, $i,j,\ldots$ refer to the flavor space, and the projectors onto the positive and negative states are defined in the standard fashion as $\Lambda^\pm(k) = (E_k^\pm + \boldsymbol{\alpha}\cdot\mathbf{k} + m\gamma_0)/2E_k^\pm$, where $E_k^\pm = \pm\sqrt{k^2 + m^2}$ and $\boldsymbol{\alpha} = \gamma_0\boldsymbol{\gamma}$. The coupled Equations (6)–(10) must be solved for the gap function, which is a function of three-momentum and the frequency. In the low-temperature limit, the relevant momenta are close to the Fermi momentum and the dependence on the magnitude of the three-momentum can be eliminated by fixing it at the Fermi momentum. The gap Equation (9) then depends only on the energy, which reflects the fact that the pairing interaction is not instantaneous—a common feature of the Fock self-energies in ordinary many-body perturbation theory. The solutions for the positive energy projection of the gap function are shown in Figure 3 as a function of frequency. The structure of the real and imaginary components of the gap function shows a maximum around frequencies at which the meson spectral functions are peaked. Thus, it is important to include the retardation effect when the color superconductor is probed at such frequencies. In the low-frequency limit, it is sufficient to use the BCS approximation where the interaction is instantaneous so that the imaginary part vanishes $\mathrm{Im}\,\Delta(\omega) = 0$ and the real part is a constant $\mathrm{Re}\,\Delta = \Delta(\omega = 0)$.

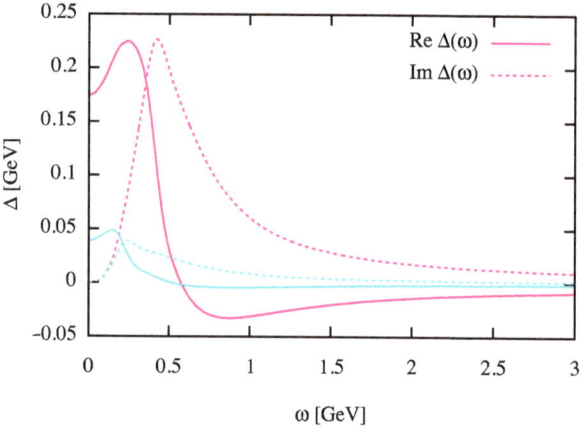

Figure 3. Dependence of the real (solid) and imaginary (dashed) components of the positive energy projection of the gap function on frequency for two different values of the coupling shown by blue and red lines [29]. The BCS theory predicts a constant on-shell value $\mathrm{Re}\,\Delta = \mathrm{Const.}$ and a vanishing $\mathrm{Im}\,\Delta(\omega)$.

Ref. [30] considered the full momentum and energy dependence of the gap in the Fock approximation within the Yukawa model but neglected the imaginary part of the anomalous self-energy. Their work shows that the retardation implies a CFL phase that is not a perfect insulator, as charge neutrality requires some electrons to be present in matter. This is not the case in the treatment based on the BCS model [55]. Thus, the phenomenology of the CFL phase is modified: its specific heat, thermal conductivity, and magnetic response will change due to the contribution of electrons. This example, in which the simple BCS ansatz for the gap is replaced by a more complete gap function, demonstrates some unexpected features of color superconductors, which may be important for their transport and dynamic response to various probes.

4. Equation of State and Mass–Radius Diagram

We have seen in the previous section that the phase diagram of quark matter may have a complicated structure. At minimum, there are two robust phases of color superconducting matter: the low-density two-flavor 2SC phase and the high-density three-flavor CFL phase. See, for example, Ref. [56] for a Nambu–Jona-Lasinio study and compact stars with two phase transitions in this model. However, additional phases are very likely because it is energetically favorable to break the rotational and translational symmetries due to the stress on the paired state induced by the finite mass of the strange quark and β-equilibrium, which induces disparity in the chemical potentials of u and d quarks. In addition, or alternatively, quarkyonic phases may interfere.

For the specific computation below, we adopt a covariant density functional EoS of nuclear matter in the nucleonic phase [39,40,57]. This EoS, in the absence of the phase transition to quark matter, produces nucleonic compact stars with a maximum $m \equiv M/M_\odot \simeq 2.6$, where M is the gravitational mass of the star and $M_\odot$ is the solar mass. Allowing for phase transition to quark matter, we consider a straightforward extension of the constant speed of sound EoS of Ref. [15] that allows for a conformal phase of quark matter at high densities with a constant speed of sound, i.e.,

$$p(\epsilon) = \begin{cases} p_1, & \epsilon_1 < \epsilon < \epsilon_1 + \Delta\epsilon_1 \\ p_1 + s_1[\epsilon - (\epsilon_1 + \Delta\epsilon_1)], & \epsilon_1 + \Delta\epsilon_1 < \epsilon < \epsilon_2 \\ p_2, & \epsilon_2 < \epsilon < \epsilon_2 + \Delta\epsilon_2 \\ p_2 + s_2[\epsilon - (\epsilon_2 + \Delta\epsilon_2)], & \epsilon_2 + \Delta\epsilon_2 < \epsilon < \epsilon_3 \\ p_3, & \epsilon_3 < \epsilon < \epsilon_3 + \Delta\epsilon_3 \\ p_3 + s_3[\epsilon - (\epsilon_3 + \Delta\epsilon_3)], & \epsilon > \epsilon_3 \end{cases} \quad (11)$$

where the three pairs of the pressure and energy density $p_{1,2,3}$ and $\epsilon_{1,2,3}$ correspond, respectively, to the transition from hadronic to quark matter, from a low-density (2SC) quark phase to a high-density (CFL) quark phase, and from the high-density quark phase to the conformal fluid. The squared sound speeds in the quark phases are denoted by s_1, s_2, and $s_3 = c^2_{\rm conf.}$. Note that we assume that the 2SC and CFL quark phases are separated by a jump at the phase boundary, as it follows from the study of Ref. [56]. At high densities, the CFL phase reaches the "conformal limit" where the interactions are dominated by the underlying conformal symmetry of QCD. In this limit, the speed of sound squared is $s_3 = 1/3$ (in units of speed of light), whereas the effects of the pairing gap of the CFL phase can be neglected in a first approximation. Note that we allow for a small jump between proper CFL and conformal zero-gap fluid, but its effect on the observables is marginal, i.e., a smooth interpolation would not change the results.

According to Equation (11), the modeling of the EoS of quark phases involves the following parameters:

- The three (energy) densities at which the sequential transitions between the nucleonic phase, 2SC phase, CFL phase, and conformal fluid take place.
- The magnitudes of the jumps in the energy density at the points of the transition from nuclear to the 2SC phase, $\Delta\epsilon_1$, from the 2SC to the CFL phase, $\Delta\epsilon_2$, and from the CFL to the conformal fluid phase $\Delta\epsilon_3$.
- The speeds of sound in the 2SC and CFL phases s_1 and s_2. The speed of sound of the conformal fluid is held fixed at $s_3 = 1/3$. Note that for any phase, $s \leq 1$ by causality.

Our model EoS is constructed using the following parameters. The transition pressure and energy density from nuclear and quark matter are $p_1 = 1.7 \times 10^{35}$ dyn/cm^2 and $\epsilon_1 = 8.4 \times 10^{14}$ g cm^{-3}, respectively. The magnitude of the first jump $\Delta\epsilon_1 = 0.6\,\epsilon_1$. The upper range of the energy density of the 2SC phase is determined as $\epsilon_1^{max} = \delta_{2SC}(\epsilon_1 + \Delta\epsilon_1)$, where δ_{2SC} is a dimensionless parameter measuring the width of the 2SC phase. The magnitude of the second jump is parametrized in terms of the ratio parameter $r = \Delta\epsilon_2/\Delta\epsilon_1$. The extent of the CFL phase is determined by limiting its energy density range to $\epsilon_2^{max} = \delta_{CFL}(\epsilon_2 + \Delta\epsilon_2)$. The transition to the conformal fluid is assumed to be of the first order with a small (compared to other scales) energy-density jump equal to $0.1r$. The transition to the conformal fluid phase occurs at densities $\epsilon_3 = 2.25$–2.57×10^{15} g cm^{-3}, i.e., by about a factor of 10 larger than the saturation density. The speeds of sound squared are fixed as

$$s_1 = 0.7, \qquad s_2 = 1, \qquad s_3 = \frac{1}{3}. \tag{12}$$

The values of s_1 and s_2 are chosen to obtain triplet configurations with large enough masses of hybrid stars. The magnitudes of jumps between the nuclear, 2SC, and CFL phases were chosen suitably to produce twin and triplet configurations [15].

Figure 4 shows a collection of EoS constructed based on Equation (11), which shares the same low-density nuclear EoS. In this collection, we vary the parameter r (as indicated in the plot) for fixed values $\delta_{2SC} = \delta_{CFL} = 0.27$. The corresponding M–R relations for static, spherically symmetrical stars obtained from solutions to Tolman–Oppenheimer–Volkoff equations are shown in Figure 5.

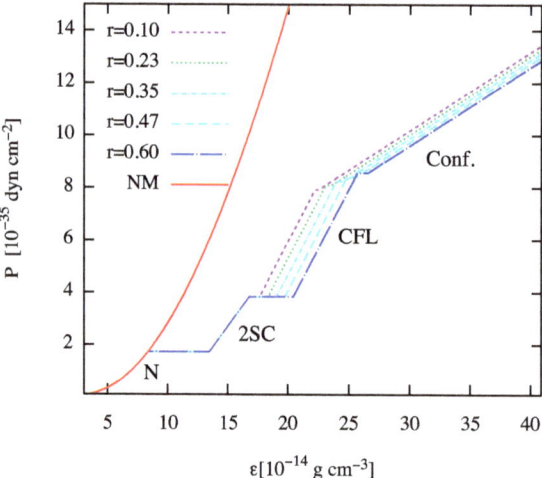

Figure 4. The pressure vs. energy density (EoS) for nucleonic matter (long-dash-dotted curve) and a series of EoSs that contain two sequential phase transitions via Maxwell construction manifest in the jumps of the energy density. The models differ by the magnitude of the second jump measured in terms of the ratio $r = \Delta\epsilon_2/\Delta\epsilon_1$.

For the chosen magnitude of the first jump $\Delta\epsilon_1$, the M–R curves show the phenomenon of twins—two stars of the same mass but different radii. The radii of twins differ by about 1 km. The more compact configuration is a hybrid star, i.e., a star with a quark core and nuclear envelope, whereas the less compact counterpart is a purely nucleonic star. The second phase transition may or may not result in a classically stable sequence depending on the value of the parameter r parameterizing the magnitude of the second jump. For small jumps $r = 0.1$ and 0.23, new stable branches arise, which are continuously connecting to the stable 2SC branch ($r = 0.1$) or are separated by a region where the stars are unstable ($r = 0.23$). It is seen that, in this case, triplets of stars with different radii but the same masses appear. The densest stars contain, in addition to the 2SC phase, a layer of the CFL phase, whereby the central density on the stable branch can exceed the onset density of the conformal fluid. This implies that the densest member of a triplet will contain in its center conformal fluid with $c_{\text{conf.}} = 1/\sqrt{3}$. For each M–R curve in Figure 5, the star with a central density at which the conformal fluid first appears is shown by a dot (this density is fixed at 10 n_{sat}). The stable branch of conformal fluid containing stars is followed by a classically unusable branch with $dM/d\rho_c < 0$. For asymptotically large central densities, the masses and radii increase again. The family of the EoSs that differ only in the value of the parameter r cross at a "special point". This type of crossing has been observed for twin star configurations with a variation in a particular parameter of the EoS [58]; however, the EoS excluded two sequential phase transitions. The behavior of M–R curves at very high central densities differs from the ones that were found in Ref. [33], where a branch of ultracompact twin stars with masses of the order of 1 $M_\odot$ and radii in the range of 6–7 km were found for a single phase transition from the nuclear matter to the quark phase. Thus, we conclude that the high-density asymptotics of the EoS modifies the behavior of the M–R curves if the conformal limit is achieved at densities of the order of 10 n_{sat}.

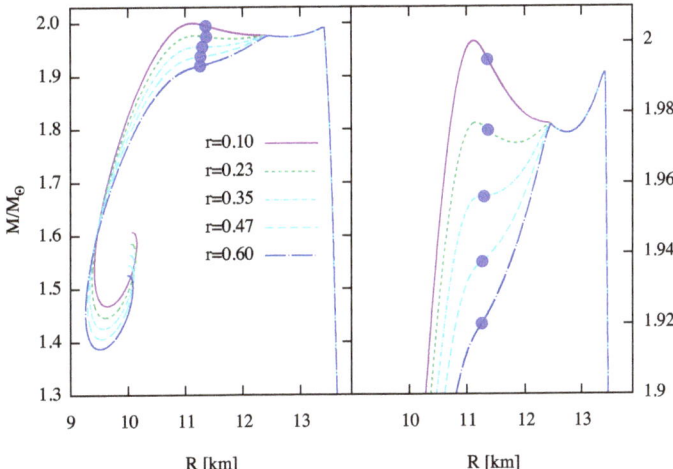

Figure 5. The M–R relations corresponding to the EoS shown in Figure 4 for several ratios of the second jump. The right panel enhances the high-mass range to demonstrate the emergence of the triplets and the fourth family of compact stars. Note that the different MR curves cross each other at the special point located in the low-mass and low-radius region, in analogy to the single-phase transition case; see Ref. [58]. The blue circles indicate the stars in which the central density corresponds to 10 n_{sat} at which the conformal fluid sets in.

The observation above may have phenomenological implications for the following reason. The stability of stellar configurations is commonly determined by the requirement that the star's mass must increase with increasing central density (or central pressure), i.e., $\partial M/\partial \rho_c > 0$. An alternative and physically more transparent method is to compute the radial modes of oscillation of a star and determine the stable configurations from the requirement that their frequencies are real. Ref. [17] showed that the classical stability conditions fail if the conversion rate is slow, i.e., if its characteristic timescale is longer than the period of oscillations. In that case, the fundamental modes are stable even when $\partial M/\partial \rho_c < 0$; i.e., stars with central densities larger than the one corresponding to the maximum-mass star (which lie to the left from the maximum on the M–R diagram in

Figure 5) will be stable. This observation also applies to configurations with two-phase transitions, as shown in Refs. [19,20]. Furthermore, Ref. [20] shows that, in this case, the classically unstable stars contribute to the count of same-mass stars, which leads to the appearance of higher-order multiplets such as quadruplets, quintuplets, and sextuplets. We will return to the stability of hybrid stars in Section 6.

5. Cooling of Compact Stars with Quark Matter Cores

The cooling of compact stars may provide indirect information about quark phases in hybrid stars. The properties of phases of dense quark matter affect both neutrino emission and the specific heat content that determine the cooling rate of a compact object in general; see Refs. [59–75].

Non-superconducting relativistic quark matter cools predominantly via the direct Urca processes involving d, u, and s quarks [76]

$$
\begin{aligned}
d &\to u + e + \bar{\nu}_e, \\
u + e &\to d + \nu_e, \\
s &\to u + e + \bar{\nu}_e, \\
u + e &\to s + \nu_e,
\end{aligned}
\tag{13}
$$

where ν_e and $\bar{\nu}_e$ are the electon neutrino and antineutrions. The neutrino emissivity through the direct Urca process for non-strange quarks is given via [77]

$$\epsilon_\beta = 8.8 \times 10^{26} \alpha_s \left(\frac{n}{n_{\text{sat}}}\right) Y_e^{1/3} T_9^6 \quad \text{ergs cm}^{-3}\ \text{s}^{-1}, \tag{14}$$

where n is the baryon density, Y_e is the electron fraction, T_9 is the temperature in units of 10^9 K, and α_s is the running strong coupling constant. The emissivity given by Equation (14) implies that the stars containing unpaired quark matter would cool quickly via this direct Urca process. The cooling would be slower if the quark spectrum contains a gap. In the case of the phenomenologically relevant 2SC phase, two alternatives are possible, depending on whether the Fermi surfaces of quarks are a full gap or they contain zero-gap segments (nodes). The latter feature arises in the case of pairing between fermions on different Fermi surfaces, as discussed in Section 2.

Ref. [78] studied a generic case where the quark spectrum is gapped if the parameter $\zeta = \Delta_0/\delta\mu$ associated with the new scale $\delta\mu = (\mu_d - \mu_u)/2$, where $\mu_{u,d}$ are the chemical potentials of light quarks and Δ_0 is the gap for $\delta\mu = 0$. The suppression of emissivity by pairing is qualitatively different in the cases in which $\zeta > 1$ and $\zeta < 1$. The novelty arises in the second case, where Fermi surfaces have nodes and particles can be excited around these nodes without any energy cost (which is not the case for gapped Fermi surfaces). Note that in the case of the FF phase, the shift in the chemical potential is replaced by a more general function—the anti-symmetric in the flavor part of the single particle spectrum of up and down quarks. This new physics can be captured by adopting a generic parameterization of the suppression factor of the quark Urca process with pairing suggested in Ref. [78]. The neutrino emissivity of the 2SC phase ϵ_{2SC}^{rg} can be related to the Urca rate in the normal phase (14) as

$$\epsilon_{2SC}^{rg}(\zeta; T \leq T_c) = 2f(\zeta)\epsilon_\beta, \quad f(\zeta) = \frac{1}{\exp\left[(\zeta - 1)\frac{\delta\mu}{T} - 1\right] + 1}, \tag{15}$$

where the parameters ζ and $\delta\mu$ were introduced above, T is the temperature, and T_c is the critical temperature of the phase transition from normal to the 2SC phase. Furthermore, the parameter $\zeta(T)$ is temperature-dependent and we adopt the parametrization

$$\zeta(T) = \zeta_i - \Delta\zeta g(T), \tag{16}$$

where ζ_i is the initial value, $\Delta\zeta$ is the constant change in this function, and the function $g(T)$ describes the transition from the initial value ζ_i to the asymptotic final value $\zeta_f = \zeta_i - \Delta\zeta$. The transition is conveniently modeled by the following function

$$g(T) = \frac{1}{\exp\left(\frac{T-T^*}{w}\right) + 1}, \quad (17)$$

which allows one to control the temperature of transition by adjusting the parameter T^* and the smoothness of the transition via the width parameter w. An additional issue to address is the role of the blue quarks that do not participate in the 2SC pairing. Blue quarks may pair among themselves due to the attractive component of the strong force as in the ordinary BCS case (as both members of the Cooper pair are on the same Fermi surface). Then, the emissivity of blue quarks in the superfluid state is given by

$$\epsilon^b_{BCS}(T \leq T_{cb}) \simeq \epsilon^b_\beta(T > T_{cb}) \exp\left(-\frac{\Delta_b}{T}\right), \quad (18)$$

where Δ_b is the gap in the blue quark spectrum, T_{cb} is the corresponding critical temperature, and ϵ^b_β is the neutrino emissivity of blue quarks in the normal state. As discussed in Section 4, the densest members of the triplets contain cores of CFL matter that is fully gapped. In this case, the excitations are the Goldstone modes of the CFL phase. Their emissivity, as well as the specific heat, is rather small compared to other phases due to their very small number density [79]. In the following discussion, we will ignore the role of the CFL phase in the cooling of hybrid stars. In the conformal fluid phase, we expect three-flavor pairing gap $\Delta \sim \mu g^{-5} \exp(-1/g)$, $g = \sqrt{4\pi\alpha_s}$, with a spin–flavor structure of the CFL phase.

Let us turn to the cooling simulations of hybrid stars with a gapless 2SC superconductor. The cooling tracks are shown in Figure 6, and the input physics beyond the emissivities is discussed elsewhere [61,62,67,68]. The key parameter regulating the behavior of the cooling curves in Figure 6 is the temperature T^*, which controls the transition from the gapped to ungapped 2SC phase. Similar results were obtained in the context of rapid cooling of the compact star in Cassiopeia (Cas) A remnant in Ref. [61,62,68]. The model has a second parameter, the gap for blue-colored quarks Δ_b, which prohibits rapid cooling via the Urca process involving only blue quarks. The third parameter w in Equation (17) accounts for the finite time scale of the phase transition—see Refs. [62,68]—but it is important only for the fine-tuning of the cooling curves close to the age of the Cas A. The various cooling tracks shown in Figure 6 correspond to various values of T^* for fixed values of w and Δ_b and stellar configuration of mass 1.93 $M_\odot$. It is seen that if T^* is small, then the quark core does not influence the cooling, because during the entire evolution $T > T^*$; therefore the neutrino emission is suppressed by the fully gapped Fermi surfaces of red-green quarks. For large T^*, early transition to the gapless phase occurs, and the star cools fast via the direct Urca process. Note that the value of T^* can be fine-tuned to reproduce not only the current temperature of Cas A but also the fast decline claimed to be observed during the last decade or so; see Ref. [80] and references therein. From the brief discussion above, one may conclude that the phase transitions within the cold QCD phase diagram may induce interesting and phenomenologically relevant changes in the cooling behavior of compact stars. Although we will not discuss in any depth the dependence of cooling tracks on the stellar mass, it should be pointed out that the onset of new phases in the interiors of compact stars, for example, hyperonization, meson condensation, and phase transition to quark matter, lead to mass hierarchy in the cooling curves [81–84]. Typically, one finds that heavier stars that have central densities beyond the threshold for the onset of the new phase cool faster than the light stars containing only nucleonic degrees of freedom. This is also the case for models of stars studied here. For example, stars with masses $M \sim 1.1$–$1.6\, M_\odot$ remain warm over longer time scales and are thus hotter than their heavy analogs, which develop large quark cores.

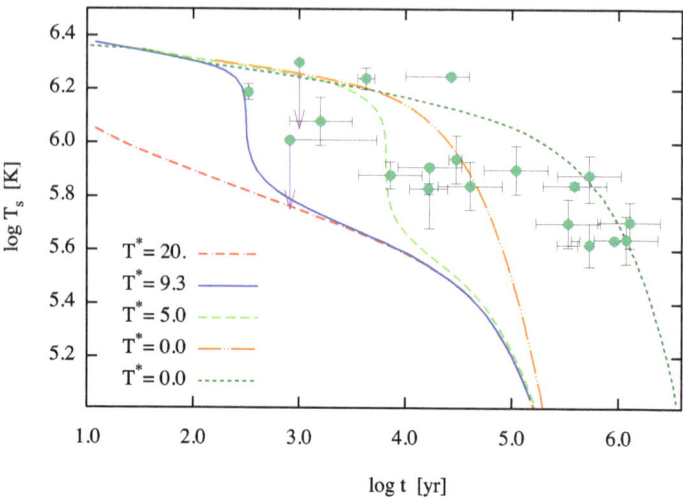

Figure 6. Cooling tracks of compact stars with quark cores in the surface-temperature–age diagram. The masses of the stars are the same, $M = 1.93\,M_\odot$, and the different curves correspond to different values of the parameter T^* in units of keV, except the dotted line, which corresponds to $1\,M_\odot$ mass nucleonic compact star without a quark core. The observational points with error bars are shown by green circles; the arrows show the upper limits on surface temperatures of known objects.

6. Stability Criteria for Hybrid Stars

The oscillation modes of a compact star are important probes of their internal structure, as has been shown in the case of g modes, which are sensitive to the size of the density jumps at a first-order phase transition between hadronic and quark matter [85–87]. They are expected to leave an imprint on the emitted gravitational wave signal during the binary inspiral of a neutron star, as well as in the post-merger phase [88–91].

As discussed briefly in Section 4, high-central density stars on the descending branch of M–R diagram can have phenomenological implications if they are stabilized by some mechanism, which we discuss in this section. The main mode of instability for non-rotating, spherically symmetrical fluid stars in general relativity is the instability against the radial f-mode of oscillations [92]. If the f-mode frequency $\omega_f^2 > 0$, the stellar configuration is dynamically stable, and it is unstable if $\omega_f^2 < 0$. The location of this instability point on the M–R diagram agrees well with the turning point of the mass–central-density ($M - \rho_c$) curve. The stars on the ascending branch are stable, whereas those on the descending branch are unusable. The maximum mass is the point of marginal stability. Numerical simulations found some violations of this criterion [93,94], but quantitative deviations are insignificant. However, recent work found that the agreement between these criteria is strongly violated for stars with first-order phase transition, as we review below.

Early work on stellar oscillations with phase transitions inside the star was carried out in the Newtonian theory assuming uniform phases [95,96]. Two possibilities arise depending on the interplay between the scales in the problem: (a) when the conversion rate from one phase to another is fast, the interface between phases oscillates as a whole when perturbed; (b) if, however, conversion is slow, then the interface is fixed over the period of characteristic oscillations. The second case is interesting because, as shown in Ref. [17], the sign of ω_f^2 does not change at the maximum mass $M_{\max}$ but stays positive over a segment where $\partial M/\partial\rho_c < 0$. This implies that the classically unstable branch becomes stable against f-mode oscillations. Several subsequent studies confirmed this feature in the case of single- [18] and two-phase transitions [19,20]. The case of two-phase transition was extended in several directions in Ref. [20] by focusing on EoS, which supported classical twin and triplet star configurations, as discussed in Section 2. It was shown that in the case of slow conversion, higher-order multiplet stars arise, since now the stars on the $\partial M/\partial \rho_c < 0$ segments of the mass–central-density curve are located on the stable branch. Also, the properties of the reaction mode of a compact star [96], which arises in case (a) with one or more rapid phase transitions, were studied.

The fundamental modes of hybrid stars are obtained from the set of equations [97,98]

$$\frac{d\xi}{dr} = \left(\frac{d\nu}{dr} - \frac{3}{r}\right)\xi - \frac{\Delta P}{r\Gamma P'} \tag{19}$$

$$\frac{d\Delta P}{dr} = \left[e^{2\lambda}\left(\omega^2 e^{-2\nu} - 8\pi P\right) + \frac{d\nu}{dr}\left(\frac{4}{r} + \frac{d\nu}{dr}\right)\right](\rho + P)r\xi$$
$$- \left[\frac{d\nu}{dr} + 4\pi(\rho + P)re^{2\lambda}\right]\Delta P, \tag{20}$$

where $\xi = \xi_{\text{dim}}/r$, with ξ_{dim} being the Lagrangian displacement, r the radial coordinate, ΔP the Lagrangian perturbation of pressure, ρ the mass–energy-density, ω the angular frequency, Γ the adiabatic index, and $e^{2\nu}$ and $e^{2\lambda}$ the metric coefficients entering the Tomann–Oppenheimer–Volkoff equations. In a first approximation, the adiabatic index for a chemically equilibrated relativistic fluid can be taken as that of the matter in β-equilibrium $\Gamma = [(\rho + P)/P](dP/d\rho)$. The set of Equations (19) and (20) can be solved provided the boundary conditions are known. These are specified by assuming that the displacement field is divergence-free at the center and that the Lagrangian variation of the pressure vanishes at the surface of the star:

$$\Delta P(r = 0) = -3\Gamma P\xi(r = 0), \qquad \Delta P(r = R) = 0. \tag{21}$$

The ω^2 values obtained in this manner are usually labeled according to the number of radial nodes in ξ and the f mode corresponds to the nodeless mode.

In the case of multiple phase transitions in the QCD phase diagram, one needs junction conditions that relate the values of Lagrangian perturbations on both sides of the interface between phases. Such junction conditions already appear in the work of Ref. [96] in the Newtonian cases, whereas the the general relativistic case is treated in Ref. [99]. For the *slow conversion rate* one has the junction condition

$$[\Delta P]_-^+ = 0, \qquad [\xi]_-^+ = 0; \tag{22}$$

for *rapid conversion rate*, one has

$$[\Delta P]_-^+ = 0, \qquad \left[\xi - \frac{\Delta P}{r}\left(\frac{dP}{dr}\right)^{-1}\right]_-^+ = 0, \tag{23}$$

where $+/-$ refer to the high- and low-density sides of the transition, respectively. At present, it is not possible to state with confidence which limit is realized in quark matter, as the conversion rate varies significantly over the parameter space; see Ref. [100] for a discussion and earlier references. Ref. [20] considered modified junction conditions that smoothly interpolate between the two limiting cases.

Phenomenologically, the most interesting implication of the modified stability criteria is the existence of new stable configurations beyond those that are classically stable. In particular, in the case where twins and triplets exist according to classical criteria of stability, additional configurations will arise when conversion between phases at the interface is slow. These can form quadruplets (the maximum number in the case of twins) and quintuplets and sextuplets in the case of triplets. A particular case that allows for classical triplet stars is illustrated in Figure 7, adapted from Ref. [20]. The fundamental mode frequency ω_f is shown as a function of the central pressure of the stars in two cases when both interfaces (i.e., nucleonic to 2SC and 2SC-CFL) feature rapid or slow conversion. (The case of rapid–slow and slow–rapid conversions are intermediate cases, and we omitted them.) To recover the classical case, one needs to assume that the conversion at each interface is rapid: in this case, the instability region is characterized by the vanishing of the real part of ω_f, as seen in Figure 7. In the case of slow conversions at both interfaces, one finds a continuous positive solution across the values of central densities of the stellar sequences, thus indicating that the stars are always stable, even on the descending branch of the mass–central-pressure curve.

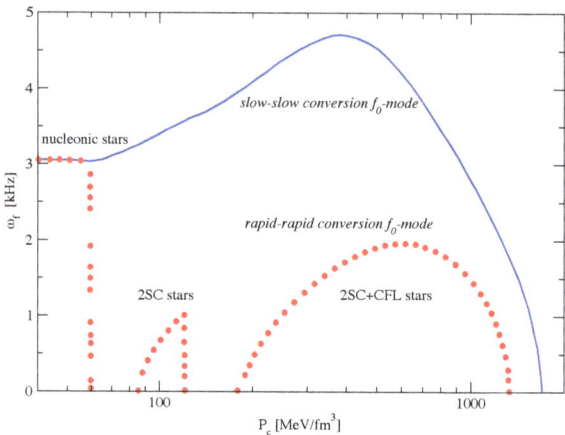

Figure 7. The fundamental mode of triplet stars as a function of the central pressure of the configuration in the cases when both nucleonic-2SC and 2SC-CFL interfaces feature slow (solid line) or rapid conversion (dotted line). In the case of rapid conversion, the classical stability criteria apply; i.e., there are no real f_0 solutions in the region where stars are unstable. The corresponding curves are not shown. For more details, see Ref. [20].

To summarize the recent findings regarding the stability of the hybrid stars, we have seen that their stability against the fundamental oscillation modes strongly depends on the junction conditions at the interfaces between the phases. These are determined by the rate of conversion between phases at the phase boundary. In the case of slow phase transitions (i.e., when the conversion time scale is larger than the characteristic period of the oscillations), the usual stability criteria are modified and new stable segments appear that were previously unstable. Alternative variants of junction conditions that are intermediate between slow and rapid conversion were also considered, but the resulting radial modes do not differ significantly from the slow conversion case, with corresponding implications for the stability of the stars [20].

7. Conclusions

The investigation of dense QCD through the astrophysics of compact stars is an actively pursued subject. This is due to the substantial observational progress, which includes measurements of the masses and radii of pulsars and gravitational wave signals from mergers of two neutron stars and neutron-star–black-hole binaries. A more thorough comprehension of the thermodynamics of dense QCD, weak interactions, and the dynamics of phase transitions would greatly enhance our ability to model astrophysical phenomena relevant to current observational programs.

This work gave an overview of the phase diagram of cold and dense QCD appropriate for compact stars. We stressed that the universality of the phase diagram of imbalanced fermionic superfluids, such as cold atomic gases and nuclear matter, provides a valuable guide to the possible arrangement of the color-superconducting phases in neutron stars, the presence of tri-critical points, and BCS–BEC crossovers. The universality allows one to conjecture the possible structures of the phase diagram in the density–temperature plane including such phases, such as the Fulde–Ferrel phase, deformed Fermi surface phase, and the phase separation.

As a novel contribution, the previously proposed parametrization of the EoS of dense quark matter with sequential phase transitions was extended to include a conformal fluid at large densities ($n \geq 10\, n_{\text{sat}}$) with the speed of sound $c_{\text{conf.}} = 1/\sqrt{3}$. The part of the M-R diagram that contains twins and triplets remains intact because the transition to conformal fluid occurs at larger central densities than those achieved in these objects. Nevertheless, for large central densities, we find behavior that is qualitatively different from earlier studies of this regime: the M–R curves spiral in; i.e., after reaching a minimum, they turn to the right (larger radius region), thus avoiding the region of ultra-compact stars. Therefore, if the conformal limit is reached for densities much larger than those considered here, the ultracompact region with radii 6–7 km can be populated [33]. In the opposite case of the early onset of the conformal limit (as discussed in Section 4), the radii will remain large, but small-mass regions can be populated if the stability criteria are modified by the slow conversion at the interface(s) between the phases. Another interesting new observation is that the change in the magnitude of the jump from 2SC to the CFL phase induces a special point on the M–R diagram at which all the curves meet in analogy to the case of single-phase transition; see Ref. [58]. The importance of studying this asymptotically large central density regime is phenomenologically relevant if the conversion between various quark and nuclear phases is slow compared to the characteristic timescale of oscillations, as discussed in Section 6. In this case, the stars on the descending branch of mass–central-density (and its counterpart on the M-R diagram) may be stable [17–20], contrary to the classical requirement $dM/d\rho_c > 0$ for the branch to be stable, which in turn leads to higher multipole (beyond triplets) stars on the M–R diagram.

Funding: This research was funded by Deutsche Forschungsgemeinschaft Grant No. SE 1836/5-2 and the Polish NCN Grant No. 2020/37/B/ST9/01937.

Data Availability Statement: The data presented in this study are available on request from the author.

Acknowledgments: The author is grateful to M. Alford, J.-J. Li, and P. B. Rau for collaboration on modeling compact stars with quark cores and the referees for helpful comments.

Conflicts of Interest: The author declares no conflict of interest.

References

1. Abbott, B.P.; Abbott, R.; Abbott, T.D.; Acernese, F.; Ackley, K.; Adams, C.; Adams, T.; Addesso, P.; Adhikari, R.X.; Adya, V.B.; et al. GW170817: Observation of Gravitational Waves from a Binary Neutron Star Inspiral. *Phys. Rev. Lett.* **2017**, *119*, 161101. [CrossRef] [PubMed]
2. Cromartie, H.T.; Fonseca, E.; Ransom, S.M.; Demorest, P.B.; Arzoumanian, Z.; Blumer, H.; Brook, P.R.; DeCesar, M.E.; Dolch, T.; Ellis, J.A.; et al. Relativistic Shapiro delay measurements of an extremely massive millisecond pulsar. *Nat. Astron.* **2020**, *4*, 72–76. [CrossRef]
3. Riley, T.E.; Watts, A.L.; Ray, P.S.; Bogdanov, S.; Guillot, S.; Morsink, S.M.; Bilous, A.V.; Arzoumanian, Z.; Choudhury, D.; Deneva, J.S.; et al. A NICER View of the Massive Pulsar PSR J0740+6620 Informed by Radio Timing and XMM-Newton Spectroscopy. *Astrophys. J. Lett.* **2021**, *918*, L27. [CrossRef]
4. Miller, M.C.; Lamb, F.K.; Dittmann, A.J.; Bogdanov, S.; Arzoumanian, Z.; Gendreau, K.C.; Guillot, S.; Ho, W.C.G.; Lattimer, J.M.; Loewenstein, M.; et al. The Radius of PSR J0740+6620 from NICER and XMM-Newton Data. *Astrophys. J. Lett.* **2021**, *918*, L28. [CrossRef]
5. Weber, F. Strange quark matter and compact stars. *Prog. Part. Nucl. Phys.* **2005**, *54*, 193–288. [CrossRef]
6. Alford, M.G.; Schmitt, A.; Rajagopal, K.; Schäfer, T. Color superconductivity in dense quark matter. *Rev. Mod. Phys.* **2008**, *80*, 1455–1515. [CrossRef]
7. Fukushima, K.; Hatsuda, T. The phase diagram of dense QCD. *Rep. Prog. Phys.* **2011**, *74*, 014001. [CrossRef]
8. Anglani, R.; Casalbuoni, R.; Ciminale, M.; Ippolito, N.; Gatto, R.; Mannarelli, M.; Ruggieri, M. Crystalline color superconductors. *Rev. Mod. Phys.* **2014**, *86*, 509–561. [CrossRef]
9. Blaschke, D.; Chamel, N. *Phases of Dense Matter in Compact Stars*; Rezzolla, L., Pizzochero, P., Jones, D.I., Rea, N., Vidaña, I., Eds.; Astrophysics and Space Science Library; Springer: Cham, Switzerland, 2018; Volume 457, p. 337.

10. Baym, G.; Hatsuda, T.; Kojo, T.; Powell, P.D.; Song, Y.; Takatsuka, T. From hadrons to quarks in neutron stars: A review. *Rep. Prog. Phys.* **2018**, *81*, 056902. [CrossRef]
11. Burgio, G.F.; Schulze, H.J.; Vidaña, I.; Wei, J.B. Neutron stars and the nuclear equation of state. *Prog. Part. Nucl. Phys.* **2021**, *120*, 103879.
12. Lattimer, J.M. Constraints on Nuclear Symmetry Energy Parameters. *Particles* **2023**, *6*, 30–56. [CrossRef]
13. Sedrakian, A.; Li, J.J.; Weber, F. Heavy baryons in compact stars. *Prog. Part. Nucl. Phys.* **2023**, *131*, 104041. [CrossRef]
14. Ding, M.; Roberts, C.D.; Schmidt, S.M. Emergence of Hadron Mass and Structure. *Particles* **2023**, *6*, 57–120. [CrossRef]
15. Alford, M.G.; Sedrakian, A. Compact stars with sequential QCD phase transitions. *Phys. Rev. Lett.* **2017**, *119*, 161104. [CrossRef]
16. Bardeen, J.M.; Friedman, J.L.; Schutz, B.F.; Sorkin, R. A new criterion for secular instability in rapidly rotating stars. *Astrophys. J.* **1977**, *217*, L49–L53. [CrossRef]
17. Pereira, J.P.; Flores, C.V.; Lugones, G. Phase transition effects on the dynamical stability of hybrid neutron stars. *Astrophys. J.* **2018**, *860*, 12. [CrossRef]
18. Curin, D.; Ranea-Sandoval, I.F.; Mariani, M.; Orsaria, M.G.; Weber, F. Hybrid stars with color superconducting cores in an extended fcm model. *Universe* **2021**, *7*, 370. [CrossRef]
19. Gonçalves, V.P.; Lazzari, L. Impact of slow conversions on hybrid stars with sequential QCD phase transitions. *Eur. Phys. J. C* **2022**, *82*, 288. [CrossRef]
20. Rau, P.B.; Sedrakian, A. Two first-order phase transitions in hybrid compact stars: Higher-order multiplet stars, reaction modes, and intermediate conversion speeds. *Phys. Rev. D* **2023**, *107*, 103042. [CrossRef]
21. McLerran, L.; Pisarski, R.D. Phases of dense quarks at large N_c. *Nucl. Phys. A* **2007**, *796*, 83–100. [CrossRef]
22. Fukushima, K.; Kojo, T. The Quarkyonic Star. *Astrophys. J.* **2016**, *817*, 180. [CrossRef]
23. Pisarski, R.D.; Skokov, V.V.; Tsvelik, A. A Pedagogical Introduction to the Lifshitz Regime. *Universe* **2019**, *5*, 48. [CrossRef]
24. McLerran, L.; Reddy, S. Quarkyonic Matter and Neutron Stars. *Phys. Rev. Lett.* **2019**, *122*, 122701. [CrossRef]
25. Margueron, J.; Hansen, H.; Proust, P.; Chanfray, G. Quarkyonic stars with isospin-flavor asymmetry. *Phys. Rev. C* **2021**, *104*, 055803. [CrossRef]
26. Strinati, G.C.; Pieri, P.; Röpke, G.; Schuck, P.; Urban, M. The BCS-BEC crossover: From ultra-cold Fermi gases to nuclear systems. *Phys. Rep.* **2018**, *738*, 1–76.
27. Stein, M.; Sedrakian, A.; Huang, X.G.; Clark, J.W. BCS-BEC crossovers and unconventional phases in dilute nuclear matter. *Phys. Rev. C* **2014**, *90*, 065804. [CrossRef]
28. Sedrakian, A.; Clark, J.W. Superfluidity in nuclear systems and neutron stars. *Eur. Phys. J. A* **2019**, *55*, 167. [CrossRef]
29. Sedrakian, A.; Tripolt, R.A.; Wambach, J. Color superconductivity from the chiral quark-meson model. *Phys. Lett. B* **2018**, *780*, 627–630. [CrossRef]
30. Alford, M.G.; Pangeni, K.; Windisch, A. Color Superconductivity and Charge Neutrality in Yukawa Theory. *Phys. Rev. Lett.* **2018**, *120*, 082701. [CrossRef]
31. Zdunik, J.L.; Haensel, P. Maximum mass of neutron stars and strange neutron-star cores. *Astron. Astrophys.* **2013**, *551*, A61. [CrossRef]
32. Alford, M.G.; Han, S.; Prakash, M. Generic conditions for stable hybrid stars. *Phys. Rev. D* **2013**, *88*, 083013. [CrossRef]
33. Li, J.J.; Sedrakian, A.; Alford, M. Ultracompact hybrid stars consistent with multimessenger astrophysics. *Phys. Rev. D* **2023**, *107*, 023018. [CrossRef]
34. Gerlach, U.H. Equation of State at Supranuclear Densities and the Existence of a Third Family of Superdense Stars. *Phys. Rev.* **1968**, *172*, 1325–1330. [CrossRef]
35. Glendenning, N.K.; Kettner, C. Possible third family of compact stars more dense than neutron stars. *Astron. Astrophys.* **2000**, *353*, L9–L12.
36. Schertler, K.; Greiner, C.; Schaffner-Bielich, J.; Thoma, M.H. Quark phases in neutron stars and a third family of compact stars as signature for phase transitions. *Nucl. Phys. A* **2000**, *677*, 463–490. [CrossRef]
37. Alvarez-Castillo, D.E.; Blaschke, D.B.; Grunfeld, A.G.; Pagura, V.P. Third family of compact stars within a nonlocal chiral quark model equation of state. *Phys. Rev. D* **2019**, *99*, 063010. [CrossRef]
38. Kojo, T. Delineating the properties of matter in cold, dense QCD. In Proceedings of the Xiamen-CUSTIPEN Workshop on the Equation of State of Dense Neutron-Rich Matter in the Era of Gravitational Wave Astronomy, Xiamen, China, 3–7 January 2019; American Institute of Physics Conference Series; Volume 2127, p. 020023.
39. Li, J.J.; Sedrakian, A.; Alford, M. Relativistic hybrid stars with sequential first-order phase transitions and heavy-baryon envelopes. *Phys. Rev. D* **2020**, *101*, 063022. [CrossRef]
40. Li, J.J.; Sedrakian, A.; Alford, M. Relativistic hybrid stars in light of the NICER PSR J0740+6620 radius measurement. *Phys. Rev. D* **2021**, *104*, L121302. [CrossRef]
41. Müther, H.; Sedrakian, A. Breaking rotational symmetry in two-flavor color superconductors. *Phys. Rev. D* **2003**, *67*, 085024. [CrossRef]
42. Bedaque, P.F.; Caldas, H.; Rupak, G. Phase Separation in Asymmetrical Fermion Superfluids. *Phys. Rev. Lett.* **2003**, *91*, 247002. [CrossRef]
43. Alford, M.; Rajagopal, K.; Wilczek, F. Color-flavor locking and chiral symmetry breaking in high density QCD. *Nucl. Phys. B* **1999**, *537*, 443–458. [CrossRef]

44. Alford, M.; Kouvaris, C.; Rajagopal, K. Evaluating the gapless color-flavor locked phase. *Phys. Rev. D* **2005**, *71*, 054009. [CrossRef]
45. Sun, G.; He, L.; Zhuang, P. BEC-BCS crossover in the Nambu-Jona-Lasinio model of QCD. *Phys. Rev. D* **2007**, *75*, 096004. [CrossRef]
46. Sedrakian, A.; Rischke, D.H. Phase diagram of chiral quark matter: From weakly to strongly coupled Fulde-Ferrell phase. *Phys. Rev. D* **2009**, *80*, 074022. [CrossRef]
47. Ferrer, E.J.; de la Incera, V.; Keith, J.P.; Portillo, I. BCS-BEC crossover and stability in a Nambu-Jona-Lasinio model with diquark-diquark repulsion. *Nucl. Phys. A* **2015**, *933*, 229–244. [CrossRef]
48. Duarte, D.C.; Farias, R.L.S.; Manso, P.H.A.; Ramos, R.O. Optimized perturbation theory applied to the study of the thermodynamics and BEC-BCS crossover in the three-color Nambu-Jona-Lasinio model. *Phys. Rev. D* **2017**, *96*, 056009. [CrossRef]
49. Lombardo, U.; Nozières, P.; Schuck, P.; Schulze, H.J.; Sedrakian, A. Transition from BCS pairing to Bose-Einstein condensation in low-density asymmetric nuclear matter. *Phys. Rev. C* **2001**, *64*, 064314. [CrossRef]
50. Sedrakian, A.; Clark, J.W. Pair condensation and bound states in fermionic systems. *Phys. Rev. C* **2006**, *73*, 035803. [CrossRef]
51. Jin, M.; Urban, M.; Schuck, P. BEC-BCS crossover and the liquid-gas phase transition in hot and dense nuclear matter. *Phys. Rev. C* **2010**, *82*, 024911. [CrossRef]
52. Giorgini, S.; Pitaevskii, L.P.; Stringari, S. Theory of ultracold atomic Fermi gases. *Rev. Mod. Phys.* **2008**, *80*, 1215–1274. [CrossRef]
53. Son, D.T. Superconductivity by long-range color magnetic interaction in high-density quark matter. *Phys. Rev. D* **1999**, *59*, 094019. [CrossRef]
54. Schäfer, T.; Wilczek, F. Superconductivity from perturbative one-gluon exchange in high density quark matter. *Phys. Rev. D* **1999**, *60*, 114033. [CrossRef]
55. Rajagopal, K.; Wilczek, F. Enforced Electrical Neutrality of the Color-Flavor Locked Phase. *Phys. Rev. Lett.* **2001**, *86*, 3492–3495. [CrossRef]
56. Bonanno, L.; Sedrakian, A. Composition and stability of hybrid stars with hyperons and quark color-superconductivity. *Astron. Astrophys.* **2012**, *539*, A16. [CrossRef]
57. Li, J.J.; Sedrakian, A. Constraining compact star properties with nuclear saturation parameters. *Phys. Rev. C* **2019**, *100*, 015809. [CrossRef]
58. Cierniak, M.; Blaschke, D. The special point on the hybrid star mass–radius diagram and its multi-messenger implications. *Eur. Phys. J. Spec. Top.* **2020**, *229*, 3663–3673. [CrossRef]
59. Lyra, F.; Moreira, L.; Negreiros, R.; Gomes, R.O.; Dexheimer, V. Compactness in the thermal evolution of twin stars. *Phys. Rev. C* **2023**, *107*, 025806. [CrossRef]
60. Zapata, J.; Sales, T.; Jaikumar, P.; Negreiros, R. Thermal relaxation and cooling of quark stars with a strangelet crust. *Astron. Astrophys.* **2022**, *663*, A19. [CrossRef]
61. Sedrakian, A. Exploring phases of dense QCD with compact stars. *EPJ Web Conf.* **2017**, *164*, 01009. [CrossRef]
62. Sedrakian, A. Cooling compact stars and phase transitions in dense QCD. *Eur. Phys. J. A* **2016**, *52*, 44. [CrossRef]
63. de Carvalho, S.M.; Negreiros, R.; Orsaria, M.; Contrera, G.A.; Weber, F.; Spinella, W. Thermal evolution of hybrid stars within the framework of a nonlocal Nambu-Jona-Lasinio model. *Phys. Rev. C* **2015**, *92*, 035810. [CrossRef]
64. Grigorian, H.; Blaschke, D.; Voskresensky, D.N. Cooling of neutron stars and hybrid stars with a stiff hadronic EoS. *Phys. Part. Nucl.* **2015**, *46*, 849. [CrossRef]
65. Noda, T.; Hashimoto, M.A.; Yasutake, N.; Maruyama, T.; Tatsumi, T.; Fujimoto, M. Cooling of Compact Stars with Color Superconducting Phase in Quark-hadron Mixed Phase. *Astrophys. J.* **2013**, *765*, 1. [CrossRef]
66. Negreiros, R.; Dexheimer, V.A.; Schramm, S. Quark core impact on hybrid star cooling. *Phys. Rev. C* **2012**, *85*, 035805. [CrossRef]
67. Hess, D.; Sedrakian, A. Thermal evolution of massive compact objects with dense quark cores. *Phys. Rev. D* **2011**, *84*, 063015. [CrossRef]
68. Sedrakian, A. Rapid cooling of Cassiopeia A as a phase transition in dense QCD. *Astron. Astrophys.* **2013**, *555*, L10. [CrossRef]
69. Stejner, M.; Weber, F.; Madsen, J. Signature of Deconfinement with Spin-Down Compression in Cooling Hybrid Stars. *Astrophys. J.* **2009**, *694*, 1019–1033. [CrossRef]
70. Blaschke, D.; Grigorian, H. Unmasking neutron star interiors using cooling simulations. *Prog. Part. Nucl. Phys.* **2007**, *59*, 139–146. [CrossRef]
71. Anglani, R.; Nardulli, G.; Ruggieri, M.; Mannarelli, M. Neutrino emission from compact stars and inhomogeneous color superconductivity. *Phys. Rev. D* **2006**, *74*, 074005. [CrossRef]
72. Popov, S.B.; Grigorian, H.; Blaschke, D. Neutron star cooling constraints for color superconductivity in hybrid stars. *Phys. Rev. C* **2006**, *74*, 025803. [CrossRef]
73. Alford, M.; Jotwani, P.; Kouvaris, C.; Kundu, J.; Rajagopal, K. Astrophysical implications of gapless color-flavor locked quark matter: A hot water bottle for aging neutron stars. *Phys. Rev. D* **2005**, *71*, 114011. [CrossRef]
74. Shovkovy, I.A.; Ellis, P.J. Impact of Cfl Quark Matter on the Cooling of Compact Stars. In *Strong Coupling Gauge Theories and Effective Field Theories*, World Scientific: Singapore, 2003; pp. 192–198.
75. Blaschke, D.; Grigorian, H.; Voskresensky, D.N. Cooling of hybrid neutron stars and hypothetical self-bound objects with superconducting quark cores. *Astron. Astrophys.* **2001**, *368*, 561–568. [CrossRef]
76. Duncan, R.C.; Wasserman, I.; Shapiro, S.L. Neutrino emissivity of interacting quark matter in neutron stars. II—Finite neutrino momentum effects. *Astrophys. J.* **1984**, *278*, 806–812. [CrossRef]

77. Iwamoto, N. Quark Beta Decay and the Cooling of Neutron Stars. *Phys. Rev. Lett.* **1980**, *44*, 1637–1640. [CrossRef]
78. Jaikumar, P.; Roberts, C.D.; Sedrakian, A. Direct Urca neutrino rate in color superconducting quark matter. *Phys. Rev. C* **2006**, *73*, 042801. [CrossRef]
79. Jaikumar, P.; Prakash, M.; Schäfer, T. Neutrino emission from Goldstone modes in dense quark matter. *Phys. Rev. D* **2002**, *66*, 063003. [CrossRef]
80. Shternin, P.S.; Ofengeim, D.D.; Heinke, C.O.; Ho, W.C.G. Constraints on neutron star superfluidity from the cooling neutron star in Cassiopeia A using all Chandra ACIS-S observations. *Mon. Not. R. Astron. Soc.* **2023**, *518*, 2775–2793. [CrossRef]
81. Raduta, A.R.; Sedrakian, A.; Weber, F. Cooling of hypernuclear compact stars. *Mon. Not. R. Astron. Soc.* **2018**, *475*, 4347–4356. [CrossRef]
82. Raduta, A.R.; Li, J.J.; Sedrakian, A.; Weber, F. Cooling of hypernuclear compact stars: Hartree-Fock models and high-density pairing. *Mon. Not. R. Astron. Soc.* **2019**, *487*, 2639–2652. [CrossRef]
83. Anzuini, F.; Melatos, A.; Dehman, C.; Viganò, D.; Pons, J.A. Fast cooling and internal heating in hyperon stars. *Mon. Not. R. Astron. Soc.* **2022**, *509*, 2609–2623. [CrossRef]
84. Tsuruta, S.; Kelly, M.J.; Nomoto, K.; Mori, K.; Teter, M.; Liebmann, A.C. Ambipolar Heating of Magnetars. *Astrophys. J.* **2023**, *945*, 151. [CrossRef]
85. Orsaria, M.G.; Malfatti, G.; Mariani, M. Phase transitions in neutron stars and their links to gravitational waves Phase transitions in neutron stars and their links to gravitational waves. *J. Phys. G Nucl. Part. Phys.* **2019**, *46*, 073002. [CrossRef]
86. Wei, W.; Salinas, M.; Klähn, T.; Jaikumar, P.; Barry, M. Lifting the Veil on Quark Matter in Compact Stars with Core g-mode Oscillations. *Astrophys. J.* **2020**, *904*, 187. [CrossRef]
87. Jaikumar, P.; Semposki, A.; Prakash, M.; Constantinou, C. G -mode oscillations in hybrid stars: A tale of two sounds. *Phys. Rev. D* **2021**, *103*, 123009. [CrossRef]
88. Bauswein, A.; Bastian, N.U.F.; Blaschke, D.B.; Chatziioannou, K.; Clark, J.A.; Fischer, T.; Oertel, M. Identifying a first-order phase transition in neutron star mergers through gravitational waves. *Phys. Rev. Lett.* **2019**, *122*, 061102. [CrossRef] [PubMed]
89. Weih, L.R.; Hanauske, M.; Rezzolla, L. Post-merger gravitational-wave signatures of phase transitions in binary mergers. *Phys. Rev. Lett.* **2020**, *124*, 171103. [CrossRef]
90. Liebling, S.L.; Palenzuela, C.; Lehner, L. Effects of high density phase transitions on neutron star dynamics. *Class. Quantum Gravity* **2021**, *38*, 115007. [CrossRef]
91. Prakash, A.; Radice, D.; Logoteta, D.; Perego, A.; Nedora, V.; Bombaci, I.; Kashyap, R.; Bernuzzi, S.; Endrizzi, A. Signatures of deconfined quark phases in binary neutron star mergers. *Phys. Rev. D* **2021**, *104*, 83029. [CrossRef]
92. Chandrasekhar, S. The Dynamical Instability of Gaseous Masses Approaching the Schwarzschild Limit in General Relativity. *Astrophys. J.* **1964**, *140*, 417. [CrossRef]
93. Gourgoulhon, E.; Haensel, P.; Gondek, D. Maximum mass instability of neutron stars and weak interaction processes in dense matter. *Astron. Astrophys.* **1995**, *294*, 747–756.
94. Takami, K.; Rezzolla, L.; Yoshida, S. A quasi-radial stability criterion for rotating relativistic stars. *Mon. Not. Roy. Astr. Soc.* **2011**, *416*, L1–L5. [CrossRef]
95. Bisnovatyi-Kogan, G.S.; Seidov, Z.F. Oscillations of a star with a phase transition. *Astrofizika* **1984**, *20*, 563–571. [CrossRef]
96. Haensel, P.; Zdunik, J.L.; Schaeffer, R. Phase transitions in dense matter and radial pulsations of neutron stars. *Astron. Astrophys.* **1989**, *217*, 137–144.
97. Chanmugam, G. Radial oscillations of zero-temperature white dwarfs and neutron stars below nuclear densities. *Astrophys. J.* **1977**, *217*, 799–808. [CrossRef]
98. Gondek, D.; Haensel, P.; Zdunik, J.L. Radial pulsations and stability of protoneutron stars. *Astron. Astrophys.* **1997**, *325*, 217–227.
99. Karlovini, M.; Samuelsson, L.; Zarroug, M. Elastic stars in general relativity: II. Radial perturbations. *Class. Quantum Gravity* **2004**, *21*, 1559–1581. [CrossRef]
100. Bombaci, I.; Logoteta, D.; Vidaña, I.; Providência, C. Quark matter nucleation in neutron stars and astrophysical implications. *Eur. Phys. J. A* **2016**, *52*, 58. [CrossRef]

Disclaimer/Publisher's Note: The statements, opinions and data contained in all publications are solely those of the individual author(s) and contributor(s) and not of MDPI and/or the editor(s). MDPI and/or the editor(s) disclaim responsibility for any injury to people or property resulting from any ideas, methods, instructions or products referred to in the content.

Article

Precision Storage Rings for Electric Dipole Moment Searches: A Tool En Route to Physics Beyond-the-Standard-Model

Hans Ströher [1,*], Sebastian M. Schmidt [2,3], Paolo Lenisa [4] and Jörg Pretz [1,3]

1. Forschungszentrum Jülich (FZJ), 52425 Jülich, Germany
2. Helmholtz-Zentrum Dresden-Rossendorf (HZDR), 01328 Dresden, Germany
3. Rheinisch-Westfälische Technische Hochschule (RWTH) Aachen, 52074 Aachen, Germany
4. Istituto Nazionale di Fisica Nucleare (INFN), University of Ferrara, 44100 Ferrara, Italy
* Correspondence: h.stroeher@fz-juelich.de

Abstract: Electric Dipole Moments (EDM) of particles (leptons, nucleons, and light nuclei) are currently deemed one of the best indicators for new physics, i.e., phenomena which lie outside the Standard Model (SM) of elementary particle physics—so-called physics "Beyond-the-Standard-Model" (BSM). Since EDMs of the SM are vanishingly small, a finite permanent EDM would indicate charge-parity (CP) symmetry violation in addition to the well-known sources of the SM, and could explain the baryon asymmetry of the Universe, while an oscillating EDM would hint at a possible Dark Matter (DM) field comprising axions or axion-like particles (ALPs). A new approach exploiting polarized charged particles (proton, deuteron, ^{3}He) in precision storage rings offers the prospect to push current experimental EDM upper limits significantly further, including the possibility of an EDM discovery. In this paper, we describe the scientific background and the steps towards the realization of a precision storage ring, which will make such measurements possible.

Keywords: Baryon asymmetry; dark matter; electric dipole moments; storage rings; polarized beams

1. Introduction

1.1. Scientific Background

During the past century mankind has made mind-blowing progress in the understanding of nature, which continues to this day and age with spectacular new discoveries, such as the observation of the Higgs-boson or the detection of gravitational waves. Nevertheless, there is a plethora of fundamental questions which wait to be answered [1]. As will be illustrated below, some of them relate the largest scales (i.e., cosmology) to the smallest ones (elementary particle physics), and answers to them will point towards physics not contained in the celebrated respective Standard Model.

Electric Dipole Moments (EDM) of elementary particles—the permanent separation of positive and negative electric charge centroids—have a very strong science case [2] as they aim to solve two of the most pressing problems in contemporary cosmology and particle physics:

1. Reason for the Baryon-asymmetric Universe (BAU) (Why is there matter and no antimatter?) [3]
2. Composition of Dark Matter (DM) in the Universe (What is Dark Matter made of?) [4,5].

EDMs do not conserve discrete symmetries [6]; they violate time-reversal invariance (T), parity (P), and—if the combined charge-parity-time reversal (CPT) symmetry holds—in addition, charge-parity (CP) is "broken". These symmetries and conservation laws play a very important role in modern physics; via Noether´s theorem [7], the invariance of a system subject to a continuous symmetry transformation (e.g., a time-translation, a spacial translation or a rotation) leads to a conservation law (energy, momentum and angular momentum, respectively). In addition to continuous symmetries,

there are discrete symmetries—parity (P) (reflection of space coordinates at the origin: (x, y, z) → (−x, −y, −z)), charge conjugation (C) (replacing a particle by its antiparticle, thus changing the sign of the electric charge: (+) → (−)), and time reversal (T) (change of the direction of the time coordinate: (t) → (−t))—as well as combinations thereof: charge-parity (CP) and charge-parity-time (CPT). Until around the middle of the 20th century, it was taken for granted that all discrete symmetry transformations are good ones—that is these operations do not make a difference in the experimental outcome—but then parity-violation and later CP-violation were discovered experimentally, implying that nature does distinguish between left and right, and between matter and antimatter. Today, only the combined CPT-symmetry is assumed to be exact (the so-called CPT theorem [8]). As a consequence, this means that CP-violation implicates T-reversal violation and vice versa to preserve CPT. CP-violation is one prerequisite to explain the observed matter–antimatter asymmetry in the Universe. The size of CP-violation, which has been found up to now and is implemented in the Standard Model (SM) of particle physics, falls short of explaining this asymmetry by many orders of magnitude.

No particle EDM has been found so far—only impressive experimental upper limits exist, e.g., for the electron [9] (indirectly deduced in atomic measurements) and for the neutron [10]. It turns out that the well-established CP-violation of the Standard Model of particle physics leads only to very tiny EDMs, below the sensitivity of current experiments. The discovery of a finite EDM value would signal the presence of additional CP-violation, which is required for an understanding of the apparent matter–antimatter asymmetry of our Universe. Observation of a finite EDM could thus solve this puzzle.

A new symmetry (called "Peccei–Quinn"(PQ) after its inventors) has originally been suggested to account for the smallness of CP-violation in the strong interaction (dubbed the "strong CP-problem") [11]. One consequence of PQ symmetry breaking is the existence of a new spin-0 ("scalar") particle called an "axion". These particles (or its generalization, axion-like particles (ALP)) [12] are considered promising candidates for Dark Matter (DM), which represents a large fraction (about 25%) of the energy content of the Universe, but its nature is still unknown. No experimental indications for axions/ALPs have been found yet. Axions/ALPs will couple to gluons (the exchange particles that bind quarks inside hadrons) and induce oscillating EDMs, e.g., of nucleons (protons and neutrons) [13]. Observation of such an oscillating EDM could thus clarify the nature of DM.

A discovery of static (not changing with time) EDMs and/or time-dependent (oscillating) EDMs (oEDM) would provide a breakthrough in the quest to understand the set-up and evolution of the Universe on a deeper, more fundamental level. The current model of elementary particle physics—the Standard Model (SM), which is spectacularly successful in many other areas of fundamental particle physics—is not able to explain both the Baryon-asymmetric Universe (BAU) and Dark Matter. Finding an EDM experimentally will therefore inevitably also imply an extension to physics beyond the SM (BSM).

Recently, it has been realized that charged particles, in particular the proton and the deuteron can be used to search for EDMs with unprecedented sensitivity, improving current limits by orders of magnitude [14]. For these experiments, a dedicated—not yet existing—new type of precision storage ring is required, which will allow the measurement of the corresponding observable, the time-development of the polarization in a stored polarized particle beam under investigation.

In the following two chapters the science case is outlined in more detail.

1.1.1. The Ambition to Solve the Matter–Antimatter Asymmetry

We live in a Universe that apparently contains an overwhelming amount of matter compared to antimatter.

Our Universe was created 13.8 billion years ago in a singular event known as the Big Bang. According to our current understanding, it created matter and antimatter in equal amounts. However, in the Universe we live in and investigate today, we only find matter (except for tiny amounts created in violent particle interactions), from our immediate neighbourhood to the largest distances. During the past decades, cosmology, assisted by nuclear and particle physics, advanced our understanding of the evolution of the Universe tremendously. The fate of the antimatter in this evolution is one of the biggest remaining mysteries. During the early evolution of the Universe, antimatter particles annihilated with corresponding matter particles into photons, leaving behind only a tiny fraction of matter—for reasons yet unknown. This tiny fraction constitutes the material foundation of today's Universe and of our very existence. Diverse research is ongoing in the quest to understand the formation of a matter-dominated Universe.

In the 1960s, the work of Andrei Sakharov [15] provided the following conditions for matter to escape the annihilation:

- Baryon number violation—a change in the difference between the number of baryons and antibaryons at the particle level,
- Charge conjugation (C) and charge-parity (CP) violation—a different behaviour of matter and antimatter—and
- Thermal non-equilibrium (i.e., effective during a phase transition of the Universe)—to prevent a subsequent compensation of the asymmetry between matter and antimatter.

The Standard Models of cosmology and particle physics meet these conditions qualitatively, but fail by 8 to 9 orders of magnitude at a quantitative description (see below). The models predict an insufficient amount of matter that survived the annihilation phase in the early Universe. It is frequently believed that the failure may arise from the second condition, i.e., from yet undiscovered violations of fundamental symmetries between matter and antimatter (particles and antiparticles), in particular additional CP-violation not yet found.

Here, EDMs enter the stage. Subatomic particles with nonzero spin (regardless of either elementary or composite nature) can only support a nonzero permanent EDM if both time-reversal (T) and parity (P) symmetries are violated explicitly, while the charge symmetry (C) can be maintained. Assuming the conservation of the combined CPT symmetry, T-violation also implies CP-violation. The CP-violation generated by the Kobayashi–Maskawa mechanism of weak interactions contributes a very small EDM that is several orders of magnitude below current experimental limits [16]. However, many models for physics beyond the Standard Model (BSM) predict EDM values near the current experimental limits (see, e.g., [17]).

Finding a non-zero EDM value of any subatomic particle would be a signal that there exists a new source of CP-violation, either induced by the strong CP-violation via the so-called "θ_{QCD} angle" or by a genuine BSM effect. In fact, the best upper limit on θ_{QCD} follows from the experimental bound on the EDM of the neutron. As indicated above, CP-violation beyond the SM is also essential for explaining the mystery of the observed baryon-antibaryon asymmetry of our Universe (BAU). A measurement of a single EDM will not be sufficient to establish the sources of any new CP-violation. Complementary observations of EDMs in multiple systems will thus prove essential.

Up to now, measurements have focused on neutral systems (neutron, atoms, molecules); Figure 1 gives a summary of the EDM upper limits (for a recent summary, see [18]). Note that the limit for the proton has been deduced from a ^{199}Hg measurement and no limit for the deuteron exists.

In order to directly measure the EDM of charged particles, the design of a dedicated precision storage ring needs to be developed [21]. The goal is to provide optimal experimental conditions to investigate the time evolution of the polarization of stored polarized beams of protons and/or deuterons.

The advantages of the storage ring method are:

- Direct observation of the EDM effect of the system (proton, deuteron, and possibly other light ions such as ^{3}He) under investigation, i.e., no need to extract particle EDM from measurement in a complex system;
- Virtually no limit to the number of particles used in the investigation, i.e., no limitation in the achievable statistical accuracy;

- There are theoretical relations between EDMs for protons and deuterons (see, e.g., [21]), which will serve as consistency checks for theory.

On a more general level, there will be different systematic errors and fake effects compared to neutral systems, and in the end—once EDMs for all of them have been obtained—they must all be internally consistent within the theoretical interpretation.

Due to the smallness of the effect, the measurements will need an excessive sensitivity, and this poses technological and metrological challenges, which will be discussed below in the section "methodology".

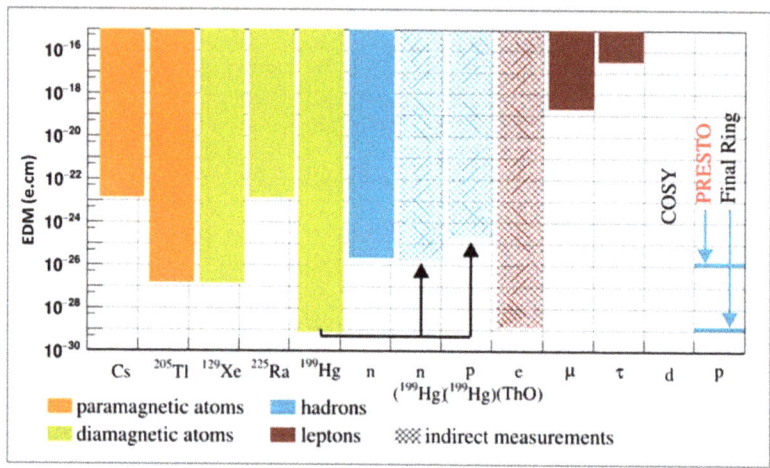

Figure 1. Current status of EDM searches (95% upper limits) of molecules, atoms, and particles: the full-colour bars represent direct measurements, while the shaded bars are deduced results for these systems. In particular, the proton result has been obtained indirectly from the experimental EDM limit for ^{199}Hg; this is indicated by the black arrows. Impressive EDM limits for the electron come from Thorium-oxide (ThO) and HfF$^+$ measurements [9]—still many orders of magnitude larger than the SM-prediction. It generally holds that the contribution from the SM is negligible at the sensitivity level accessible with current experimental techniques. For the deuteron (d), for which neither a direct nor an indirect limit is known, an EDM limit from the measurement at the Cooler Synchrotron (COSY) of Forschungszentrum Jülich (Germany) will soon be available (see below). The expected sensitivity is comparable to the muon (µ) [19] and tau (τ) limits [20]. The goal of the proton EDM (p) of the final ring is to reach a sensitivity of 10^{-29} e·cm, and the goal for the intermediate step—a prototype storage ring (called PRESTO in the figure)—is to reach 10^{-26} e·cm.

1.1.2. The Intent to Understand the Nature of Dark Matter

Since the earliest times of mankind, people have looked at the night sky with awe, wondering what these tiny luminous objects in heaven might be. Today, though few of us have had a chance to see even our own galaxy (Milky Way) in full glory, it is known to most people that we are only able to see a small fraction of the Milky Way and an even tinier part of the vast visible Universe—as, e.g., observed through electromagnetic radiation, ranging from radio waves to visible light and beyond. It may come as no surprise that astronomical observations over the past almost 100 years have accumulated more and more evidence that this luminous part (stars, galaxies, and clusters) only constitutes a portion of the mass in our Universe. In fact only about one fifth of the matter is made of fundamental particles that we know and understand; the majority of the matter is not visible and thus dubbed "Dark Matter" (DM) [22].

DM has been postulated by the need for much more mass than actually seen in rotating galaxies in order to bind them by the gravitational force, compensating the centrifugal force. This is shown schematically in Figure 2.

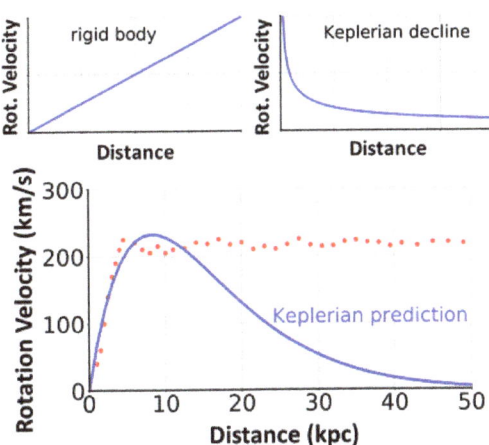

Figure 2. Rotational speed for a rigid body (top left) and for Keplerian motion, e.g., our solar system, with most of the mass centred at zero (top right). Below, the result (red dots) for a typical spiral galaxy is shown. The disagreement of the rotational speed of galaxies from the Keplerian prediction for R > 10 kpc (i.e., the constant speed) was first observed by F. Zwicky [23,24] in the 1930s in the Coma Cluster, and indicates that more mass than visible must be present.

There are further compelling observations indicating the existence of DM [25], e.g., via gravitational lensing and its role in the evolution of structure in the Universe. However, what DM actually is remains a mystery. It is important to note that none of the known building blocks of the (otherwise remarkably successful) Standard Model (SM) of elementary particle physics can serve as DM. This implies that the SM is incomplete and that a discovery of DM particles would be conclusive evidence for physics beyond the SM (BSM). There is mounting evidence for the need of DM, and the search is ongoing by a suite of dedicated experiments, but no DM particle(s) has(have) been observed to date, although there have been claims. The allowed DM-parameters of mass and coupling to SM particles spans a huge range, which requires very different techniques for discovery (see, e.g., [26]).

The following principal search strategies for Dark Matter are currently exploited:

- Direct production of DM particles at accelerators;
- Detection of galactic DM particles through interactions with SM matter; and
- Detection of DM as coherently oscillating waves.

EDMs will exploit the third option, assuming a high occupation number per unit volume, i.e., a field rather than individual particles. The effect of DM on the spin motion of an ensemble of particles (a beam) in a storage ring would then be observed as oscillating Electric Dipole Moments (oEDM) in nucleons and nuclei [13].

To conclude a discussion of motivations, the science case is summarized in Figure 3.

1.2. State-of-the-Art Knowledge

EDMs can be searched for in elementary particles such as leptons (electrons, muons, taus), composite particles (e.g., neutrons and protons), and—indirectly—in complex systems, for example atoms and molecules.

At first, two important remarks seem to be indicated:

1. The connection between any finite experimental EDM result and its sources is less complex for simpler systems—this is schematically shown in Figure 4. A more detailed picture of such connections is given, e.g., in [27].

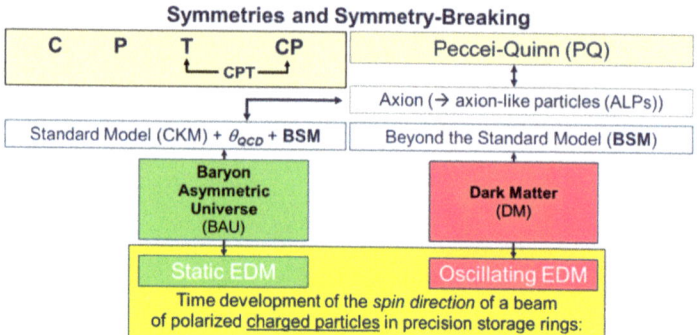

Figure 3. Summary of the scientific background: discrete symmetries (C, P, T, and combinations thereof) and the presumed Peccei–Quinn symmetry (PQ) together with their breaking (CP and PQ symmetry violation) can be related to static and oscillating EDMs. Static EDMs may be responsible for the Baryon-asymmetric Universe (BAU), while oscillating EDMs could be due to the axion and/or axion-like particles (ALP) as Dark Matter candidates. Experimentally, both can be investigated by a measurement of the time development of the spins of a beam of charged particles circulating in a storage ring.

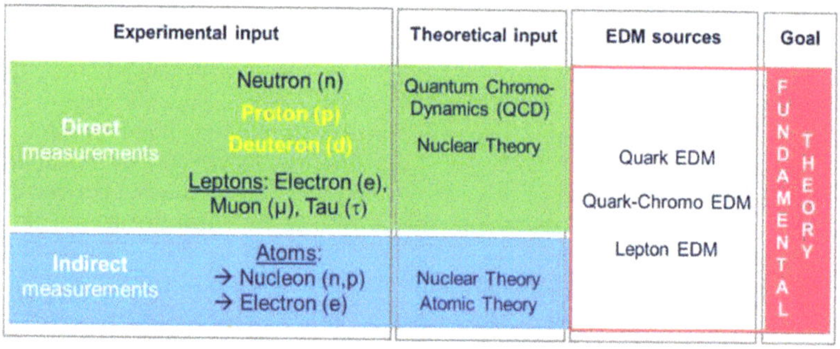

Figure 4. Schematic connection between experiment and a fundamental theory as the ultimate goal of EDM investigations. There are two major paths to experimentally search for EDMs: direct measurements, which study a sole particle species, and indirect measurements, which use atoms or molecules (or even bulk matter). Although indirect experiments have resulted in impressive upper limits, direct investigations are clearly preferable, since they will be much easier to interpret.

While results from indirect searches—i.e., experiments with systems which contain the respective particle—are available for the electron and the proton (see Figure 1), in the end, one has to strive for the observation of the pure (naked) object under investigation. For charged particles, this will imply the use of stored beams to extend the observation times to such a level that the miniscule expected effects of an EDM can be observed with available experimental techniques.

2. EDM results from different systems will be needed to pin down the (set of) EDM sources.

Experimental results—EDM upper limits of the order of 10^{-26} e·cm—are available for the free neutron [10] (see Figure 1), and it can be expected that they will be further improved by ongoing and upcoming new experiments, although the limitations for the provision of ultracold neutrons (and the finite neutron lifetime) will finally limit the achievable sensitivity. In order to complement these results, experimental EDM searches for free protons must be conducted, and this is now in reach with a new class of measurements of the spin motion of polarized protons, which are circulating in a precision trap called a storage ring. In such experiments, the number of stored protons is hardly limited, which implies that the statistical uncertainty (relating to the achievable sensitivity) can be higher than for the neutron. The challenge will be, as outlined below, to reduce the systematic uncertainties to a comparable level.

2. Methodology of Charged-Particle EDM Searches

The basis for observing an EDM is its interaction with an external electric field. The possible observables are given by the fact that the EDM vector (which can be thought of as connecting the negative and the positive electric charge that forms the EDM) and the intrinsic angular momentum ("spin") vector are parallel to each other. Thus, one way to detect an EDM is to observe the impact it has on the spin of a particle (in practice: an ensemble of particles with spins oriented in a preferred direction, a polarized beam). In the following, we focus on charged particles (proton, deuteron), which need to be confined in a trap—in this case a storage ring—since otherwise they would be accelerated out-of-sight by the electric field. While a full description of the spin-motion of particles in the electric and magnetic fields of the ring requires the use of the Thomas–BMT equation [28,29], the following discussion focuses on a generic description of the method.

In order to introduce the basic principle of an EDM measurement of polarized charged particles in a storage ring, it is reasonable to start by describing what is happening to them as they go around the ring, when no EDM of the particles would exist. To simplify the situation further, it is assumed that at the beginning the spins are pointing in the direction of motion (i.e., parallel to the particle's momentum vector):

- Since the spin s of a particle is linked to a magnetic moment $\mu = (G+1)(q/m)\,s$, where G is the magnetic anomaly, q the charge, and m the mass, the particles will precess in the electric and/or magnetic field of the storage ring, which are necessary to keep them on a closed orbit. The total angular precession frequency of the spin is given by $ds/dt = \Omega_s \times s$, where Ω_s has components due to the magnetic and electric dipole moments of the particle.
- For any particle and energy, the electric and magnetic fields bending the beam can be chosen such that the angular rotation of the spin vector s from turn to turn vanishes, the spin is then said to "be frozen", and, e.g., an initial longitudinal polarization of the particle beam is maintained.
- For particles with $G > 0$, such as protons, the above frozen-spin condition can be fulfilled with electric fields only by operating the ring at a specific energy, the so-called "magic" energy.

Now consider that the particles possess also an EDM $d = ds$ (parallel to the spin s) and are exposed to a radial electric field E:

- The interaction between electric field and EDM will produce a torque $d \times E$ that rotates the spin out of the ring plane, as schematically shown in Figure 5.

As a result, the principle for searching/observing EDMs of charged particles is as follows:

1. Inject a polarized charged particle beam with its spin vectors pointing in the direction of the momentum vector into a storage ring fulfilling the "frozen spin" condition. Thus, when the particles possess only a magnetic dipole moment (MDM), the beam will remain polarized in longitudinal direction.
2. Let the particles interact with an electric field that couples to the possible EDM.
3. Observe the rotation of the beam polarization out of the storage ring plane due to a non-zero EDM as a function of time. This can be observed with a dedicated detector system inside the ring, called a polarimeter.

At this point, it is important to note that the MDM of the particle completely dominates the spin motion if no special provisions are taken. In order to maximise the sensitivity to EDM, the MDM contributions must vanish.

This can be achieved by the following measures:

- Switch off the external magnetic field: B = 0 (i.e., use an all-electric storage ring and shield all external magnetic fields);
- Arrange that bunches are polarized in the longitudinal direction during the duration of the measurement by choosing a "magic momentum"; this is possible for particles with a positive magnetic anomaly (G > 0), e.g., for the proton for a momentum of p = 700.7 MeV/c (corresponding to a kinetic energy of 232.8 MeV).

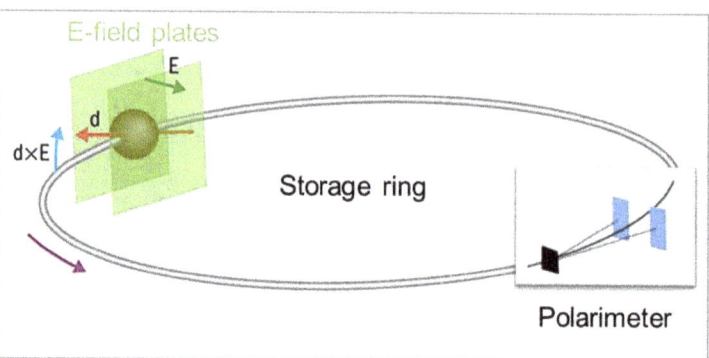

Figure 5. Basic principle of an EDM measurement. For clarity, only a single particle is shown and the magnetic moment precession is neglected. The particle is kept on its orbit by the transverse electric field E, generated by the electrostatic deflector plates that are depicted in green. Initially, all particle spins s in the beam point along the direction of motion. An electric dipole moment $d = ds$ tilts the spin s out of the horizontal plane due to the torque $d \times E$. This effect is experimentally observable, because the vertical polarization component of the beam can be detected in the polarimeter.

For particles with G < 0 (such as the deuteron), an all-electric solution is not possible and one needs to use a combination of radial electric and vertical magnetic fields to make the spin "frozen" along the particle momentum. Though this measurement procedure sounds straightforward and simple, a closer look indicates substantial intricacy. As mentioned earlier, this can be investigated using the Thomas–BMT equation, which describes the temporal behaviour of the spin of a particle with a magnetic dipole moment (MDM) and an electric dipole moment (EDM) in the presence of electric (E) and magnetic (B) fields.

In the remainder of this chapter, challenges will be discussed which come along with the vastly different magnitude of response of a particle's spin to electric (via an EDM) and magnetic (via its MDM) fields. In order to set a benchmark, it may suffice to mention that the minuscule (radial) magnetic field of $\approx 10^{-17}$ T (10 aT (atto-Tesla)) mimics the effect of an EDM of 10^{-29} e·cm—the sensitivity goal for the final EDM ring. It has been shown [21] that with reasonable assumptions for beam intensity, beam polarization, measurement time, applied electric field, and polarimetry, a statistical accuracy of $d = 10^{-29}$ e·cm can be achieved in one year of measurement time; the task, therefore. is to conceive and realise a research infrastructure which allows to bring the systematic uncertainty to a similar level. This implies to identify all sources of false effects to be able to remove all of them or to mitigate their effects to the highest possible level.

The research infrastructure—an all-electric precision storage ring for polarized proton beams of 232.8 MeV—is in many fundamental aspects a blank area; many questions will have to be answered and technological solutions will need to be scrutinised before one can start to realise the facility.

This comprises the following list of items:

- Beam characteristics: storage time of the intensive beams, spin coherence time of the beam polarization and beam emittance; need of phase space cooling;

- Hardware: electric bends, storage ring vacuum; need of cryogenics, instrumentation for beam and spin manipulation, monitoring and control, polarimetry/polarimeter, injection scheme into the storage ring, and stability of power supplies;
- Other: alignment and metrology of ring elements, systematics investigations, and beam and spin tracking simulations.

3. Strategy towards Realization

Research and development towards a precision storage ring has been vigorously advanced at the storage ring COSY, which currently is the only facility worldwide where corresponding tests and measurements can be conducted. A schematic layout of the COSY facility is given in Figure 6.

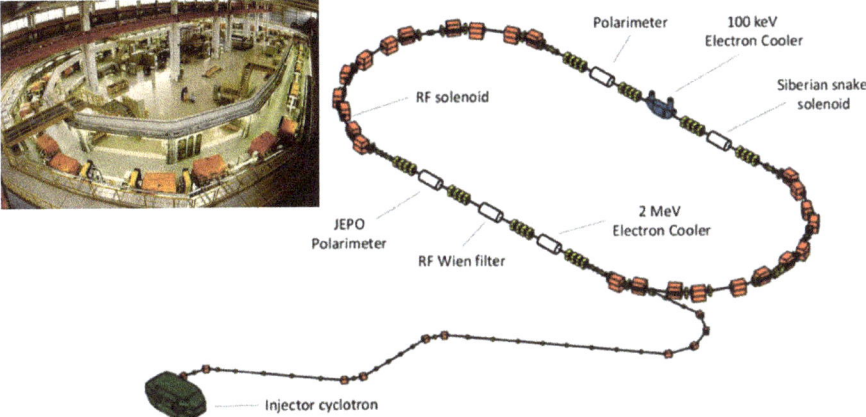

Figure 6. Layout of the Cooler Synchrotron (COSY) storage ring at Forschungszentrum Jülich (Germany) together with an early photograph: the ring has a circumference of 184 m and can accelerate and store unpolarized and polarized beams of protons and deuterons from the injector cyclotron with momenta up to 3.7 GeV/c. The equipment comprises electron and stochastic cooling apparatuses, as well as spin manipulation and detection devices. The unique capabilities of COSY have been used by the JEDI collaboration to investigate manipulations of polarized particle beams and its response to external electric and magnetic fields as preparation for EDM measurements—both at COSY and future dedicated storage rings.

COSY [30] was conceived and built in the late 1980s, and originally used for hadron physics experiments with both unpolarized and polarized proton and deuteron beams at different internal and external target stations until about 2015. Starting from 2010, the JEDI collaboration used COSY beams for the development of hardware (e.g., a new beam polarimeter, beam-position monitors, and a radio-frequency Wien-filter), measurement techniques (e.g., optimization of the so-called "spin-coherence time", the "pilot-bunch method", and feedback loops) and—last, but not the least—first experiments with polarized deuteron beams.

3.1. Stage 1: R&D and Measurements at COSY as "Proof-of-Capability"

Over the years, the following achievements have been accomplished at COSY [31–34]:

- Development and implementation of techniques to preserve, manipulate, and observe the polarization of polarized beam bunches in a storage ring (optimization of the spin coherence time, use of a radio-frequency Wien filter, including the pilot-bunch method, design and set-up of an in-beam polarimeter), since they are essential prerequisites for a successful measurement of electric dipole moments of charged particles.

- A Rogowski beam-position monitor has been developed, optimized, installed into COSY, and used in the deuteron EDM (dEDM) measurements.
- A new polarimeter to determine the polarization direction of a circulating polarized beam, based on LYSO scintillators with a silicon photosensor readout, has been developed, optimized, installed into COSY, and used in the dEDM measurements [35].
- A radio-frequency Wien filter has been designed, built, and installed in COSY [36]. In a magnetic (B-field) storage ring such as COSY, the effect of a possible EDM on the particle spin (i.e., on the polarization of an ensemble of polarized particles) will exactly cancel, since the torque due to EDM changes sign—this is why the final EDM storage ring will be an all-electric ring with E-fields only. However, with the concept of such a Wien filter, it is possible in principle to observe the EDM effect (at reduced sensitivity). This device has been exploited for the so-called "precursor experiments" [37] with a polarized deuteron beam.
- In order to study the influence of the RF WF on the polarization (and possibly the orbit), two beam-bunches were injected into the COSY ring, and the RF WF was upgraded such that it could be switched on/off when one of the two bunches passes through the device. After successful tests, this so-called "pilot bunch" method was exploited in the second dEDM precursor experiment.

Scientific Results

- A successful measurement of an oscillating electric dipole moment (limit) for a polarized deuteron beam has been performed in COSY—this was not anticipated originally, but it demonstrates that small changes of the beam polarization can be observed reliably [38]. A preliminary exclusion plot for the size of the oscillating EDM is given in Figure 7.

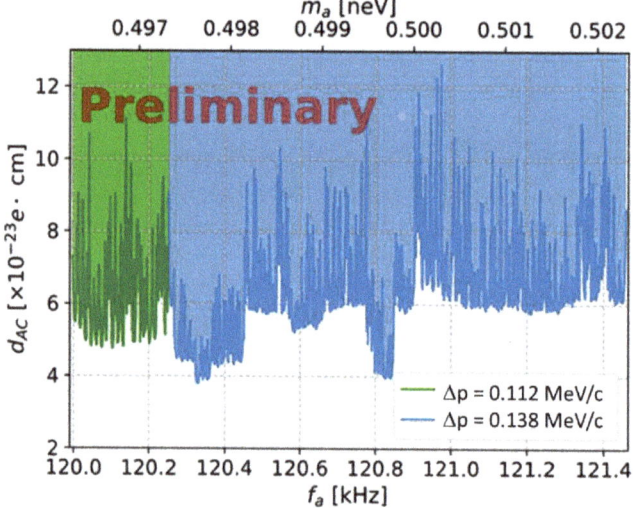

Figure 7. Preliminary result for the 90% confidence level sensitivity for excluding an ALP's induced oscillating EDM (oEDM) in the frequency range 120.0–121.4 kHz (corresponding to a mass range m_a = 0.496–0.502 neV/c^2). A dedicated paper has been submitted [38].

- A proof-of-capability of measuring a limit for a deuteron electric dipole moment in a magnetic storage ring (COSY) has also been achieved; the EDM result has not been finalized up to now due to very complex investigations required to quantify the systematic uncertainty.

COSY has been an indispensable development tool and test device for the storage ring EDM venture. The efforts have been supported by a grant of the European Research Council (ERC AdG srEDM with Principal Investigator H.S., with additional beneficiaries from RWTH Aachen University (Germany), J.P., and the University of Ferrara (Italy), P.L.) from 2016 to 2022. Unfortunately, the operation of COSY will be terminated by the end of 2024 within the context of the transfer of the Institute for Nuclear Physics (IKP) of Forschungszentrum Jülich to the GSI Helmholtz Center Darmstadt, which forfeits the community its current centre-of-gravity.

3.2. Stage 2: A Prototype Storage Ring (PSR) for EDM Searches

COSY is a magnetic storage ring, thus many issues listed above cannot be investigated there. Before starting the construction of the final ring, which is supposed to be one with a circumference of the order of 500 m for a purely electric ring [14], or even larger for a hybrid machine [39,40], and thus constitutes a major investment, members of the JEDI and the CPEDM (Charged Particle EDM) collaboration have concluded that the next step must be the design and construction of a prototype storage ring (PSR) [21]. Because COSY will no longer be available, the PSR cannot rely on the facility, e.g., as a tool for beam preparation and injector.

The PSR should operate at a beam kinetic energy between 30 and 45 MeV in two modes:

1. as an all-electric ring for CW/CCW operation, but not at the magic momentum; and
2. in the "frozen spin" mode after complementing the ring with B-fields; this will allow to perform a first competitive proton (pEDM) experiment with a sensitivity similar to the neutron EDM, i.e., about 10^{-26} e·cm.

Activities have started for a design study of the PSR, including discussions of the host institution as well as its financing and timeline.

3.3. Stage 3: Design and Implementation of the Final Precision Storage Ring

There are suggestions to build a precision storage ring to search for charged-particle EDMs with a sensitivity of the order of 10^{-29} e·cm with the knowledge acquired by now— mainly based on paper work [39,40]. If there is a lesson that we have learned over the past 10 years or so in our experimental investigations of production, preservation, manipulation and detection of polarized beams in a storage ring, it is that one is always facing unexpected challenges. Thus, a final ring which should yield an unprecedented sensitivity must be very carefully approached—also having in mind that the construction cost is significant. The JEDI/CPEDM collaborations are convinced that starting to construct this ring now is premature, and consequently are suggesting to design and build a PSR first. Only after the experience gained in operating the PSR, including a first competitive EDM measurement, the *holy grail* EDM ring should be approached.

4. Summary and Outlook

According to our present understanding, the early Universe contained the same amount of matter and antimatter and, if the Universe had behaved symmetrically as it developed, every particle would have been annihilated by one of its antiparticles. One of the great mysteries is therefore why matter dominates over antimatter in the visible Universe. Furthermore, among the deepest mysteries in natural sciences stands the question of what the Universe is made of. Visible matter, comprising Standard Model particles, accounts only for a small fraction. As inferred from cosmological observations, about 5 times more matter is invisible and is called Dark Matter (DM).

Electric Dipole Moments of particles offer the possibility of providing answers to both enigmas: static EDMs would imply new sources of CP-violation, required for the baryon asymmetry, and oscillating EDMs could be due to axions/ALPs, a prime candidate for DM.

EDM searches with charged particles in storage rings are a new approach, which may be a game-changer due to its projected sensitivity of 10^{-29} e·cm. However, technological and metrological challenges abound and require a staged approach, i.e., a PSR as next step before embarking on the final ring—as outlined in Figure 8.

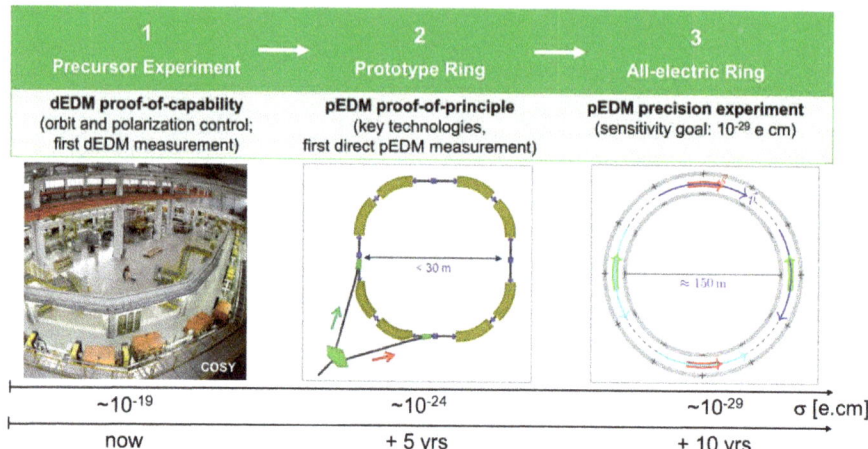

Figure 8. Stages of the strategy for charged particle EDM searches as outlined by the CPEDM collaboration. The EDM sensitivity goals are given for each step. Stage 1 is currently ongoing, while the timeline for stages 2 and 3 is of the order of 5 and 10 years, respectively.

The storage ring EDM enterprise is a long-term project. For further information, we refer the reader to a "CERN Yellow Report" [21], in which the current status of the accumulated knowledge for searches of charge-particle EDMs in storage rings has been summarized. An additional source of detailed information is the web-site given in Supplementary Materials below.

Supplementary Materials: Additional information can be downloaded at: http://collaborations.fz-juelich.de/ikp/jedi/.

Author Contributions: Conceptualization, H.S., S.M.S., P.L. and J.P.; methodology, H.S., P.L. and J.P.; writing—original draft preparation, H.S.; writing—review and editing, S.M.S., P.L. and J.P. All authors have read and agreed to the published version of the manuscript.

Funding: H.S., P.L. and J.P. acknowledge the support of the European Commission via the European Research Council (ERC) by the Advanced Grant "Search for electric dipole moments using storage rings" (ERC-2015-AdG, Grant Agreement number 694340).

Data Availability Statement: The study does not report any finalized data yet.

Acknowledgments: The authors want to acknowledge the long-term cooperation within the JEDI and the CPEDM collaborations, as well as the outstanding support of the scientists and technicians operating the COSY facility.

Conflicts of Interest: The authors declare no conflict of interest.

References

1. Allen, R.E.; Lidström, S. Life, the Universe, and everything—42 fundamental questions. *Phys. Scr.* **2017**, *92*, 12501. [CrossRef]
2. Alarcon, R.; Alexander, J.; Anastassopoulos, V.; Aoki, T.; Baartman, R.; Baeßler, S.; Bartoszek, L.; Beck, D.H.; Bedeschi, F.; Berger, R.; et al. Electric dipole moments and the search for new physics. *arXiv* **2022**, arXiv:2203.08103. [CrossRef]
3. Canetti, L.; Drewes, M.; Shaposhnikov, M. Matter and Antimatter in the Universe. *New J. Phys.* **2012**, *14*, 95012. [CrossRef]
4. Chadha-Day, F.; Ellis, J.; Marsh, D.J. Axion dark matter: What is it and why now? *arXiv* **2021**, arXiv:2105.01406v2. [CrossRef] [PubMed]
5. Boveia, A.; Berkat, M.; Chen, T.Y.; Desai, A.; Doglioni, C.; Drlica-Wagner, A.; Gardner, S.; Gori, S.; Greaves, J.; Harding, P.; et al. Snowmass 2021 Dark Matter Complementarity Report. *arXiv* **2022**, arXiv:2211.07027v2. 17. [CrossRef]

6. Ramsey, N.F. Electric-dipole moments of elementary particles. *Rep. Prog. Phys.* **1982**, *45*, 95. [CrossRef]
7. Leach, P.; Paliathanasis, A. (Eds.) *Noether's Theorem and Symmetry*; MDPI: Basel, Switzerland, 2020; ISBN 978-3-03928-234-0/978-3-03928-235-7. [CrossRef]
8. Blum, A.S.; de Velasco, A. The genesis of the CPT theorem. *Eur. Phys. J. H* **2022**, *47*, 5. [CrossRef]
9. Roussy, T.S.; Caldwell, L.; Wright, T.; Cairncross, W.B.; Shagam, Y.; Ng, K.B.; Schlossberger, N.; Park, S.Y.; Wang, A.; Ye, J.; et al. A new bound on the electron's electric dipole moment. *arXiv* **2022**, arXiv:2212.11841v3. [CrossRef]
10. Abel, C.; Afach, S.; Ayres, N.J.; Baker, C.A.; Ban, G.; Bison, G.; Bodek, K.; Bondar, V.; Burghoff, M.; Chanel, E.; et al. Measurement of the Permanent Electric Dipole Moment of the Neutron. *Phys. Rev. Lett.* **2020**, *124*, 81803. [CrossRef]
11. Peccei, R.D. *The Strong CP Problem and Axions*; Lecture Notes in Physics; Springer: Berlin/Heidelberg, Germany, 2008; Volume 741, pp. 3–17. [CrossRef]
12. Ringwald, A. Axions and Axion-Like Particles. *arXiv* **2014**, arXiv:1407.0546v1. [CrossRef]
13. Kim, O.; Semertzidis, Y.K. New method of probing an oscillating EDM induced by axionlike dark matter using an RF Wien Filter in storage rings. *Phys. Rev. D* **2021**, *104*, 96006. [CrossRef]
14. Anastassopoulos, V.; Andrianov, S.; Baartman, R.; Baessler, S.; Bai, M.; Benante, J.; Berz, M.; Blaskiewicz, M.; Bowcock, T.; Brown, K.; et al. A storage ring experiment to detect a proton electric dipole moment. *Rev. Sci. Instr.* **2016**, *87*, 115116. [CrossRef] [PubMed]
15. Sakharov, A.D. Violation of CP invariance, C asymmetry, and baryon asymmetry of the universe. *Sov. Phys. Usp.* **1991**, *34*, 392. [CrossRef]
16. Mohanmurthy, P.; Winger, J.A. Estimation of CP violating EDMs from known mechanisms in the SM. *arXiv* **2020**, arXiv:2009.00852v3. 66. [CrossRef]
17. Fukuyama, T. Searching for New Physics beyond the Standard Model in Electric Dipole Moment. *Int. J. Mod. Phys. A* **2012**, *27*, 1230015. [CrossRef]
18. Kirch, K.; Schmidt-Wellenburg, P. Search for electric dipole moments. *EPJ Web Conf.* **2020**, *234*, 1007. [CrossRef]
19. Ema, Y.; Gao, T.; Pospelov, M. Improved indirect limits on muon EDM. *arXiv* **2021**, arXiv:2108.05398.
20. Bernreuther, W.; Chen, L.; Nachtmann, O. Electric dipole moment of the tau lepton revisited. *Phys. Rev. D* **2021**, *103*, 96011. [CrossRef]
21. Abusaif, F.; Aggarwal, A.; Aksentev, A.; Alberdi-Esuain, B.; Andres, A.; Atanasov, A.; Barion, L.; Basile, S.; Berz, C.; Böhme, C.; et al. *Storage Ring to Search for Electric Dipole Moments of Charged Particles: Feasibility Study*; CERN Yellow Reports: Monographs; CERN: Meyrin, Switzerland, 2021; Volume 3. [CrossRef]
22. Trimble, V. Existence and Nature of Dark Matter in the Universe. *Ann. Rev. Astron. Astrophys.* **1987**, *25*, 425–472. [CrossRef]
23. Zwicky, F. Die Rotverschiebung von extragalaktischen Nebeln. *Helv. Phys. Acta* **1933**, *6*, 110–127.
24. Johnson, J., Jr. *Zwicky: The Outcast Genius Who Unmasked the Universe*; Harvard University Press: Cambridge, MA, USA, 2019; ISBN 9780674979673.
25. Bahcall, N.A. Dark matter universe. *Proc. Natl. Acad. Sci. USA* **2015**, *112*, 12243–12245. [CrossRef] [PubMed]
26. Rajendran, S. New Directions in the Search for Dark Matter. *arXiv* **2022**, arXiv:2204.03085v1.84. [CrossRef]
27. Chupp, T.E.; Fierlinger, P.; Ramsey-Musolf, M.J.; Singh, J.T. Electric dipole moments of atoms, molecules, nuclei, and particles. *Rev. Mod. Phys.* **2019**, *91*, 15001. [CrossRef]
28. Lee, S.Y. The Thomas–BMT equation. In *Spin Dynamics and Snakes in Synchrotrons*; World Scientific: Singapore, 1997; pp. 9–24.
29. Fukuyama, T.; Silenko, A.J. Derivation of Generalized Thomas-Bargmann-Michel-Telegdi Equation for a Particle with Electric Dipole Moment. *Int. J. Mod. Phys. A* **2013**, *28*, 1350147. [CrossRef]
30. Wilkin, C. The legacy of the experimental hadron physics programme at COSY. *Eur. Phys. J. A* **2017**, *53*, 114. [CrossRef]
31. Annual Report 2018, Institut für Kernphysik COSY, Jül-4418, ISSN 0944-2952 (Forschungszentrum Jülich). pp. 15–21. Available online: https://www.fz-juelich.de/de/ikp/service/downloads (accessed on 25 February 2023).
32. Annual Report 2019, Institut für Kernphysik COSY, Jül-4423, ISSN 0944-2952 (Forschungszentrum Jülich). pp. 8–13. Available online: https://www.fz-juelich.de/de/ikp/service/downloads (accessed on 25 February 2023).
33. Annual Report 2020, Institut für Kernphysik COSY, Jül-4427, ISSN 0944-2952 (Forschungszentrum Jülich). pp. 7–13. Available online: https://www.fz-juelich.de/de/ikp/service/downloads (accessed on 25 February 2023).
34. Annual Report 2021, Institut für Kernphysik COSY, Jül-4429, ISSN 0944-2952 (Forschungszentrum Jülich). pp. 6–11. Available online: https://www.fz-juelich.de/de/ikp/service/downloads (accessed on 25 February 2023).
35. Müller, F.; Javakhishvili, O.; Shergelashvili, D.; Keshelashvili, I.; Mchedlishvili, D.; Abusaif, F.; Aggarwal, A.; Barion, L.; Basile, S.; Böker, J.; et al. A new beam polarimeter at COSY to search for electric dipole moments of charged particles. *J. Instrum.* **2020**, *15*, P12005. [CrossRef]
36. Slim, J.; Nikolaev, N.N.; Rathmann, F.; Wirzba, A.; Nass, A.; Hejny, V.; Pretz, J.; Soltner, H.; Abusaif, F.; Aggarwal, A.; et al. (JEDI Collaboration), First detection of collective oscillations of a stored deuteron beam with an amplitude close to the quantum limit. *Phys. Rev. Accel. Beams* **2021**, *24*, 124601. [CrossRef]
37. Lehrach, A.; Lorentz, B.; Morse, W.; Nikolaev, N.; Rathmann, F. Precursor Experiments to Search for Permanent Electric Dipole Moments (EDMs) of Protons and Deuterons at COSY. *arXiv* **2012**, arXiv:1201.5773.

38. Karanth, S.; Stephenson, E.J.; Chang, S.P.; Hejny, V.; Park, S.; Pretz, J.; Semertzidis, Y.; Wrońska, A.; Abusaif, F.; Aksentev, A.; et al. First Search for Axion-Like Particles in a Storage Ring Using a Polarized Deuteron Beam. *arXiv* **2022**, arXiv:2208.07293v2. [CrossRef]
39. Hacıömeroğlu, S.; Semertzidis, Y.K. Hybrid ring design in the storage-ring proton electric dipole moment experiment. *Phys. Rev. Accel. Beams* **2019**, *22*, 34001. [CrossRef]
40. Omarov, Z.; Davoudiasl, H.; Hacıömeroğlu, S.; Lebedev, V.; Morse, W.M.; Semertzidis, Y.K.; Silenko, A.J.; Stephenson, E.J.; Suleiman, R. Comprehensive symmetric-hybrid ring design for a proton EDM experiment at below 10^{-29} e·cm. *Phys. Rev. D* **2022**, *105*, 32001. [CrossRef]

Disclaimer/Publisher's Note: The statements, opinions and data contained in all publications are solely those of the individual author(s) and contributor(s) and not of MDPI and/or the editor(s). MDPI and/or the editor(s) disclaim responsibility for any injury to people or property resulting from any ideas, methods, instructions or products referred to in the content.

Article

Masses of Compact (Neutron) Stars with Distinguished Cores

Rico Zöllner [1,*], Minghui Ding [2] and Burkhard Kämpfer [2,3]

[1] Institut für Technische Logistik und Arbeitssysteme, TU Dresden, 01062 Dresden, Germany
[2] Helmholtz-Zentrum Dresden-Rossendorf, 01314 Dresden, Germany
[3] Institut für Theoretische Physik, TU Dresden, 01062 Dresden, Germany
* Correspondence: rico.zoellner@tu-dresden.de

Abstract: In this paper, the impact of core mass on the compact/neutron-star mass-radius relation is studied. Besides the mass, the core is parameterized by its radius and surface pressure, which supports the outside one-component Standard Model (SM) matter. The core may accommodate SM matter with unspecified (or poorly known) equation-of-state or several components, e.g., consisting of admixtures of Dark Matter and/or Mirror World matter etc. beyond the SM. Thus, the admissible range of masses and radii of compact stars can be considerably extended.

Keywords: compact stars; core-corona decomposition; impact of core mass; Dark-Matter admixture

1. Introduction

Strong interaction rules a variety of systems, ranging from hadrons to nuclei up to neutron stars. The related mass scales are typically $\geq m_\pi = 0.13$ GeV for mesons, $\geq m_p = 0.938$ GeV for baryons, $(1 \cdots 250) m_p$ for nuclei, and $\mathcal{O}(10^{57}) m_p$ for neutron stars. ($m_{\pi,p}$ stand for pion and proton masses.) Further systems awaiting their confirmation are glueballs ($\mathcal{O}(\text{GeV})$) and non-baryon stars, such as pion stars [1]. Besides weak interaction, it is the long-range Coulomb interaction that limits the size (or baryon number) of nuclei, and gravity is the binding force of matter in neutron stars. The phenomenon of hadron mass emergence is a fundamental issue tightly related to non-perturbative effects in the realm of QCD, cf. Refs. [2,3] and citations therein. Once the masses and interactions among hadrons are understood, one can make the journey to address the masses and binding energies of nuclei and then jump to constituents of neutron stars. Despite the notion, cool neutron stars accommodate, in the crust, various nuclei immersed in a degenerate electron-muon environment (maybe as "pasta" or "spaghetti" or crystalline medium). In the deeper interior, above the neutron drip density, neutrons and light clusters begin to dominate the matter composition. These constituents and their interactions govern the mass (or energy density) of the medium. At nuclear saturation density, $n_0 \approx 0.15 \, \text{fm}^{-3}$, one meets conditions similar to the interior of heavy nuclei but with crucial impact of the symmetry energy when extrapolating from nuclear matter with comparable proton and neutron numbers to a very asymmetric proton-neutron mixture. Above saturation density, various effects hamper a reliable computation of properties of the strong-interaction medium: three-body interactions may become even more important than at n_0, and further baryon species become excited, e.g., strangeness is lifted from vacuum into baryons forming hyperons whose interaction could be a miracle w.r.t. the hyperon puzzle [4], and the relevant degrees of freedom become relativistic. Eventually, at asymptotically high density, the strong-interaction medium is converted into quarks and gluons; color-flavor locking and color superconductivity can essentially determine the medium's properties. The turn of massive hadronic degrees of freedom into quark-gluon excitations is thereby particularly challenging.

While lattice QCD represents, in principle, an *ab initio* approach to strong-interaction systems in all their facets, the "sign problem" prevents the access to non-zero baryon

Citation: Zöllner, R.; Ding, M.; Kämpfer, B. Masses of Compact (Neutron) stars with Distinguished Cores. *Particles* **2023**, *6*, 217–238. https://doi.org/10.3390/particles6010012

Academic Editor: Armen Sedrakian

Received: 17 December 2022
Revised: 30 January 2023
Accepted: 31 January 2023
Published: 2 February 2023

Copyright: © 2023 by the authors. Licensee MDPI, Basel, Switzerland. This article is an open access article distributed under the terms and conditions of the Creative Commons Attribution (CC BY) license (https://creativecommons.org/licenses/by/4.0/).

number systems. Thus, the exploration of compact/neutron stars, in particular their possible mass range, by stand-alone theory is presently not feasible. Instead, an intimate connection of astrophysical data and compact-star modeling is required.

The advent of detecting gravitational waves from merging neutron stars, the related multimessenger astrophysics [5–12] and the improving mass-radius determinations of neutron stars, in particular by NICER data [13–17], stimulated a wealth of activities. Besides masses and radii, moments of inertia and tidal deformabilities become experimentally accessible and can be confronted with theoretical models [18–24]. The baseline of the latter ones is provided by non-rotating, spherically symmetric cold dense matter configurations. The sequence of white dwarfs (first island of stability) and neutron stars (second island of stability) and possibly [25] a third island of stability [26–30] shows up when going to more compact objects, with details depending sensitively on the actual equation of state (EoS). The quest for a fourth island has been addressed as well [31,32]. "Stability" means here the damping of radial disturbances, at least. Since the radii of configurations of the second (neutron stars) and third (hypothetical quark/hybrid stars [33–40]) islands are very similar, the notion of twin stars [31–33,41,42] has been coined for equal-mass configurations; "masquerade" was another related term [43].

We emphasize the relation of ultra-relativistic heavy-ion collision physics, probing the EoS $p(T, \mu_B \approx 0)$, and compact star physics, probing $p(T \approx 0, \mu_B)$ when focusing on static compact-star properties [44,45]. Of course, in binary or ternary compact-star merging-events, also finite temperatures T and a large range of baryon-chemical potential μ_B are probed, which are accessible in medium-energy heavy-ion collisions [46]. Implications of the conjecture of a first-order phase transition at small temperatures and large baryo-chemical potentials or densities [47–50] can also be studied by neutron-hybrid-quark stars [51–54]. It is known since some time [26–28,55,56] that a cold EoS with special pressure-energy density relation $p(e)$, e.g., a strong local softening up to first-order phase transition with a density jump, can give rise to a "third family" of compact stars, beyond white dwarfs and neutron stars. In special cases, the third-family stars appear as twins of neutron stars [43,57,58]. Various scenarios of the transition dynamics to the denser configuration as mini-supernova have also been discussed quite early [59,60].

While the Standard Model (SM) of particle physics seems to accommodate nearly all of the observed phenomena of the micro-world, severe issues remain. Among them is the $(g-2)_\mu$ puzzle or the proton's charge radius. Another fundamental problem is the very nature of Dark Matter (DM): Astrophysical and cosmological observations seem to require inevitably its existence, but details remain elusive despite many concerted attempts, e.g., Refs. [61–64]. Supposing that DM behaves like massive particles, it could be captured gravitationally in the centers of compact stars [65–67], thus providing a non-SM component there. This would be an uncertainty on top of the less reliably known SM-matter state. Beyond the SM, other feebly interacting particles could also populate compact stars. A candidate scenario is provided, for instance, by Mirror World (MW) [68–72], i.e., a parity-symmetric complement to our SM-world with very tiny non-gravity interaction. There are many proposals of portals from our SM-world to such beyond-SM scenarios, cf. Ref. [73].

Guided by these remarks we follow here an access to static cold compact stars already launched in [74]: We describe the core by a minimum of parameters and determine the resulting compact-star masses and radii by assuming the knowledge of the equation of state of the SM-matter enveloping the core. A motivation is the quest of a mass gap between compact stars and black holes.

Our paper is organized as follows. In Section 2 we recall the Tolman–Oppenheimer–Volkoff equations, their scaling property, and introduce our core-corona decomposition. Small cores with and without MW/DM admixtures are considered in Section 3. Section 4 is devoted to the core-corona decomposition, where a specific EoS is deployed for the explicit construction. We summarize in Section 5. We supplement our paper in Appendix A by a brief retreat to the emergence of hadron masses as a key issue in understanding the typical scales of compact (neutron) star masses. Appendix B sketches a complementary approach:

the construction of an EoS by holographic means, thus transporting information of a hot (quark-gluon) QCD EoS to a cool EoS and connecting heavy-ion collisions and compact star physics. These appendices survey (Appendix A) and exemplify (Appendix B) symmetry and the governing principles where the access to astrophysical objects is based upon.

2. One-Component Static Cool Compact Stars: TOV Equations

The standard modeling of compact star configurations is based on the Tolman–Oppenheimer–Volkoff (TOV) equations

$$\frac{dp}{dr} = -G_N \frac{[e(r) + p(r)][m(r) + 4\pi r^3 p(r)]}{r^2[1 - \frac{2G_N m(r)}{r}]}, \tag{1}$$

$$\frac{dm}{dr} = 4\pi r^2 e(r), \tag{2}$$

resulting from the energy-momentum tensor of a one-component static isotropic fluid (described locally by pressure p and energy density e as only quantities relevant for the medium) and spherical symmetry of both space-time and matter, within the framework of Einstein gravity without cosmological term [75]. Newton's constant is denoted by G_N, and natural units with $c = 1$ are used, unless when relating mass and length and energy density, where $\hbar c$ is needed.

2.1. Scaling of TOV Equations and Compact/Neutron Star Masses and Radii

The TOV equations become free of any dimension by the scalings

$$e = \mathfrak{s}\bar{e}, \quad p = \mathfrak{s}\bar{p}, \quad r = (G_N\mathfrak{s})^{-1/2}\bar{r}, \quad m = (G_N^3\mathfrak{s})^{-1/2}\bar{m}, \tag{3}$$

where $\mathfrak{s}$ is a mass dimension-four quantity or has dimension of energy density. It may be a critical pressure of a phase transition or a limiting energy density, e.g., at the boundary matter-vacuum [76]. Splitting up $\mathfrak{s} = nm_p$ into a number density n and the energy scale m_p, one gets $(G_N\mathfrak{s})^{-1/2} = 7$ km$/\sqrt{10^{-2}n/n_0}$ and $(G_N^3\mathfrak{s})^{-1/2} = 4.8 M_\odot/\sqrt{10^{-2}n/n_0}$, where the nuclear saturation density $n_0 = 0.15$ fm^{-3} is used as reference density. The scales $m_{\pi,p}$ facilitate the densities $n = m_\pi^3 \to 2.3 n_0$ and $n = m_p^3 \to 833 n_0$. That is, the scale solely set by the nucleon mass, i.e., $\mathfrak{s} = m_p^4$ [77], yields $(G_N\mathfrak{s})^{-1/2} = 2.74$ km and $(G_N^3\mathfrak{s})^{-1/2} = 1.86 M_\odot$, suggesting that the nucleon mass (see Appendix A on its emergence in QCD) determines the gross properties of neutron stars, such as mass and radius (modulus factor 2π) in an order-of magnitude estimate. In contrast, the density estimate via $M = \frac{4\pi}{3}\langle n\rangle m_p R^3$ yields, independently of G_N, $\langle n\rangle = 2n_0 \frac{M}{M_\odot}\frac{10^3}{(R/\text{km})^3} \approx 4n_0$ for $M = 2M_\odot$ and $R = 10$ km, thus pointing to the importance of the actual numerical values of $\bar{r}$ and $\bar{m}$.

In fact, recent measurements and supplementary work report averaged mean and individual heavy compact/neutron stars masses and radii (often on 67% credible level) as follows

PSR	M [$M_\odot$]	R [km]
	1.4	$11.94^{+0.76}_{-0.87}$ (1) [12], 12.45 ± 0.65 [13], $12.33^{+0.76}_{-0.81}$ [15]
J0030+0451	$1.34^{+0.15}_{-0.16}$	$12.71^{+1.14}_{-1.19}$ [16]
	$1.44^{+0.15}_{-0.14}$	$13.02^{+1.24}_{-1.06}$ [14], $12.18^{+0.56}_{-0.79}$ [17]
J1614–2230	1.908 ± 0.016 [78]	
J0348+0432	2.01 ± 0.04 [79]	
J0740+6620	$2.072^{+0.067}_{-0.066}$	$12.39^{+1.30}_{-0.98}$ [15]
	2.08 ± 0.07 [80]	$13.7^{+2.6}_{-1.5}$ (2) [13], $11.96^{+0.86}_{-0.81}$ [12]
0952-0607 (3)	2.35 ± 0.17 [81]	

(1) 90% confidence. (2) With nuclear physics constraints at low density and gravitational radiation data from GW170817 added in, the inferred radius drops to (12.35 ± 0.75) km [13]. (3) Black-widow binary pulsar PSR 0952-0607.

One has to add GW190814: gravitational waves from the coalescence of a 23 solar mass black hole with a $2.6 M_\odot$ compact object [82]. An intriguing question concerns a possible mass gap between compact-star maximum-mass [83] and light black holes, cf., [84,85].

2.2. Solving TOV Equations

Given a unique relationship of pressure p and energy density e as EoS $e(p)$, in particular at zero temperature, the TOV equations are integrated customarily with boundary conditions $p(r=0) = p_c$ and $m(r=0) = 0$ (implying $p(r) = p_c - \mathcal{O}(r^2)$ and $m(r) = 0 + \mathcal{O}(r^3)$ at small radii r), and $p(R) = 0$ and $m(R) = M$ with R as circumferential radius and M as gravitational mass (acting as parameter in the external (vacuum) Schwarzschild solution at $r > R$). The quantity p_c is the central pressure. The solutions $R(p_c)$ and $M(p_c)$ provide the mass-radius relation $M(R)$ in parametric form.

A great deal of effort is presently concerned about the EoS at supra-nuclear densities [86]. For instance, Figure 1 in Ref. [87] exhibits the currently admitted uncertainty: up to a factor of 10 in pressure as a function of energy density. At asymptotically large energy density, perturbative QCD constrains the EoS, though it is just the non-asymptotic supra-nuclear density region that crucially determines the maximum mass and whether twin stars may exist or quark-matter cores appear in neutron stars. Accordingly, one can fill this gap by a big number (e.g., millions [88]) of test EoSs to scan through the possibly resulting manifold of mass-radius curves, see Refs. [89–92]. However, the possibility that neutron stars may accommodate other components than Standard Model matter, e.g., exotic material as Dark Matter [93–96], can be an obstacle for the safe theoretical modeling of a concise mass-radius relation in such a manner. Of course, inverting the posed problem with sufficiently precise data of masses and radii as input offers a promising avenue towards determining the EoS [17,89,97–101].

Here, we pursue another perspective. We parameterize the supra-nuclear core by a radius r_x and the included mass m_x and integrate the above TOV equations only within the corona (our notion "corona" is a synonym for "mantel" or "crust" or "envelope" or "shell", it refers to the complete part of the compact star outside the core, $r_x \leq r \leq R$), i.e., from pressure p_x to the surface, where $p = 0$. This yields the total mass $M(r_x, m_x; p_x)$ and the total radius $R(r_x, m_x; p_x)$ by assuming that the corona EoS $e(p)$ is reliably known at $p \leq p_x$ and that only SM matter occupies that region. Clearly, without knowledge of the matter composition at $p > p_x$ (may it be SM matter with an uncertainly known EoS or may it contain a Dark-Matter admixture, for instance, or monopoles or some other type of "exotic" matter) one does not get a simple mass-radius relation by such a procedure, but admissible area(s) over the mass-radius plane, depending on the core parameters r_x and m_x and the matching pressure p_x and related energy density e_x. This is the price of avoiding a special model of the core matter composition.

If the core is occupied by a one-component SM medium, the region $p > p_x$ and $e > e_x$ can be mapped out by many test EoSs which locally obey the constraint $v_s^2 \in [0, 1]$ to obtain the corresponding region in the mass-radius plane, cf., Figure 2 in Ref. [102] for an example processed by Bayesian inference. This is equivalent, to some extent, to our core-corona decomposition for SM matter-only.

3. Small-Core Approximation and Beyond
3.1. One-Component Core

For small one-component distinguished cores one can utilize the EoS parameterization from a truncated Taylor expansion of $p(e) \geq p_x$ at λe_x, $p(e) = p(\lambda e_x) + \frac{\partial p}{\partial e}|_{\lambda e_x}(e - \lambda e_x) + \cdots$,

$$p(e) = p_x + v_s^2(e - \lambda e_x), \qquad (4)$$

where $\lambda > 1$ is a density jump at the core boundary and $\lambda = 1$ continuously continues (but may be kinky) the corona EoS at $p \geq p_x$; p_x and e_x mark the "end point" of the corona EoS. A Taylor expansion for small cores,

$$p(r) = \sum_{i=0}^{\infty} p_{2i} r^{2i}, \tag{5}$$

$$m(r) = \sum_{i=0}^{\infty} m_{2i+1} r^{2i+1}, \tag{6}$$

gives by means of Equations (1) and (2) with (4) (that is $e = \lambda e_x - v_s^{-2} p_x + v_s^{-2} \sum_{i=0}^{\infty} p_{2i} r^{2i}$) and $p_{2i+1} = 0$ and $m_{2i} = 0$ for all $i \in \mathbb{N}_0$

$$m_1 = 0, \; m_3 = \frac{4\pi}{3} e_c, \; m_{2i+1} = \frac{4\pi v_s^{-2}}{2i+1} p_{2i-2} \text{ for } i \geq 2, \tag{7}$$

where $e_c \equiv e(p_c) = \lambda e_x + v_s^{-2}(p_c - p_x)$, and the recurrence

$$p_0 = p_c, \quad p_2 = -\frac{2\pi}{3} G_N (e_c + p_c)(e_c + 3p_c), \tag{8}$$

$$p_{2i} = \frac{G_N}{2i} \left(2A_i - [\lambda e_x - v_s^{-2} p_x](m_{2i+1} + 4\pi p_{2i-2}) - (1 + v_s^{-2}) B_{i-1} \right) \text{ for } i \geq 1, \tag{9}$$

$$A_i = \sum_{j=0}^{i} 2(i-j) m_{2j} p_{2i-2j}, \tag{9a}$$

$$B_i = \sum_{j=0}^{i} (m_{2j+3} + 4\pi p_{2j}) p_{2i-2j}. \tag{9b}$$

In leading order one obtains

$$r_x \approx \frac{\delta^{1/2}(1 - \mathcal{O}(\delta))}{\sqrt{\frac{2\pi}{3} G_N p_x W}} \approx \frac{60.1 \text{ km}}{\sqrt{W}} \sqrt{\delta \frac{100 \text{ MeV/fm}^3}{p_x}}, \tag{10}$$

$$m_x \approx 2\lambda \frac{e_x}{p_x} \frac{\delta^{3/2}(1 - \mathcal{O}(\delta))}{\sqrt{\frac{2\pi}{3} G_N^3 p_x W^3}} \approx \frac{81.2 M_\odot}{W^{3/2}} \lambda \frac{e_x}{100 \text{ MeV/fm}^3} \times \left(\delta \frac{100 \text{ MeV/fm}^3}{p_x} \right)^{3/2}, \tag{11}$$

where $\delta := p_c / p_x - 1$ and $W := 3 + 4\lambda \frac{e_x}{p_x} + \lambda^2 \left(\frac{e_x}{p_x} \right)^2$. The scale setting is by p_x and $\lambda e_x / p_x$. The core-mass–core-radius relation is $m_x(\delta) \approx \frac{4\pi}{3} p_x \lambda \left(\frac{e_x}{p_x} \right) r_x(\delta)^3 (1 + \mathcal{O}(\delta))$.

To control and extend the above approximations we numerically solve the scaled TOV equations by assuming that Equation (4) holds true in the core. (Of course, this is an ad hoc assumption aimed at providing an explicit example of mass-radius relations of the core.) The core-corona matching is at p_x, i.e., the maximum (minimum) pressure of the corona (core) EoS. The related energy density is $e_x = 3p_x / (1 - 3\Delta^{corona})$, where Δ^{corona} denotes the trace anomaly measure discussed below in Equation (12). It is used here for a suitable parameterization of the double (e_x, p_x) by means of the corona EoS. The core EoS Equation (4), $e(p) = \lambda e_x + v_s^{-2}(p - p_x)$, enters, after scaling according to Equation (3), the dimensionless TOV equations

$$\frac{d\bar{p}}{d\bar{r}} = -\frac{[\bar{e}(\bar{r}) + \bar{p}(\bar{r})][\bar{m}(\bar{r}) + 4\pi \bar{r}^3 \bar{p}(\bar{r})]}{\bar{r}^2 [1 - \frac{2\bar{m}(\bar{r})}{\bar{r}}]}, \tag{11a}$$

$$\frac{d\bar{m}}{d\bar{r}} = 4\pi \bar{r}^2 \bar{e}(\bar{r}), \tag{11b}$$

for $\bar{p} \in [\bar{p}_c, \bar{p}_x]$ to get $\bar{r}_x(\bar{p}_c)$ and $\bar{m}_x(\bar{p}_c)$. The corresponding scaled core mass vs. core radius relations are exhibited in Figure 1, where the scaling quantity $\mathfrak{s} = p_x$ is employed, i.e., $\bar{p}_x = 1$. The figure offers a glimpse on the systematic of the parameter dependence. Note that it applies to all values of $p_x > 0$. The finite pressure and energy density at the core boundary facilitates a pattern of mass-relations and dependence on sound velocity as known from bag model EoS, cf., Figure 7 in Ref. [76]. $\lambda > 1$ causes an overall shrinking of the pattern plus a slight up-shift; see the right panel for $\lambda = 1.5$. Considering, e.g., $e_x \in [150, 1500]$ MeV/fm^3 and the scaling $\propto 1/\sqrt{e_x}$, cf., (3), the core masses and radii change by a factor up to three, depending on the actual value of e_x.

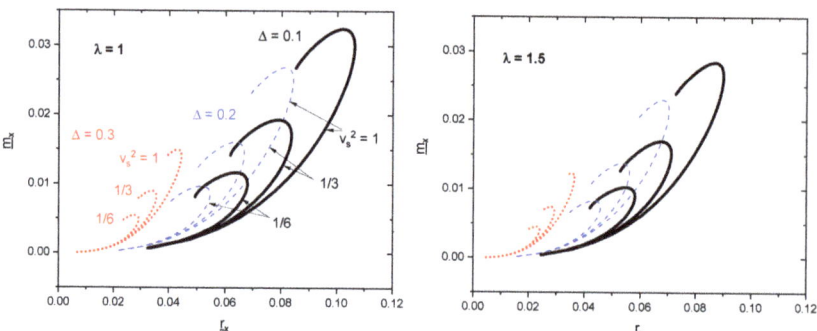

Figure 1. Scaled core masses $\bar{m}_x$ as a function of the scaled core radii $\bar{r}_x$ for various values of $\Delta^{corona} = \Delta$ and v_s^2 for a one-component medium with EoS (4). The **left** (**right**) panel is for $\lambda = 1$ (1.5). The scaling quantity is $\mathfrak{s} = p_x$. For central pressures $\bar{p}_c = 1.1^n$, $n = 1 \cdots 55$. The displayed curves are limited by $\bar{m}_x < \bar{r}_x/2$ (black hole limit) and $\bar{m}_x > \frac{4\pi}{3} \lambda \frac{3}{1-3\Delta^{corona}} \bar{r}_x^3$. The latter expression is for the respective asymptotic curve in the small-$\bar{r}_x$ region. To convert to usual dimensions, one employs $r_x = \bar{r}_x \frac{86.9 \text{ km}}{\sqrt{p_x/100 \text{ [MeV/fm}^3\text{]}}}$ and $m_x = \bar{m}_x \frac{58.8 \, M_\odot}{\sqrt{p_x/100 \text{ [MeV/fm}^3\text{]}}}$, where "$p_x/100 \text{ [MeV/fm}^3\text{]}$" denotes the scaling pressure p_x in units of 100 MeV/fm^3. The approximations (10) and (11) apply only in the small-$\bar{m}_x$ and small-$\bar{r}_x$ regions.

3.2. Multi-Component Cores

In a multi-component medium (note that our considerations apply to components which do not mutually interact in a direct microscopical manner, but are coupled solely by the common gravity field) with known EoSs one has to add the contributions by $[m(r) + 4\pi r^3 p(r)] \to \sum_i [m_{(i)}(r) + 4\pi r^3 p_{(i)}(r)]$ and $[1 - \frac{2G_N m(r)}{r}] \to [1 - \frac{2G_N \sum_i m_{(i)}(r)}{r}]$ and solve in parallel the multitude of TOV equations for the components i. A particular minimum-parameter model is Mirror World matter (component $i = 2$) which is completely symmetric to SM matter (component $i = 1$), i.e., $p_{(1)}(r) = p_{(2)}(r)$ etc., turning $\sum_{(i)} \to 2$. The results are exhibited in Figure 2, left. The increased energy density by the two superimposed fluids cause a shrinking of the core-mass–core-radius curves similar to the increase of λ in Figure 1. The extension to three components is exhibited in Figure 2, right. Here, $\sum_{(i)} \to 3$ applies. Due to further increased energy density and pressure, the configurations become even more "compact", i.e., core masses and core radii shrink further on.

One may quantify the core compactness by $\bar{C}_x := 2\bar{m}_x^{max}/\bar{r}_x|_{\bar{m}_x^{max}}$ and note that, within the scanned parameter patch, it (i) is independent of the number of fluids, (ii) increases slightly with Δ^{corona} at $v_s^2 = const$ and (iii) decreases with decreasing $v_s^2 = const$ at $\Delta^{corona} = const$, see Table 1. Note that the usual definition of compactness is without the factor of two.

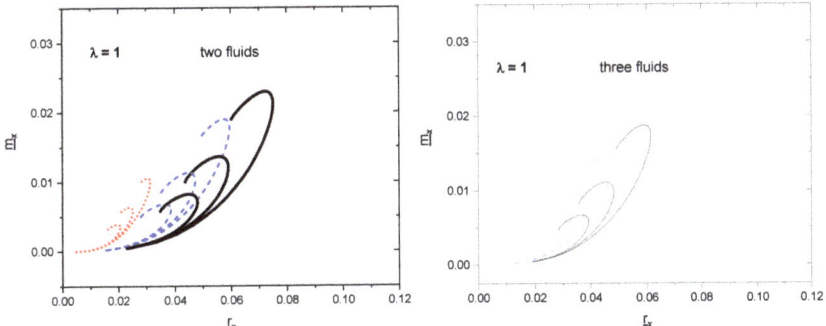

Figure 2. **Left** panel: As the left panel of Figure 1, but for a two-component medium of SM + MW matter with $p_{(1)}(r) = p_{(2)}(r)$ and EoS (4) for both components. Note the difference to the two-fluid core-shell construction in Ref. [103]. **Right** panel: As a left panel but for a three-component medium.

Table 1. Core compactness $\bar{C}_x = 2\bar{m}_x^{max}/\bar{r}_x|_{\bar{m}_x^{max}}$ for various values of v_s^2 (sound velocity squared in the core) and $\Delta^{corona} = \Delta$. For $\lambda = 1$.

v_s^2 \ Δ	0.1	0.2	0.3
1	0.64	0.67	0.70
1/3	0.49	0.51	0.53
1/6	0.37	0.38	0.40

Of course, more general is an asymmetric Mirror World component facilitating $p_{(1)}(r) \neq p_{(2)}(r)$, especially $p_{(1)}(r = 0) \neq p_{(2)}(r = 0)$, even for the common EoS (4), dealt with in Appendix A in Ref. [74]. We do not dive into various conceivable scenarios here and refer the interested reader to Refs. [103,104].

4. Core-Corona Decomposition with NYΔ DD-EM2 EoS

4.1. Trace Anomaly

An example of an EoS largely compatible with neutron star data is NYΔ based on the DD-ME2 density functional [31]. Despite some peculiarities (see the left panel in Figure 3), it shares features recently advocated as essential w.r.t. QCD tracy anomaly and conformality [105,106]. A suitable measure of the trace anomaly is

$$\Delta := \frac{1}{3} - \frac{p}{e} \tag{12}$$

which is related to the sound velocity squared

$$v_s^2 := \frac{\partial p}{\partial e} = \frac{n_B}{\mu_b \chi_B} = \frac{1}{3} - \Delta - e\frac{\partial \Delta}{\partial e}, \tag{13}$$

where n_B stands for the baryon density, μ_B the baryo-chemical potential, and $\chi_B = \partial^2 p/\partial \mu_B^2 = \partial n_B/\partial \mu_B$ is the second-order cumulant of the net-baryon number density. Thermodynamic stability and causality constrain $\Delta \in [-2/3, 1/3]$. Restoration of scale invariance means $\Delta \to 0$ and $v_s^2 \to 1/3$. Although Δ approaches monotonically to zero, v_s^2 can develop a peak at lower energy densities. In fact, v_s^2 as a function of $\eta \equiv \ln e/(n_0 m_p)$ displays such a peak at $\eta \approx 1.3$ (see Figure 2 in Ref. [105] or Figure 1 in Ref. [88] and discussion in Ref. [107]), which should be considered a signature of conformality even at strong coupling. Continuing with the adjustment of Δ beyond the values tabulated in

Ref. [31] (see the symbols in Figure 3, left), v_s^2 drops first slightly below $1/3$ and approaches then slowly $1/3$ from below, in agreement with QCD asymptotic.

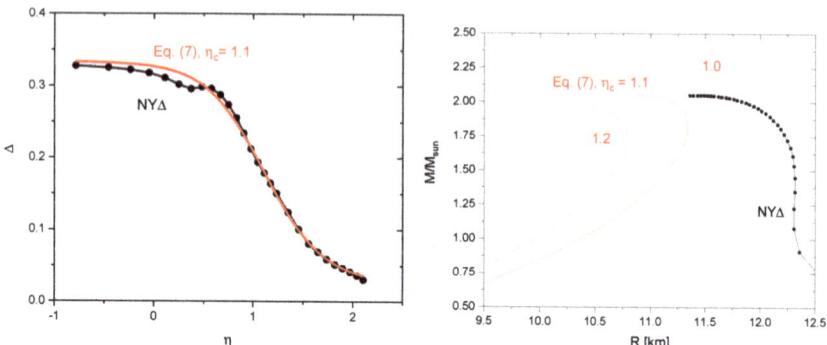

Figure 3. **Left** panel: Trace anomaly measure $\Delta = 1/3 - p/e$ as a function of $\eta \equiv \ln e/(n_0 m_p)$ for the NYΔ DD-EM2 EoS of [31] (black curve with symbols) and the adjusted parameterization by Equation (7) in Ref. [105] (red curve; $\eta_c = 1.1$, other parameters as in Ref. [105], see Equation (14) below). NYΔ DD-EM2 shows a softening at $p \approx 10$ MeV/fm^3, $\eta \approx 0.5$, caused by the onset of Δ proliferation, similar to Ref. [108]. **Right** panel: Mass-radius relation for NYΔ DD-EM2 EoS of Ref. [31] (solid black curve) in comparison with the parameterization by Equation (7) in Ref. [105] (see Equation (14) below for three values of the parameter η_c (read dashed/solid/dotted curves for $\eta_c = 1.2, 1.1$ and 1.0).

The comparison of the resulting mass-radius relations is exhibited in Figure 3, right. We follow the parameterization [105]

$$\Delta = \frac{1}{3} - \frac{1}{3}\left(1 - \frac{A}{B + \eta^2}\right) \frac{1}{1 + e^{-\kappa(\eta - \eta_c)}}, \quad (14)$$

however, with parameters adjusted to [31]: $\kappa = 3.45$, $\eta_c = 1.1$, $A = 2$ and $B = 20$, see Figure 3. Despite the tiny differences of the red curve (for adjusted Δ) and the solid black curve with markers (for NYΔ DD-EM2 EoS of [31]), the overall shapes of $M(R)$ differ in detail, see Figure 3, right. While the maximum masses agree, the radii for NYΔ DD-EM2 EoS of [31] are greater by about 1 km at $M \approx 1.4\,M_\odot$. A small dropping of the parameter η_c from 1.2 to 1.0 in the Δ EoS lets us increase the radius by 1 km. All these differences can be traced back essentially to differences in the low-density part of the employed EoSs.

4.2. Distinguished Cores with NYΔ Envelope

Focusing now on NYΔ DD-EM2 EoS of Ref. [31] and the core-corona decomposition, one obtains the resulting mass-radius radius relation exhibited in Figure 4. We find it convenient to keep the respective core radius constant and vary the core masses as $m_x = 10^{-4} 1.5^n M_\odot$, $n = 1, 2, 3 \cdots$. The very small core masses (of course, a large core with several-km radius and small included mass is a very exotic thing; it may be referred to as bubble or void with surface pressure p_x and the stable interface to SM matter) occupy the right end of the solid blue curves, where the dots refer to the increasing values of n. Heavy core masses occupy the left sections of the blue curves, i.e., larger values of n. For considerably smaller values of the core radii (not displayed), the blue curves approach the conventional mass-radius curve of the NYΔ DD-EM2 EoS of Ref. [31] (fat black curve). The limit $r_x \to 0$ and $m_x \to 0$ of the core-corona decomposition curve is depicted by the asterisk, which also agrees with the mass and radius of NYΔ DD-EM2 EoS of Ref. [31] with $p_c = p_x$. Increasing values of p_x make an up-shift of the core-corona mass-radius curves for a constant value of the core radius r_x, and a larger range of masses is occupied.

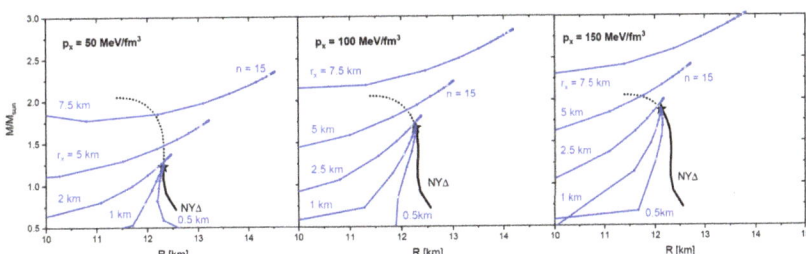

Figure 4. Mass-radius plane and its occupancy by compact stars with given core radii r_x (blue curves with dots at core masses $m_x = 10^{-4} \, 1.5^n M_\odot$, $n = 1, 2, 3 \cdots$ from right to left; the open circles depict points for $n = 15$, and the right-most endpoints are for $n = 1$). The values of p_x are 50 (**left** panel), 100 (**middle** panel), and 150 MeV/fm^3 (**right** panel). The fat solid curve is obtained by standard integration of the TOV equations using the NYΔ EoS tabulated in Ref. [31] with linear interpolation both in between the mesh and from the tabulated minimum energy density to the $p = 0$ point at $e_0 = 1$ MeV/fm^3. The asterisks display the mass-radius values for $p_c = p_x$. That is, the sections above the asterisks (dotted curves) are for a particular continuation of NYΔ at $p > p_x$, which is, (trivially) in this case, NYΔ itself. One could instead employ the parameterization Equation (4) which would deliver another dotted curve. For other examples, in particular the small-R region near black hole and Buchdahl limits, the interested reader is referred to Ref. [74].

All the features discussed in Section 2 in Ref. [74] are recovered, e.g., the convex shape of the curves $r_x = const$, which however becomes visible only by extending the plot towards smaller values of R (not shown). The right end points of the core-corona curves refer to very small values of m_x, while the not displayed left end points are on the limiting black hole curve $2G_N M = R$. However, the core-corona decomposition shows that the maximum-mass region of NYΔ-DD-ME2 (see fat solid curves) is easily uncovered too, interestingly with sizeable core radii and noticeably smaller up to larger total radii.

We refrain from displaying the core-corona mass-radius curves for $m_x = const$. They can be easily inferred by connecting the points $n = const$ in each of the panels in Figure 4.

To stress the relation to the usual mass-radius relation $M(R)$ obtained from the parametric representations $M(p_c)$ and $R(p_c)$, let us mention that the respective dotted curve sections above the asterisk on the fat solid curves are just an example—here simply NYΔ DD-EM2 EoS of Ref. [31] for $p > p_x$. Other continuations of the EoS at $p > p_x$ are within the region filled by the blue core-corona curves, which however may not be completely mapped out by many conceivable EoSs: The core can contain more than just SM matter, such as Dark Matter or Mirror Matter or other exotic material. In the simplest case, a multi-component composition with hypothesized EoS for each component can be used for the explicit construction. All that counts in our core-corona decomposition is that a core with radius r_x and included mass m_x is present and supports the pressure p_x at its boundary. By definition, outside the core, only SM matter with known EoS is there. (For a dedicated study of crust properties, cf., Ref. [109].)

4.3. Example of Radial Pressure and Mass Profiles

To illustrate that feature in some detail let us consider an example and select $p_x = 100$ MeV/fm^3. Assuming that the continuation of the EoS above this p_x is, hypothetically, by NYΔ itself, one obtains the pressure and mass profiles as displayed by black solid curves in Figure 5 for $p_c = 200$ MeV/fm^3. Clearly, taking r_x, m_x and p_x as start values and integrating in the corona up to the surface at $p = 0$, one obtains the same values of M and R as for the standard integration from $p(r = 0) = p_c$ to $p(R) = 0$ (see the blue circles on top of the black solid curve sections). However, keeping r_x but using other core masses m_x, e.g., $m_x^{(1)} = 0.5 m_x$ or $m_x^{(2)} = 2 m_x$, one obtains different pressure and mass profiles and, consequently, also different values of M and R, see Figure 5. This is the very construction of the core-corona decomposition leading to the results exhibited in Figure 4.

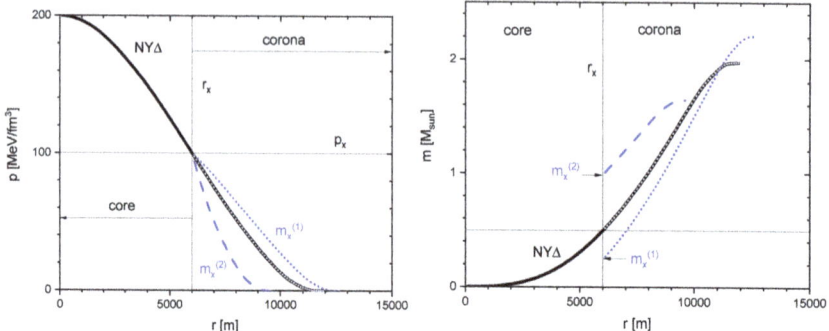

Figure 5. Pressure $p(r)$ (**left** panel) and mass $m(r)$ (**right** panel) as a function of the radius r for the special value $p_x = 100$ MeV/fm^3, selected here as the end point of the "reliable EoS" NYΔ at $p \leq p_x$. Assuming a possible continuation at $p > p_x$ by NYΔ itself as a particular example, it yields the fat solid curves for the (ad hoc) choice $p_c = 200$ MeV/fm^3. Keeping the resulting value r_x from $p(r_x) = p_x$ and integrating the TOV equations in the corona, $r \geq r_x$ with p_x and $m_x^{(1,2)}$ as initial values, one gets the dashed blue ($m_x^{(1)} = 0.5 m_x$) and dotted blue ($m_x^{(2)} = 2 m_x$) curves, where the respective values of R and M can be read off. Using m_x as core mass, the blue circles (on top of the fat black curve at $r \geq r_x$) are obtained. Using a multitude of values $m_x^{(n)}$ would generates one additional blue curve in Figure 4. In the present example, for $r_x = 6$ km.

5. Summary

The core-corona decomposition relies on the assumption that the EoS of compact/neutron star matter (i) is reliably known up to energy density e_x and pressure p_x and (ii) occupies the star as the only component at radii $r \geq r_x$. The base line for static, spherically symmetric configurations is then provided by the TOV equations, which are integrated, for $r \in [r_x, R]$, to find the circumferential radius R (where $p(R) = 0$) and the gravitational mass $M = m(R)$. We call that region $r \in [r_x, R]$ "corona", but "crust" or "mantle" or "envelope" or "shell" are also suitable synonyms. The region $r \in [0, r_x]$ is the "core", which is parameterized by the included mass m_x. The core must support the corona pressure at the interface, i.e., $p(r_x^-) = p(r_x^+)$. The core can contain any material compatible with the symmetry requirements. In particular, it could be modeled by a multi-component fluid with SM matter plus Dark Matter and/or Mirror World matter or anything beyond the SM. The TOV equations then deliver $R(r_x, m_x, p_x)$ and $M(r_x, m_x, p_x)$.

It helps our intuition to think of a SM matter material in the core with assumed fiducial EoS, which determines, for given value of p_x, $r_x(p_c)$ and $m_x(p_c)$, thus $R(p_c)$ and $M(p_c)$ resulting in $M(R)$, as conventionally done. Using many fiducial test EoSs at $p > p_x$, one maps out a certain region in the M-R plane accessible by SM matter. Our core-corona decomposition extends this region, since the core can contain much more than SM matter only. With reference to a special EoS, applied tentatively in core and corona, the accessible masses and radii become smaller (larger) for heavy (light) cores. That effect is noticeable for large (km size) and heavy (fraction of solar mass) cores. Small and light cores hardly have an impact.

This analysis should be refined in follow-up work by employing improved EoSs in the low-density region and by taking care of mass-radius values for the heaviest neutron star(s), still keeping precise radius values for the $1.4 M_\odot$ neutron stars. In addition, the holographic model in Appendix B, which is aimed at mapping hot QCD thermodynamics to a cool EoS, deserves further investigations, before arriving quantitatively at an EoS suitable for compact star properties and (merging) dynamics. The emphasis in Appendix B is the illustration of the extrapolation scheme beyond a patch of QCD thermodynamic state space.

Taking up the suggestion in Ref. [77] that the scaling property of the TOV equations with a mass dimension-4 quantity, e.g., by m_p^4, where m_p is the proton mass, essentially determines the gross parameters of compact/neutron stars, we supply in Appendix A a brief contemplation on the emergence of hadron masses within QCD-related approaches, thus bridging from micro to macro physics: $M_{PSR\,J0740+6620} \approx 2 M_\odot \approx 2.38 \times 10^{57} m_p$, $R_{PSR\,J0740+6620} \approx 12.3$ km $\approx 1.5 \times 10^{19} r_p$ with proton charge radius $r_p \approx 0.88$ fm $\approx 4.2 m_p^{-1}$ which is somewhat larger than its "natural" scale $\hbar c / m_p$. Our focus, however, is here on the emergence of the gravitational mass (the gravitational mass in our core-corona decomposition is $M = m_x + 4\pi \int_{r_x}^{R} \mathrm{d}r\, r^2 e(r)$ which, due to gravitational binding, is different from the total mass $\int \mathrm{d}^3 V e(r)$ where the integral measure $\mathrm{d}^3 V$ accounts for the gravitational spatial deformation, cf., Ref. [110]) determining the trajectories of light rays and test particles outside of compact astrophysical objects.

Author Contributions: Conceptualization, B.K.; methodology, R.Z., M.D. and B.K.; validation, R.Z., M.D. and B.K.; investigation, R.Z., M.D. and B.K.; writing—original draft preparation, B.K.; writing—review and editing, B.K. and R.Z.; visualization, B.K. and R.Z.; supervision, B.K.; project administration, B.K. All authors have read and agreed to the published version of the manuscript.

Funding: The work is supported in part by the European Union's Horizon 2020 research and innovation program STRONG-2020 under grant agreement No 824093.

Acknowledgments: We thank S. M. Schmidt and C. D. Roberts for inviting our contribution to the topic of the emergent mass phenomenon. M. Ding is grateful for support by Helmholtz-Zentrum Dresden-Rossendorf High Potential Programme. One of the authors (BK) acknowledges continuous discussions with J. Schaffner-Bielich and K. Redlich and for the encouragement to deal with the current topic.

Conflicts of Interest: The authors declare no conflict of interest.

Appendix A. Emergence of Hadron Masses

Astronomical observations of neutron stars have provided us with a way to understand these mysterious objects. The literature suggests that the outer and inner core of neutron stars may consist of nucleons or cold, dense quark matter. The question one can ask is how almost massless light quarks could combine and produce such massive nucleons to give rise to neutron star mass, as emphasized in Section 2.1. A widely recognized mechanism is the Higgs mechanism, which gives rise to the current quark masses. However, as was immediately realized, the current quark mass contributes only a few MeV, while the mass of a proton or neutron is much larger. The mystery of the missing masses lies in the emergence of hadron mass (EHM).

Appendix A.1. Three Pillars of EHM

Contemporary studies of continuum Schwinger methods (CSMs) have shown that the emergence of hadron mass rests on three pillars: (a) the running quark mass, (b) the running gluon mass, and (c) the process-independent effective charge. These three pillars provide the basis for giving observable results of hadron properties [111].

Appendix A.1.1. Running Quark Mass

The first pillar is the most familiar, namely the running quark mass. The dressed-quark propagator can be represented by a special quantity, the dressed-quark mass function. Modern CSMs calculations of the quark gap equation show that the dressed-quark mass function is a finite value of $M_0(k^2 = 0) = 0.41$ GeV in the far infrared, even in the chiral limit. This is an explicit expression of the dynamical chiral symmetry breaking (DCSB), and the infrared scale is responsible for the masses of all hadrons, and the running quark mass is therefore regarded as an expression of the EHM. In a sense, the quark in the far infrared can be seen as a quasiparticle, produced by the interaction of the high-energy quark parton with the gluon. The scale of the dressed-quark mass function in the far infrared is comparable to the scale in the constituent quark model. The difference is that the dressed-quark mass function is momentum-dependent, and it runs from the far infrared to the ultraviolet, where it matches the current quark mass in perturbation theory. The dressed-quark mass function is also flavor-dependent, with the Higgs mechanism gradually dominating in describing it from light to heavy quarks, while the dressed-quark mass function decreases at a slower rate. It is worth emphasizing that the ultraviolet behavior of the dressed-quark mass function in the chiral limit is related to the well known chiral condensates.

Appendix A.1.2. Running Gluon Mass

The second pillar is the running gluon mass. Gauge invariance requires that the gluon parton be massless. However, in the Schwinger mechanism of QCD, it was conjectured long ago that the two-point gluon Schwinger function might give rise to a massive dressed gluon. Thus, the dressed gluon gains mass and the gluon parton is transformed into the gluon quasiparticle by interaction with themselves. Consequently, gluons can have a momentum-dependent mass function, and this non-zero function has been confirmed by both the continuum and lattice QCD. The dressed gluon mass function is power-law suppressed in the ultraviolet, so it is invisible in perturbation theory. However, it is non-zero in the infrared momentum range, and in the far infrared it yields a value of $m_g = 0.43$ GeV. This is purely a manifestation of "mass from nothing" and is responsible for all hadron masses. It is worth emphasizing that the running gluon mass may also be associated with confinement. The gluon two-point Schwinger function has an inflection point, so it has no Källén–Lehmann representation, and therefore the relevant states cannot appear in the Hilbert space of physical states.

Appendix A.1.3. Process-Independent Effective Charge

The third pillar is the process-independent effective charge. QCD has a running coupling which expresses the feature of asymptotic freedom in the ultraviolet at one-loop order, but it also shows a Landau pole in the infrared, where the coupling becomes divergent when $k^2 = \Lambda_{QCD}^2$. However, recent advances in CSMs based on pinch techniques and background field methods show that QCD has a unique, nonperturbatively well-defined, computable, process-independent effective charge. Its large-momentum behavior is smoothly connected to the QCD running coupling at one-loop order, while it is convergent at small momentum. The Landau pole is eliminated due to the appearance of the gluon mass scale in the infrared. It is noteworthy that, at small momentum, the effective charge run ceases and it enters a domain that can be regarded as an effective conformal. In this domain, the gluons are screened so that the valence quasiparticle can be seen as carrying all the properties of a hadron. Furthermore, the process-independent effective charge matches the Bjorken process-dependent charge, and in practical use they can be considered to be indistinguishable. Additionally, the process-independent effective charge is a well-defined smoothing function at all available momenta, from the far infrared to the ultraviolet, so it can be a good candidate for applications, for example, in evolving parton distribution functions to different scales.

Appendix A.2. Hadrons in Vacuum

A direct correlation with the running quark mass is its effect on pseudoscalar mesons, especially the lightest meson, the pion [112,113]. As a chiral symmetry breaking Nambu–Goldstone boson, the well-known Goldberger–Treiman relation relates the dressed-quark mass function to pion's Bethe–Salpeter amplitude, so that, to some extent, it can be seen that, once the one-body problem is solved, the two-body problem is also solved. Furthermore, because of its direct connection to the dressed quark mass function, the properties of the pion can be seen as the cleanest window to glimpse the emergence of hadron mass. In the quark model, the vector meson is a spin-flip state of the pseudoscalar meson, and thus the ρ meson is the closest relative to pions. However, the properties of ρ mesons are significantly different from those of pion. Compared to the surprisingly light pion, the ρ is much heavier. Therefore, one can ask a straightforward question: how does the spin flip produce such different masses for the two mesons? Since QCD is a well-defined theory of strong interactions, it must answer such a fundamental question.

The meson spectrum includes not only ground states, but also excited states with high orbital angular momenta between the quark and antiquark constituents of the meson. The study of excited mesons is more difficult in both lattice QCD and CSMs. Experience tells us that the well-known rainbow ladder approximation does not provide the correct ordering of the meson masses between ground and excited states when using CSMs [114]. Recent improvements to the Bethe-Salpeter kernel have made it possible for practitioners of CSMs to reproduce the empirical results. The new feature of modern Bethe-Salpeter kernel is the inclusion of a term closely related to the anomalous chromomagnetic moment, which reflects the influence of the EHM on the quark gluon vertex and which can be used to effectively describe the excited meson state very well.

In addition to studying mesons made up of light quarks, it is also worth exploring how the properties of mesons vary with meson mass. In the quark model view, if a valence light quark in pion is replaced by a strange quark, a kaon is formed. A kaon is also a Nambu–Goldstone boson in the chiral limit, however, in the real world, the strange quark, produced through the Higgs mechanism, is 27 times more massive than the current light quarks. Thus, the properties of the kaon are the result of the combined effect of the Higgs mechanism and the emergence of hadron mass. This has been revealed from the study of the parton distribution functions of kaons. The skewness of the distribution is caused by the heavier strange quark, while the overall broadening of the distribution is caused by the EHM. If mesons consisted of heavier quarks, charm, and bottom quarks, the Higgs mechanism would dominate and be the largest source of heavy meson properties. In particular, pseudoscalar mesons and heavy vector mesons are of particular interest as mesons with zero orbital angular momentum in the quark model [115,116]. Their distributions are usually narrower than those of light mesons, since it has been pointed out that the distribution in the heavy-quark limit is a Dirac delta function. The difference in distributions provides a clear picture of how the properties of mesons evolve with increasing meson mass, and the CSMs is a unified framework for describing all mesons, from pion to Upsilon.

It is worth mentioning that progress in mesons has also been extended to baryons. The properties of baryons are calculated using the Faddeev equation describing the three-quark scattering problem. Since the complete three-body problem in nature is much more complicated, the quark dynamical-diquark method is usually introduced, which is useful as a means of elucidating many qualitative features. Central to this approach is the incorporation of five different diquark correlations, of which the axial-vector diquark is of outstanding importance, to produce the correct baryon spectrum as well as baryon structure functions such as distribution functions, electromagnetic, axial, and pseudoscalar form factors.

In addition to the existing traditional hadrons, mesons, and baryons, many other new hadron states have been proposed experimentally and theoretically, such as exotic states, pentaquark, tetraquark, hybrid states, and glueballs. In the field of research on mesons consisting of heavy quarks, a comprehensive study of the charm family has been carried out thanks to a large amount of experimental data from the B-factory. As a result, states such as XYZ states, pentaquark, and tetraquark have extensively extended our knowledge of QCD bound states. In the field of research on mesons consisting of light quarks, there is a tendency to think that gluon degrees of freedom may also play a role in the formation of bound states, and thus there is speculation about the existence of states such as hybrid states and even glueballs [117]. This has been found from calculations of lattice QCD and CSMs, and future experiments, such as the 12 GeV upgrade experiment at Jefferson Lab, which will provide an opportunity to test these theoretical predictions.

Appendix A.3. Hadrons in Cold Dense Medium

In a cold and dense medium, i.e., for non-zero baryo-chemical potentials, quarks and emergent hadrons suffer the impact of the ambient matter. From low to high chemical potential domains, QCD will go through a phase of confinement and dynamical chiral symmetry breaking to a phase of deconfinement and chiral symmetry restoration. The key to the study is to determine the critical point/region at which the transition occurs and to which category the "phase transition" belongs.

Since the full domain of the finite chemical potential is currently not fully available in lattice QCD simulations, a complementary approach is CSMs [118–120], which expresses the dynamical chiral symmetry breaking and confinement in QCD, so that it can therefore be used as a tool to explore the properties of quark and hadron matter in cold dense medium, thus revealing relevant features of objects such as neutron stars. In the CSMs, the medium-induced dressed-quark propagator can be obtained by solving the quark gap equation in the medium, and the quark condensates are proportional to the matrix trace of the dressed quark propagator in the chiral limit. The quark condensate is crucial because it is commonly seen as the order parameter of the deconfinement (phase) transition. In some studies it has been shown that the chiral quark condensate is discontinuous with chemical potential, so that in the chiral limit the phase transition is of first order. Furthermore, it has been suggested that the QCD quark condensate may be completely contained in hadrons [121]. In addition to quarks, hadrons are also affected by the ambient medium at a non-zero chemical potential and, consequently, their masses change, showing noticeable deviations from the masses in vacuum. The appearance of a turning point in the chemical potential dependence of the hadron mass can also characterize the occurrence of chiral symmetry transition. It has also been proposed that increasing the chemical potential promotes the possibility of quark-quark Cooper pairing, i.e., diquark condensation. Quark-quark Cooper pairs are composite bosons with both electric and color charge, and thus superfluidity in quark matter entails superconductivity and color superconductivity.

These hadronic in-medium information is important not only for imaging the QCD phase diagram, but also for constructing a unified EoS from low-density nucleonic matter to high-density quark matter, and thus becomes crucial for determining the properties of neutron stars. Analog reasoning applies to other forms of strong-interaction matter which could govern new classes of compact stars, e.g., pion stars [1].

Appendix A.4. Supplementary Remarks

Before supplementing the above reflections by another approach to EHM w.r.t. condensates as fundamental QCD quantities, let us mention that the *ab initio* access to hadron vacuum masses is directly based on QCD as the theoretical basis of strong interaction. The current status is reviewed in Ref. [122]: a compelling description of the mass spectrum and various other hadron parameters is achieved by lattice QCD.

The operator product expansion relates parameters of a hadron model of the spectral function to a series of QCD condensates, most notably the above mentioned chiral quark condensate $\langle \bar{q}q \rangle$, the gluon condensate $\langle \frac{\alpha_s}{\pi} G^2 \rangle$, the mixed quark-gluon condensate $\langle \bar{q} g_s \sigma G q \rangle$, the triple gluon condensate $\langle g_s^3 G^3 \rangle$, the four-quark condensates $\langle \bar{q}\Gamma q \bar{q} \Gamma q \rangle$ (Γ denotes all possible structures formed by Dirac and Gell–Mann matrices) [123] and the poorly known condensates with higher mass-dimension. Further ingredients are the Wilson coefficients, which are accessible by perturbation theory [124]. In a strong-interaction medium, the condensates change: $\langle \cdots \rangle_0 \to \langle \cdots \rangle_{T,\mu_B}$; one could say that they are expelled from vacuum (label "0") by a higher spatial occupancy due to non-zero temperature and density (parametrized by μ_B) (similar to concerns as on the Higgs field condensate challenge such a picture of the vacuum, cf., [125] and follow-up citations). The induced dropping of condensates facilitates in-medium modifications of hadrons, in particular due the coupling to the chiral condensate, which is often considered as an order parameter of chiral symmetry. A recent review is Ref. [126].

In the baryon sector, the impact of the four-quark condensates and other in-medium-only condensates make unambiguous analyses of QCD sum rules somewhat vague. Previous factorizations relate the many four-quark condensates to the chiral condensate modulo some uncertain factor. For a comprehensive discussion, see Ref. [127], which focuses on the emergence of the nucleon mass, most notably m_p, from various quark and gluon condensates.

This brief journey should supplement our perspective of how the hadronic constituents of matter in compact stars acquire their masses and that the ambient medium modifies them. The very first step of gaining the above mentioned bare input masses by the Higgs mechanism is therefore left out, see Figure A1. The various bare quark masses appear in several QCD approaches as parameters adjusted to data. This is evidenced, e.g., in the approach [114]: A formula of two-quark meson masses is developed, $m(m_1, m_2)$, where $m_{1,2}$ refer to the bare quark masses. As shown in Figure 3 in Ref. [114], inspection of the contour lines $m(m_1, m_2) = const.$ can be used to pin down numerical values of bare masses $m_{1,2}$ which reproduce empirical values of ground states in the pseudo-scalar and vector channels including light u, d, strange and charm quarks at once; redundancy is used for cross checking. In addition, Regge trajectories become accessible to some extent [114,128].

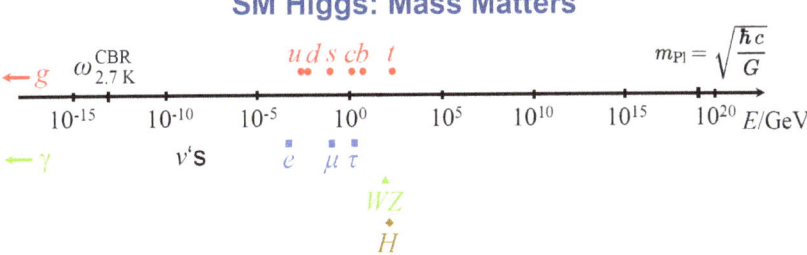

Figure A1. Masses within the SM. Mysterious concentration of bare SM masses (quarks $[u, d, s, c, t, b]$, leptons $[e, \mu, \tau]$, gauge bosons $[W^\pm, Z^0]$, Higgs $[H]$) and separation of neutrinos ($\nu's = [\nu_{e,\mu,\tau}]$) on a large energy scale ranging from present-day cosmic background radiation $\omega_{2.7\,K}^{CBR} \approx 0.233 \times 10^{-3}$ eV to Planck mass $m_{Pl} = \sqrt{\hbar c/G_N} \approx 1.22 \times 10^{19}$ GeV. Only the QED and QCD gauge Bosons $[\gamma, g]$ remain massless.

Appendix B. Holographic Approach to the EoS

Here we present a particular model of strong-interaction matter that is based on the famous AdS/CFT correspondence. In line with Refs. [129–131] we employ the action in a fiducial five-dimensional space-time with asymptotic AdS symmetry:

$$S_{EdM} = \frac{1}{2\kappa_5^2} \int d^4x\, dz\, \sqrt{-g_5} \left(R - \frac{1}{2} \partial^M \phi\, \partial_M \phi - V(\phi) - \frac{1}{4} \mathcal{G}(\phi) F_B^2 \right), \quad \text{(A1)}$$

where R is the Einstein–Hilbert gravity part, $F_B^{MN} = \partial^M \mathcal{B}^N - \partial^N \mathcal{B}^M$ stands for the field strength tensor of an Abelian gauge field $\mathcal{B}$ à la Maxwell with $\mathcal{B}_M dx^M = \Phi(z)\,dt$ defining the electro-static potential. An embedded black hole facilitates the description of a hot and dense medium, since the black hole has a Hawking surface temperature and sources an electric field, thus encoding holographically a temperature and a density of the system. Dynamical objects are a dilatonic (scalar) field ϕ and a Maxwell-type field Φ that are governed by a dilaton potential $V(\phi)$ and a dynamical coupling $\mathcal{G}(\phi)$ and geometry-related quantities. For a more general holographic approach to compact star physics, cf., Ref. [132]. Space-time is required to be described by the line element squared

$$ds^2 = g_{MN} dx^M dx^N := \exp\{A(z, z_H)\} \left[f(z, z_H)\, dt^2 - d\vec{x}^2 - \frac{dz^2}{f(z, z_H)} \right], \quad \text{(A2)}$$

where z_H is the horizon position, $z \in [z_H, \infty]$ is the radial coordinate, A is the warp factor, and f the blackness function.

The resulting Einstein equations are a set of coupled second-order ODEs to be solved with appropriate boundary conditions. The quantities V and $\mathcal{G}$ are tuned (in contrast to former work, we here put emphasis on side conditions which ensure that, at $\mu_B = 0$, no phase transition is facilitated outside the temperature range uncovered by the lattice data) to quantitatively describe the lattice QCD results [133]: Meanwhile, at $\mu_B = 0$, the datasets [134,135] are consistent, special combinations of quantities, e.g., e/p, enhance the small differences. This is, in particular, is striking at $T \in [130, 140]$ MeV, where a +4% (-10%) deviation in energy density (pressure) even changes the shape of the e/p curve when ignoring the error bars

$$\partial_\phi \ln V(\phi) = (p_1\phi + p_2\phi^2 + p_3\phi^3)\exp\{-\gamma\phi\}, \tag{A3}$$

$$\mathcal{G}(\phi) = \frac{1}{1+c_5}\left(\frac{1}{\cosh(c_1\phi + c_2\phi^2)} + \frac{c_3}{\cosh c_4\phi}\right) \tag{A4}$$

with parameters $\{p_{1,2,3}, \gamma\} = \{0.165919, 0.269459, -0.017133, 0.471384\}$ and $\{c_{1,2,3,4,5}\} = \{-0.276851, 0.394100, 0.651725, 101.6378, -0.939473\}$. These parameters and the scales implicitly refer to the QCD input, see Figure A2. The trace anomaly measure Δ, Equation (12), is exhibited in Figure A2, left, for various temperatures T and at baryon-chemical potential $\mu_B = 0$; the middle panel is for the ratio e/p. As argued in Ref. [105], the high-temperature, small-density and low-temperature, high-density behavior is strikingly different. A further crucial input quantity is the susceptibility $\chi_2 = \partial_{\mu_B} n_B|_{\mu_B=0}$, see the right panel in Figure A2. The length scale of the z coordinate is set by $L^{-1} = 1465$ MeV, which relates the horizon position z_H and temperature via $T = -1/(4\pi\partial_z f(z, z_H))|_{z=z_H}$, and by $\kappa_5^2 = 8.841$ fm^{-3}, which determines entropy density $s(T, \mu_B) = \frac{2\pi}{\kappa_5^2}\exp\{\frac{3}{2}A(z_H, z_H)\}$ and baryon density $n_B(T, \mu_B) = -\frac{\pi L^2}{\kappa_5^2}\partial_z^2\Phi(z, z_H)|_{z=0}$.

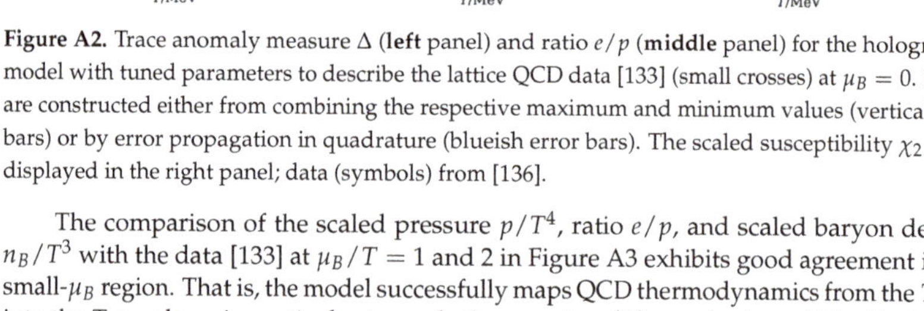

Figure A2. Trace anomaly measure Δ (**left** panel) and ratio e/p (**middle** panel) for the holographic model with tuned parameters to describe the lattice QCD data [133] (small crosses) at $\mu_B = 0$. Errors are constructed either from combining the respective maximum and minimum values (vertical error bars) or by error propagation in quadrature (blueish error bars). The scaled susceptibility χ_2/T^2 is displayed in the right panel; data (symbols) from [136].

The comparison of the scaled pressure p/T^4, ratio e/p, and scaled baryon density n_B/T^3 with the data [133] at $\mu_B/T = 1$ and 2 in Figure A3 exhibits good agreement in the small-μ_B region. That is, the model successfully maps QCD thermodynamics from the T axis into the T-μ_B plane, in particular towards the μ_B axis: $p(T) \mapsto p(\mu_B) = p(T(\mu_B))$. Curves of constant pressure, $T(\mu_B)|_{p=const}$, are determined by $dT/d\mu_B = -n_B(T, \mu_B)/s(T, \mu_B)$, (from $p(T(x), \mu_B(x)) = const$ and assuming a parametric representation $T(x)$ and $\mu_B(x)$ with x being the arc length. $dp/dx = (\partial p/\partial T)(dT/dx) + (\partial p/\partial \mu_B)(d\mu_B/dx) = 0$ then uses $s = \partial p/\partial T$ and $n_B = \partial p/\partial \mu_B$. The technicalities of holographically determining these quantities and the properties of the resulting EoS are relegated to a separate paper.) The left panel in Figure A4 is for a toy model demonstrating how the EoS given on the temperature axis is mapped to the chemical potential axis exhibited in Figure A4. The lattice data [133] are for $T \in [125, 240]$ MeV at small values of μ_B (see the right panel). The "transport" into the restricted region of the T-μ_B plane by the holographic model is already an extrapolation, which requires a smooth pattern of $T(\mu_B)|_{p=const}$ curves without junctions, branchings, crossings, etc. The extrapolation of the data to $T < 125$ MeV and $T > 240$ MeV at $\mu_B = 0$ by the parameterizations (A3) and (A4) is another type of extrapolation. Thus, the region beyond the displayed $p = const$ curves could be hampered by both types of uncertainties, but nevertheless may serve as educated guess of the cool EoS in the high-density realm. The benefit of such an approach is the direct coupling of the hot EoS, being an essential input in describing ultra-relativistic heavy-ion collisions by fluid dynamics, and the cool EoS, being the input for compact (neutron) star structure and (merger) dynamics.

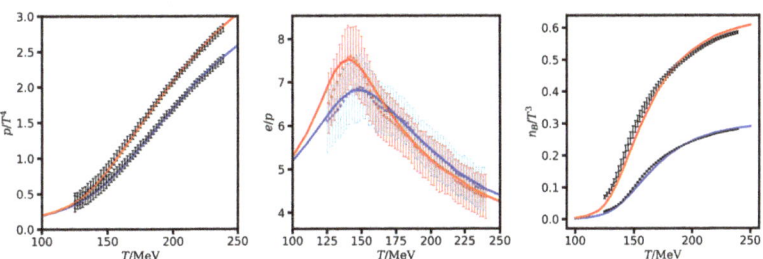

Figure A3. Scaled pressure p/T^4 (**left** panel), ratio e/p (**middle** panel), and scaled baryon density n_B/T^3 (**right** panel) as a function of temperature for $\mu_B/T = 1$ (blue) and 2 (red) in comparison with the data [133] (symbols with error bars).

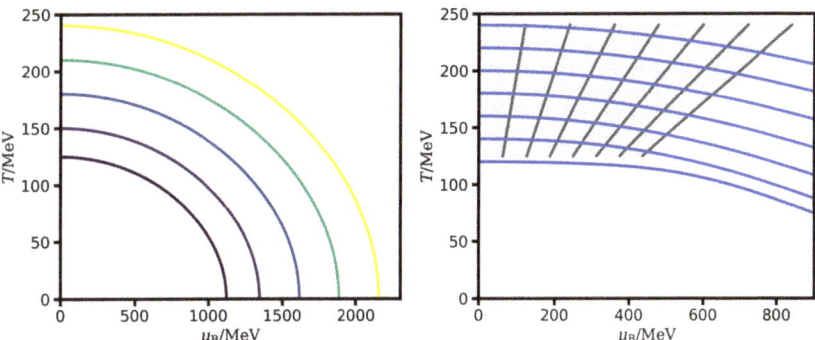

Figure A4. Curves of constant pressure over the T-μ_B plane. In such a way, the pressure data on the T axis are directly "transported" towards the μ_B axis, in particular $p(T = 0, \mu_B^{(0)}) = p_0 := p(T_0, \mu_B = 0)$ along the constant-pressure curve $T(\mu_B)|_{p=p_0}$ starting at $T(\mu_B = 0) = T_0$ and terminating at $T(\mu_B^{(0)}) = 0$. The energy density, with $s(T, \mu_B)$ and $n_B(T, \mu_B)$ given, then follows from $e = -p + Ts + \mu_B n_B$ (Gibbs–Duhem). **Left** panel: A simple toy model is employed here for the purpose of demonstration ($s = 4aT^3 + 2bT\mu_B^2$, $n_B = 4c\mu_B^3 + 2bT^2\mu_B$, and numerical values $b/a = 0.027384$, $c/a = 0.000154$ referring to a two-flavor ideal quark-gluon plasma). **Right** panel: For the holographic model (A1), (A3) and (A4) in a region (grey hatched) controlled by lattice QCD data [133] on the dark-grey beam sections.

Instead of explicitly using hadronic degrees of freedom to devise the holographic EoS, here the EoS as relation of energy density and pressure is solely deployed. Of course, the underlying hot QCD EoS is anchored in the common quark-gluon dynamics with intimate contact to the hadron observables.

References

1. Brandt, B.B.; Endrodi, G.; Fraga, E.S.; Hippert, M.; Schaffner-Bielich, J.; Schmalzbauer, S. New class of compact stars: Pion stars. *Phys. Rev. D* **2018**, *98*, 094510. [CrossRef]
2. Roberts, C.D.; Richards, D.G.; Horn, T.; Chang, L. Insights into the emergence of mass from studies of pion and kaon structure. *Prog. Part. Nucl. Phys.* **2021**, *120*, 103883. [CrossRef]
3. Roberts, C.D.; Schmidt, S.M. Reflections upon the emergence of hadronic mass. *Eur. Phys. J. ST* **2020**, *229*, 3319–3340. [CrossRef]
4. Tolos, L.; Fabbietti, L. Strangeness in Nuclei and Neutron Stars. *Prog. Part. Nucl. Phys.* **2020**, *112*, 103770. [CrossRef]
5. Silva, H.O.; Holgado, A.M.; Cárdenas-Avendaño, A.; Yunes, N. Astrophysical and theoretical physics implications from multi-messenger neutron star observations. *Phys. Rev. Lett.* **2021**, *126*, 181101. [CrossRef]
6. Tews, I.; Pang, P.T.H.; Dietrich, T.; Coughlin, M.W.; Antier, S.; Bulla, M.; Heinzel, J.; Issa, L. On the Nature of GW190814 and Its Impact on the Understanding of Supranuclear Matter. *Astrophys. J. Lett.* **2021**, *908*, L1. [CrossRef]
7. Tang, S.P.; Jiang, J.L.; Gao, W.H.; Fan, Y.Z.; Wei, D.M. Constraint on phase transition with the multimessenger data of neutron stars. *Phys. Rev. D* **2021**, *103*, 063026. [CrossRef]
8. Margutti, R.; Chornock, R. First Multimessenger Observations of a Neutron Star Merger. *Ann. Rev. Astron. Astrophys.* **2021**, *59*, 155–202. [CrossRef]
9. Nicholl, M.; Margalit, B.; Schmidt, P.; Smith, G.P.; Ridley, E.J.; Nuttall, J. Tight multimessenger constraints on the neutron star equation of state from GW170817 and a forward model for kilonova light-curve synthesis. *Mon. Not. R. Astron. Soc.* **2021**, *505*, 3016–3032. [CrossRef]
10. Yu, J.; Song, H.; Ai, S.; Gao, H.; Wang, F.; Wang, Y.; Lu, Y.; Fang, W.; Zhao, W. Multimessenger Detection Rates and Distributions of Binary Neutron Star Mergers and Their Cosmological Implications. *Astrophys. J.* **2021**, *916*, 54. [CrossRef]
11. Annala, E.; Gorda, T.; Katerini, E.; Kurkela, A.; Nättilä, J.; Paschalidis, V.; Vuorinen, A. Multimessenger Constraints for Ultradense Matter. *Phys. Rev. X* **2022**, *12*, 011058. [CrossRef]
12. Pang, P.T.H.; Tews, I.; Coughlin, M.W.; Bulla, M.; Broeck, C.V.D.; Dietrich, T. Nuclear Physics Multimessenger Astrophysics Constraints on the Neutron Star Equation of State: Adding NICER's PSR J0740+6620 Measurement. *Astrophys. J.* **2021**, *922*, 14. [CrossRef]
13. Miller, M.C.; Lamb, F.K.; Dittmann, A.J.; Bogdanov, S.; Arzoumanian, Z.; Gendreau, K.C.; Guillot, S.; Ho, W.C.G.; Lattimer, J.M.; Loewenstein, M.; et al. The Radius of PSR J0740+6620 from NICER and XMM-Newton Data. *Astrophys. J. Lett.* **2021**, *918*, L28. [CrossRef]
14. Miller, M.C.; Lamb, F.K.; Dittmann, A.J.; Bogdanov, S.; Arzoumanian, Z.; Gendreau, K.C.; Guillot, S.; Harding, A.K.; Ho, W.C.G.; Lattimer, J.M.; et al. PSR J0030+0451 Mass and Radius from $NICER$ Data and Implications for the Properties of Neutron Star Matter. *Astrophys. J. Lett.* **2019**, *887*, L24. [CrossRef]
15. Riley, T.E.; Watts, A.L.; Ray, P.S.; Bogdanov, S.; Guillot, S.; Morsink, S.M.; Bilous, A.V.; Arzoumanian, Z.; Choudhury, D.; Deneva, J.S.; et al. A NICER View of the Massive Pulsar PSR J0740+6620 Informed by Radio Timing and XMM-Newton Spectroscopy. *Astrophys. J. Lett.* **2021**, *918*, L27. [CrossRef]
16. Riley, T.E.; Watts, A.L.; Bogdanov, S.; Ray, P.S.; Ludlam, R.M.; Guillot, S.; Arzoumanian, Z.; Baker, C.L.; Bilous, A.V.; Chakrabarty, D.; et al. A $NICER$ View of PSR J0030+0451: Millisecond Pulsar Parameter Estimation. *Astrophys. J. Lett.* **2019**, *887*, L21. [CrossRef]
17. Raaijmakers, G.; Greif, S.K.; Hebeler, K.; Hinderer, T.; Nissanke, S.; Schwenk, A.; Riley, T.E.; Watts, A.L.; Lattimer, J.M.; Ho, W.C.G. Constraints on the Dense Matter Equation of State and Neutron Star Properties from NICER's Mass–Radius Estimate of PSR J0740+6620 and Multimessenger Observations. *Astrophys. J. Lett.* **2021**, *918*, L29. [CrossRef]
18. Pereira, J.P.; Bejger, M.; Tonetto, L.; Lugones, G.; Haensel, P.; Zdunik, J.L.; Sieniawska, M. Probing elastic quark phases in hybrid stars with radius measurements. *Astrophys. J.* **2021**, *910*, 145. [CrossRef]
19. Christian, J.E.; Schaffner-Bielich, J. Twin Stars and the Stiffness of the Nuclear Equation of State: Ruling Out Strong Phase Transitions below $1.7n_0$ with the New NICER Radius Measurements. *Astrophys. J. Lett.* **2020**, *894*, L8. [CrossRef]
20. Chatziioannou, K. Neutron star tidal deformability and equation of state constraints. *Gen. Rel. Grav.* **2020**, *52*, 109. [CrossRef]
21. Motta, T.F.; Thomas, A.W. The role of baryon structure in neutron stars. *Mod. Phys. Lett. A* **2022**, *37*, 2230001. [CrossRef]
22. Jokela, N.; Järvinen, M.; Remes, J. Holographic QCD in the NICER era. *Phys. Rev. D* **2022**, *105*, 086005. [CrossRef]
23. Kovensky, N.; Poole, A.; Schmitt, A. Building a realistic neutron star from holography. *Phys. Rev. D* **2022**, *105*, 034022. [CrossRef]
24. Zhang, N.B.; Li, B.A. Impact of NICER's Radius Measurement of PSR J0740+6620 on Nuclear Symmetry Energy at Suprasaturation Densities. *Astrophys. J.* **2021**, *921*, 111. [CrossRef]
25. Christian, J.E.; Schaffner-Bielich, J. Supermassive Neutron Stars Rule Out Twin Stars. *Phys. Rev. D* **2021**, *103*, 063042. [CrossRef]
26. Gerlach, U.H. Equation of State at Supranuclear Densities and the Existence of a Third Family of Superdense Stars. *Phys. Rev.* **1968**, *172*, 1325. [CrossRef]
27. Kämpfer, B. On the Possibility of Stable Quark and Pion Condensed Stars. *J. Phys. A* **1981**, *14*, L471. [CrossRef]
28. Kämpfer, B. On stabilizing effects of relativity in cold spheric stars with a phase transition in the interior. *Phys. Lett. B* **1981**, *101*, 366–368. [CrossRef]
29. Zdunik, J.L.; Haensel, P.; Schaeffer, R. Phase transitons in stellar cores, II Equilibrium configurations in general relativity. *Astron. Astrophys.* **1987**, *172*, 95–110.
30. Zdunik, J.L.; Haensel, P.; Schaeffer, R. Phase transitons in stellar cores, I Equilibrium configurations. *Astron. Astrophys.* **1983**, *126*, 121–145.

31. Li, J.J.; Sedrakian, A.; Alford, M. Relativistic hybrid stars with sequential first-order phase transitions and heavy-baryon envelopes. *Phys. Rev. D* **2020**, *101*, 063022. [CrossRef]
32. Alford, M.G.; Sedrakian, A. Compact stars with sequential QCD phase transitions. *Phys. Rev. Lett.* **2017**, *119*, 161104. [CrossRef] [PubMed]
33. Malfatti, G.; Orsaria, M.G.; Ranea-Sandoval, I.F.; Contrera, G.A.; Weber, F. Delta baryons and diquark formation in the cores of neutron stars. *Phys. Rev. D* **2020**, *102*, 063008. [CrossRef]
34. Pereira, J.P.; Bejger, M.; Zdunik, J.L.; Haensel, P. Differentiating sharp phase transitions from mixed states in neutron stars. *arXiv* **2022**, arXiv:2201.01217.
35. Bejger, M.; Blaschke, D.; Haensel, P.; Zdunik, J.L.; Fortin, M. Consequences of a strong phase transition in the dense matter equation of state for the rotational evolution of neutron stars. *Astron. Astrophys.* **2017**, *600*, A39. [CrossRef]
36. Li, J.J.; Sedrakian, A.; Alford, M. Relativistic hybrid stars in light of the NICER PSR J0740+6620 radius measurement. *Phys. Rev. D* **2021**, *104*, L121302. [CrossRef]
37. Cierniak, M.; Blaschke, D. The special point on the hybrid star mass–radius diagram and its multi–messenger implications. *Eur. Phys. J. ST* **2020**, *229*, 3663–3673. [CrossRef]
38. Ranea-Sandoval, I.F.; Han, S.; Orsaria, M.G.; Contrera, G.A.; Weber, F.; Alford, M.G. Constant-sound-speed parametrization for Nambu–Jona-Lasinio models of quark matter in hybrid stars. *Phys. Rev. C* **2016**, *93*, 045812. [CrossRef]
39. Alford, M.G.; Han, S. Characteristics of hybrid compact stars with a sharp hadron-quark interface. *Eur. Phys. J. A* **2016**, *52*, 62. [CrossRef]
40. Tan, H.; Dore, T.; Dexheimer, V.; Noronha-Hostler, J.; Yunes, N. Extreme matter meets extreme gravity: Ultraheavy neutron stars with phase transitions. *Phys. Rev. D* **2022**, *105*, 023018. [CrossRef]
41. Glendenning, N.K.; Kettner, C. Nonidentical neutron star twins. *Astron. Astrophys.* **2000**, *353*, L9.
42. Jakobus, P.; Motornenko, A.; Gomes, R.O.; Steinheimer, J.; Stoecker, H. The possibility of twin star solutions in a model based on lattice QCD thermodynamics. *Eur. Phys. J. C* **2021**, *81*, 41. [CrossRef]
43. Alford, M.; Braby, M.; Paris, M.W.; Reddy, S. Hybrid stars that masquerade as neutron stars. *Astrophys. J.* **2005**, *629*, 969. [CrossRef]
44. Klahn, T.; Blaschke, D.; Typel, S.; van Dalen, E.N.E.; Faessler, A.; Fuchs, C.; Gaitanos, T.; Grigorian, H.; Ho, A.; Kolomeitsev, E.E.; et al. Constraints on the high-density nuclear equation of state from the phenomenology of compact stars and heavy-ion collisions. *Phys. Rev. C* **2006**, *74*, 035802. [CrossRef]
45. Most, E.R.; Motornenko, A.; Steinheimer, J.; Dexheimer, V.; Hanauske, M.; Rezzolla, L.; Stoecker, H. Probing neutron-star matter in the lab: Connecting binary mergers to heavy-ion collisions. *arXiv* **2022**, arXiv:2201.13150.
46. Adamczewski-Musch, J.; Arnold, O.; Behnke, C.; Belounnas, A.; Belyaev, A.; Berger-Chen, J.C.; Biernat, J.; Blanco, A.; Blume, C.; Böhmer, M.; et al. Probing dense baryon-rich matter with virtual photons. *Nat. Phys.* **2019**, *15*, 1040–1045.
47. Stephanov, M.A.; Rajagopal, K.; Shuryak, E.V. Event-by-event fluctuations in heavy ion collisions and the QCD critical point. *Phys. Rev. D* **1999**, *60*, 114028. [CrossRef]
48. Karsch, F. Lattice QCD at high temperature and density. *Lect. Notes Phys.* **2002**, *583*, 209–249.
49. Fukushima, K.; Hatsuda, T. The phase diagram of dense QCD. *Rept. Prog. Phys.* **2011**, *74*, 014001. [CrossRef]
50. Halasz, A.M.; Jackson, A.D.; Shrock, R.E.; Stephanov, M.A.; Verbaarschot, J.J.M. On the phase diagram of QCD. *Phys. Rev. D* **1998**, *58*, 096007. [CrossRef]
51. Blacker, S.; Bastian, N.U.F.; Bauswein, A.; Blaschke, D.B.; Fischer, T.; Oertel, M.; Soultanis, T.; Typel, S. Constraining the onset density of the hadron-quark phase transition with gravitational-wave observations. *Phys. Rev. D* **2020**, *102*, 123023. [CrossRef]
52. Blaschke, D.; Cierniak, M. Studying the onset of deconfinement with multi-messenger astronomy of neutron stars. *Astron. Nachr.* **2021**, *342*, 227–233. [CrossRef]
53. Orsaria, M.G.; Malfatti, G.; Mariani, M.; Ranea-Sandoval, I.F.; García, F.; Spinella, W.M.; Contrera, G.A.; Lugones, G.; Weber, F. Phase transitions in neutron stars and their links to gravitational waves. *J. Phys. G* **2019**, *46*, 073002. [CrossRef]
54. Cierniak, M.; Blaschke, D. Hybrid neutron stars in the mass-radius diagram. *Astron. Nachr.* **2021**, *342*, 819. [CrossRef]
55. Kämpfer, B. Phase transitions in dense nuclear matter and explosive neutron star phenomena. *Phys. Lett. B* **1985**, *153*, 121–123. [CrossRef]
56. Kämpfer, B. Phase transitions in nuclear matter and consequences for neutron stars. *J. Phys. G* **1983**, *9*, 1487. [CrossRef]
57. Schertler, K.; Greiner, C.; Schaffner-Bielich, J.; Thoma, M.H. Quark phases in neutron stars and a 'third family' of compact stars as a signature for phase transitions. *Nucl. Phys. A* **2000**, *677*, 463–490. [CrossRef]
58. Christian, J.E.; Zacchi, A.; Schaffner-Bielich, J. Classifications of Twin Star Solutions for a Constant Speed of Sound Parameterized Equation of State. *Eur. Phys. J. A* **2018**, *54*, 28. [CrossRef]
59. Migdal, A.B.; Chernoutsan, A.I.; Mishustin, I.N. Pion condensation and dynamics of neutron stars. *Phys. Lett. B* **1979**, *83*, 158–160. [CrossRef]
60. Kämpfer, B. On the collapse of neutron stars and stellar cores to pion-condensed stars. *Astrophys. Space Sci.* **1983**, *93*, 185–197. [CrossRef]
61. Bertone, G.; Hooper, D.; Silk, J. Particle dark matter: Evidence, candidates and constraints. *Phys. Rept.* **2005**, *405*, 279–390. [CrossRef]
62. Luzio, L.D.; Gavela, B.; Quilez, P.; Ringwald, A. Dark matter from an even lighter QCD axion: Trapped misalignment. *JCAP* **2021**, *10*, 001. [CrossRef]

63. Hodges, H.M. Mirror baryons as the dark matter. *Phys. Rev. D* **1993**, *47*, 456. [CrossRef] [PubMed]
64. Tulin, S.; Yu, H.B. Dark Matter Self-interactions and Small Scale Structure. *Phys. Rept.* **2018**, *730*, 1–57. [CrossRef]
65. Karkevandi, D.R.; Shakeri, S.; Sagun, V.; Ivanytskyi, O. Bosonic dark matter in neutron stars and its effect on gravitational wave signal. *Phys. Rev. D* **2022**, *105*, 023001. [CrossRef]
66. Dengler, Y.; Schaffner-Bielich, J.; Tolos, L. Second Love number of dark compact planets and neutron stars with dark matter. *Phys. Rev. D* **2022**, *105*, 043013. [CrossRef]
67. Hippert, M.; Dillingham, E.; Tan, H.; Curtin, D.; Noronha-Hostler, J.; Yunes, N. Dark Matter or Regular Matter in Neutron Stars? How to tell the difference from the coalescence of compact objects. *arXiv* **2022**, arXiv:2211.08590.
68. Beradze, R.; Gogberashvili, M.; Sakharov, A.S. Binary Neutron Star Mergers with Missing Electromagnetic Counterparts as Manifestations of Mirror World. *Phys. Lett. B* **2020**, *804*, 135402. [CrossRef]
69. Alizzi, A.; Silagadze, Z.K. Dark photon portal into mirror world. *Mod. Phys. Lett. A* **2021**, *36*, 2150215. [CrossRef]
70. Goldman, I.; Mohapatra, R.N.; Nussinov, S. Bounds on neutron-mirror neutron mixing from pulsar timing. *Phys. Rev. D* **2019**, *100*, 123021. [CrossRef]
71. Berezhiani, Z. Antistars or antimatter cores in mirror neutron stars? *Universe* **2022**, *8*, 313. [CrossRef]
72. Berezhiani, Z.; Biondi, R.; Mannarelli, M.; Tonelli, F. Neutron-mirror neutron mixing and neutron stars. *Eur. Phys. J. C* **2021**, *81*, 1036. [CrossRef]
73. Beacham, J.; Burrage, C.; Curtin, D.; Roeck, A.D.; Evans, J.; Feng, J.L.; Gatto, C.; Gninenko, S.; Hartin, A.; Irastorza, I.; et al. Physics Beyond Colliders at CERN: Beyond the Standard Model Working Group Report. *J. Phys. G* **2020**, *47*, 010501. [CrossRef]
74. Zöllner, R.; Kämpfer, B. Exotic Cores with and without Dark-Matter Admixtures in Compact Stars. *Astronomy* **2022**, *1*, 36–48. [CrossRef]
75. Schaffner-Bielich, J. *Compact Star Physics*; Cambridge University Press: Cambridge UK, 2020.
76. Schulze, R.; Kämpfer, B. Cold quark stars from hot lattice QCD. *arXiv* **2009**, arXiv:0912.2827.
77. Baym, G.; Hatsuda, T.; Kojo, T.; Powell, P.D.; Song, Y.; Takatsuka, T. From hadrons to quarks in neutron stars: A review. *Rept. Prog. Phys.* **2018**, *81*, 056902. [CrossRef]
78. Demorest, P.; Pennucci, T.; Ransom, S.; Roberts, M.; Hessels, J. Shapiro Delay Measurement of A Two Solar Mass Neutron Star. *Nature* **2010**, *467*, 1081. [CrossRef]
79. Antoniadis, J.; Freire, P.C.C.; Wex, N.; Tauris, T.M.; Lynch, R.S.; van Kerkwijk, M.H.; Kramer, M.; Bassa, C.; Dhillon, V.S.; Driebe, T.; et al. A Massive Pulsar in a Compact Relativistic Binary. *Science* **2013**, *340*, 6131. [CrossRef]
80. Fonseca, E.; Cromartie, H.T.; Pennucci, T.T.; Ray, P.S.; Kirichenko, A.Y.; Ransom, S.M.; Demorest, P.B.; Stairs, I.H.; Arzoumanian, Z.; Guillemot, L.; et al. Refined Mass and Geometric Measurements of the High-mass PSR J0740+6620. *Astrophys. J. Lett.* **2021**, *915*, L12. [CrossRef]
81. Romani, R.W.; Kandel, D.; Filippenko, A.V.; Brink, T.G.; Zheng, W. PSR J0952−0607: The Fastest and Heaviest Known Galactic Neutron Star. *Astrophys. J. Lett.* **2022**, *934*, L18. [CrossRef]
82. Abbott, R.; Abbott, T.D.; Abraham, S.; Acernese, F.; Ackley, K.; Adams, C.; Adhikari, R.X.; Adya, V.B.; Affeldt, C.; Agathos, M.; et al. GW190814: Gravitational Waves from the Coalescence of a 23 Solar Mass Black Hole with a 2.6 Solar Mass Compact Object. *Astrophys. J. Lett.* **2020**, *896*, L44. [CrossRef]
83. Ecker, C.; Rezzolla, L. Impact of large-mass constraints on the properties of neutron stars. *arXiv* **2022**, arXiv:2209.08101.
84. Shao, Y. On the Neutron Star/Black Hole Mass Gap and Black Hole Searches. *Res. Astron. Astrophys.* **2022**, *22*, 122002. [CrossRef]
85. Farah, A.M.; Fishbach, M.; Essick, R.; Holz, D.E.; Galaudage, S. Bridging the Gap: Categorizing Gravitational-wave Events at the Transition between Neutron Stars and Black Holes. *Astrophys. J.* **2022**, *931*, 108. [CrossRef]
86. Reed, B.T.; Fattoyev, F.J.; Horowitz, C.J.; Piekarewicz, J. Implications of PREX-2 on the Equation of State of Neutron-Rich Matter. *Phys. Rev. Lett.* **2021**, *126*, 172503. [CrossRef] [PubMed]
87. Annala, E.; Gorda, T.; Kurkela, A.; Nättilä, J.; Vuorinen, A. Evidence for quark-matter cores in massive neutron stars. *Nat. Phys.* **2020**, *16*, 907–910. [CrossRef]
88. Altiparmak, S.; Ecker, C.; Rezzolla, L. On the Sound Speed in Neutron Stars. *Astrophys. J. Lett.* **2022**, *939*, L34. [CrossRef]
89. Ayriyan, A.; Blaschke, D.; Grunfeld, A.G.; Alvarez-Castillo, D.; Grigorian, H.; Abgaryan, V. Bayesian analysis of multimessenger M-R data with interpolated hybrid EoS. *Eur. Phys. J. A* **2021**, *57*, 318. [CrossRef]
90. Oter, E.L.; Windisch, A.; Llanes-Estrada, F.J.; Alford, M. nEoS: Neutron Star Equation of State from hadron physics alone. *J. Phys. G* **2019**, *46*, 084001. [CrossRef]
91. Lattimer, J.M.; Prakash, M. The Equation of State of Hot, Dense Matter and Neutron Stars. *Phys. Rept.* **2016**, *621*, 127–164. [CrossRef]
92. Greif, S.K.; Hebeler, K.; Lattimer, J.M.; Pethick, C.J.; Schwenk, A. Equation of state constraints from nuclear physics, neutron star masses, and future moment of inertia measurements. *Astrophys. J.* **2020**, *901*, 155. [CrossRef]
93. Anzuini, F.; Bell, N.F.; Busoni, G.; Motta, T.F.; Robles, S.; Thomas, A.W.; Virgato, M. Improved treatment of dark matter capture in neutron stars III: Nucleon and exotic targets. *JCAP* **2021**, *11*, 056. [CrossRef]
94. Bell, N.F.; Busoni, G.; Robles, S. Capture of Leptophilic Dark Matter in Neutron Stars. *JCAP* **2019**, *06*, 054. [CrossRef]
95. Das, H.C.; Kumar, A.; Kumar, B.; Patra, S.K. Dark Matter Effects on the Compact Star Properties. *Galaxies* **2022**, *10*, 14. [CrossRef]
96. Das, H.C.; Kumar, A.; Patra, S.K. Dark matter admixed neutron star as a possible compact component in the GW190814 merger event. *Phys. Rev. D* **2021**, *104*, 063028. [CrossRef]

97. Blaschke, D.; Ayriyan, A.; Alvarez-Castillo, D.E.; Grigorian, H. Was GW170817 a Canonical Neutron Star Merger? Bayesian Analysis with a Third Family of Compact Stars. *Universe* **2020**, *6*, 81. [CrossRef]
98. Newton, W.G.; Balliet, L.; Budimir, S.; Crocombe, G.; Douglas, B.; Head, T.B.; Langford, Z.; Rivera, L.; Sanford, J. Ensembles of unified crust and core equations of state in a nuclear-multimessenger astrophysics environment. *Eur. Phys. J. A* **2022**, *58*, 69. [CrossRef]
99. Huth, S.; Pang, P.T.H.; Tews, I.; Dietrich, T.; Fèvre, A.L.; Schwenk, A.; Trautmann, W.; Agarwal, K.; Bulla, M.; Coughlin, M.W.; et al. Constraining Neutron-Star Matter with Microscopic and Macroscopic Collisions. *Nature* **2022**, *606*, 276–280. [CrossRef]
100. Raaijmakers, G.; Riley, T.E.; Watts, A.L.; Greif, S.K.; Morsink, S.M.; Hebeler, K.; Schwenk, A.; Hinderer, T.; Nissanke, S.; Guillot, S.; et al. A *NICER* view of PSR J0030+0451: Implications for the dense matter equation of state. *Astrophys. J. Lett.* **2019**, *887*, L22. [CrossRef]
101. Raaijmakers, G.; Greif, S.K.; Riley, T.E.; Hinderer, T.; Hebeler, K.; Schwenk, A.; Watts, A.L.; Nissanke, S.; Guillot, S.; Lattimer, J.M.; et al. Constraining the dense matter equation of state with joint analysis of NICER and LIGO/Virgo measurements. *Astrophys. J. Lett.* **2020**, *893*, L21. [CrossRef]
102. Gorda, T.; Komoltsev, O.; Kurkela, A. Ab-initio QCD calculations impact the inference of the neutron-star-matter equation of state. *arXiv* **2022**, arXiv:2204.11877.
103. Cassing, M.; Brisebois, A.; Azeem, M.; Schaffner-Bielich, J. Exotic Compact Objects with Two Dark Matter Fluids. *arXiv* **2022**, arXiv:2210.13697.
104. Kain, B. Dark matter admixed neutron stars. *Phys. Rev. D* **2021**, *103*, 043009. [CrossRef]
105. Fujimoto, Y.; Fukushima, K.; McLerran, L.D.; Praszalowicz, M. Trace anomaly as signature of conformality in neutron stars. *Phys. Rev. Lett.* **2022**, *129*, 252702. [CrossRef]
106. Marczenko, M.; McLerran, L.; Redlich, K.; Sasaki, C. Reaching percolation and conformal limits in neutron stars. *arXiv* **2022**, arXiv:2207.13059.
107. Hippert, M.; Fraga, E.S.; Noronha, J. Insights on the peak in the speed of sound of ultradense matter. *Phys. Rev. D* **2021**, *104*, 034011. [CrossRef]
108. Marczenko, M.; Redlich, K.; Sasaki, C. Reconciling Multi-messenger Constraints with Chiral Symmetry Restoration. *arXiv* **2022**, arXiv:2208.03933.
109. Suleiman, L.; Fortin, M.; Zdunik, J.L.; Haensel, P. Influence of the crust on the neutron star macrophysical quantities and universal relations. *Phys. Rev. C* **2021**, *104*, 015801. [CrossRef]
110. Gao, H.; Ai, S.K.; Cao, Z.J.; Zhang, B.; Zhu, Z.Y.; Li, A.; Zhang, N.B.; Bauswein, A. Relation between gravitational mass and baryonic mass for non-rotating and rapidly rotating neutron stars. *Front. Phys. (Beijing)* **2020**, *15*, 24603. [CrossRef]
111. Ding, M.; Roberts, C.D.; Schmidt, S.M. Emergence of Hadron Mass and Structure. *Particles* **2023**, *6*, 57–120. [CrossRef]
112. Dorkin, S.M.; Kaptari, L.P.; Hilger, T.; Kampfer, B. Analytical properties of the quark propagator from a truncated Dyson-Schwinger equation in complex Euclidean space. *Phys. Rev. C* **2014**, *89*, 034005. [CrossRef]
113. Dorkin, S.M.; Kaptari, L.P.; Kämpfer, B. Accounting for the analytical properties of the quark propagator from the Dyson-Schwinger equation. *Phys. Rev. C* **2015**, *91*, 055201. [CrossRef]
114. Greifenhagen, R.; Kämpfer, B.; Kaptari, L.P. Regge Trajectories of Radial Meson Excitations: Exploring the Dyson–Schwinger and Bethe–Salpeter Approach. In *Discoveries at the Frontiers of Science*; FIAS Interdisciplinary Science Series; Springer Nature: Berlin/Heidelberg, Germany, 2020; p. 55.
115. Dorkin, S.M.; Hilger, T.; Kaptari, L.P.; Kampfer, B. Heavy pseudoscalar mesons in a Schwinger-Dyson–Bethe-Salpeter approach. *Few Body Syst.* **2011**, *49*, 247–254. [CrossRef]
116. Dorkin, S.M.; Hilger, T.; Kampfer, B.; Kaptari, L.P. A Combined Solution of the Schwinger-Dyson and Bethe-Salpeter Equations for Mesons as $q\bar{q}$ Bound States. *arXiv* **2010**, arXiv:1012.5372.
117. Kaptari, L.P.; Kämpfer, B. Mass Spectrum of Pseudo-Scalar Glueballs from a Bethe–Salpeter Approach with the Rainbow–Ladder Truncation. *Few Body Syst.* **2020**, *61*, 28. [CrossRef]
118. Maris, P.; Roberts, C.D.; Schmidt, S.M. Chemical potential dependence of pi and rho properties. *Phys. Rev. C* **1998**, *57*, R28215. [CrossRef]
119. Bender, A.; Poulis, G.I.; Roberts, C.D.; Schmidt, S.M.; Thomas, A.W. Deconfinement at finite chemical potential. *Phys. Lett. B* **1998**, *431*, 263–269. [CrossRef]
120. Roberts, C.D.; Schmidt, S.M. Dyson-Schwinger equations: Density, temperature and continuum strong QCD. *Prog. Part. Nucl. Phys.* **2000**, *45*, S1. [CrossRef]
121. Brodsky, S.J.; Roberts, C.D.; Shrock, R.; Tandy, P.C. Confinement contains condensates. *Phys. Rev. C* **2012**, *85*, 065202. [CrossRef]
122. Brambilla, N.; Eidelman, S.; Foka, P.; Gardner, S.; Kronfeld, A.S.; Alford, M.G.; Alkofer, R.; Butenschoen, M.; Cohen, T.D.; Erdmenger, J.; et al. QCD and Strongly Coupled Gauge Theories: Challenges and Perspectives. *Eur. Phys. J. C* **2014**, *74*, 2981.
123. Thomas, R.; Zschocke, S.; Kampfer, B. Evidence for in-medium changes of four-quark condensates. *Phys. Rev. Lett.* **2005**, *95*, 232301. [CrossRef] [PubMed]
124. Buchheim, T.; Hilger, T.; Kampfer, B. Wilson coefficients and four-quark condensates in QCD sum rules for medium modifications of D mesons. *Phys. Rev. C* **2015**, *91*, 015205. [CrossRef]
125. Brodsky, S.J.; Shrock, R. Condensates in Quantum Chromodynamics and the Cosmological Constant. *Proc. Nat. Acad. Sci. USA* **2011**, *108*, 45–50. [CrossRef]

126. Gubler, P.; Satow, D. Recent Progress in QCD Condensate Evaluations and Sum Rules. *Prog. Part. Nucl. Phys.* **2019**, *106*, 1–67. [CrossRef]
127. Thomas, R.; Hilger, T.; Kampfer, B. Four-quark condensates in nucleon QCD sum rules. *Nucl. Phys. A* **2007**, *795*, 19–46. [CrossRef]
128. Fischer, C.S.; Kubrak, S.; Williams, R. Mass spectra and Regge trajectories of light mesons in the Bethe-Salpeter approach. *Eur. Phys. J. A* **2014**, *50*, 126. [CrossRef]
129. DeWolfe, O.; Gubser, S.S.; Rosen, C. A holographic critical point. *Phys. Rev. D* **2011**, *83*, 086005. [CrossRef]
130. Grefa, J.; Noronha, J.; Noronha-Hostler, J.; Portillo, I.; Ratti, C.; Rougemont, R. Hot and dense quark-gluon plasma thermodynamics from holographic black holes. *Phys. Rev. D* **2021**, *104*, 034002. [CrossRef]
131. Critelli, R.; Noronha, J.; Noronha-Hostler, J.; Portillo, I.; Ratti, C.; Rougemont, R. Critical point in the phase diagram of primordial quark-gluon matter from black hole physics. *Phys. Rev. D* **2017**, *96*, 096026. [CrossRef]
132. Järvinen, M. Holographic modeling of nuclear matter and neutron stars. *Eur. Phys. J. C* **2022**, *82*, 282. [CrossRef]
133. Borsányi, S.; Fodor, Z.; Guenther, J.N.; Kara, R.; Katz, S.D.; Parotto, P.; Pásztor, A.; Ratti, C.; Szabó, K.K. Lattice QCD equation of state at finite chemical potential from an alternative expansion scheme. *Phys. Rev. Lett.* **2021**, *126*, 232001. [CrossRef] [PubMed]
134. Borsanyi, S.; Fodor, Z.; Hoelbling, C.; Katz, S.D.; Krieg, S.; Szabo, K.K. Full result for the QCD equation of state with 2+1 flavors. *Phys. Lett. B* **2014**, *730*, 99–104. [CrossRef]
135. Bazavov, A.; Bhattacharya, T.; DeTar, C.; Ding, H.T.; Gottlieb, S.; Gupta, R.; Hegde, P.; Heller, U.M.; Karsch, F.; Laermann, E.; et al. Equation of state in (2+1)-flavor QCD. *Phys. Rev. D* **2014**, *90*, 094503. [CrossRef]
136. Bellwied, R.; Borsanyi, S.; Fodor, Z.; Katz, S.D.; Pasztor, A.; Ratti, C.; Szabo, K.K. Fluctuations and correlations in high temperature QCD. *Phys. Rev. D* **2015**, *92*, 114505. [CrossRef]

Disclaimer/Publisher's Note: The statements, opinions and data contained in all publications are solely those of the individual author(s) and contributor(s) and not of MDPI and/or the editor(s). MDPI and/or the editor(s) disclaim responsibility for any injury to people or property resulting from any ideas, methods, instructions or products referred to in the content.

MDPI AG
Grosspeteranlage 5
4052 Basel
Switzerland
Tel.: +41 61 683 77 34

Particles Editorial Office
E-mail: particles@mdpi.com
www.mdpi.com/journal/particles

Disclaimer/Publisher's Note: The statements, opinions and data contained in all publications are solely those of the individual author(s) and contributor(s) and not of MDPI and/or the editor(s). MDPI and/or the editor(s) disclaim responsibility for any injury to people or property resulting from any ideas, methods, instructions or products referred to in the content.

www.ingramcontent.com/pod-product-compliance
Lightning Source LLC
LaVergne TN
LVHW070151100526
838202LV00015B/1930